Northeastern Regional
Aquaculture
Center

Recirculating Aquaculture Systems 2nd Edition

by

MICHAEL B. TIMMONS
JAMES M. EBELING
FRED W. WHEATON
STEVEN T. SUMMERFELT
BRIAN J. VINCI

NRAC Publication No. 01-002

NRAC Publication No. 01-002

Recirculating Aquaculture Systems 2nd Edition

Michael B. Timmons
Cornell University, Biological and Environmental Engineering Department, 302 Riley-Robb Hall, Ithaca, NY 14853

James M. Ebeling
Freshwater Institute, PO Box 1889, Shepherdstown, WV 25443

Fred W. Wheaton
University of Maryland, Biological Resources Engineering Department, 1439 Animal Science Building, College Park, MD 20742

Steven T. Summerfelt
Freshwater Institute, PO Box 1889, Shepherdstown, WV 25443

Brian J. Vinci
Freshwater Institute, PO Box 1889, Shepherdstown, WV 25443

Northeastern Regional Aquaculture Center

USDA CSREES

NRAC Publication No. 01-002

CAYUGA AQUA VENTURES
126 Sunset Drive
Ithaca, NY 14850

Northeastern Regional Aquaculture Center

ISBN 0-9712646-1-9

Cayuga Aqua Ventures, 2002

Printed in the United States

NRAC Publication No. 01-002

DEDICATION

This book effort is dedicated to Dr. William D. Youngs, Professor Emeritus, Department of Natural Resources, Cornell University. Dr. Youngs spent over 30 years at Cornell teaching and mentoring students in Fishery science and aquaculture. Dr. Youngs is most recognized for one of the seminal texts in fisheries, "Principles of Fishery Science", that he co-authored with Dr. W. Harry Everhart, a book that many keep in their personal libraries (published 1981, Cornell University Press, p. 349).

For me it was a phone call from Bill in 1984 that was my introduction to the world of aquaculture and, in particular, recirculating aquaculture. The phone call went something like "What size pump do I need to move 1,000 gpm of water against 10 feet of head?" I responded with my own question, "Why do you want to know that?" Bill took me under his wing and we spent many hours together building the first recirculating aquaculture system at the Cornell Dairy Research farm. Bill was also the first General Manager of my commercial aquaculture venture, Fingerlakes Aquaculture (see Chapter 15).

Bill has always been a constant inspiration to me. I truly value Bill's advice on a wide variety of subjects, but I consider him unequaled in experience and scientific knowledge on fisheries management as applied to recirculating aquaculture. I also have the honor of being the only person that Bill ever took fishing and failed to even have a fish strike!

Thanks Bill!

Michael Timmons
June 4, 2002

NRAC Publication No. 01-002

NRAC Publication No. 01-002

FOREWORD

Aquaculture has a long history with its origins dating back to at least 475 BC in China (Milne, 1973). Trout culture started in Germany in 1741 (Leitritz and Lewis, 1980) but it wasn't until the 1880's that trout culture came to the U.S. This was the first U.S. aquaculture effort. However, aquaculture was not of much importance until the late 1940's when it was discovered that aquaculture methods could be used to raise fish for planting in natural waters as a means to supplement natural spawning. At this time the U.S. fish and Wildlife Service began growing trout, bluegill, largemouth bass and other species for planting. From its U. S. initiation in the 1880's the U.S. the trout industry grew slowly until the late 1940's or early 1950's when it began to expand more rapidly. In the 1960's the U.S. catfish industry got its start and began to grow rapidly. Although the catfish industry had its ups and downs over the years, it is an excellent success story. The catfish industry started from essentially no commercial production in 1960 to over 600 million pounds of production in 2000. If the imports (many of which are wild caught) and those not processed are added, the total supply comes to over 1 billion pounds in 2000 (Aquaculture Magazine, 2001). With the rise of the trout and catfish industries in the U.S. there has been experimentation and now commercial production of many other species of fish including but not limited to striped bass, salmon, yellow perch, tilapia, blue gill, small mouth bass, several species of bait fish, goldfish, koi, red fish, sturgeon, a wide variety of tropical fish, and a variety of other fish species. Tremendous progress has been made in fish culture in development of production systems, nutrition, genetics, engineering, disease control, physiology, basic understanding of fish biology and other areas. However, much remains to be done in all of these areas.

Shellfish aquaculture has been practiced in the U.S. since about the 1850's, but serious culture really became more popular only in the last 30 years. Species such as blue mussels, various species of oysters and clams, shrimp, lobsters, crabs, and others have been raised in aquaculture settings. Most shellfish aquaculture in the U.S. uses leased bottom systems, raft and rack culture, and/or some floating systems using either lines or cages. U.S. shellfish culture is almost exclusively done in salt or brackish water while, except for salmon, most fish culture is in fresh or brackish water. However, salt water fish culture is rapidly increasing. Development of shellfish culture is lagging behind fish culture in areas of nutrition, engineering, disease control, genetics, basic biology and other areas.

Aquaculture has been the fastest growing segment of U.S. agriculture for more than 10 years and is projected to remain that way for the foreseeable future. This rapid growth rate is driven by several factors including: 1) many fisheries have reached their sustainable yield, 2) food safety concerns of consumers, and 3) consumer demand for high quality, safe aquatic products that are low in fat and high in protein. The consumer trend that sees more meals eaten away from the home also contributes to aquaculture production as most seafood is eaten in restaurants and other eating places. These businesses need a reliable supply that can provide products on a regular basis year around, something a natural fishery can rarely provide. Although the future of aquaculture is bright in the U.S. there also are risks. Regulation of water supplies (both quantity and quality), waste discharges, and health regulations are becoming more and more onerous and costly to the industry. Competition for coastal sites, the public's concerns about environmental factors ranging from pollution to concerns about the visual "pollution" of aquaculture facilities in front of vacation homes on the shore, and the recreational use of waters that are also suitable for aquaculture are only a few of the clouds on the aquaculture horizon. These and other concerns are encouraging the aquaculture industry to move from open pond and cage culture systems to the more closely controlled recirculating systems. Typically recirculating (closed) aquatic production systems have higher capital and operating costs than many of the extensive systems such as cage culture in natural waters and raceway and/or pond culture systems. However, when the control provided by recirculating systems and the benefits this environmental control provides in terms of marketing, waste control, product quality, product availability and other factors are considered recirculating systems become much more attractive. Thus, this text is designed primarily for recirculating systems, which the authors feel will be the systems of choice for most new aquaculture ventures. The information provided in this text does, however, also apply to open, semi-closed, and closed systems.

The objectives of this text are the practical application of aquacultural engineering and how to develop an aquatic production system. It provides the reader with essential information necessary to get started in aquaculture production and it emphases practical information rather than in-depth theoretical discussions. It does not provide the reader with information on nutrition, genetics, basic biology, marketing, and all of the other areas important to development of a successful aquaculture operation. Many of these topics are touched on in the text, but are presented only in sufficient detail to allow the reader to understand the relationship of each of these aspects to production of fish. There is no attempt to present in-depth discussions of these topics. Rather the object is to provide sufficient information so the reader can: 1) look at a system and make a good judgment as to how well the systems will operate, 2) work with a systems designer to develop an aquatic production system of your own, and 3) know what to look for when shopping for aquacultural production systems.

The authors of this text combined have over 90 years of experience in aquacultural engineering.

Michael B. Timmons, Ph.D. Dr. Michael Timmons received his B.S. in Agricultural Engineering from the Ohio State University, his M.S. in Agricultural Engineering from the University of Hawaii, and his Ph.D. from Cornell University. Dr. Timmons has worked in aquacultural engineering for 20 years as a researcher and extension specialist. He has published widely and has served as primary editor on many of the Aquacultural Engineering Society meeting proceedings and for the series of bi-annual meetings sponsored by Virginia Tech on Water Recirculation Systems. He was one of the founders of the Aquacultural Engineering Society and has served in several officer positions including being the current President-elect. Dr. Timmons currently occupies the J. Thomas Clark Professorship of Entrepreneurship and Personal Enterprise, Biological and Environmental Engineering Department, Cornell University. Dr. Timmons has been a principle in the design, construction, and operation of a commercial recirculating fish production system and thus provides the viewpoint of a commercial aquaculturist in addition to his experience as a researcher and extension specialist.

James M. Ebeling, Ph.D. Dr. James Ebeling has a B.S. and M.S. in physics from Albion College in Albion, Michigan and Washington State University in Pullman, Washington, respectively. He has a second M.S. in agricultural engineering from Washington State University and has three years of formal training at the University of California, Davis in Aquacultural Engineering. He has his Ph.D. in Biological Resources Engineering from the University of Maryland, College Park, Maryland, where he worked on the kinetics of biofilters operating on aquacultural systems. James has been involved in aquaculture for over 18 years and has cultured over 20 species of fish. He spent three years at the Mariculture Research and Training Center, University of Hawaii as a research coordinator, and two years as project manager for the design and construction of the "Fish Barn" at the North Carolina State University. James also spent five years as a research and extension associate at the Piketon Research and Extension Center, Ohio State University, Piketon, Ohio where he was responsible for design, construction, and maintenance of the aquaculture facilities and for maintaining the fish as well as for the Center Aquaculture Extension Program. Dr. Ebeling joined the Freshwater Institute as an Environmental Research Engineer in 2000, working in basic and applied research as well as the application of monitoring and computer control to biological and aquatic ecosystems.

Fred Wheaton, Ph.D. Dr. Fred Wheaton has been involved in aquacultural engineering for over 30 years. He developed the first aquacultural engineering curriculum in the U.S., which was offered at the University of Maryland, and authored the first aquacultural engineering textbook "Aquacultural Engineering, Krieger Publishing Company, Melbourne, Florida, original edition (1977). Graduates of this curriculum are located throughout the U.S. and in many international locations. Dr. Wheaton developed one of the first aquacultural engineering research and extension programs in the U.S. His research has included recirculating systems, seafood processing, automation of oyster shucking, and a variety of other topics related to aquacultural engineering. He has published widely producing over 100 articles and three books. He was one of the founding members of and has served as president of the Aquacultural Engineering Society. He is currently employed as Professor and Chairman, Department of Biological Resources Engineering, University of Maryland, College Park, Maryland.

Steven T. Summerfelt, Ph.D. Dr. Steve Summerfelt is a research engineer with the Freshwater Institute. He received a Bachelor's Degree in Chemical Engineering from Iowa State University, a Master's Degree in Chemical Engineering from Michigan State University, and a Doctorate in Civil

Engineering from Iowa State University. Dr. Summerfelt is currently the President of the Aquacultural Engineering Society, and is the Editor of *AES News*. He researches processes for wastewater treatment and aquaculture systems intensification. During his career, he has authored or co-authored numerous scientific publications.

Brian J. Vinci. Mr. Brian Vinci is an environmental engineer with the Freshwater Institute. He is also the Secretary/Treasurer for the Aquacultural Engineering Society, and a member of the American Society of Agricultural Engineers. He received a Bachelor's Degree and a Master's Degree in Agricultural and Biological Engineering from Cornell University, and is a Ph.D. candidate in Biological and Environmental Engineering at Cornell. Mr. Vinci has been working in aquaculture for the last 10 years including an appointment with the US Fish & Wildlife Service, Tunison Laboratory (Cortland, NY).

All five authors participate in the Cornell University/Freshwater Institute "Hands-On" short course that focuses on Recirculating Aquaculture Systems. This short course has been conducted for the last 8 years and has drawn aquaculturalists from all over the world.

Dr. Fred Wheaton
May 1, 2002

ACKNOWLEDGEMENTS

The book authors would like to express their sincere appreciation to Mr. Joe Hankins, Director and the Freshwater Institute Conservation Fund (Shepherdstown, WV) who supported the development and writing of this book. Also, to the Northeastern Regional Aquaculture Center who provided funding for this book. Also to the NRAES (Natural Resources, Agriculture, and Engineering Services, a Land Grant Outreach program, Ithaca, NY; see www.NRAES.org for a complete list of available publications and resource materials) who permitted the use of some materials previously published in NRAES-49 "Engineering Aspects of Intensive Aquaculture". Finally, we would like to express our thanks to our review panel of experts who were assembled to provide initial guidance and review of the book progress. Our thanks go to each of them:

John Ewart	Extension	Aquaculture Extension Specialist, Delaware Sea Grant Marine Advisory Service, Graduate College of Marine Studies, University of Delaware.
Donald Webster	Extension	Marine Science Educator, University of Maryland Cooperative Extension
Michael Iannello	Industry	General Manager, Fingerlakes Aquaculture, Groton, NY
Jerry Redden	Industry	Director, Worcester County Economic Development Office, Snow Hill, MD 21863 Email: ecodevo@ezy.net Past Manager, AquaMar, Pocomoke, MD
Glenn Snapp and Terry McCarthy	Industry	Owners, Water Management Technologies, Baton Rouge, LA
Thomas Losordo	Research	Professor, North Carolina State University

Gordon Durant	Government	Fish Culture Coordinator-Technical Development, Fish Culture Section, Fish and Wildlife Branch, Ontario Ministry of Natural Resources
William Foulkrod	Secondary Education	Teacher, Secondary Education, Syracuse, NY

Our highest sense of appreciation goes to Mr. James E. Timmons (Engineer Management, Boeing Aircraft Division) who was our text editor who actually read every line of text to make sure it was non-sensical! And to Mrs. Brenda L. Marchewka, who did the final formatting of the book prior to printing. These as we all know are extremely tedious jobs that were done in the highest professional manner. Thank you Jim and Brenda.

We would also like to thank several other contributors to the book, specifically:

Chapter 11 (System Management and Operations):

- **Mr. Don Webster,** Marine Science Educator, Sea Grant Extension Program, University of Maryland Cooperative Extension, Wye Research & Education Center, PO Box 169, Queenstown MD 21658
- **Dr. Joe M. Regenstein,** Professor, Department of Food Science, Cornell University.

Chapter 12 (Sterilization and Disinfection):

- **Dr. Helge Liltved** (Scientist, Norwegian Institute for Water Research, Branch Office South, Televeien 3, N-4879 Grimstad, Norway)

Chapter 13 (Fish Health Management):

- **Dr. Julie Bebak-Williams** (Research Veterinarian with the Freshwater Institute. She received her Veterinary Degree (V.M.D.) and Ph.D. from the University of Pennsylvania School of Veterinary Medicine.)
- **Dr. Alicia C. Noble** (Clinical Veterinarian with the Freshwater Institute. Dr. Noble received her D.V.M. from the Universidad de Chile in Santiago, Chile.)

- **Dr. Paul R. Bowser** (Fish Pathologist and Professor of Aquatic Animal Medicine, College of Veterinary Medicine, Cornell University; PhD Auburn University). Dr. Bowser is also one of the project leaders of the Aquatic Animal Health Program at Cornell University:

  Aquatic Animal Health Program:
  http://web.vet.cornell.edu/Public/FishDisease/AquaticProg/

  AQUAVET Program:
  http://web.vet.cornell.edu/public/aquavet/
  Email: prb4@cornell.edu
- **Mr. Greg Wooster** (Research Technician, College of Veterinary Medicine)

Chapter 16 (Fish Nutrition and Feeding):

- **Dr. H. George Ketola** (Research Physiologist, Tunison Laboratory of Aquatic Science, USGS, Great Lakes Science Center, Cortland, NY and Adjunct Assistant Professor, Department of Natural Resources, Cornell University, Ithaca, NY; Dr. Ketola received his Ph.D. from Cornell University)
- **Dr. Paul D. Maugle** (P.D. M. and Associates, Norwich CT; B.S. University of West Florida, M.S. University of Rhode Island, Ph. D. Kagoshima University JAPAN - University of Rhode Island US

Chapter 17 (Software and Computing):

- **Ms. Rosa Aguirre**
  Aquaculture Consultant
  Pescados del Caribe, S.A.
  Managua, Nicaragua
  E-mail: rda@starosa.com.ni

Chapter 18 (Aquaponics):

- **Dr. James Rakocy**
  University of the Virgin Islands
  Agricultural Experiment Station
  RR2, Box 10000
  Kingshill, St. Croix, VI 00850
  E-mail: james.rakocy@uvi.edu

This publication is supported by the **Cooperative State Research, Education, and Extension Service** (CSREES), U.S. Department of Agriculture, under Agreement No. 97-38500-4641, awarded to the Northeastern Regional Aquaculture Center at the University of Massachusetts Dartmouth. Any opinions, findings, conclusions, or recommendations expressed in this publication are those of the authors and do not necessarily reflect the views of the U.S. Department of Agriculture, the Northeastern Regional Aquaculture Center, or the University of Massachusetts.

TABLE OF CONTENTS

CHAPTER 1

INTRODUCTION TO RECIRCULATING AQUACULTURE TECHNOLOGY

1.0 BACKGROUND

Recirculating Aquaculture System (RAS) technology has been under development and refinement for the past thirty years, and the authors have been at the forefront of this activity. The document that has become known as the original aquacultural engineering bible was written by our senior author in 1977 and published by John Wiley & Sons, Inc., with the title "Aquacultural Engineering". Collectively, our primary focus has always been to create, develop, and refine RAS technologies that can produce food fish on an economically competitive basis. The US is known for being a leader in agriculture and agriculture is the leading economic business sector in the economy of several states. However, the number of farms and the associated labor force has continued to shrink as individual farms become larger and larger and animal productivity continues to increase. For example, in New York there are 8,700 dairy farms today, compared to 13,000 dairy farms just 10 years ago (1988). Some of these former dairy farmers have focused on aquaculture as a method of generating new opportunities. Similar stories exist in North Carolina where tobacco farmers and hog farmers are looking for alternative enterprises. How realistic is this change of focus and can RAS play a role in what is possibly the next transformation of worldwide farming techniques?

1.1 AQUACULTURE IS NOT FOR EVERYONE

There seems to be an assumption that given alternate forms of animal husbandry, a currently unsuccessful dairy or hog farmer could become a successful fish farmer. Let us call that General Falsehood #1. Fish farming and RAS aquaculture in particular are generally more critically dependent upon expert management and precise activity than are other forms of farming. Thus, it is unlikely that a farmer with "average" management skills would be successful in fish farming. On the other hand, a successful dairy farmer could probably be a successful user of RAS technology. The importance of good management to a successful RAS farm cannot be overemphasized. Poor management is almost always the primary reason for failure in aquaculture ventures. Experience with flow-through raceways or outdoor ponds has almost nothing to do with understanding how to effectively and efficiently manage an RAS farm.

1.2 Some Quick Case Histories

The authors have been involved in several RAS startup ventures in their public institution and private roles. Some have resulted in failure. Failure is valuable in that it accentuates problems that must be recognized and addressed. To date, Cornell University has initiated two unsuccessful technology transfer efforts and is currently indirectly involved in a third effort via the first author's private involvement in a large tilapia venture (600 tons or 1.2 million lbs per year production level; reported on in more detail in Chapter 15). The first technology transfer failure (mid 1980's) involved a very successful integrated turkey farm. This farm had 130 employees, a processing plant, an extensive distribution network for fresh and frozen products, and a restaurant. The second failure in the early 1990's was a cooperative trout farm that consisted of seven members named the Northern Fresh Fish Cooperative (NFFC). The cooperative members were smart, talented, experienced, and successful people – all with college degrees and some with advanced degrees. Failure of the first two efforts had the following common elements:

- Individuals involved had no previous experience with indoor fish culture
- Technology that was labor intensive
- Engineering Technology was poor
- Sensitive species (brook and rainbow trout)
- Enterprise was essentially an expensive hobby venture

Cornell designed the turkey farm and assisted them with their implementation: four 10,000 gallon (38 m^3) culture tanks designed for 0.7 lbs per gallon (84 kg/m^3) at harvest density. The turkey farm ran their operation for 2 years before abandoning the effort. Solids accumulation was probably the primary reason for failure, even though there was a reasonably adequate solids settling chamber in the system.

The turkey farm failure and subsequent analysis led the Cornell efforts to focus on the development of new biofilter designs and more energy efficient methods to move water between system components. Several years were spent refining the application of rotating biological contactors (RBC's), which replaced the submerged rock filter designs, and the use of airlift pumps to move the water between system components.

The efforts with the Northern Fresh Fish Cooperative including layouts of their system components are described by Timmons et al. (1993). Again, this particular system had been successfully operated by Cornell for three years before the technology transfer effort and achieved high carrying capacity (0.7 lb/gal, 84 kg/m^3), excellent feed conversions (1 to 1) and high fish growth rate (an inch per month, 25 mm/month). This system was quite amazing and efficient in that the airlift pumps moved water from the culture tank to the solids settling chamber to the biofiltration chamber and back to the culture tank on a total water level differential of 1.5 cm. The system also employed foam fractionators (Weeks et al. 1992; Chen et al. 1992; Chen et al. 1994B; Timmons et al. 1995) in the culture tank to remove fine solids, provide oxygen and CO_2 removal and increase the circular motion in the culture tank to assist in solids removal.

1.3 Other Experiences

A few years ago, Peter Redmayne (Editor of Seafood Leader, January/February issue 1992) did a story on the financial failures in the aquaculture industry. He mentioned several notable failures and their reason for demise:

- Idaho-based J.R. Simplot Co. closed the doors on its two-year old, intensive tilapia operation, losing more than $20 million in the process. Reason: inadequate biofilter.

- Bodega Farms shut down its $9.5 million steelhead, coho salmon and abalone farm near Bodega Bay, CA. Reason: State of CA would not allow two million fingerlings across the border, and they had no place else to go (no fish, no cash).

- Aquaculture Technologies of Louisiana (ATL) went bankrupt leaving 2,000 acres of catfish ponds in St. Landry Parish and $9 million in debts to some 300 creditors. Reason: bad management.

- NAIAD Corp, largest catfish farming and processing venture in Texas filed Chapter 11 (August of 1991) after starting processing its own catfish. Reason: lack of operating cash related to poor cash flow management.

- Blue Ridge Fisheries (Martinsville, VA) the largest indoor catfish operation in the world (at that time, 1991) lost its assets to bank foreclosure. Reason: the RAS was not cost effective (this facility was resurrected as a tilapia facility and is currently a major producer in the Northeast US, essentially under the same management structure).

More recently, the following scenarios have unfolded:

- Fish & Dakota lost several hundred thousand lbs of fish and did not reopen its doors. Reason: New management eliminated some of the 24 hour coverage and a power outage and failure of the "automatic" stand-by generator killed the fish.

- Sunflower Aquaculture (Kansas) lost their fish (not the enterprise) when their roof caved in; this was a military facility that had been unused for many years and had clear span timber trusses; when the trusses regained some moisture, they structurally failed.

- Northern Fresh Fish Cooperative (central NY), last member to lose fish was because his dialer had not been hooked up to alert them of a lack of water (had left a drain open during a cleaning operation); the 2nd to last member went out of business when his well went dry.

- Perch operation in Western Pennsylvania closed their doors when their new system had finally reached near full design carrying capacity and then the liner in their culture vessel "broke".

- Southern Pennsylvania perch grower finally gave up after their initial stocking of perch showed growth rates a fraction of what was anticipated.

Be careful. Remember the following rule of thumb.

"RULE OF THUMB"

Only invest what you can afford to **LOSE**!

1.4 THE OPTIMISTIC VIEW

Peter F. Drucker, a world recognized business leader and economic forecaster, predicts that aquaculture will be one of three major economic opportunities in the new millennium. Everyone everywhere is either eating more fish or thinking they should! Changes in dietary patterns and the fact that the US Surgeon General now recommends eating fish as a significant protein source for the diet are strong indicators that opportunities in aquaculture will continue to expand. The US catfish industry is a ready example of how fast aquaculture fish markets can grow; i.e., this market has grown by 100,000 tons (220 million lbs) in just the last seven years. The Atlantic salmon industry has been adding 50,000 tons per year (110 million lbs) of new production for the last 10 years. The production of tilapia has been exponential in the last several years to the point that the US market demand for tilapia has gone from essentially nothing to importing the equivalent of 50,000 tons (110 million lbs) in year 2000.

We believe that aquaculture – the farming of aquatic organisms - is the most probable and feasible solution to providing the seafood products for this ever increasing market demand and shrinking supply of product from the oceans. Aquaculture is an environmentally responsible alternative to fishing. It provides a consistent and reliable source of high quality, fresh seafood that is nutritious, safe to eat, and reasonably priced. Nearly a third of all seafood consumed in the world today is the product of aquaculture, and the industry continues to grow as consumers demand more and more seafood and wild supplies continue to dwindle. Much of the salmon, catfish, tilapia, and shrimp seen on restaurant menus are farmed, and analysts predict that hybrid striped bass, sturgeon caviar, yellow perch, cod, flounder, snapper, halibut, char, and scallop farm production will continue to increase.

Seafood is only as good as the water in which it lives. Aquaculturists control the quality of the water, so the seafood they produce can be free of environmental contaminants. Consumers have demonstrated a marked preference for cultured seafood because it is more consistent in quality and presentation, and it tends to have a milder taste than wild seafood.

Aquaculture systems can be extensive, semi-intensive, or intensive, depending upon the number of organisms grown per volume of water and the water source and supply. Pond culture is extensive, cage culture is semi-intensive but intensive within the cage, and RAS are intensive. Pond and cage systems are open-air, and therefore there is always a risk of air or water-borne contaminants. Because water quality control is more difficult in pond and cage systems, the number of organisms that can be grown effectively is limited. In this book as noted by the title, we are concentrating on RAS. The principles of RAS for the water environment can be employed in the open air, but you lose total control of the environment.

1.5 Overview of Recirculating Aquaculture Systems (RAS)

Fish can be stocked most intensively in RAS. In RAS, the total environment is controlled. Fish are raised in tanks and in the most secure environment are placed within an enclosed building to control the aerial environment as well. Water circulates throughout the system, and only a small percentage of the water is changed daily. Temperature, salinity, pH, alkalinity, chemical composition, and oxygen are all monitored and continuously controlled. Solid wastes are filtered and removed, oxygen is added to maintain sufficient dissolved oxygen levels for the stocking density, and effluent water is passed through a biofilter for a biological conversion of ammonia-nitrogen to nitrate-nitrogen. Designing and operating RAS requires a solid understanding of the many unit processes or operations involved, (see Fig. 1.1). The failure of any one of these operations can cause the whole system to fail, usually killing the fish in the process. These operations will be covered in detail in the following chapters.

RAS are more capital-intensive than most other types of traditional aquaculture systems, and must rely on economic productivity per unit volume of rearing space for profitability. An RAS farm will have support features such as backup generators whose cost is also part of the overall production costs for the farm. Sufficient production quantity must be achieved in order that these support component costs are reduced to a level that the overall production costs are competitive. Large production levels are achieved by replicating a successful modular unit that insulates itself from failure of other production modules. This feature permits controlled, incremental increases of seafood production in response to rising market demand. It also facilitates raising capital and spreading investment risk through business arrangements such as franchising, contract growing, and production/marketing cooperatives. Finally, RAS technology is species-adaptable, allowing operators to follow market trends for seafood preference.

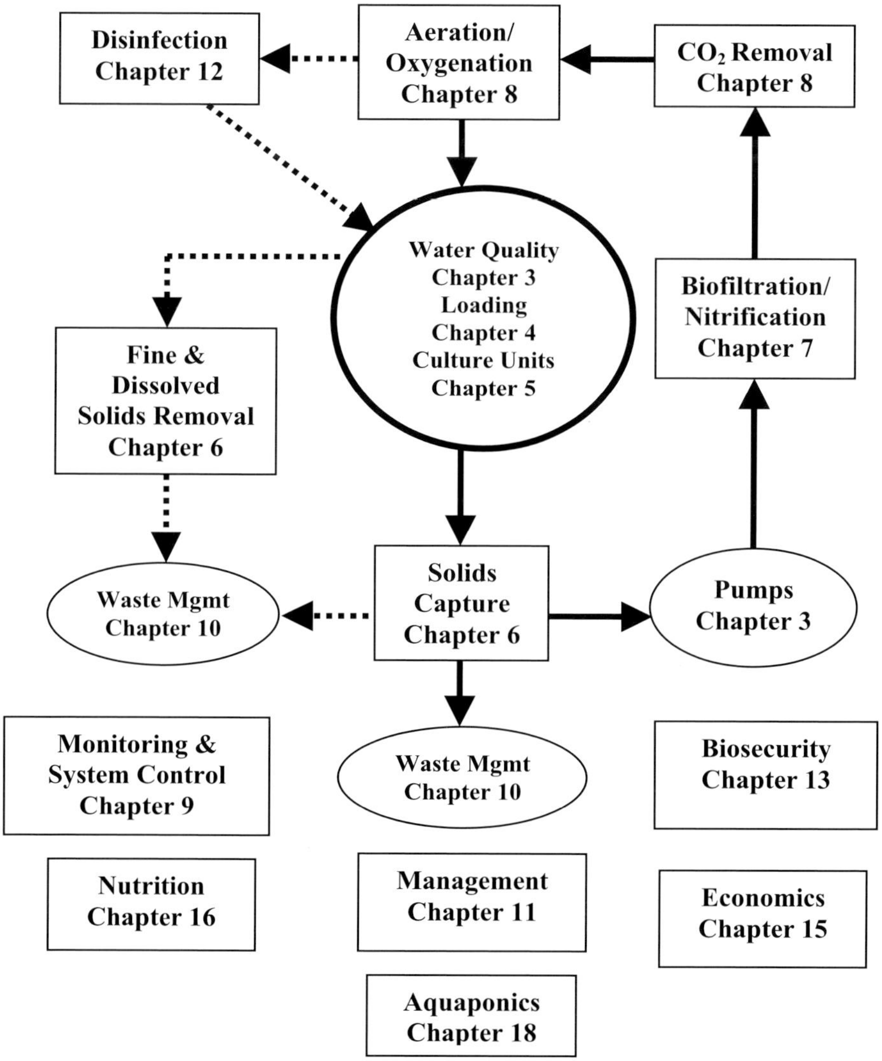

Figure 1.1 Unit RAS Operations.
Reference given to the book chapters that cover these topics.

1.6 WATER REQUIREMENTS, USE AND CONSERVATION

RAS requires less than 10% of the water and much less land than does extensive systems like ponds to grow the same quantity of fish. Table 1.1 provides a comparison of water used per kg of fish produced. The RAS assumes a tilapia culture system with a density of 100 kg/m^3, a 1% feeding rate and a conversion of 1 to 1. Some current commercial RAS are using less water, higher densities and similar feed conversions. RAS allow year-round production of consistent volumes of product, and complete climate control of the environment. Because RAS can be set up to produce the same volume of fish every week, week in and week out, they have a competitive advantage over outdoor tank and pond systems, which are seasonal and sporadic in harvest.

Table 1.1 Water Use Per kg of Production of Aquaculture Products (after Phillips et al. 1991)

Species and System	Production Intensity (kg/ha/y)	Water required (Liter/kg)	Ratio = Land or Water Use to RAS Use	
			Land	Water
O. niloticus (Nile tilapia) ponds	17,400	21,000	77	210
I. punctatus (Channel catfish) ponds	3,000	3,000–5,000	448	400
S. gairdneri (Rainbow trout) raceways	150,000	210,000	9	2,100
Panaeid shrimp pond (Taiwan)	4,200–11,000	11,000–21,340	177	160
O. niloticus (Nile tilapia) RAS	1,340,000[a]	100	1	1

[a]does not account for land used external to building space

RAS offers a high degree of environmental control. This not only mitigates the risks of outdoor aquaculture (natural disaster, pollution and disease) but also allows for optimized species growth on a year-round basis. Similar optimization can be observed in the domestic poultry industry, where chickens were brought indoors and the added cost of environmental control was easily overcome by higher growth rates. In addition to the growth advantages of RAS technology, the low environmental impact of these systems means that they can be built closer to the consumer. RAS has the clear advantage of being able to place production on the doorstep of the consumer, and deliver a fresher, safer product at lower transportation costs.

In the past, intensive fish farming used flowing water resources for two purposes: (1) to transport oxygen to the fish, and (2) to carry the waste produced in the system (metabolic by-products and other materials) away so that they did not accumulate in/around the fish farm to undesirable levels (Phillips et al. 1991). More recently, the carrying capacity of a flowing water fish farm has also become limited by imposition of state or federal discharge regulations. For example, traditional trout culture requires relatively large volume water resources to produce fish in a single-pass tank and serial-reuse raceway system. Clear Springs Trout Company (Buhl, Idaho), established in 1966, produced 8.2 million kilograms of trout in 1990 and is the world's single largest producer of rainbow trout for human consumption (MacMillan, 1992). According to MacMillan (1992), Clear Springs uses 22.6 m^3/s (360,000 gal/min) of water flow for up to five serial-reuses through concrete raceways. The only effluent treatment is either quiescent zone settling within the raceways or sometimes full-flow settling following the raceways (MacMillan, 1992; IDEQ, 1998).

Fortunately, serial-reuse systems may not have to actually capture large percentages of waste solids for their average discharge to meet concentration-based effluent limits on TSS (total suspended solids), even in Idaho where permits issued by NPDES (National Pollution Discharge Elimination System-EPA promulgated) limit fish farm effluents to monthly average suspended solids concentration of 5 mg/L (net). Similarly, serial-reuse systems may not have to capture large percentages of waste solids under typical fish production levels, i.e., 50 lb production per year per 1 gal/min or 6 kg/yr per Lpm, because a simple mass balance shows that only about 5 mg/L of total suspended solids concentration would be added to the flow if averaged over the entire day. Also, nutrient limits (is this basically residue of uneaten fish food?) are becoming a discharge issue. However, because of the large water volumes used in single-pass and serial-reuse production systems, it is not realistic for farms such as Clear Springs Trout Company to remove nutrients from their effluents (MacMillan, 1992; IDEQ, 1998). Therefore, some treatment by solids settling is used to ensure that spikes in TSS are not discharged, although overall waste capture efficiencies may only be 25–50% within serial-reuse systems (Mudrak, 1981); this topic is discussed in detail in Chapter 6: Solids Capture.

To abate the environmental impact of aquaculture, production practices and technologies are being adopted to minimize waste production, conserve water, and concentrate wastes into smaller flows during fish culture, thus the heightened interest in RAS. Traditional flowing water systems can produce approximately 6 kg of fish annually for every 1 Lpm of water flow, i.e., 50 lb of fish annually for every 1 gpm of water flow. By reusing or recycling 80 to 90% of the water prior to discharge, partial-reuse systems can produce as much as 48 kg of fish annually for every 1 Lpm of make-up water flow, i.e., 400 lb of fish annually for every gallon per minute of water flow (Summerfelt et al. 2000A). And of course in the extreme case where a 100% RAS is employed, the production is based upon rates of evaporation which means that several 100 fold increases in production per unit of water can be achieved compared to pond production.

Fully recirculating systems, because of their extremely low makeup water requirements, can readily capture from 96% (Heinen et al. 1996) to 100% of the waste produced, depending upon the amount of percentage of make-up water passed through the system. In comparison, a well operated serial-reuse raceway system can typically achieve overall waste capture efficiencies of only 25–50% (Mudrak, 1981). Additionally, by using "Cornell-type" dual-drain circular culture tanks and microscreen filters, recirculating systems can produce a much smaller and more concentrated waste stream, which can be treated more economically and efficiently (Timmons et al. 1998; Summerfelt et al. 2000A; 2000B Summerfelt, 1996; 1999). Thus, partial-reuse and fully-recirculating systems offer key advantages over traditional fish culture in serial-reuse raceway systems, including 80–100% reduced water resource requirements (respectively) and an overall waste capture efficiency of 80–100% (respectively). Also, solids removal from dual-drain circular tanks is so rapid and effective that partial-reuse and fully-recirculating systems can treat and return water to the culture tanks with <2.0 mg/L TSS (Summerfelt et al. 2000A; 2000B). Coldwater recirculating technology has now advanced to the point that these coldwater systems can provide a controlled environment with optimized water temperature, quality, and culture tank velocities (Summerfelt, 1996; Summerfelt et al. 2000A; Summerfelt et al. 2001).

1.7 MARKET DYNAMICS

With the fishing industry suffering from dramatic reductions in the supply of wild caught traditional species, alternatives to wild catches now have a competitive advantage in the marketplace. Where aquaculture was once viewed as the wave of the future, it is now generally accepted as a significant source of product. More than one third (33%) of all the seafood consumed around the world today is a product of aquaculture.

The leading forms of dietary protein consumed in the US include beef, poultry, pork, seafood, and turkey, which have consumption rates of 64 (29), 52 (24), 49 (22), 15 (7), and 14 (6) pound (kg) per capita, respectively (USDA). Inconsistent supply and subsequent higher retail costs has kept US consumption of seafood steady at approximately 15 pounds (7 kg) per capita during the last decade. As the wild catch of certain species such as cod has declined during the past decade, the supply of aquacultured seafood products has risen to meet the demand. Due to the highly favorable opinion of seafood (most Americans view seafood as being healthy), it is thought that the increased supply of cost-effective aquacultured products may increase overall seafood consumption rates. For example, the tilapia market is growing because consumers find its flavor to be delicate and its texture flaky compared to other species at a time when reduction in the supply of wild caught traditional species such as cod, haddock, halibut, and pollock, is forcing utilization of other species. Tilapia are of particular interest because they adapt readily to RAS and most of the US production of tilapia is from RAS, the subject of this book.

United States per capita consumption rates of individual seafood species is presented in Table 1.2. As indicated in bold typeface, ***aquacultured*** seafood species (catfish, shrimp and salmon) are becoming a significant portion of the $8.8 billion United States seafood market and have increased market share in contrast to ocean caught species such as cod which has dropped from then number 3 to number 6 position in consumption.

According to the January 2000 edition of Seafood Business, domestic imports of fresh and frozen tilapia fillets increased to 10,000 tons in 1999, an increase of 55% from 1998. Combined with domestic tilapia production, estimated to be less than 9,500 tons whole weight basis, United States per capita consumption of tilapia is estimated to have been approximately 0.15 pounds per capita in 1999. Thus, tilapia should be in the "top ten" within one or two years, replacing oysters. Increased importation of tilapia fillets is predicted to continue, with growth to

15,000 tons in CY 2000 (30,000,000 lbs of fillets or 100 million lbs whole fish basis). As mainstream domestic seafood customers become more familiar with tilapia and wild catch supplies of other whitefish continue to decrease, the tilapia market share should continue to exhibit strong growth. Although the tilapia market has doubled in the past few years, domestic production accounts for less than 20% of the total supply and is shrinking. Only shrimp and Atlantic salmon exceed tilapia on the list of imports of aquacultured seafood products.

Table 1.2 Current U.S. Seafood Consumption Rates by Species in Pounds (Bold signifies partial contribution from aquacultured products, National Marine Institute)

Species	2000 Rank	2000	1998	1995	1990	1990 Rank
Tuna	1	3.60	3.40	3.40	3.70	1
Shrimp	2	3.20	2.80	2.50	2.20	2
Pollock	3	1.69	1.65	1.52	1.27	4
Salmon	4	1.60	1.38	1.19	0.73	5
Catfish	5	1.13	1.06	0.86	0.70	6
Cod	6	0.77	0.97	0.98	1.38	3
Clams	7	0.49	0.39	0.57	0.61	7
Crabs	8	0.54	0.57	0.32	0.29	10
Flatfish	9	0.43	0.39	0.30	0.57	8
Scallops	10	0.28			0.30	9
Oysters	11	0.20	0.23	0.19	0.22	
Tilapia	Not ranked	0.15[a]			0.05	

The authors predict that the US tilapia industry will surpass the catfish industry within next 20 years. Eventually, RAS produced tilapia will be comparable to the US broiler industry that produces 15 billion lbs per year of chicken. This is because RAS will continue to become more cost efficient, and RAS technology allows production to be infinitely expanded. Generally, the primary constraint to building a fish farm or other animal production unit is finding a site that can handle the waste disposal needs. Natural environments that will support traditional

[a] Fillet basis

aquaculture such as raceways, net-pens, and outdoor ponds are extremely limited and subject to a political process. For example, in Canada recently, a net pen operation was denied their site renewal permit without explanation after 10 years of trouble free operation. RAS technology has a primary and exclusive advantage by minimizing the effluent waste stream by a factor of 500 to 1,000! Even so, RAS sites still need to address the issue of waste disposal, but at the reduced effluent load and ability to concentrate the waste stream, there are many cost effective waste treatment options.

Another key advantage of using RAS is that these production sites can be located near the markets for the products being produced. Fresh product becomes the immediate focus, since *fresh seafood commands twice the product value* of frozen seafood. There is a strong consumer preference for fresh and RAS are ideally suited to provide this product. Locating near the market also minimizes product-hauling costs and maximizes available shelf life. For reference only, Table 1.3 provides a breakdown of current US consumer preference and wholesale pricing on fresh seafood:

Table 1.3 US Market for Fresh Seafood (US NMFS, 2001)

Per capita Consumption (2000) Species	lb per capita	Wholesale Price, $/lb
Tuna	3.60	$5.50
Shrimp	3.20	
Alaska Pollock	1.69	$1.75
Salmon	1.60	$2.65
Cod	1.13	$2.65
Catfish (fillets)	0.77	$2.75
Clams	0.49	
Crabs	0.46	
Flounder/Sole	0.43	$7.50
Scallops	0.28	
Subtotal	13.65	
Sum of "White fish" only	4.02	
Average price of White fish, $/lb		$2.86
Wholesale Market Value for USA $3.2 billion of White Fish "only"		
Total US seafood consumption 15.6 lb/capita		

1.8 Interactions Between Objectives, Resources, Business Strategy, and Design

The authors frequently receive calls from individuals who desire to enter aquaculture with the callers having the inherent assumption that "aquaculture" means growing finfish. And of course, everyone's objective is to have a profitable enterprise. The economics of RAS are thoroughly reviewed in Chapter 15 and the reader is encouraged to read this chapter before starting an aquaculture venture. But, the first rule in aquaculture RAS is to never invest more than you can afford to lose. There are many hard-luck stories for every tale of success. Some families have lost everything in their pursuit of something they just wanted to do. One thing is certain: poor skills from one profession do not become great skills in aquaculture. You have to be good at it to make a profit and this generally entails several years of hands-on experience. This book gives you a head start, but is no substitute for real experience.

There are countless ways to become involved in aquaculture and perhaps the last choice should be to actually enter into the culture of fish. One must review their short term and long term objectives, the natural resources available to the enterprise, the management and business skills available for this enterprise (referred to as business strategy), and finally the design of a system to achieve these objectives. If any one of these steps is ignored or components are missing or insufficient, then the success of the planned operation is unlikely. Design of a physical system should be the last step in the planning process. Depending on the reader's own perspective, design is constrained by objectives, resources, and business strategy, or the design expresses the object resulting from considering them.

Although there is interaction between all components, the objectives must be determined first, and the resources available to pursue the objectives must be identified.

Table 1.4 lists some simple examples that will help to convey our thoughts behind the interactions among the resources, business strategy and resulting design. The objectives for each case are similar and not specific enough to define a design other than stating a possible structure. Selecting different management strategies that result in alternate designs, as shown in the second and third case, may approach the identical objective. The same resources are given for each case and are again too broad to be of value in design but show dependency upon components. For example, a person need not possess water as a personal resource in order to work in the aquaculture industry. Why not consider buying fish from a grower who is not interested in marketing their product? These examples show that objectives, resources and strategy must be specific to affect an appropriate design.

Table 1.4 Example Interactions of Objectives, Resources, Strategy and Design

Objectives	Resources	Business Strategy	Design
Have an annual income of $50,000	$500,000; financial institutions, water, market, knowledge	Invest at a net profit of 10%	Not necessary
Have an annual income of $50,000 from Aquaculture	Same	Buy and Sell	Not necessary
Same	Same	Buy, hold briefly & sell	Holding tanks, transport capability
Have an annual income of $50,000 from buying, processing tilapia fillets	Same	Buy, process, & sell	Processing facility
Have an annual income of $50,000 from selling tilapia fingerlings	Same	Maintain brood stock, incubate and hatch eggs, and maintain hatchery	Breeding tanks, egg incubator, fingerling production
Produce 114,000 kg per year of tilapia	Same	Maintain advanced growout system (needs more specific objectives, e.g., harvest size)	RAS or pond

1.9 TERMINOLOGY AND NOMENCLATURE

A primary requirement for discussing RAS technology is a common base of terms. We have a fairly well accepted set of terms now in use, which is based upon a report from the European Inland Fisheries

Advisory Commission (EIFAC, 1987). The following are the definitions and terminology that will be used throughout the chapters to come. Commonly used conversion factors are provided in the Appendix.

- **Carrying Capacity**: The maximum mass of aquacultured product that can be maintained within a culture system; usually expressed as mass per unit volume of the culture system

- **Flow Through Rate**: The volume of new water per unit time passing through a culture tank; refers to the make-up water specifically.

- **Mean Hydraulic Residence Time (HRT)**: Refers to the time required at a given rate of flow for a complete volume of water in a tank to be exchanged, e.g., Q(flow rate) / V (volume of tank)

- **Percentage Replacement**: The percentage of the total system volume replaced per day

- **Percentage Recycle**: The percentage of the total system volume that is retained, on a daily basis

- **Specific Surface Area**: Surface area of the media per unit volume; usually referring to the surface area of a particular media used in filtration or settling components

- **Stocking Density**: Mass of cultured product per volume of tank (ignores the effects of fish displacing part of the water volume)

- **Total Biomass**: Mass of cultured product in the culture system

- **Total System Volume**: Volume of water in the culture tank, pipes, reservoirs, treatment tanks, and pumps

- **Water Reuse**: Water is reused in multiple tanks, moving in one direction never used in the same tank twice; often referred to as serial reuse

- **Water Recycle**: Water flows from a tank to a treatment process then back to the tank

1.10 WEBSITES FOR REFERENCE

The reader is also referred to the following list of websites that contain vast amounts of information that can be useful for both the new and experienced aquaculturalist:

Annual Recirculating Aquaculture Systems Short Course
http://www.bee.cornell.edu/extension/aquaculture

Aquacultural Engineering Society
http://www.aesweb.org/
Newsletter - The latest quarterly newsletter of the AES is now on-line. Philosophy & Purpose - The official philosophy, purpose and operation of AES for your reference. Officers & Board - The current AES officers and board of directors. On-line Application - Use this electronic version of the member application to join AES!

American Tilapia Association
http://www.tilapia.org
Access to information about the fish, which is the fastest growing aquaculture crop in the United States and around the world. We are pleased to provide this information to those already producing tilapia for the food industry, for those interested in joining the industry and to potential customers and consumers of farm-raised tilapia.

AquaNIC - Aquaculture Network Information Center
http://aquanic.org/
An excellent clearinghouse of aquaculture information available on the Internet. It also provides information on how to obtain aquaculture information that is not on the Internet.

Alternative Farming Systems Information Center (AFSIC)
http://www.nal.usda.gov/afsic/
The AFSIC is one of 10 information centers at the National Agricultural Library. AFSIC serves as a national clearinghouse for aquaculture information and provides materials for aquafarmers, consumers, industry personnel, educators, government agencies, associations, libraries, the media, students, scientists, and prospective farmers.

1.11 SUMMARY

The authors have spent significant or all parts of their careers in aquaculture. Cornell has periodically hosted a one-week short course that hundreds of professionals have attended and this course is now (year 2001 and forward) being offered jointly by the Freshwater Institute and Cornell University. The course is also offered in a Distance Learning format (see details at: http://www.bee.cornell.edu/extension/aquaculture. The materials developed over the years for this short course are the primary materials being used in this book. The RAS short course emphasizes practical "how to" applications and concludes with a series of spreadsheets and computer programs to make the many tedious calculations necessary to design and manage an RAS easier. However, the computer programs are much more useful to the user if the basic fundamentals are first understood. This book attempts to provide a solid foundation for understanding the basics and also includes the computer software in Chapter 17. Example problems are given throughout the Chapters and these example problems (with solutions) can be used as a first try when using the software. Chapter 18, Aquaponics, was added to this revised version of the first edition. Have fun and enjoy the book.

CHAPTER 2

WATER QUALITY

2.0 INTRODUCTION

The success of a commercial aquaculture enterprise depends on providing the optimum environment for rapid growth at the minimum cost of resources and capital. One of the major advantages of intensive recirculation systems is the ability to control the environment and its numerous water quality parameters to optimize fish health and growth rates. Although the aquatic environment is a complex eco-system consisting of several water quality variables, it is fortunate that only a few of these parameters play decisive roles. The critical parameters are temperature, suspended solids, and pH and concentrations of dissolved oxygen, ammonia, nitrite, CO_2, and alkalinity. Each individual parameter is important, but it is the aggregate and interrelationship of all the parameters that influence the health and growth rate of the fish.

Each water quality parameter interacts with and influences other parameters, sometimes in complex ways. Concentrations of any one parameter that would be harmless in one situation can be toxic in another. For example, when aeration and degassing problems occur, carbon dioxide levels will generally become high while at the same time dissolved oxygen levels become low. The result of this particular situation is that not only is there less oxygen available to the fish, the fish are less able to use the oxygen that is available. The high carbon dioxide level of the water affects the fishes' blood capacity to transport oxygen, aggravating the stress imposed by low dissolved oxygen levels. Another excellent example of the complex interaction among water quality elements is the relationship between pH and the toxicity of ammonia. As will be discussed later, only the unionized fraction of the total ammonia concentration is toxic, and at low pH, most of the ammonia in the water is in the non-toxic ionized form. However, increasing the pH by only one unit, i.e., from 6.5 to 7.5, increases the concentration of the toxic unionized ammonia concentration by a factor of ten. Simply adding baking soda to a system to increase its alkalinity can inadvertently but easily create this extremely undesirable condition. The result, for one of the authors, was 32 tanks of angelfish swimming upside down. This catastrophic result provided a lesson learned the hard way about why it is important to understand the interrelationships of these parameters, and to routinely monitor all the critical water quality parameters and when adjustments are needed, to make them carefully and slowly.

The relationship between water quality factors and their effect on fish growth rate and health is complicated. For example, fish lack the means to control their body temperature and maintain it independent of the environment. Environmental temperature changes affect the fishes' rate of biochemical reactions, which leads to different metabolic and oxygen consumption rates. At the lower ranges of the species tolerable temperature range, these rates decrease. As water temperatures increase, fish become more active and consume more dissolved oxygen, while simultaneously producing more carbon dioxide and other excretory products, such as ammonia. These increasing rates of consumption of necessary elements and production of detrimental elements can have a direct effect on overall fish health and survival if these parameters are allowed to exceed nominal values. If not corrected, the fish will become stressed to some degree. Even low levels of stress can have adverse long-term consequences in the form of reduced growth rates or mortality due to opportunistic organisms that take advantage of the stressed fish.

2.1 PHYSICAL PROPERTIES

There are several physical properties of water that are important for the understanding of some of the engineering concepts that follow. Water has several very unique properties that make life on the planet possible. One of these is its density as a function of temperature, Table 2.1. Maximum density of pure water is reached at 3.98°C. A large discontinuity occurs at the freezing point, where the density drops drastically. Thus ice is lighter than water and floats. If water did not exhibit this unusual density pattern associated with temperature, lakes and oceans would freeze from the bottom up and life, as we know it, would be impossible. The addition of salt and other impurities also increases the density of water. For example, seawater at 35 ppt salt has a density of 1.028 g/cm^3, as compared to 1.000 g/cm^3 for pure water. Thus the density of water increases as salinity increases and/or temperature decreases. These density differences are the major driving force for overturn of freshwater systems and the vertical circulation patterns in the oceans.

Water viscosity is a measure of a fluid's resistance to shear, and is called absolute or dynamic viscosity. Highly viscous fluids, like molasses or motor oil, flow very slowly when subject to a shear stress. Absolute viscosity is measured in centipoises, and water at 20°C has a value of 1.0 centipoise, compared to air, which has a value of 0.17 centipoise, at this temperature. The absolute viscosity divided by the liquid density is the given the name kinematic viscosity. Table 2.1 shows the variation of density, absolute viscosity, and kinematic viscosity of water at a range of temperatures. Viscosity increases as temperature decreases. Thus, pumping costs will increase slightly as water temperature decreases because viscosity and density both increase.

Vapor pressure is the pressure exerted by the gaseous phase of a material when in equilibrium with its solid or liquid phase. Another way of expressing vapor pressure is the pressure at which a liquid just begins to boil and change to a vapor (Lawson, 1995). Vapor pressure of pure water is a function of temperature and increases with an increase in temperature, Table 2.1. The addition of salt to water lowers its vapor pressure. In a closed pipe, water may change phase from a liquid to a vapor because of a reduction in pressure, even though the temperature remains unchanged. This can occur on the suction side of pipe, where vapor bubbles form in local regions of very low pressure and then collapse when they move into regions of higher pressure downstream. This process is called cavitation. When a pump is cavitating, it is easily heard and causes severe vibrations in the pump impeller. These collapsing bubbles contain a great deal of energy and can cause considerable damage to pump impellers and pipes.

Formulae for generating water physical properties (T is in °C):

$$\text{Density (kg/m}^3\text{)} = 999.842594 + 6.793952\text{E-}2 \cdot T - 9.095290\text{E-}3 \cdot T^2 + 1.001685\text{E-}4 \cdot T^3 - 1.120083\text{E-}6 \cdot T^4 + 6.536336\text{E-}9 \cdot T^5 \quad (2.1)$$

$$\text{Vapor Pressure (mm Hg)} = 4.7603\ e^{0.0645T} \quad (2.2)$$

$$\text{Kinematic Viscosity (m}^2\text{/s } 10^{-6}\text{)} = -9.9653\text{E-}6 \cdot T^3 + 0.001143T^2 - 0.05807T + 1.7851 \quad (2.3)$$

2.2 WATER QUALITY STANDARDS

The very nature of aquaculture makes it almost impossible to formulate a definitive "one size fits all" list of water quality standards. The wide range of species, temperature regimes, and production techniques makes any such list only “recommendations”, at best. In intensive recirculation systems, the species of choice is raised in a kind of chemical soup consisting of numerous physical, chemical, and biological factors that are interrelated in a complex series of physico-bio-chemical reactions. These reactions affect every aspect of the culture from fish survival and growth rates to biofilter performance and solids removal. A basic understanding of water chemistry is critical for the success of any intensive system. Table 2.2 lists the most critical water quality parameters required to fully characterize the potential water supply and provides very general recommendations of water quality criteria for each parameter.

2.3 WATER QUANTITY REQUIREMENTS

During the initial site selection process, one of the most critical factors to consider is the availability of an adequate water supply for both the initial facility and any planned (or imagined) expansion. When it comes to the availability of water, too much is definitely better than too little. After all, one is constructing an **AQUA**culture facility. The amount of water needed will depend on several factors such as species, density, management practices, production technology, and the degree of risk one is willing to accept. At a minimum, sufficient quantities of water are needed to routinely fill production tanks within a reasonable time (24–48 hrs), provide for routine and emergency flushing of tanks, filter backwashing, facility wash down and clean-up, and domestic requirements. A good rule of thumb is to have sufficient water available to provide a 100% water exchange of total system volume per day (Timmons, 2000). Thus for a total system volume of 100,000 gallons (379 m^3), a water supply able to provide 100,000 gallons per day or 70 gpm (0.26 m^3/min) as the suggested requirement. This may be reduced some for a warm water species.

Beyond the minimum exchange volume, the additional quantity of new water required for any given system is directly dependent on the degree of reuse or recycling of the "old" water that is already available in the system. There are numerous reasons to reuse the supply water as much as possible, besides the obvious decrease in water demand and reduced discharge water that must be treated. These include the reduction of water heating or cooling requirements, which are major factors in the economics of many warm-water species production systems. Additionally, and becoming increasingly important, recycling of system waters results in the reduction in wastewater discharge and the corresponding costs involved in the treatment of large discharge flow streams.

Table 2.1 Physical Properties of Water

Temp. C	Density kg/m^3	Kinematic Viscosity (m^2/s)·E-06	Vapor Pressure mm Hg
0	999.84	1.79	4.8
1	999.90	1.73	5.1
2	999.94	1.68	5.4
3	999.97	1.62	5.8
4	1000.00	1.57	6.2
5	999.97	1.52	6.6
6	999.94	1.48	7.0
7	999.90	1.43	7.5
8	999.85	1.39	8.0
9	999.78	1.35	8.5
10	999.70	1.31	9.1
11	999.61	1.27	9.7
12	999.50	1.23	10.3
13	999.38	1.20	11.0
14	999.25	1.17	11.7
15	999.10	1.13	12.5
16	998.94	1.10	13.4
17	998.78	1.07	14.3
18	998.60	1.05	15.2
19	998.41	1.02	16.2
20	998.21	0.99	17.3
21	997.99	0.97	18.4
22	997.77	0.95	19.7
23	997.54	0.93	21.0
24	997.30	0.90	22.4
25	997.05	0.88	23.9
26	996.79	0.87	25.5
27	996.52	0.85	27.2
28	996.24	0.83	29.0
29	995.95	0.81	30.9
30	995.65	0.80	33.0
31	995.34	0.78	35.2
32	995.03	0.77	37.5
33	994.71	0.75	40.0
34	994.38	0.74	42.7
35	994.04	0.72	45.5
36	993.69	0.71	48.5
37	993.33	0.70	51.8
38	992.97	0.68	55.2
39	992.60	0.67	58.9
40	992.22	0.66	62.8

Table 2.2 Water Quality Criteria for Aquaculture

Parameter	Concentration (mg/L)
Alkalinity (as $CaCO_3$)	50–300
Aluminum (Al)	<0.01
Ammonia (NH_3-N unionized)	<0.0125 (Salmonids)
Ammonia (TAN) Cool-water fish	<1.0
Ammonia (TAN) Warm-water fish	<3.0
Arsenic (As)	<0.05
Barium (Ba)	<5
Cadmium (Cd)	
Alkalinity < 100 mg/L	<0.0005
Alkalinity > 100 mg/L	<0.005
Calcium (Ca)	4–160
Carbon Dioxide (CO_2)	
Tolerant Species (tilapia)	<60
Sensitive Species (salmonids)	<20
Chlorine (Cl)	<0.003
Copper (Cu)	
Alkalinity < 100 mg/L	0.006
Alkalinity > 100 mg/L	0.03
Hardness, Total (as $CaCO_3$)	>100
Hydrogen cyanide (HCN)	<0.005
Hydrogen sulfide (H_2S)	<0.002
Iron (Fe)	<0.15
Lead (Pb)	<0.02
Magnesium (Mg)	<15
Manganese (Mn)	<0.01
Mercury (Hg)	<0.02
Nitrogen (N_2)	<110% total gas pressure <103 % as nitrogen gas
Nitrite (NO_2)	<1, 0.1 in soft water
Nitrate (NO_3)	0–400 or higher
Nickel (Ni)	<0.1
Oxygen Dissolved (DO) (see Chapter 4 for more detail)	>5 > 90 mm Hg partial pressure)
Ozone (O_3)	<0.005
PCB's	<0.002
pH	6.5–8.5
Phosphorous (P)	0.01–3.0
Potassium (K)	<5
Salinity	depends on salt or fresh species
Selenium (Se)	<0.01
Silver (Ag)	<0.003
Sodium (Na)	<75
Sulfate (SO_4)	<50
TGP (total gas pressure)	<105% (species dependent)
Sulfur (S)	<1
Total dissolved solids (TDS)	<400 (site specific and species specific; use as rough guideline)
Total suspended solids (TSS)	<80
Uranium (U)	<0.1
Vanadium (V)	<0.1
Zinc (Zn)	<0.005

Source: Meade, 1991; Piper et al. 1982; Lawson, 1995

There are three categories of reuse systems: serial-reuse systems, partial-reuse systems, and fully recirculating systems. Partial-reuse systems reuse a greater percentage of total system volume than do serial-reuse systems, and the fully recirculating systems reuse a greater percentage of total system volume than do the partial-reuse systems. The degree of water reuse affects the depletion/accumulation rate of important water quality parameters. The greater the reuse rate, the more conditioning the reused water must have in order to restore the water quality to target parameters. Typically, the most important and first encountered limiting factor that governs the density of fish reared in a system is the concentration of dissolved oxygen that is available to the fish. The next most important factors are the amounts of un-ionized ammonia and dissolved carbon dioxide levels. These two parameters are interconnected. This is due to the direct effect that dissolved carbon dioxide has on pH and the relationship of pH to the toxicity of ammonia-nitrogen. As the dissolved carbon dioxide levels decrease, the pH increases, which in turn increases the toxicity of the total ammonia-nitrogen in the system. For example, if salmonids are the species being grown, the maximum upper safe limit for chronic exposure to carbon dioxide is from <9 to 30 mg/L, and for un-ionized ammonia-nitrogen <0.0125 to 0.03 mg/L. The chosen reuse system must be able to sustain the necessary levels of dissolved oxygen while also keeping the amounts of dissolved carbon dioxide and pH below their respective limiting levels.

Serial-reuse systems have been used extensively in trout and salmonid raceway production systems, where the limiting water quality factor is usually the dissolved oxygen concentration between raceway sections. By adding oxygen, the water can be used over again in the next section of the raceway, until ultimately the accumulated ammonia levels become too high. This simple concept has significantly increased raceway production, but at a higher system cost due to the requirements for more sophisticated oxygenation systems and monitoring systems. Economic risk is also higher, because there are more fish in the raceway that are subject to a catastrophic event.

Partial-reuse systems are an alternative culture system that can sustain high production densities on less than 20% of the total flow that would be required to grow the same quantity of fish in a serial-reuse system. Partial-reuse systems separate the solid wastes from the main water recirculation loop by using a dual-drain system. In this system, the circular production tank is used as a "swirl" separator to concentrate the solids into the center drain. From there, they are removed with a relatively small discharge flow stream (15–20% of the quantity of water in the tank). The majority of the water flow leaving the tank, i.e., the remaining 80–85%, is discharged from the tank through a fish-excluding port located approximately midway on the tank sidewall. This flow stream is relatively free of settleable solids and can be easily treated with a high-capacity microscreen filter, carbon dioxide stripping system, and an oxygenation system. The ammonia levels are controlled by dilution with make-up water, typically 10–20% of the flow (a volume approximately equal to the solids waste discharge flow from the center drain), and by controlling the system pH. The key operating water quality parameter in this type of system is the amount of dissolved carbon dioxide. Adjusting the amount of carbon dioxide stripped by the degassing system will control the system's pH. When the system is operated so that the dissolved carbon dioxide level is the limiting water quality parameter, the water will have a low pH, and thus the corresponding maximum total ammonia-nitrogen level will also be well below the level at which it would become critical.

The nomograph in Fig. 2.1 provides an estimate of the maximum total ammonia-nitrogen that can be allowed to accumulate in a partial-reuse system, depending upon the water's alkalinity and assumed maximum limits on the concentrations of dissolved carbon dioxide and un-ionized ammonia (Summerfelt et al. 2000A). Thus setting the maximum limits on the concentrations of dissolved carbon dioxide and un-ionized ammonia, at a given alkalinity, will set the minimum operating pH and the maximum allowable total ammonia-nitrogen (TAN).

2.4 WATER SOURCES

One of the most important requirements for a successful aquaculture facility site is a high quality water source, with sufficient capacity to provide for initial needs and contemplated future expansions. Thus, when choosing an aquaculture site, the water supply must be thoroughly investigated and quantified in terms of both quality and quantity. Aquaculture is unlike other agriculture activities, in that it requires a continuous supply of high quality water. Most wells operate only periodically, so a traditional well pump test is usually conducted for only a short time, to measure the well's ability to deliver a certain quantity of water within a certain time. However, under normal aquaculture operations, the well is pumped continuously. The result can be that the actual productivity of the well is much less than indicated by the short duration test, and that the originally expected quantities of water will not be available. This will result in serious difficulties down the road.

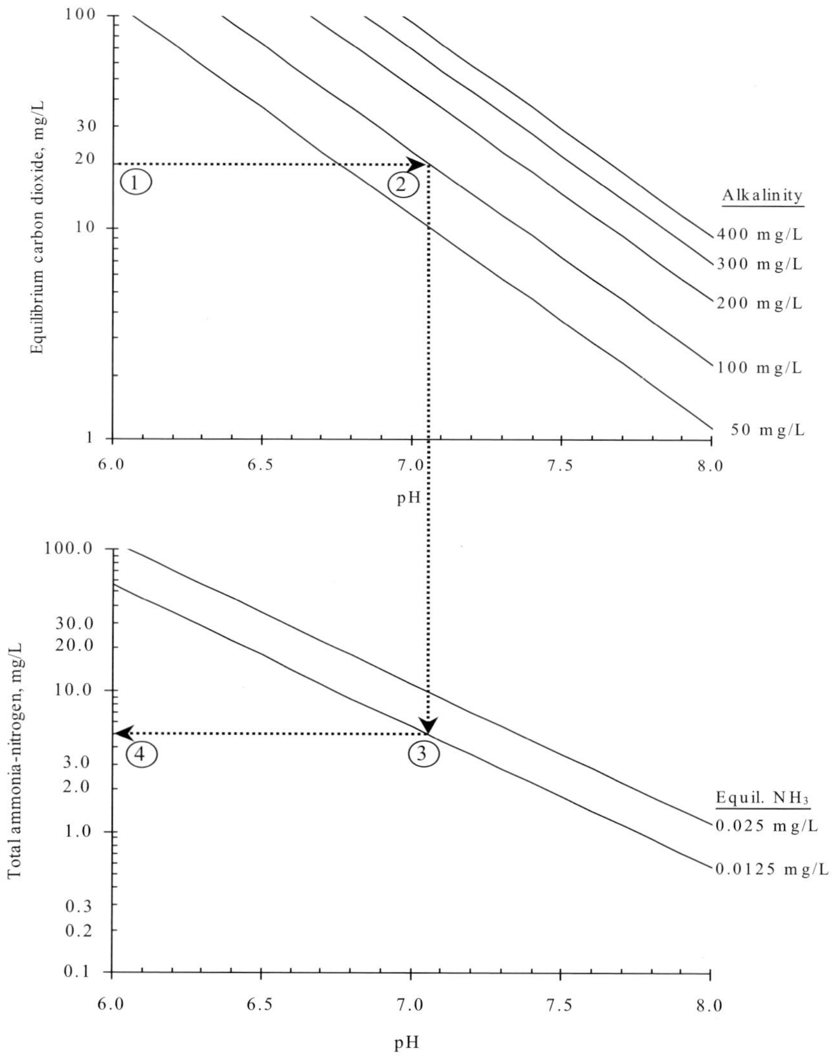

Figure 2.1 Dependence of Dissolved Carbon Dioxide Concentrations on pH and Alkalinity (top graph), and Un-ionized Ammonia Concentrations and Total Ammonia-nitrogen (bottom graph), According to Acid-base Equilibrium at 15°C. Dashed lines indicate how setting a maximum limit of 20 mg/L on the concentrations of dissolved carbon dioxide (1) and of 0.0125 mg/L on un-ionized ammonia (3), at an alkalinity of 100 mg/L (2), will set the minimum operating pH (about 7.06) and the maximum allowable TAN (4) (about 5.0 mg/L), from Summerfelt et al. 2000A.

There are several sources of water for aquaculture operations, each with distinct advantages and disadvantages. Since the whole point of a recirculating aquaculture system (RAS) is to minimize water usage, the two most often used water sources are groundwater and municipal water supplies. Both normally have the quality, quantity, and reliability required, so the actual choice between them is based on availability and economics. Surface waters are generally not used, due to the higher risks of contamination by pollutants, fish eggs, insect larva, disease microorganisms, and wide seasonal temperature variations.

One of the major advantages of groundwater sources is their constant temperature throughout the year. Shallow sources of groundwater approximate the mean air temperature of the area. The chemistry of groundwater is directly dependent on the geology of the area surrounding the source. In limestone areas, groundwater is hard, and high in calcium and carbon dioxide. In areas of granite formation, the groundwater tends to be soft, low in dissolved minerals and carbon dioxide. As will be discussed later, there are advantages and disadvantages to both, emphasizing the need for early extensive water quality testing.

The disadvantages of groundwater, especially from deep wells, are the high concentrations of dissolved toxic gases such as hydrogen sulfide (H_2S), methane (CH_4), and carbon dioxide (CO_2). Most groundwater contains little or no dissolved oxygen because of the biological processes occurring in the aquifer recharge zone. In addition, once pumped to the surface, groundwater often is supersaturated with nitrogen gas. Thus, the groundwater must be initially conditioned by some form of aeration and degassing to remove the excess nitrogen gas and add oxygen. High concentrations of dissolved iron can also be a problem. Once aerated, the iron precipitates out of solution and often must be removed by some form of granular filtration or settling basin.

Another potential source of water is from a municipal water supply. However, it must be kept in mind that these sources are designed and treated to safeguard the health of humans, i.e., chemicals such as chlorine (1.0 mg/L) and fluorine are added. To be useable in RAS; these chemicals must be neutralized either chemically (sodium thiosulfate to remove chlorine and chloramines) or by filtration (activated carbon). Chlorine is very detrimental to fish and levels as low as 0.02 mg/L are stressful.

2.5 WATER QUALITY PARAMETERS

DISSOLVED OXYGEN

Of all the water quality parameters, dissolved oxygen is the most important and most critical parameter, requiring continuous monitoring in intensive production systems. Nature played a cruel joke on aquaculture when she decided that the saturation concentration of dissolved oxygen would be highest at low temperature and lowest at high temperatures. This condition is exactly the opposite of what fish require for basic metabolism and food conversion, which is highest at high temperatures and lowest at low temperatures. Although the air we breathe contains 21% oxygen, oxygen is only slightly soluble in water. As a result, aquatic species must spend a great deal of energy to remove the dissolved oxygen from water, as compared to the energy that land dwelling species expend to obtain oxygen from the air. As previously mentioned, oxygen solubility decreases as temperature and salinity increase. Both barometric pressure and altitude directly affect oxygen concentration. Table 2.3 lists the oxygen concentration of water as function of salinity and temperature.

It is difficult to specify critical dissolved oxygen concentrations, because the response to low dissolved oxygen is not life-or-death, but a continuum of physiological effects. Also, these effects are influenced by the exposure time, the size and health of the fish, water temperature, concentration of carbon dioxide, and other environmental conditions. In general though, warm water fish feed best, grow fastest, and are healthiest when dissolved oxygen concentrations are above about 5 mg/L. However, concentrations of dissolved oxygen greater than this level of saturation appear to provide no additional benefit to fish. The gills can only transfer so much oxygen to the blood, and are very near or at the maximum transfer capability when the environmental dissolved oxygen concentrations are at the recommended concentrations. Higher oxygen concentrations in the water do not result in any additional oxygen carried by the blood stream.

TEMPERATURE

Water temperature is second only to dissolved oxygen in importance and impact on the economic viability of a commercial aquaculture venture. Temperature directly affects the physiological processes, such as respiration rate, efficiency of feeding and assimilation, growth, behavior, and reproduction. Fish have traditionally been grouped into three classifications depending on their temperature preference: coldwater, cool-water, and warm-water. Cold-water species prefer water temperatures in the range of 15°C (60°F), cool-water species between 15°–20°C (60°–68°F), and warm-water species above 20°C (68°F). These are not exact definitions and several factors are involved in determining the tolerance of fish to different temperatures; these include species, age, size, and past thermal history.

Table 2.3 Dissolved Oxygen (mg O_2 per Liter, ppm) at Saturation in Freshwater, Brackish Water, and Seawater at Different Temperatures

Temperature	Chlorinity (0/00)[1]					
(C)	0	4	8	12	16	20
1	14.24	13.54	12.91	12.29	11.70	11.15
2	13.84	13.18	12.56	11.98	11.40	10.86
3	13.45	12.84	12.25	11.68	11.12	10.59
4	13.09	12.51	11.93	11.38	10.83	10.34
5	12.75	12.18	11.63	11.09	10.57	10.10
6	12.44	11.86	11.33	10.82	10.32	9.86
7	12.13	11.58	11.06	10.56	10.07	9.63
8	11.85	11.29	10.80	10.32	9.84	9.40
9	11.56	11.02	10.54	10.07	9.61	9.20
10	11.29	10.77	10.30	9.84	9.40	9.00
11	11.05	10.53	10.07	9.63	9.20	8.80
12	10.80	10.29	9.84	9.41	9.00	8.61
13	10.56	10.07	9.63	9.21	8.81	8.42
14	10.33	9.86	9.41	9.01	8.62	8.25
15	10.10	9.64	9.23	8.83	8.44	8.09
16	9.89	9.44	9.03	8.64	8.28	7.94
17	9.67	9.26	8.85	8.47	8.11	7.78
18	9.47	9.07	8.67	8.31	7.97	7.64
19	9.28	8.88	8.50	8.15	7.08	7.49
20	9.11	8.70	8.32	7.99	7.66	7.36
21	8.93	8.54	8.17	7.84	7.52	7.23
22	8.75	8.38	8.02	7.69	7.39	7.11
23	8.60	8.22	7.87	7.55	7.26	6.99
24	8.44	8.07	7.72	7.42	7.13	6.86
25	8.27	7.92	7.58	7.29	7.01	6.75
26	8.12	7.78	7.45	7.16	6.89	6.63
27	7.98	7.64	7.32	7.03	6.78	6.52
28	7.84	7.51	7.19	6.92	6.66	6.40
29	7.69	7.38	7.08	6.82	6.55	6.29
30	7.56	7.25	6.96	6.70	6.45	6.19

[1]Conversion Chlorinity (CI) to Salinity (S) is: S = 1.80655 x CI
Normal seawater is 35 0/00 or 35 ppt S (19.37 ppt CI)

Fish are classified as poikilothermic or cold-blooded, which means that their body temperature is approximately the same as their surrounding environment. Therefore, each species has an optimum temperature range that maximizes growth and an upper and lower limit beyond which they cannot survive. Within the species tolerable temperature range, growth rates increase as the water temperature increases, until the optimum temperature is reached. Above the optimum temperature, the increased energy requirements for food conversion and other metabolic processes ensure that the law of diminishing returns applies. Also, at higher than optimum temperatures, the fish food conversion ratios are lower. Further temperature increases beyond optimum are of no benefit, and may in fact approach lethal levels. Thus to insure maximum growth and minimize stress, system temperatures need to be maintained as close as possible to optimum value. Optimum temperature range for several representative species is presented in Table 2.4.

Table 2.4 Optimum Temperature Ranges (°C) for Representative Aquaculture Species

Species	Temp. Range	Source
brown trout (*Salmo trutte*)	12–14°	Aston, 1981
rainbow trout (*O. mykiss*)	14–16°	Aston, 1981
brook trout *(Salvelinus fontinalus)*	7–13°	Piper, et al. 1982
Chinook salmon *(Oncorhynchus tshawytscha)*	10–14°	Piper, et al. 1982
Coho salmon *(Oncorhynchus kisutch)*	9–14°	Piper, et al. 1982
Atlantic salmon (*S. salar*)	15°	Aston, 1981
sockeye salmon (*Oncorhynchus nerda*)	15°	Aston, 1981
turbot (*Scophthalmus maximus*)	19°	Aston, 1981
sole (*Solea solea*)	15°	Aston, 1981
channel catfish (*Ictalurus punctatus*)	25–30°	Tucker & Robinson, 1990
striped bass (*Moronoe saxatilis*)	13–24°	Piper, et al. 1982
red drum (*Sciaenops ocellatus*)	22–28°	Piper, et al. 1982
largemouth bass (*Micropterus salmoides*)	25–30°	Piper, et al. 1982
European eel (*Anguilla anguilla*)	22–26°	Aston, 1981
Japanese eel (*A. jabonica*)	24–28°	Aston, 1981
carp (*Cypriums carpio*)	25–30°	Aston, 1981
mullet (*Mugil cephalus*)	28°	Aston, 1981
tilapia (*Sarotheorodon, sp.)*	28–32°	Aston, 1981

AMMONIA / NITRITE / NITRATE

These derivatives of nitrogen are contaminants in aquaculture system water columns, and must be contained within acceptable concentrations. Nitrogen is an essential nutrient for all living organisms, and is found in proteins, nucleic acids, adenosine phosphates, pyridine nucleotides and pigments. However, nitrogen is required in relatively small quantities, and physiological needs are easily satisfied. Excess quantities then become nitrogenous wastes, and removal is necessary. The fish create and expel various nitrogenous waste products through gill diffusion, gill cation exchange, and urine and feces excretion. In addition to the urea, uric acid, and amino acid excreted by the fish, nitrogenous wastes accumulate from the organic debris of dead and dying organisms, uneaten feed, and from nitrogen gas in the atmosphere. Decomposing these nitrogenous compounds is particularly important in intensive RAS because of the toxicity of ammonia, nitrite, and to some extent, nitrate.

Ammonia, nitrite, and nitrate are all highly soluble in water. Ammonia exists in two forms: un-ionized NH_3, and ionized NH_4^+. The relative concentration of ammonia is primarily a function of water pH, salinity, and temperature. The sum of the two ($NH_4^+ + NH_3$) is called total ammonia or simply ammonia. It is common in chemistry to express inorganic nitrogen compounds in terms of the nitrogen they contain, i.e., NH_4^+–N (ionized ammonia nitrogen), NH_3–N (un-ionized ammonia nitrogen), NO_2^-–N (nitrite nitrogen) and NO_3^-–N (nitrate nitrogen). This allows for easier computation of total ammonia-nitrogen (TAN = NH_4^+–N + NH_3–N) and easy conversion between the various stages of nitrification.

Un-ionized ammonia (NH_3–N) is the most toxic form of ammonia, so the toxicity of TAN is dependent on the percentage of the un-ionized form in the TAN concentration. An increase in pH, temperature, or salinity increases the proportion of the un-ionized form of ammonia nitrogen. For example, at 20°C and a pH of 7.0, the mole fraction of un-ionized ammonia is only 0.004, but at a pH of 10, the un-ionized ammonia increases to 0.8. The fraction of un-ionized ammonia at different temperatures and pH are included in the Appendix. Un-ionized ammonia is toxic to fish at low concentrations, with 96-hour LC_{50}'s varying widely by species from as low as 0.08 mg/L NH_3-N for pink salmon to 2.2 mg/L NH_3-N for common carp. In general, warm-water fish are more tolerant to ammonia toxicity than cold-water fish, and freshwater fish are more tolerant than saltwater fish. Un-ionized ammonia concentrations should be held below 0.05 mg/L and TAN concentrations below 1.0 mg/L for long-term exposure. Table 2.5

summarizes the effects of temperature and pH on the percentage of free ammonia in freshwater.

Table 2.5 Percentage of Free Ammonia (as NH_3) in Freshwater at Varying pH and Water Temperature, (Spotte, 1979)

pH	10°C (50°F)	15°C (59°F)	20°C (68°F)	25°C (77°F)
7.0	0.19	0.27	0.40	0.55
7.1	0.23	0.34	0.50	0.70
7.2	0.29	0.43	0.63	0.88
7.3	0.37	0.54	0.79	1.10
7.4	0.47	0.68	0.99	1.38
7.5	0.59	0.85	1.24	1.73
7.6	0.74	1.07	1.56	2.17
7.7	0.92	1.35	1.96	2.72
7.8	1.16	1.69	2.45	3.39
7.9	1.46	2.12	3.06	4.24
8.0	1.83	2.65	3.83	5.28
8.1	2.29	3.32	4.77	6.55
8.2	2.86	4.14	5.94	8.11
8.3	3.58	5.16	7.36	10.00
8.4	4.46	6.41	9.09	12.27
8.5	5.55	7.98	11.18	14.97

Nitrite is the intermediate product in the process of nitrification of ammonia to nitrate. Although nitrite is converted to nitrate relatively quickly by ozone and the nitrifying bacteria in a properly balanced biofilter, it is a problem in recirculating systems because it is being created on a constant basis, so the fish are continually exposed to certain concentrations of this chemical. Therefore, it is an important water quality parameter to monitor and correct if acceptable limits are exceeded. Nitrite is toxic because it affects the blood hemoglobin's ability to carry oxygen. When nitrite enters the bloodstream, it oxidizes the iron in the hemoglobin molecule from the ferrous state to the ferric state. The resulting product is called methemoglobin, which has a characteristic brown color, hence the common name "brown-blood disease". The amount of nitrite entering the blood depends on the ratio of nitrite to chloride in the water, in that increased levels of chloride reduce the amount of nitrite absorption. Chloride levels can be increased by adding ordinary salt (sodium chloride) or calcium chloride. At least a 20:1 ratio of chloride to nitrite-nitrogen (Cl: NO_2–N) is recommended for channel catfish in ponds, tilapia, and rainbow trout.

Nitrate is the end product of nitrification and is the least toxic of the nitrogen compounds, with a 96-h LC values usually exceeding 1000 mg NO_3-N /L. In recirculation systems, nitrate levels are usually controlled by daily water exchanges. In systems with low water exchange or high hydraulic retention times, denitrification has become increasingly important.

PH

The pH value expresses the intensity of the acidic or basic characteristic of water. In chemical terms, pH is the negative logarithm of the hydrogen ion concentration. The pH scale ranges from 0 to 14, with a pH of 7.0 corresponding to the neutral point. Values of pH below 7.0 are acidic (the H^+ ion predominates), and above 7.0, values are basic or alkaline (the OH^+ ion predominates). The pH of most groundwaters and surface waters are buffered by the bicarbonate-carbonate system and have pH values between 5 and 9. Exceptions to this are groundwater with high concentrations of dissolved carbon dioxide, and water draining from some mines (acid mine drainage (AMD)). AMD is the result of the oxidation of mineral sulfides, which becomes sulfuric acid. Seawater is buffered by the bicarbonate-borate system and has a relatively stable pH between 8.0 and 8.5. The optimum pH for the growth and health of most freshwater aquatic animals is in the range of 6.5 to 9.0.

Exposure to extreme pH can be stressful or lethal, but it is the indirect effects resulting from the interactions of pH with other variables that are more important in aquaculture. pH controls a wide variety of solubility and equilibria reactions, the most important of which is the relationship between the un-ionized and the ionized form of ammonia and nitrite. pH also affects the toxicity of hydrogen sulfide and metals such as copper, cadmium, zinc and aluminum.

ALKALINITY/HARDNESS

In general terms, alkalinity is a measure of the pH-buffering capacity or the acid-neutralizing capacity of water. In chemical terms, alkalinity is defined as the total amount of titratable bases in water expressed as mg/L equivalent calcium carbonate ($CaCO_3$). Sometimes alkalinity is expressed as milliequivalents/liter, where 1 meq/L equals 50 mg/L as $CaCO_3$. The principle ions that contribute to alkalinity are carbonate

(CO_3^{-}) and bicarbonate (HCO_3^{-}). Table 2.6 provides a list of commonly used alkalinity supplements, their relative solubility, and their equivalents basis (also see Chapter 7 for discussion of nitrification and alkalinity consumption). See Table A-7 for hardness conversions to a variety of units.

Table 2.6 Alkalinity Supplement Properties (Bisogni and Timmons, 1991)

Chemical Formula	Common Name(s)	Equivalent Wt. (gm/eq.)	Solubility	Rate of solubilization
$NaOH$	sodium hydroxide	40	high	rapid
Na_2CO_3	sodium carbonate soda ash	53	high	rapid
$NaHCO_3$	sodium bicarbonate baking soda	83	high	rapid
$CaCO_3$	calcium carbonate calcite	50	moderate	moderate
CaO	slaked lime	28	high	moderate
$Ca(OH)_2$	calcium hydroxide hydrated lime	37	high	moderate
$CaMg(CO_3)_2$	dolomite	46	moderate	slow
$MgCO_3$	magnesium carbonate magnesite	42	moderate	slow
$Mg(OH)_2$	magnesium hydroxide brucite	29	moderate	slow

Note: Na compounds are highly soluble in water while Mg compounds have poor solubility. Ca compounds are intermediate. Mg compounds tend to dissolve very slowly, so may have application in situations requiring a long-term application. Na compounds may prove to be the most expensive.

Based on 100% pure compound. To calculate for impurities, divide the tabulated value by the pure fraction ({100%-impurities %}/100 gives pure fraction) to get the true value.

In practical terms, alkalinity is measured by titration with sulfuric or hydrochloric acid to the methyl orange endpoint (pH of 4.5). Alkalinity of freshwater ranges from less than 5 mg/L in soft water to over 500 mg/L, and is determined by the geology of the aquifer or watershed. The alkalinity of seawater is about 120 mg/L $CaCO_3$. Required alkalinity concentrations are directly linked to system pH and carbon dioxide concentrations. Maintaining carbon dioxide concentrations at less than 15 mg/L and pH between 7.0 and 7.4 requires an alkalinity concentration less than 70 (high pH condition) to 190 mg/L $CaCO_3$ (low pH condition). The relationship between pH, alkalinity, and CO_2 concentrations is shown in Fig. 2.2.

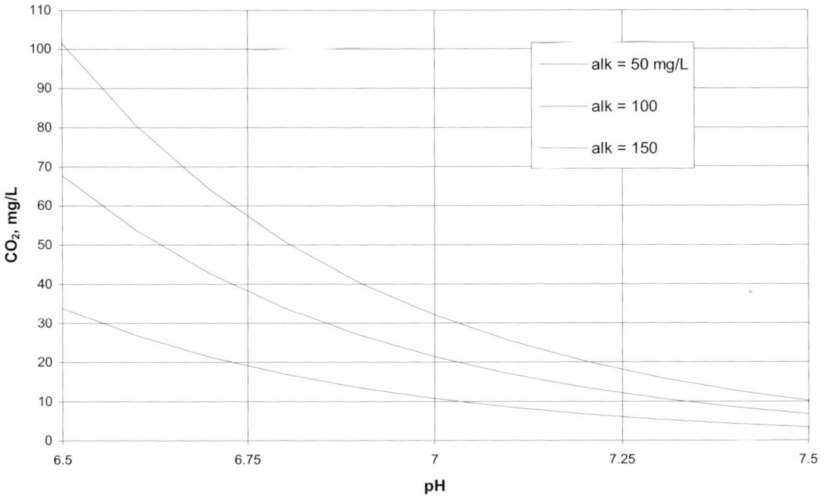

Figure 2.2 The Relationship Between pH, Alkalinity, and CO_2 Concentrations When Alkalinity Is 50, 100, 150 mg/L (Note CO_2 Concentration Is Proportional to Alkalinity).

In recent years, as aquaculture system stocking density and hydraulic retention time has increased, the relationship between pH and alkalinity has become a significant issue. This relationship requires careful monitoring and adjustment of both alkalinity and carbon dioxide levels to maintain optimum pH for both the aquatic species being grown and the biofilters. Alkalinity is easily adjusted through the addition of sodium bicarbonate ($NaHCO_3$), common baking soda. Other materials can be used, but sodium bicarbonate is commercially available in 50 to 100 lb (23 to 45 kg) bags, safe, inexpensive, and easy to apply. It has very high water solubility and rapidly dissolves in water at ambient temperature. A general rule of thumb is that for every pound of feed, approximately 0.25 lbs (113 g) of sodium bicarbonate should be added to the water. Carbon dioxide concentrations are routinely controlled through degasification systems, such as counter-flow gas stripping towers.

Hardness is the term used to describe the ability of water to precipitate soap. The harder the water, the greater the amount of soap that must be added to a given volume of water to get the same cleaning action. In chemical terms, hardness is defined as the total concentration of primarily calcium (Ca^{2+} and magnesium (Mg^{2+}), iron, and manganese in terms of mg/L equivalent calcium carbonate ($CaCO_3$). In practical terms, hardness is measured by chemical titration. The total hardness of natural waters ranges from less than 5 to over 10,000 mg/L $CaCO_3$. Waters have traditionally been classified as soft (0–75 mg/L), moderately hard (75–150 mg/L), hard (150–300 mg/L), or very hard (> 300 mg/L).

Hardness is often confused with alkalinity, probability because both are commonly defined in terms of mg/L $CaCO_3$. In fact, if the alkalinity of the water originates from limestone, the concentrations of hardness and alkalinity can be similar if not identical. Conversely, many coastal-plain area groundwaters have high alkalinity and very low hardness concentrations. Aquifers in regions of basalt and granite often have waters of low total hardness and low alkalinity, because of the relative low solubility of these minerals. If this low hardness and low alkalinity water is used for aquaculture, the water must be "hardened" by adding dissolved calcium to support newly fertilized freshwater fish eggs and for calcification of larval skeletal structures. Calcium and magnesium also decrease the toxicity of dissolved metals. Recommendations for total hardness range from 20 to 300 mg/L.

Carbon Dioxide and the Carbonate Cycle

Carbon dioxide is highly soluble in water, but concentrations in pure water are low (0.54 mg/L at 20°C), due to its low concentration in the atmosphere (about 0.035% by volume). Most of the carbon dioxide in an aquaculture water column is produced by animal respiration and the decomposition of organic matter, with a small percentage coming from atmospheric diffusion. The concentration of carbon dioxide in groundwater can range from 0 to 100 mg/L, depending on the biological activity in the aquifer recharge zone. The concentration of dissolved carbon dioxide in surface waters depends on the rate of respiration, photosynthesis, and gas exchange with the atmosphere.

Exposure to high carbon dioxide concentrations reduces respiration efficiency and decreases the tolerance to low dissolved oxygen concentrations. High levels of carbon dioxide in the water reduce carbon dioxide excretion at fish gills. This, in turn, causes the CO_2 concentration in the fish blood to increase, lowering the blood plasma pH, which creates a condition called respiratory acidosis. When a fish is in this condition, the amount of oxygen that the blood hemoglobin can carry is reduced and respiration distress can occur, even with high concentrations of dissolved oxygen in the water. This is referred to as the Bohr-Root effect. An upper limit of 15–20 mg/L carbon dioxide is recommended as steady state maximum for finfish (see Table 2.2), although this recommendation is poorly supported by research. However, as a management technique, higher concentrations (60–80 mg/L) have a narcotic effect on aquatic animals, and are sometimes used on a temporary basis as an anesthetic to reduce handling stress and during treatment procedures.

Carbon dioxide differs from oxygen, nitrogen, and other gases, because its concentration in water is determined both by a gas-liquid equilibrium relationship and also by a series of acid-base reactions. Gas-liquid equilibrium influences the transfer of carbon dioxide between air and water and the acid-base reactions determine the chemical form in which dissolved inorganic carbon is present in water. Dissolved carbon dioxide concentration is therefore a function of pH and the total dissolved inorganic carbon ($CtCO_3$) present in water, i.e., carbon dioxide (CO_2), carbonic acid (H_2CO_3), bicarbonate ions (HCO_3^-), and carbonate ions (CO_3^{2-}). It is more correct to write $H_2CO_3^*$ for carbonic acid where $H_2CO_3^*$ is a composite carbonic acid concentration that includes true carbonic acid plus dissolved carbon dioxide or

$$[H_2CO_3^*] = [H_2CO_3] + [CO_{2aq}] \tag{2.4}$$

The acid-base equilibrium relationships between pH and carbon dioxide in freshwater and seawater aquaculture systems have been extensively reviewed and a summary of the carbonate acid-base system is presented in Table 2.7 (Grace & Piedrahita, 1994, Stumm and Morgan, 1981).

The carbonate system can be considered to be a volatile system or a non-volatile system depending on whether or not aqueous carbon dioxide is allowed to exchange and equilibrate with atmospheric carbon dioxide, volatile being one that has equilibrated with the atmosphere. RAS's with medium or higher fish densities (> 40 kg fish biomass per m^3) are considered non-volatile due to:

- high production of CO_2
- high solubility of CO_2
- fish tank hydraulic retention times are generally 20 minutes or more (primary reason).

However, as densities decrease and hydraulic retention times reduce to less than 20 minutes, the RAS will begin to approach a volatile system. In a fully volatile system, the dissolved CO_2 will approach ~0.5 mg/L (example given below). These relationships are summarized in Figs. 2.3 and 2.4 (from Bisogni and Timmons, 1994). Using Fig. 2.3 for an open system, the dissolved CO_2 can be calculated as follows assuming a system pH of 7.0:

$$pH = 7.0 \Rightarrow [H_2CO_3] = 0.00001 \frac{mole_{CO_2}}{L} \cdot \frac{44g}{mole_{CO_2}}$$

$$= 0.00044 \frac{g}{L} \cdot 1000 \frac{mg}{g} = 0.44 \frac{mg}{L}$$

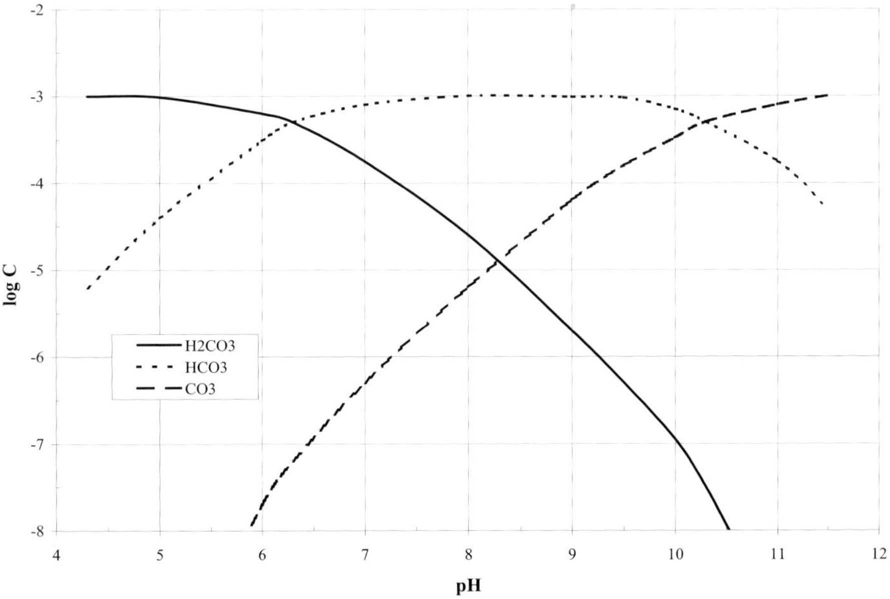

Figure 2.3 Non-volatile Log C-pH Diagram (from Bisogni and Timmons, 1994.

Determination of equilibrium carbon dioxide concentrations requires that two of following three quantities be known: pH, total carbonate carbon, alkalinity. For example, total dissolved carbon dioxide, $[H_2CO_3^*]$, can be calculated from the total carbonate carbon, $[CtCO_3]$, and pH, since pH defines $[H^+]$:

$$[H_2CO_3^*] = [CtCO_3] \cdot \frac{1}{\left(1 + \frac{K_0 K_1}{[H^+]} + \frac{K_0 K_1 K_2}{[H^+]^2}\right)} \quad (2.5)$$

as well as from the alkalinity (ALK) and pH:

$$[H_2CO_3^*] = \left(\frac{ALK}{50000} - \frac{K_w}{[H^+]} - [H^+]\right) \cdot \left(\frac{1}{\frac{K_0 K_1}{[H^+]} + \frac{K_0 K_1 K_2}{[H^+]^2}}\right) \quad (2.6)$$

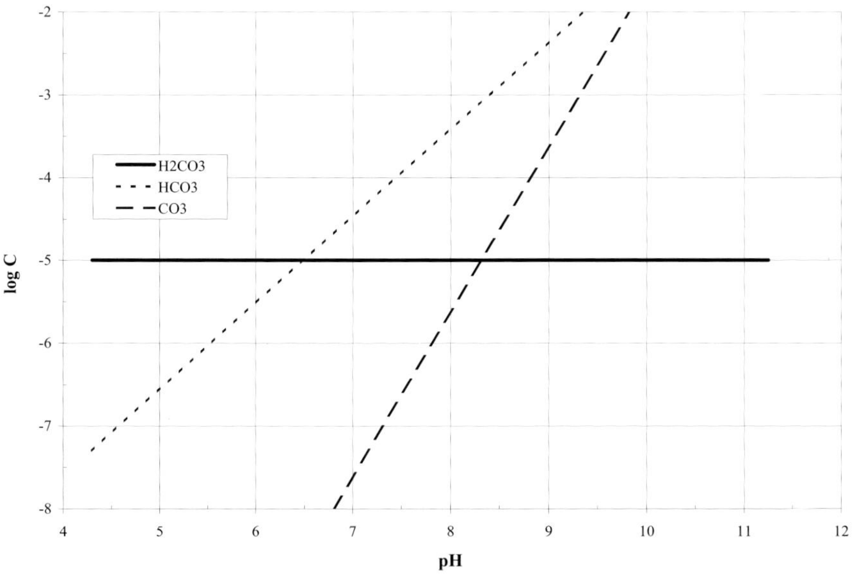

Figure 2.4 Volatile Log C-pH Diagram (from Bisogni and Timmons, 1994)

The equilibrium constants K_0, K_1, and K_2 are defined in Table 2.7, the square brackets indicate molar concentration, and alkalinity is given as mg $CaCO_3$ per liter.

Applying the equilibrium constants given in Table 2.7 for CO_2 and assuming a gas space concentration of CO_2 around the stripping device of 5,000 ppm (OSHA 8-hr limit, see Table A-15), we can calculate the concentration of CO_2 in the water as follows:

$$X_{CO_2} = K_H \cdot P_{CO_2} = \frac{0.0006111\ mol\ CO_2}{atm \cdot mol\ water} \cdot 0.0050\ atm \cdot \frac{55.6\ mol\ water}{L} \cdot$$

$$\frac{44 g}{mol\ CO_2} \cdot \frac{10^3 mg}{g} = 7.5 \frac{mg}{L} CO_2$$

Table 2.7 Carbonate system acid-base equilibrium reactions and equilibrium constants in freshwater (Stumm and Morgan, 1996) (square brackets signify molar concentrations (moles/L); the units of the partial pressure of carbon dioxide (p_{CO_2}) are (atm) (from Summerfelt et al. 2000; refer to Piedrahita and Seland, 1995, for temperature dependence of the equilibrium constants.)

Equilibrium type	Equilibrium relationships	Equilibrium constants @25°C	
gas-liquid	$CO_2\ (g) \leftrightarrow CO_2\ (aq)$	$K_H = p_{CO_2} / X_{CO_2}$	$\cong 6.11x10^{-4}\ atm^{-1}$
hydration/dehydration	$CO_2\ (aq) + H_2O \leftrightarrow H_2CO_3$	$K_\upsilon = [H_2CO_3]/[CO_2]$	$\cong 1.58x10^{-3}$
acid-base	$H_2CO_3 \leftrightarrow HCO_3^- + H^+$	$K_1 = [H^+][HCO_3^-]/[H_2CO_3]$	$\cong 2.83x10^{-4}$ mol/L
acid-base	$HCO_3^- \leftrightarrow CO_3^= + H^+$	$K_2 = [CO_3^=][H^+]/[HCO_3^=]$	$\cong 4.68x10^{-11}$ mol/L
acid-base	$H_2O \leftrightarrow OH^- + H^+$	$K_w = [OH^-][H^+]$	$\cong 1.00x10^{-14}\ mol^2/L^2$
dissolution/precipitation	$CaCO_3 \leftrightarrow CO_3^= + Ca^{++}$	$K_{sp} = [CO_3^=][Ca^{++}]$	$\cong 4.57x10^{-9}\ mol^2/L^2$

These equations illustrate that the dissolved carbon dioxide concentration can be effected by changes in the total amount of carbonate carbon in solution, the alkalinity, or the pH of the water. Changes in temperature and salinity can also have minor effects, since they impact the equilibrium constant values (Piedrahita and Seland, 1995). In addition to these equations, an approximation within 5–10% can be used to calculate the dissolved carbon dioxide concentration given the alkalinity and pH (restricted to a pH range from 6.5–9.5; Summerfelt, 1996):

$$C_{CO_2} = ALK \cdot 10^{(6.3 - pH)} \tag{2.7}$$

C_{CO_2} Dissolved carbon dioxide concentration (mg/L)

ALK Alkalinity (as $CaCO_3$ mg/L)

CO_2 Control

Methods of CO_2 degassing are presented and discussed in Chapter 8. The addition of bases to increase the pH and shift the carbonate chemistry equilibrium will control carbon dioxide in aquaculture systems (Grace and Piedrahita, 1994; Summerfelt, 1996; Loyless and Malone, 1997). Typically adding bases to the culture water does not result in the removal of dissolved inorganic carbon from solution, but merely causes a decrease in the carbon dioxide concentration by shifting the carbonate carbon balance to bicarbonate and carbonate ions as the pH is increased.

Two general classes of chemicals are used in aquaculture to control pH and decrease carbon dioxide concentrations:

- strong bases which do not contain any carbon, e.g., sodium hydroxide, and
- bases that contain carbon, e.g., sodium bicarbonate.

In either case, the addition of bases causes an increase in the alkalinity of the culture water. In the case of the carbon-containing bases, the increase in alkalinity is accompanied by an increase in total carbonate carbon. The pH increase resulting from chemical base addition is determined by the initial characteristics of the water (pH, alkalinity, and total carbonate carbon) and by the type and amount of base added. Equations derived from the carbonate chemistry system, Table 2.7, can be used to target a design carbon dioxide concentration and the corresponding alkalinity and carbonate carbon. The target of levels of alkalinity and carbonate carbon can then be used to estimate the amount of base needed for maintaining the design carbon dioxide level.

SALINITY

Water is commonly described as fresh, brackish, or saltwater. Each of these terms refers to the salinity of the water, and there are no clearly identified crossover points between these characterizations. Salinity is defined as the total concentration of dissolved ions in water, and is usually reported as parts per thousand (ppt) or grams of salt per kilogram of water. The major contributors of dissolved ions are calcium, sodium, potassium, bicarbonate, chloride and sulfate. The salinity of natural waters tends to reflect the climate, geography, and hydrological conditions of the immediate surroundings. Each aquatic species has an optimum range of salinity for reproduction and growth, although the salinity tolerance of most aquaculture species is rather broad. For example, rainbow trout fingerlings are produced in freshwater, acclimated to saltwater and then grown out to marketable size in sea cages at salinities as high as 20 ppt. Most freshwater fish of commercial importance reproduce and grow well at salinities up to at least 4–5 ppt.

Fish maintain the concentration of dissolved salts in their body fluids by regulating the uptake of ions from the environment and through the restriction of ion loss. This process is called "osmoregulation". Freshwater fish, for example, tend to accumulate water because they have body fluids more concentrated in ions than the surrounding water. When exposed to salinity values outside of their optimum range, aquatic species must spend considerable energy for osmoregulation at the expense of other functions, such as growth. If salinity deviates too far from optimum, the animal cannot maintain homeostasis and dies. The blood of freshwater fishes has an osmotic pressure approximately equal to the osmotic pressure of a 7 ppt sodium chloride solution. Freshwater aquaculture systems are generally maintained at 2–3 ppt salinity to reduce stress levels and the amount of energy required for osmoregulation, thereby increasing growth rates.

SOLIDS – SETTLEABLE, SUSPENDED, DISSOLVED

Waste solids accumulating in an aquaculture system come from uneaten feed, feed fines, fish fecal matter, algae, and biofilm cell mass sloughed from biological filters. Studies indicate that fish produce between 0.3 to 0.4 kg total suspended solids (TSS) for every 1 kg of feed fed. Waste solids influence the efficiency of all other unit processes in a recirculating system. They are a major source of carbonaceous oxygen demand and nutrient input into the water, and can directly affect fish health within recirculating systems by damaging fish gills and harboring pathogens. The tentative upper limit for freshwater fish is 25 mg TSS/L, with 10 mg TSS/L usually recommended for normal operation. Therefore, solids removal is one of the most critical processes in aquaculture systems. Optimally, solids need to be removed from the fish culture tank as soon as possible, while creating as little turbulence and mechanical shearing as possible.

Solids are generally classified into three categories: settleable, suspended, and fine or dissolved solids. The difference between settleable and suspended solids is simply the time it takes for them to settle to the bottom of an Imhoff cone. Settleable solids settle out in less than an hour. Suspended solids do not, and therefore require a treatment process other than conventional gravity settling basins. Fine and dissolved solids are by their very nature difficult to remove.

2.6 MEASUREMENTS

METERS & INSTRUMENTS

The three water quality parameters described below are the three most critical to successful aquaculture, in that they can change dramatically in a relatively short time (DO can drop to lethal levels in minutes at high loading densities!). As a result, they should be monitored continuously, with backup systems and alarms to indicate out of tolerance conditions.

Dissolved Oxygen

Although dissolved oxygen (DO) can be measured analytically by using the Winkler Method and a simple titration, the dissolved oxygen meters available today allow for rapid and accurate analysis over a range from zero to supersaturation. Two types of electrodes are commonly used: polarographic and galvanic. In simple terms, both of these meters consists of an electrode, which produces a signal proportional to the concentration of the oxygen in the water, and instrumentation to convert the signal reading to a visual display or recording device. A typical polarographic DO electrode or probe consists of a gold electrode and a silver-silver oxide reference electrode. The electrodes are bathed in 4 M KCL and separated from the sample by a membrane, usually made of Teflon, polyethylene, or fluorocarbon. The membrane is permeable to gases and the rate at which oxygen crosses the membrane is directly proportional to the dissolved oxygen concentration in the sample. When an electrical voltage is applied to the probe, molecular oxygen diffusing across the membrane reacts with the cathode (gold ring) and a small current flows to the anode electrode (silver). In a galvanic system, the electrode generates a small voltage (mV range) proportional to the dissolved oxygen concentration. Both systems require temperature, atmospheric pressure, and salinity compensation, usually accomplished by a combination of calibration and instrumentation hardware and software. Currently there are numerous DO meters available, over a wide range of price scales. As usual, you usually will "get what you pay for", so invest in a good oxygen meter (~ $500 minimum).

Temperature

Traditionally, temperature was measured with a simple mercury thermometer. Recently, these are becoming harder and harder to obtain, due to the significant environmental harm (and potential economic hardship) it will do if broken and mercury is released into the environment (or in some cases, the production tank!). Thankfully, there are a host of alternative measurement devices, and most of them are more convenient and flexible in their use. To start with, most DO meters and pH meters have some form of built-in temperature measurement for calibration and temperature compensation. Second, simple handheld temperature meters are relatively cheap, accurate, and easy to use. Finally, these types of temperature measurements allow for easy data logging and control.

pH

Like the dissolved oxygen meters, easy to read pH meters are available over a wide range of specifications and price. In addition to pH, most pH meters are also capable of measuring other ion specific electrodes, which include such parameters as ammonia, nitrate, ORP, dissolved oxygen, conductivity, among others. However, please keep in mind that most of these ion specific electrodes where designed to be used under laboratory conditions and require some sample pretreatment, and they are limited to specific ranges of concentration. It is recommended that a combination, gel-filled pH probe be used. This instrument is inexpensive, less susceptible to clogging, needs no refilling with electrolytes, and requires little or no maintenance.

The following parameters are usually slower to change than the parameters addressed above, and therefore can be monitored on a periodic (usually daily or weekly) basis. Of course, monitoring frequency can be increased if necessary due to a significant operational problem.

CO_2

Several types of dissolved carbon dioxide meters have recently become available. These work by first measuring the pH of the solution, and then calculating the CO_2 concentration based on the alkalinity (which must be known) and the carbonate equilibrium relationships. If the alkalinity is relatively constant, then CO_2 levels can be easily monitored and controlled by simply measuring the pH.

Carbon dioxide is usually measured by titration of a water sample to the phenolphthalene endpoint (pH 8.3) with a standard base. Approximate CO_2 concentrations can be determined from known values of pH, temperature, and alkalinity using either a formula or table values (see Fig. 2.2).

Salinity

Salinity can be indirectly measured using the measurement of a physical property such as refractive index, conductivity or density. Of these, the two most commonly used in aquaculture are the refractive index and conductivity measurements. A refractometer utilizes a change in the refractive index as function of the salt concentration to directly display the salinity level. It is very simple to use. Simply place a drop of water on the window and look through the eyepiece to directly read the salinity in parts per thousand. Refractometers provide fast, accurate salinity measurements, but some form of temperature-compensation is recommended when used outside their normal operating range of 20°C. One problem they do have is that most commercially available models have a wide salinity range of 0–100 ppt, making accurate measurements at low salinities difficult.

Electrical conductivity measurements can also provide a simple and quick measurement of salinity. Conductivity meters measure the flow of electrical current between two electrodes submersed in water, which is directly proportional to the concentration of ions, i.e., salts, in the water. They can be very accurate over a wide range of concentrations; these meters are direct reading and very portable.

Hydrometers measure salinity based on changes in water density (specific gravity) as a function of salinity. However, accurate and reliable hydrometers are more expensive than refractometers or electrical conductivity meters that provide the same degree of accuracy and ease of use, so they are typically not used. Inexpensive hydrometers are not recommended at all except for very gross estimates.

Chemical Analysis

Almost all of the following chemical analyses are easily performed colorimetically with off-the-shelf analysis kits manufactured by several companies. In most cases, using these kits is an easy process. Reagents come premixed in small vials, capsules, or bags, which requires only opening, mixing, and waiting for the specified period of time. Then, the mixture is measured, usually with a spectrophotometer. Analysis is quick, cheap, requires very little expensive laboratory equipment, and does not generate large quantities of hazardous wastes. Table 2.8 provides a summary of methods.

Dissolved Oxygen (DO)

In the basic Winkler method, a sample of water is treated with manganous sulfate, potassium iodide, and sodium hydroxide. Under these highly alkaline conditions, the manganous ion is oxidized by the dissolved oxygen to manganous dioxide. Sulfuric acid is then added to dissolve the precipitate and produce acid conditions for the oxidation of iodide to iodine by manganous dioxide. The quantity of iodine released is proportional to the amount of dissolved oxygen (DO), and is estimated by titration with standard sodium thiosulfate with a starch indicator as the endpoint. This sounds difficult, but in practice, all of the reagents are available in a premixed form in many commercially available test kits and thus DO measurement is a fairly routine activity. With the availability of relatively low cost dissolved oxygen meters, this method is usually used only for cross checking and calibration purposes.

Testing Procedures for DO Transfer. Standard tests for aerators are conducted using clean tap water at standard temperature and pressure (20°C and 760 mm Hg). Tests are based upon using deoxygenated water that has had the DO removed using a sodium sulfite (Na_2SO_3) solution using cobalt chloride ($CoCl_2 \cdot 6H_2O$) as a catalyst. The theoretical requirement for deoxygenation is 7.88 mg/L of sulfite (as Na_2SO_3) per 1.0 mg/L of O_2 removed. The cobalt chloride should be used at a concentration of 0.10–0.50 mg/L (as Co). A slightly higher concentration may be required if the water temperature is below 10°C. For more details, see APHA (1989).

CO_2

The simplest method of determining carbon dioxide concentration in the water column is to use a nomograph when the pH and total alkalinity are known. Then, these values can be compared to the nomographic data to determine the CO_2 concentration of the sample.

If pH and alkalinity are not known, they must first be determined. Then, CO_2 can be determined by adding a titrating solution of sodium hydroxide (0.0227 Normality or N) to a specified volume of sample water (normally, 250 mL). Since water at a pH over 8.34 does not contain appreciable amounts of carbon dioxide, the amount of a base (normally, sodium hydroxide) needed to raise the sample's pH to the 8.34 pH endpoint is approximately equivalent to the carbon dioxide content of the sample. Free carbon dioxide reacts with sodium hydroxide (0.0227 N read strength titrating solution) to form sodium bicarbonate, which raises the sample's pH. When sufficient amounts of the titrating solution of sodium hydroxide have been added to raise the sample's pH to 8.34, and then the quantity of sodium hydroxide that was required to reach this endpoint is comparable to the amount of CO_2 in the sample, at a relationship of 1 mL of sodium hydroxide (NaOH) equaling 1 mg/L CO_2 concentration of the sample.

Alkalinity

Alkalinity is a measure of the acid neutralizing capacity of the water. Alkalinity can be measured by titration with 0.02 N sulfuric acid and titrating to a pH end point or using a phenophalein indicator to show the color change as the indicator to stop adding titrant. With a 0.02 N sulfuric acid, one mL of titrant equals 10 mg/L of alkalinity (defined as calcium carbonate, $CaCO_3$). Only a few items are required to perform these measurements: a 125 mL beaker, a 50 mL buret, 0.02 N sulfuric acid (available premixed), methyl orange indicator, and a dropper.

Ammonia/Nitrite/Nitrate

The easiest method for measuring ammonia, nitrite, and nitrate is to use a colorimetric technique and either a color wheel comparison or a spectro-photometer.

For ammonia, the colorimetric Nesslerization method is the most popular. In this method, a nessler-ammonia reaction produces a yellow to brown color that is proportional to the ammonia-nitrogen concentration. This is a simple process to use, requiring only the addition of one commercially available reagent to the water sample, and then taking a measurement of the ammonia-nitrogen concentration with a spectrophotometer at 425 nm.

Nitrite is usually measured using a Colorimetric method. Diazotized sulfanilamide and NED dihydrochloride are added to the water sample, which creates a reddish purple azo dye, which is then measured with a spectrophotometer at 543 nm. The degree of transmittance in the spectrophotometer correlates to nitrite concentration data as presented in the form of standardized curves.

Finally, nitrate is measured by reducing it to nitrite with a Cadmium catalyst and then measuring nitrite concentration.

Reagents for all of these analysis methods are readily available in premixed or packet form. There are also several manufacturers of spectrophotometers, including Hach, YSI, and LaMotte. Although the use of a color wheel comparison is applicable in some cases, for commercial systems the cost of a spectrophotometer and the flexibility and accuracy it provides is worth the expense.

Table 2.8 Water Quality Parameters: Standard Methods

Water Quality Parameter	Standard Methods Monitoring method
Temperature	2550 Temperature Temperatures are traditional made with mercury-filled thermometers, but non-mercury or electronic temperature meters are preferred due to the toxicity of mercury in the environment.
pH	pH Meters Follow manufacturer's instructions for calibration using known standards 4.0, 7.02 and 10.
Total Suspended Solids	2540 Solids A well-mixed sample is filtered through a weighed standard glass-fiber filter and the residue retained on the filter is dried to a constant weight at 103 to 105 °C. The increase in the weight of the filter represents the total suspended solids.
Ammonia	4500 – NH_3 Nitrogen (Ammonia) Colorimetric Nesslerization Method: nessler-ammonia reaction produces a yellow to brown color, measured with spectrophotometer at 425 nm.
Nitrite	4500-NO_2^- Nitrogen (Nitrite) Colorimetric method with diazotized sulfanilamide with NED dihydrochloride creating a reddish purple azo dye measure with a spectrophotometer at 543 nm.
Nitrate	4500-NO_3^- Nitrogen (Nitrate) Cadmium reduction to nitrite and measurement of NO_2^-.
Phosphorus	4500-P Phosphorus (Orthophosphate) Ammonium molybdate and potassium antimonyl tartrate react to form a heteropoly acid, which is reduced with to intensely colored molybdenum blue by ascorbic acid.
Alkalinity	2320 – Titration Method Titration with 0.02 N Sulfuric Acid with methyl orange indicator end point (4.5 pH), 1 mL titrant equals 10 mg/L $CaCO_3$.
Hardness	2340 Hardness EDTA titrimetric method using EDTA and ethylenediaminetetraacetic acid and Calmagite. Titrated with EDTA from a red to a blue color.
Carbon Dioxide	4500-CO_2 Carbon Dioxide Free CO_2 reacts with sodium hydroxide (0.0227 N) to form sodium bicarbonate; completion indicated using a pH meter (8.3) or phenolphthalein indicator. 1 mL of NaOH equals 1 mg/LCO_2.
Salinity	2520 Salinity Measurement of conductivity, density, or refractive index. Commercially available conductivity meters and refractometers.
Chlorine	4500-Cl Chlorine (Residual) Iodometric method, chlorine liberates iodine from potassium iodide solutions and the iodine is tritrated with a standard solution of sodium thiosulfate with starch as the indicator.
Total Gas Pressure	2810 Dissolved Gas Supersaturation Commercially available membrane-diffusion instruments.
ORP – Oxidation-Reduction Potential	2580 Oxidation-Reduction Potential (ORP) ORP Electrode and standard reference electrode Follow manufacturer's instructions for calibration using known standards.
Conductivity	2510 Conductivity Conductivity is a measure of the ability of an aqueous solution to carry and electric current. Standard meters available, calibrate with a standard KCL solution.

Ion Specific Electrodes (ISE) allows measurement of ammonia, nitrate, and several other parameters using the pH/mV input of a standard pH meter. Although these have been used in the wastewater treatment industry and in a laboratory setting, they are not popular in the aquaculture industry. This is probably because of the difficulty in maintaining calibration, the need to pre-treat the water sample, and high equipment costs associated with this type of measurement method.

Phosphorus

Phosphorus is measured by using a colorimetric technique similar to ammonia-nitrogen. In this method, ammonium molybdate and potassium antimonyl tartrate are added to the sample, and they react to form a heteropoly acid. Then, ascorbic acid is added, which reduces the heteropoly acid to intensely colored molybdenum blue. The reactive phosphorus concentration (orthophosphorus) is then measured with a spectrophotometer at 880 nm.

Solids

Total solids (TS) are the weight of the material residue left after evaporation of a sample at a temperature of 105°C. Total suspended solids (TSS) is the portion of material retained by a filter, and total dissolved solids (TDS) is the portion of material that passes through the filter. Total volatile solids (TVS) is the difference in weight of the total solids after burning them to ashes, and Settleable solids (SS) is the material that settles out of suspension within a defined period of time.

The most commonly used measurement of solids is the total suspended solids and the settleable solids. The amount of total suspended solids is simple to determine, although a fair amount of laboratory equipment is required. The equipment requirements are: a drying oven (105°C), a filtration setup including vacuum pump, an analytical balance (0.1 mg), and glass fiber filters (Whatman GF/C). To take a TSS measurement, first weigh the glass fiber filter, and then filter a known volume of sample through it. Next, dry the filter at 105°C for a least one hour. When dry, weigh the filter again and divide the difference in weight by the volume of sample. The result is the total suspended solids concentration (mg/L).

Settleable solids are measured with an Imhoff cone, which is a cone shaped container with volume gradations at its base. A sample of 1 L is poured into the Imhoff cone and allowed to settle for one hour, after which the volume of settleable solids in the base of the cone are read in mL/L.

CHAPTER 3

FLUID MECHANICS AND PUMPS

3.0 FLUID MECHANICS

Fluid mechanics is the study of the impact that forces have on fluids, and it includes fluid statics (stationary fluids) and fluid dynamics (fluids in motion). In engineering terms, a fluid is a substance that deforms continuously when subjected to a shear force. In aquaculture, water and air are the two fluids of prime interest. Water is a liquid at ordinary temperatures and pressures, although it also exists in small quantities as a gas under these conditions. As a liquid, it can be treated as an incompressible fluid. Air, on the other hand, is a compressible gas. The two important physical properties of fluids are density and specific weight. Density is defined as mass per unit volume (kg/m^3), and specific weight is defined as weight per unit volume (lb/ft^3). The specific gravity of a fluid is the ratio of its density to the density of water, and is unitless.

The pressure of a fluid is defined as the normal force exerted over a unit area, and is usually expressed as pounds per square inch (psi or Pascals). For example, the pressure at the bottom of a column of water is equal to the specific weight of the fluid (lb/ft^3 or kg/m^3) times the height of the fluid column (ft or m) above the point in question. Pressure measurements can also be referenced to either absolute zero or atmospheric pressure. If the pressure reading is referenced to absolute zero or a vacuum, the measurement is referred to as absolute pressure or psia. Thus, atmospheric pressure at mean sea level is given as 101 kN/m^2, 14.7 psia, 760 mm of mercury (Hg) or 101.3 kPa. Common pressure gauges measure the difference between the pressure of the fluid being measured and the pressure of the surrounding atmosphere or psig. Thus, a tire gauge measures the air pressure in comparison to the existing barometric pressure. Thus, gauge pressure is equal to the difference in absolute pressure and actual atmospheric pressure. Since gauge pressure is referenced to atmospheric pressure, it can be either positive or negative. Negative gauge pressure is also referred to as vacuum pressure.

Pressure is normally measured using either a manometer or by measuring the deformation of an elastic member, whose deflection is directly proportional to the applied pressure (tradition pressure gauges).

A manometer uses the change in height of a column of liquid to measure pressure. In its simplest form, a manometer or piezometer is a straight vertical tube attached to a pipe. The gauge pressure at the center of the pipe is equal to the product of the specific weight of the fluid (lb/ft^3) times the height of the fluid column (inch or cm). In practice, a section of tygon tubing or clear acrylic tubing attached to a threaded barbed pipe adapter makes a simple, flexible piezometer. These are often placed on the discharge pipe into a tank to monitor the backpressure and thus indirectly the flow into the tank.

A U-tube manometer is used to measure a differential pressure. In a simple manometer, pressure forces the liquid in one leg of the U-tube to rise until the head pressure exerted by the column of fluid is in equilibrium with the pressure at the other leg. The difference in height between the two legs times the specific weight of the fluid yields the pressure, either positive or negative. A standard pressure gauge employs a tube having an elliptical cross section, shaped into a circular arc. When pressure is applied to the gauge, the tube tends to straighten, and a pointer that directly corresponds to pressure is actuated. These are simple and inexpensive to use and are available over a wide range of pressures and vacuums.

The study of fluids in motion is fluid dynamics and is based on several fundamental concepts including conservation of mass, energy, and Newton's Laws of motion. The law of conservation of mass, which says that mass can neither be created nor destroyed, yields the concept of the continuity equation. For pipe flow, this means that the mass of fluid flowing past one section of a closed pipe must also flow past all other section, since there is no where else for it to go. Using the continuity equation, the pipe velocities can be easily calculated knowing the flow rate through the pipe:

$$Q = \rho_1 A_1 V_1 = \rho_2 A_2 V_2 \qquad (3.1^{a})$$

In addition, the velocity through pipe sections of different diameters can be determined since the fluid density is the same, hence:

$$V_1 A_1 = V_2 A_2 \qquad (3.2)$$

[a] Symbols are defined at end of chapter.

$$A = \frac{\pi \, D^2}{4} \tag{3.3}$$

The magnitude of the water velocity is a crucial variable, as it affects the pressure losses through pipe runs, turns, and fittings.

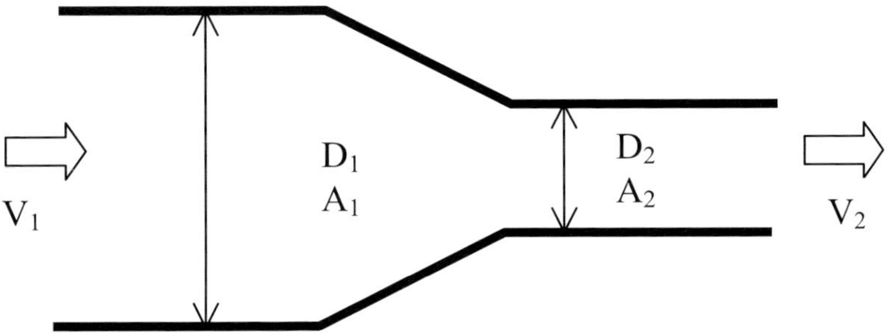

Figure 3.1 Pipe Cross-section and Pertinent Terms.

Another important concept is the law of conservation of energy. The total energy at any point in a fluid consists of three components: potential energy due to location, potential energy due to pressure, and kinetic energy due to the motion of the fluid. Potential energy is due to elevation above some point of reference. For example, water in an overhead pipe has potential energy in relationship to the tank located at floor level. The potential energy is a function of the weight of the water and the vertical distance from the overhead pipe to the water level in the tank. An example of potential energy due to pressure is the bottom discharge of a tank. In this case, the potential energy due to pressure is a function of the weight of fluid and the height of the fluid above the reference plane. Finally, the kinetic energy due to the motion of the fluid is directly related to its velocity using Newton's laws of motion. The total energy of the fluid at some point is the sum of the energies due to elevation, pressure, and velocity.

From the law of conservation of energy, the total energy at any point in a fluid system must be the same. This yields Bernoulli's equation, which is usually expressed as:

$$E_1 = Z_1 + \frac{P_1}{\gamma} + \frac{V_1^2}{2g} = Z_2 + \frac{P_2}{\gamma} + \frac{V_2^2}{2g} = E_2 \tag{3.4}$$

The term Z is often referred to as the pressure head and the term ($V^2/2g$) as the velocity head or dynamic head, since both can be expressed in feet (meters). As an example how Bernoulli's equation is used, calculate the velocity of the discharge from a storage reservoir tank that is 8 feet (2.44 m) deep. We know that the energy at the top of the tank and at the discharge point must be the same. In addition, since the system is open to the atmosphere at the top and at the discharge point, the gauge pressures are zero. If we assume that the tank is very large, the velocity of the water at the surface as the tank drains (V_1) is very small and since it is also squared (making it smaller), it can be ignored. Therefore, the velocity at the discharge reduces to:

$$V = \sqrt{2 \cdot 32.2\,ft/s^2 \cdot 8\,ft} = 22.7\ ft/s$$

SI

$$V = \sqrt{2 \cdot 9.81m/s^2 \cdot 2.4m} = 6.8m/s$$

Bernoulli's equation can thus be used to calculate the specific energy components at a particular point in the system and relate how one form of energy, potential or kinetic, is converted into another. A broader form of the conservation of energy law would take into account other forms of energy that might be added or removed from a system, including energy added by a pump, energy extracted via a turbine, or energy lost due to friction in pipes and fittings:

$$E_1 + E_{added} = E_2 + E_{lost} + E_{extracted} \tag{3.5}$$

Energy inputs to a system are normally included on the left side of equation as a positive value. Energy losses due to pipe friction and minor losses in fittings are added to the right side of the equation, since this energy is usually lost between point 1 and 2. Pipe friction losses are caused by the friction generated by the movement of the fluid against the walls of the pipes. Minor pipe loses are those associated with friction occurring when the fluid encounters restriction in the systems (valves), changes in direction (elbows, bends, tees), changes in pipe size (reducers, expanders), and losses when the fluid enters or leaves a conduit. Sometimes minor losses are referred to as dynamic losses, since "minor" can be a misnomer in that these pressure losses can be large.

3.1 FRICTIONAL LOSSES

The magnitude of pipe friction losses is a function of the internal pipe diameter, length, fluid velocity, roughness of the internal pipe surfaces and certain physical properties of the fluid such as density and viscosity. Over the years, there have been several empirical relations developed to relate these factors to the pipe friction head loss. These usually include the calculation of Reynolds number (Re) to determine the type of flow and relative roughness, and the use of Moody's diagram (see Fig. 3.2) to find the friction factor and finally the head loss using Darcy-Weisbach equation. One of problems with this method is that all of the coefficients are estimated for average or normal conditions, without considering the effects of biofouling for example. Thus the choice of the coefficient values are at best an educated guess as to the current or projected conditions, which probably will and do change with time. The head losses resulting from various water flow rates in plastic pipe can be calculated by means of the Hazen-Williams formula as opposed to the Darcy-Weisbach equation. Both methods will be demonstrated in this chapter.

Darcy-Weisbach

$$H_L = f\left(\frac{L}{D}\right)\left(\frac{V^2}{2g}\right) \tag{3.6}$$

Hazen-Williams

$$H_L = 0.2083\left(\frac{100}{C_{H-W}}\right)^{1.852}\left(\frac{Q^{1.852}}{D_{inch}^{4.87}}\right) \tag{3.7}$$

where H_L in the Hazen-Williams equation, Eq. 3.7, is per 100 ft of pipe and $C_{H\text{-}W}$ is 150 for PVC pipe and 100 for cast iron pipe.

The Reynolds number, Re, is the product of velocity and pipe diameter (inertial forces) divided by kinematic viscosity (viscous forces):

$$Re = \frac{VD}{\nu} \tag{3.8}$$

The Re is used with a Moody diagram, Fig. 3.2, to determine "f". For smooth pipes, e.g., plastic, copper, glass, etc., Table 3.1 can be used to obtain the friction factor, f, given the Reynolds number (Re) taken from Table 3.2 or calculated using Eq. 3.8. Table 3.2 can be used to read Re directly for flows between 1 and 4,000 gpm (0.2 to 900 m^3/hr) and pipe diameters from 2 to 12 inches (5 to 30 cm).

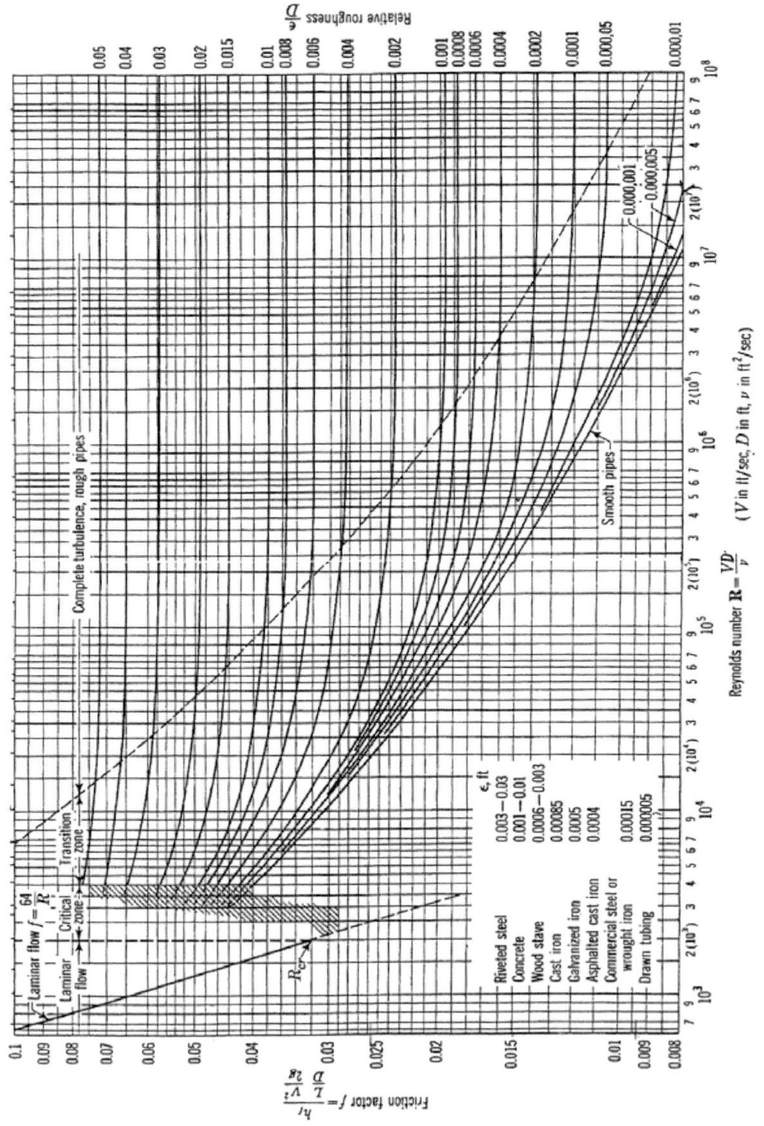

Figure 3.2 Moody Diagram.

Table 3.1 Friction Factor, f, for Various Reynolds Numbers (Re)

Re	**f**
2,300	0.042
10,000	0.030
20,000	0.025
50,000	0.021
75,000	0.019
100,000	0.018
200,000	0.016
500,000	0.013
1,000,000	0.012
2,000,000	0.011

note: for Re < 2,300, f = 64/Re

As an easy alternative, many pipe manufacturers and engineering texts (this one included) present frictional head losses for several different types of piping in tabular form, Table 3.3 for PVC pipe. The input variables are nominal pipe diameter and the flow rate desired. The table than lists the average pipe velocity and the frictional head loss per unit length, usually 100 ft (30 m). The total frictional loss for a system is than calculated by multiplying the table value by the number of unit lengths of pipe run. Again, these tables are based on some assumptions as to pipe characteristics, temperature, and conditions. Since most tables do not consider biofouling, values should be used conservatively.

Example 1

Calculate the friction loss in 200 lineal feet of 4 inch PVC pipe (smooth) flowing at 1,000 gpm:

Constants: g = 32.2 ft/s^2 and ν = 0.0000122 ft^2/s (water at 60°F; see Table 2.7)

1. $$A = \pi r^2 = \pi \cdot (0.167\ ft)^2 = 0.0876\ ft^2$$

2. $$V = \frac{Q}{A} = \frac{\frac{1{,}000\ gal}{min} \cdot \frac{1\ ft^3}{7.48\ gal} \cdot \frac{1\ min}{60\ sec}}{0.0876\ ft^2}$$

$$V = 25.4\ ft/s$$

Table 3.2 Velocity, Dynamic Head and Reynolds Number

Velocity Look up Table in ft/s

	Diameter of Pipe in Inches					
Flow, gpm	2	3	4	6	8	12
1	0.102	0.045	0.026	0.011	0.006	0.003
5	0.511	0.227	0.128	0.057	0.032	0.014
10	1.021	0.454	0.255	0.113	0.064	0.028
25	2.553	1.135	0.638	0.284	0.160	0.071
50	5.105	2.269	1.276	0.567	0.319	0.142
100	10.211	4.538	2.553	1.135	0.638	0.284
250	25.527	11.345	6.382	2.836	1.595	0.709
500	51.054	22.690	12.763	5.673	3.191	1.418
1,000	102.107	45.381	25.527	11.345	6.382	2.836
2,000	204.214	90.762	51.054	22.690	12.763	5.673
4,000	408.428	181.524	102.107	45.381	25.527	11.345

Dynamic Head Look up Table in ft

	Diameter of Pipe in Inches					
Flow, gpm	2	3	4	6	8	12
1	1.62E-04	3.20E-05	1.01E-05	2.00E-06	6.32E-07	1.25E-07
5	4.05E-03	7.99E-04	2.53E-04	5.00E-05	1.58E-05	3.12E-06
10	1.62E-02	3.20E-03	1.01E-03	2.00E-04	6.32E-05	1.25E-05
25	1.01E-01	2.00E-02	6.32E-03	1.25E-03	3.95E-04	7.81E-05
50	4.05E-01	7.99E-02	2.53E-02	5.00E-03	1.58E-03	3.12E-04
100	1.62E+00	3.20E-01	1.01E-01	2.00E-02	6.32E-03	1.25E-03
250	1.01E+01	2.00E+00	6.32E-01	1.25E-01	3.95E-02	7.81E-03
500	4.05E+01	7.99E+00	2.53E+00	5.00E-01	1.58E-01	3.12E-02
1,000	1.62E+02	3.20E+01	1.01E+01	2.00E+00	6.32E-01	1.25E-01
2,000	6.48E+02	1.28E+02	4.05E+01	7.99E+00	2.53E+00	5.00E-01
4,000	2.59E+03	5.12E+02	1.62E+02	3.20E+01	1.01E+01	2.00E+00

Reynolds Number Look up Table, dimensionless

	Diameter of Pipes in Inches					
Flow, gpm	2	3	4	6	8	12
1	1.40E+03	9.32E+02	6.99E+02	4.66E+02	3.50E+02	2.33E+02
5	6.99E+03	4.66E+03	3.50E+03	2.33E+03	1.75E+03	1.17E+03
10	1.40E+04	9.32E+03	6.99E+03	4.66E+03	3.50E+03	2.33E+03
25	3.50E+04	2.33E+04	1.75E+04	1.17E+04	8.74E+03	5.83E+03
50	6.99E+04	4.66E+04	3.50E+04	2.33E+04	1.75E+04	1.17E+04
100	1.40E+05	9.32E+04	6.99E+04	4.66E+04	3.50E+04	2.33E+04
250	3.50E+05	2.33E+05	1.75E+05	1.17E+05	8.74E+04	5.83E+04
500	6.99E+05	4.66E+05	3.50E+05	2.33E+05	1.75E+05	1.17E+05
1,000	1.40E+06	9.32E+05	6.99E+05	4.66E+05	3.50E+05	2.33E+05
2,000	2.80E+06	1.86E+06	1.40E+06	9.32E+05	6.99E+05	4.66E+05
4,000	5.59E+06	3.73E+06	2.80E+06	1.86E+06	1.40E+06	9.32E+05

3. $$Re = \frac{VD}{\nu} = \frac{25.4\ ft/s \cdot 0.333\ ft}{0.0000122\ ft^2/s} = 693{,}000$$

4. For a Re = 693,000 → f = 0.013

5. $$H_L = f\left(\frac{V^2}{2g}\right)\left(\frac{L}{D}\right)$$

$$H_L = 0.013 \cdot \left(\frac{(25.4\ ft/s)^2}{2(32.2\ ft/s^2)}\right)\left(\frac{200\ ft}{0.333\ ft}\right)$$

$$H_L = 78.2\ ft$$

3.2 Dynamic Losses

Energy losses due to elbows and fittings are proportional to the dynamic head $(V^2/2g)$, just as the friction losses were. Returning to Eqs. 3.6 (Darcy-Weisbach equation) and 3.8, we can see that:

$$K = f\frac{L}{D} \tag{3.9}$$

The frictional head loss for any given fitting or transition is found by multiplying the velocity head or dynamic head times the loss coefficient K:

$$H_L = K\left(\frac{V^2}{2g}\right) \tag{3.10}$$

Fittings are assigned a "K" value and the energy loss is simply the product of "K" and the dynamic head. The resistance coefficient K is considered to be independent of the Reynolds (Re) number, as opposed to pipe friction losses, which are dependent upon Re number.

The flow coefficient of a valve/fitting, C_v, is related to the friction factor K as follows:

$$C_V = \frac{29.9d^2}{\sqrt{K}} \tag{3.11}$$

$$K = \frac{891d^4}{C_V^2} \tag{3.12}$$

K values for common fittings are listed in Table 3.4. These K values were produced from data given by Crane (1988) that gives specific frictional loss factors based upon the size of a fitting (f_T) and loss coefficients that are based upon the type of fitting ($FITTING_{constant}$) as follows:

$$K = f_T \cdot FITTING_{constant} \tag{3.13}$$

Values for f_T are given in Table 3.5. Friction loss coefficients can be found in product guides and specification manuals, e.g., "Thermoplastic Valves & Accessories: Product Guide and Engineering Specifications, 18th Edition", Spears Manufacturing Company, Sylmar, CA; http://www.spearsmfg.com.

Estimating K values accurately depends on the type of fitting, the abruptness of the flow change, the material, and the condition of the internal surfaces. For some fittings, such as valves, there may be considerable variation in losses from one type and manufacture to another. Even for standard fittings, the variation in frictional head losses can be as large as 50%, and thus, pressure loss data from a fitting manufacturer is preferred. The head loss for other types of inline components, such as heat exchangers and UV systems, are usually provided by the manufacturer for the design flow rate.

A general rule of thumb is that you can simply assume every fitting and/or turn in a pipe is equal to one dynamic head, e.g., K = 1.

> **"RULE OF THUMB"**
>
> Each fitting will cause one dynamic head of pressure loss, $V^2/2g$.

A second method commonly used for calculating minor head losses uses the concept of Equivalent Length. In this method, the head loss for the fitting is expressed in terms of the length of straight pipe with the same head losses. Tabulated data on fittings is presented in Table 3.6 as a factor (L/D) that is multiplied by the associated pipe diameter to produce and equivalent length of straight pipe.

Thus, the process of estimating total friction head loss in a system consists of first using a diagram of the system to determine the total lengths of each section of pipe with the same diameter and flow rate. Then using the tables provided or the Hazen-William formula, calculate the head loss per 100 ft and then the head loss in each section of corresponding pipe. Add to this the equivalent length of all valves and fitting in each section, based on the velocity and loss coefficient. Finally, add up the friction losses.

Example 2

Calculate the dynamic loss due to five 90° elbows of 4 inch PVC pipe (smooth) flowing at 1,000 gpm:

Constant: $g = 32.2\ ft/s^2$

1. $A = \pi r^2 = \pi \cdot (0.167\ ft)^2 = 0.0876\ ft^2$

2. $$V = \frac{Q}{A} = \frac{\dfrac{1{,}000\ gal}{min} \cdot \dfrac{1\ ft^3}{7.48\ gal} \cdot \dfrac{1\ min}{60\ sec}}{0.0876\ ft^2}$$

$V = 25.4\ ft/s$

3. 90° elbow → K = 1.0 (or you could obtain more accurate values of K for this size pipe from Table 3.4)

4. $$H_L = K\left(\frac{V^2}{2g}\right)$$

$$H_L = 1.0 \cdot \left(\frac{(25.4\ ft/s)^2}{2(32.2\ ft/s^2)}\right)$$

$$H_L = 10.0\ ft\ per\ elbow \times 5\ elbows$$

$$H_L = 50.0\ ft$$

Table 3.3 Frictional Head Losses per 100 ft Schedule 40 PVC Pipe (friction losses for Schedule 80 pipe are provided in the Appendix Table A-19.)

GPM	Velocity (fps)	Head Loss (ft)	Head Loss (psi)	Velocity (fps)	Head Loss (ft)	Head Loss (psi)	Velocity (fps)	Head Loss (ft)	Head Loss (psi)	Velocity (fps)	Head Loss (ft)	Head Loss (psi)	Velocity (fps)	Head Loss (ft)	Head Loss (psi)	Velocity (fps)	Head Loss (ft)	Head Loss (psi)	Velocity (fps)	Head Loss (ft)	Head Loss (psi)
		1/2 inch			**3/4 inch**			**1 inch**			**1-1/4 inch**			**1-1/2 inch**			**2 inch**			**3 inch**	
1	1.13	2.08	0.90	0.63	0.51	0.22															
2	2.26	4.16	1.80	1.26	1.02	0.44	0.77	0.55	0.24	0.44	0.14	0.06									
5	5.64	23.44	10.15	3.16	5.73	2.48	1.93	1.72	0.75	1.11	0.44	0.19	0.81	0.22	0.09						
7	7.90	43.06	13.64	4.43	10.52	4.56	2.72	3.17	1.37	1.55	0.81	0.35	1.13	0.38	0.17						
10				6.32	20.04	8.68	3.86	6.02	2.61	2.21	1.55	0.67	1.62	0.72	0.31	0.98	0.21	0.09			
15				9.48	42.46	18.39	5.79	12.77	5.53	3.31	3.28	1.42	2.42	1.53	0.66	1.46	0.45	0.19			
20							7.72	21.75	9.42	4.42	5.59	2.42	3.23	2.61	1.13	1.95	0.76	0.33	0.88	0.11	0.05
25										5.52	8.45	3.66	4.04	3.95	1.71	2.44	1.15	0.50	1.10	0.17	0.07
30		**4 inch**								6.63	11.85	5.13	4.85	5.53	2.39	2.93	1.62	0.70	1.33	0.23	0.10
35										7.73	15.76	6.82	5.66	7.36	3.19	3.41	2.15	0.93	1.55	0.31	0.13
40	1.02	0.11	0.05							8.84	20.18	8.74	6.47	9.43	4.08	3.90	2.75	1.19	1.77	0.40	0.17
45	1.15	0.13	0.06		**5 inch**								7.27	11.73	5.08	4.39	3.43	1.49	1.99	0.50	0.22
50	1.28	0.16	0.07										8.08	14.25	6.17	4.88	4.16	1.80	2.21	0.60	0.26
60	1.53	0.22	0.10	0.97	0.07	0.03										5.85	5.84	2.53	2.65	0.85	0.37
70	1.79	0.30	0.13	1.14	0.10	0.04		**6 inch**								6.83	7.76	3.36	3.09	1.13	0.49
75	1.92	0.34	0.15	1.22	0.11	0.05										7.32	8.82	3.82	3.31	1.28	0.55
80	2.05	0.38	0.16	1.30	0.13	0.06	0.90	0.05	0.02							7.80	9.94	4.30	3.53	1.44	0.62
90	2.30	0.47	0.20	1.46	0.16	0.07	1.01	0.06	0.03							8.78	12.37	5.36	3.98	1.80	1.78
100	2.56	0.58	0.25	1.62	0.19	0.08	1.12	0.08	0.04		**8 inch**					9.75	15.03	6.51	4.42	2.18	1.94
125	3.20	0.88	0.38	2.03	0.29	0.13	1.41	0.12	0.05										5.52	3.31	1.43
150	3.84	1.22	0.53	2.44	0.40	0.17	1.69	0.16	0.07	0.97	0.04	0.02							6.63	4.63	2.00
175	4.48	1.63	0.71	2.84	0.54	0.24	1.97	0.22	0.10	1.14	0.06	0.02		**10 inch**					7.73	6.16	2.67
200	5.11	2.08	0.90	3.25	0.69	0.30	2.25	0.28	0.12	1.30	0.07	0.03							8.83	7.88	3.41
250	6.40	3.15	1.36	4.06	1.05	0.45	2.81	0.43	0.19	1.63	0.11	0.05	1.03	0.04	0.02		**12 inch**				
300	7.67	4.41	1.91	4.87	1.46	0.63	3.37	0.60	0.26	1.94	0.16	0.07	1.23	0.05	0.02						
350	8.95	5.87	2.55	5.69	1.95	0.85	3.94	0.79	0.34	2.27	0.21	0.09	1.44	0.07	0.03	1.01	0.03	0.01			
400	10.23	7.52	3.26	6.50	2.49	1.08	4.49	1.01	0.44	2.59	0.27	0.12	1.64	0.09	0.04	1.16	0.04	0.02			
450				7.31	3.09	1.34	5.06	1.26	0.55	2.92	0.33	0.14	1.85	0.11	0.05	1.30	0.05	0.02			
500				8.12	3.76	1.63	5.62	1.53	0.66	3.24	0.40	0.17	2.05	0.13	0.06	1.45	0.06	0.03			
750							8.43	3.25	1.41	4.86	0.85	0.37	3.08	0.28	0.12	2.17	0.12	0.05			
1000							11.24	5.54	2.40	6.48	1.45	0.63	4.11	0.48	0.21	2.89	0.20	0.09			
1250										8.11	2.20	0.95	5.14	0.73	0.32	3.62	0.31	0.13			
1500										9.72	3.07	1.33	6.16	1.01	0.44	4.34	0.43	0.19			
2000													8.21	1.72	0.74	5.78	0.73	0.32			
2500													10.27	2.61	1.13	7.23	1.11	0.49			
3000																8.68	1.55	0.67			

(Shaded area recommended flow rates to minimize settlement of solids and avoid scouring of walls and junctions)

Table 3.4 K Values for Variety of Fitting Types

Type of Fitting	Nominal Size of Pipe, inches ½	¾	1	1.25	1.5	2	2½ & 3.0	4.0	5.0	6.0	8.0 to 10.0	12 to 16	18 to 24
Standard Fittings													
Elbows-90	0.81	0.75	0.69	0.66	0.63	0.57	0.54	0.51	0.48	0.45	0.42	0.39	0.36
Elbows-45	0.432	0.4	0.368	0.352	0.336	0.304	0.288	0.272	0.256	0.24	0.224	0.208	0.192
T-flow through	0.54	0.5	0.46	0.44	0.42	0.38	0.36	0.34	0.32	0.3	0.28	0.26	0.24
T-branch off	1.62	1.5	1.38	1.32	1.26	1.14	1.08	1.02	0.96	0.9	0.84	0.78	0.72
Valves													
Gate	0.22	0.20	0.18	0.18	0.17	0.15	0.14	0.14	0.13	0.12	0.11	0.10	0.10
Globe & Angle Valve	9.18	8.50	7.82	7.48	7.14	6.46	6.12	5.78	5.44	5.10	4.76	4.42	4.08
Swing Check	2.03	1.88	1.73	1.65	1.58	1.43	1.35	1.28	1.20	1.13	1.05	0.98	0.90
Hinged Foot	2.03	1.88	1.73	1.65	1.58	1.43	1.35	1.28	1.20	1.13	1.05	0.98	0.90
Ball	0.08	0.08	0.07	0.07	0.06	0.06	0.05	0.05	0.05	0.05	0.04	0.04	0.04
Butterfly (D < 8")	1.22	1.13	1.04	0.99	0.95	0.86	0.81	0.77	0.72	0.68	0.63	0.59	0.54
Butterfly (D = 10 to 14")	0.95	0.88	0.81	0.77	0.74	0.67	0.63	0.60	0.56	0.53	0.49	0.46	0.42
90 Degree Pipe Bend	0.54	0.50	0.46	0.44	0.42	0.38	0.36	0.34	0.32	0.30	0.28	0.26	0.24
r/D = 1													
r/D = 10	0.81	0.75	0.69	0.66	0.63	0.57	0.54	0.51	0.48	0.45	0.42	0.39	0.36
r/D = 20	1.35	1.25	1.15	1.10	1.05	0.95	0.90	0.85	0.80	0.75	0.70	0.65	0.60
180-Return Bend	1.35	1.25	1.15	1.10	1.05	0.95	0.90	0.85	0.80	0.75	0.70	0.65	0.60
Mitre Bends													
30 degree	0.22	0.20	0.18	0.18	0.17	0.15	0.14	0.14	0.13	0.12	0.11	0.10	0.10
45 degree	0.41	0.38	0.35	0.33	0.32	0.29	0.27	0.26	0.24	0.23	0.21	0.20	0.18
90 degree	1.62	1.50	1.38	1.32	1.26	1.14	1.08	1.02	0.96	0.90	0.84	0.78	0.72
Non-size dependent													
Pipe Entrance, flush	0.50	0.50	0.50	0.50	0.50	0.50	0.50	0.50	0.50	0.50	0.50	0.50	0.50
Pipe Entrance, projecting	0.78	0.78	0.78	0.78	0.78	0.78	0.78	0.78	0.78	0.78	0.78	0.78	0.78
Pipe Exit, all types	1.00	1.00	1.00	1.00	1.00	1.00	1.00	1.00	1.00	1.00	1.00	1.00	1.00

Table 3.5 Friction Factors (f_T) for Valves and Fittings (clean steel)

Nominal Size	½"	¾"	1"	1.25"	1.5"	2"	2½, 3"	4"	5"	6"	8-10"	12-16"	18-24"
Friction Factor (f_T)	.027	.025	.023	.022	.021	.019	.018	.017	.016	.015	.014	.013	.012

Table 3.6 L/D Ratios for Selected Fittings for Determining Frictional Losses as Equivalent Length of Pipes

Fitting	L/D
Couplings and unions	
45° elbow	16
90° elbow	30
Closed return bend	50
"Tee" flow along run	20
"Tee" used as elbow (entering run)	—
"Tee" used as elbow (entering branch)	60
Medium sweep elbow	
Long sweep elbow	20
Globe valve (fully open)	340
Angle valve (fully open)	145
Butterfly valve	20
Check Valve (fully open)	
Swing Type	135
Ball type	150
Lift type	
Gate Valve	
Fully open	13
¾ open	35
½ open	160
¼ open	900
Strainer bucket	
With poppet lift-type disc	420
With leather-hinged disc	75

Example 3

Assume a required flow of 1,000 gpm. Calculate the total energy loss for a flow from the rearing tank to the water treatment system and gravity flow return to the tank. This system has 200 lineal feet of pipe, five 90° elbows (L), an entrance loss into the pump (E) and a gravity return (G); for simplicity assume each elbow has a K = 1.0; system depicted below:

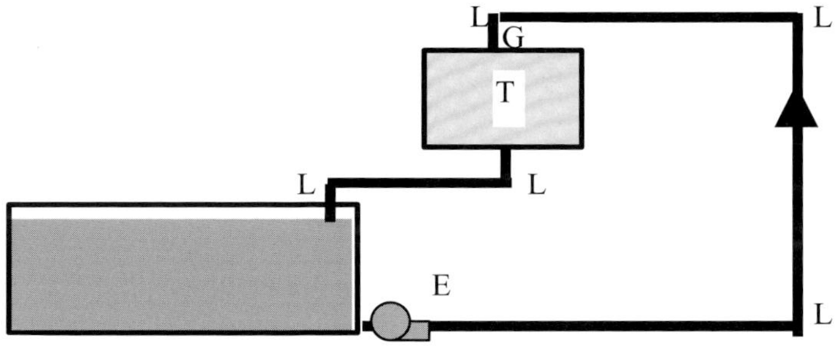

<u>For 4 inch PVC pipe:</u>

- **200 lineal feet of pipe**

  $H_L = 78.2\ ft$ (Example 1)

- **Five 90° elbows**

  $H_L = 50.0\ f$ (Example 2)

- **One entrance into the pump**

  Entrance → $K = 0.5$

$$H_L = K\left(\frac{V^2}{2g}\right)$$

$$H_L = 0.5 \cdot \left(\frac{(25.4\ ft/s)^2}{2(32.2\ ft/s^2)}\right)$$

$$H_L = 5.01\ ft$$

- **One gravity return**

  Return → $K = 1.0$

$$H_L = K\left(\frac{V^2}{2g}\right)$$

$$H_L = 1.0 \cdot \left(\frac{(25.4\ ft/s)^2}{2(32.2\ ft/s^2)}\right)$$

$$H_L = 10.0\ ft$$

Total energy loss $= \sum H_L$

$= \sum (78.2 + 50.0 + 5.01 + 10.0) ft$

$= 143\ ft\ (62.1\ psi)$

<u>For 8 inch PVC pipe:</u>

- **200 lineal feet of pipe**

1. $A = \pi r^2 = \pi \cdot (0.333\ ft)^2 = 0.348\ ft^2$

2. $$V = \frac{Q}{A} = \frac{\dfrac{1{,}000\ gal}{min} \cdot \dfrac{1\ ft^3}{7.48\ gal} \cdot \dfrac{1\ min}{60\ sec}}{0.348\ ft^2}$$

$$V = 6.40\ ft/s$$

3. $$Re = \frac{VD}{\nu} = \frac{6.40\ ft/s \cdot 0.666\ ft}{0.0000122\ ft^2/s} = 349{,}000$$

4. For a Re = 349,000 $\rightarrow f = 0.014$

5. $$H_L = f\left(\frac{V^2}{2g}\right)\left(\frac{L}{D}\right)$$

$$H_L = 0.014 \cdot \left(\frac{(6.40\ ft/s)^2}{2(32.2\ ft/s^2)}\right) \cdot \left(\frac{200\ ft}{0.666\ ft}\right)$$

$$H_L = 2.67\ ft$$

- **Five 90° elbows**

1. $A = \pi r^2 = \pi \cdot (0.333\ ft)^2 = 0.348\ ft^2$

2. $$V = \frac{Q}{A} = \frac{\frac{1{,}000\ gal}{min} \cdot \frac{1\ ft^3}{7.48\ gal} \cdot \frac{1\ min}{60\ sec}}{0.348\ ft^2}$$

$$V = 6.40\ ft/s$$

3. $90\,^{\circ} elbow \rightarrow K = 1.0$

4. $$H_L = K\left(\frac{V^2}{2g}\right) \qquad H_L = 1.0 \cdot \left(\frac{(6.40\ ft/s)^2}{2(32.2\ ft/s^2)}\right)$$

$$H_L = 0.636\ \frac{ft}{elbow} \cdot 5\ elbows$$

$$H_L = 3.18\ ft$$

- **One entrance into the pump**

$Entrance \rightarrow K = 0.5$

$$H_L = K\left(\frac{V^2}{2g}\right)$$

$$H_L = 0.5 \cdot \left(\frac{(6.40\ ft/s)^2}{2(32.2\ ft/s^2)}\right)$$

$$H_L = 0.318\ ft$$

- **One gravity return**

$Return \rightarrow K = 1.0$

$$H_L = K\left(\frac{V^2}{2g}\right)$$

$$H_L = 1.0 \cdot \left(\frac{(6.40\ ft/s)^2}{2(32.2\ ft/s^2)}\right)$$

$$H_L = 0.636\ ft$$

Total energy loss $= \sum H_L$

$= \sum (2.67 + 3.18 + 0.318 + 0.636) ft$

$= 6.80\ ft\ (2.95\ psi)$

Also, for the same velocity throughout the system, one can sum the "K" values and then multiply times the dynamic head; note that the friction loss can be represented as a "K" value by multiplying the friction factor, f, times the relative length (L/D).

Continuing with a 1,000 gpm flow with all 4 inch PVC pipe:

$$\sum K = \left[5\ elbows \cdot \left(\frac{1.0}{elbow}\right)\right] + \left[1\ entrance \cdot \left(\frac{0.5}{entrance}\right)\right] +$$

$$\left[1\ return \cdot \left(\frac{1.0}{return}\right)\right] + \left[0.013 \cdot \left(\frac{200}{0.333}\right)\right]$$

$$\sum K = 14.3$$

Total energy loss $= H_L = K\left(\frac{V^2}{2g}\right) = 14.3\left(\frac{(25.4\ ft/s)^2}{2(32.2\ ft/s^2)}\right) = 143\ ft$

3.3 MEASUREMENT

Fluid flow can be measured either directly to determine actual flow for a given time interval, such as with a bucket and stopwatch, or indirectly by measuring some variable that is directly related to the flow rate, such as pressure differential across an orifice plate. There are numerous devices available for measuring flow rate in pipelines, such as orifice plate meters, propeller meters, electromagnetic flow meters, ultrasonic flow meters, and rotameters. One can obtain a device that is appropriate for specific circumstances and which provides the degree of accuracy required at a reasonable cost. Flow rates can also be estimated from open pipe discharges by relating the height of the water jet from a vertical pipe or the trajectory of a horizontal pipe.

PITOT-STATIC GAUGE OR TUBE

You can determine the velocity in a flowing pipe by inserting a Pitot-Static Gauge. This device consists of a wall tap into the pipe that measures the static pressure energy (elevation + pressure energy) and a small tube placed to face into the flow that measures total pressure (elevation + pressure + kinetic energy). Generally the wall or static energy pressure tap is built into the pitot tube itself. The difference in these two energy measurements is the kinetic energy only. This can be seen be rearranging and solving for velocity from Eq. 3.4:

$$V = C \cdot 2.315\sqrt{H\ inches} \quad (3.14a)$$

$$V = C \cdot 0.443\sqrt{H\ cm} \quad (3.14b)$$

where V is in ft/s (m/s) and H is in inches water gauge (cm water) and

C = 0.95 to 1.00 for pitot tube
= 0.60 for sharp edged orifice

Example 4

What is the velocity in a pipe where the pitot tube indicates a pressure difference (Total – static pressure) of 12 inches (30.5 cm) water gauge.

IP units

$$V = 1.0 \cdot 2.315 \cdot \sqrt{12} = 8.02\ ft/s$$

SI units

$$V = 1.0 \cdot 0.443 \cdot \sqrt{30.5} = 2.45\ m/s$$

VENTURI TUBE

Venturi tube meters estimate flow based upon the change in pressure along a pipe that is specifically reduced in cross sectional diameter so that the fluid velocity is increased which causes a corresponding decrease in static pressure (see Eq. 3.4 and Fig. 3.3). There are commercially available venturi meters that can be installed in a straight section of pipe to obtain the pressure measurements. Knowing the pressures at the upstream (P_1) and in the throat section (P_2), the velocity can be calculated:

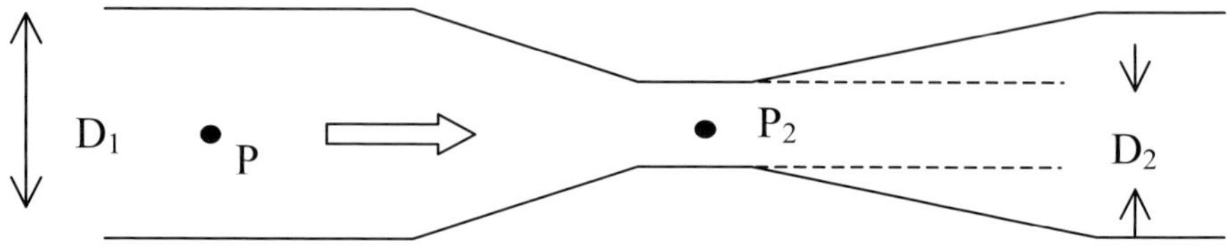

Figure 3.3 Venturi Meter Showing Pertinent Variables.

$$V_2 = \frac{C\,\beta\sqrt{(P_1 - P_2)}}{\sqrt{1 - \left(\frac{D_2}{D_1}\right)^4}} \tag{3.15}$$

where C is 1.00 for perfect venturi (provided by manufacturer, generally around 0.98 to 0.99) and dimensions are in cm or inches, pressure terms are in kPa or psi or inches water gauge, velocity is predicted in cm/s or ft/s and the closure constants, β, are:

β = 2.312 for US units with pressure in inches water gauge
12.18 for US units with pressure in psi gauge
1.414 for SI units with pressure in kPa[a]

Example 5

Calculate the velocity in a venturi meter (assume venturi coefficient of 1.00) if the venturi has a throat diameter of d = 2 inches (5.08 cm) and the main pipe has a diameter of D = 4 inches (10.2 cm) if the pressure differential between the main pipe and the venturi is 12 inch water gauge (2.983 kPa).

IP Units

$$V_2 = \frac{1.00 \cdot 2.312\sqrt{12}}{\sqrt{1 - \left(\frac{2}{4}\right)^4}} = 8.27\ ft/s$$

[a] 1 inch water gauge pressure is 0.2486 kPa

SI Units

$$V_2 = \frac{1.00 \cdot 1.414\sqrt{2.983}}{\sqrt{1-\left(\frac{5.08}{10.2}\right)^4}} = 2.52\ m/s$$

ORIFICE PLATE METER

Equation 3.15 can also be used to predict the velocity in a pipe where an orifice plate with a smaller hole of diameter "D_2" has been inserted. The C coefficient is approximately 0.60 for a sharp-edged orifices when the ratio D_2/D_1 is less than 0.3 (see ASHRAE Fundamentals for more specific information where C is related to Reynolds number and D_2/D_1).

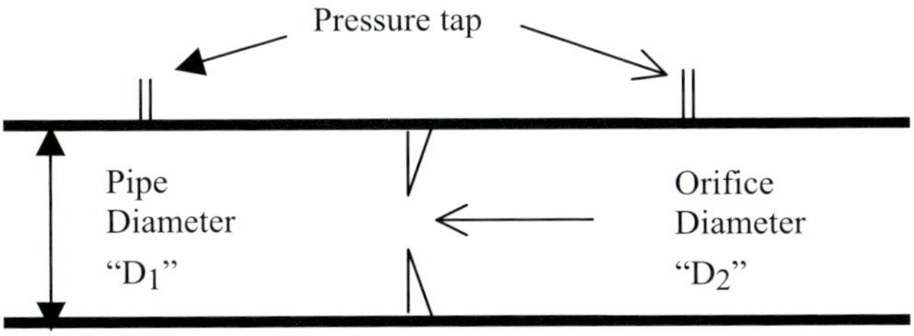

Figure 3.4 Depiction of a Pipe with a Sharp Edged Orifice Plate.

WEIRS

Weirs are essentially "dams" that are placed in a channel that obstruct the flow, thereby creating a crested flow over the weir. The height of this crest of fluid can be measured and then correlated with a channel flow rate. Use of open channels in recirculating aquaculture systems (RAS) are handy as they are easy to clean and do not plug. The two most common weirs are either rectangular flat weirs or Cippoletti weirs (see Fig. 3.5). Other weir calculations are presented by Lawson (1995) and Piper et al. (1982).

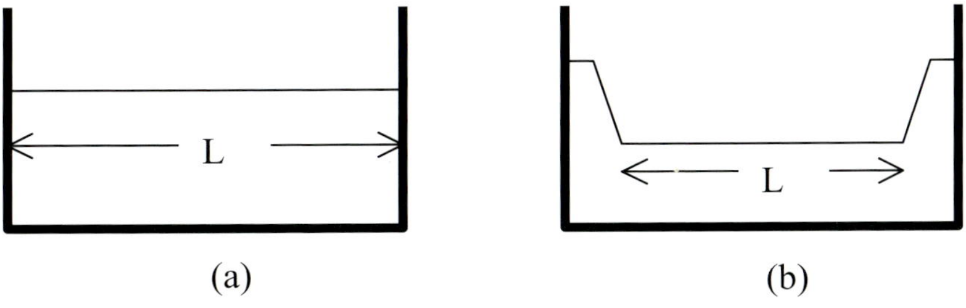

Figure 3.5 Weir Types: A) Rectangular and B) Cipolletti.

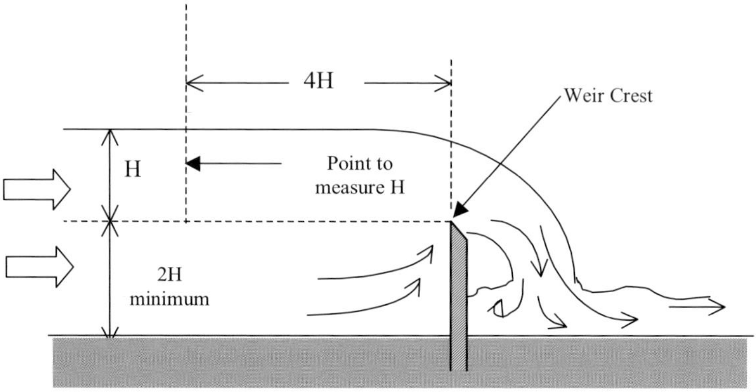

Figure 3.6 Channel Profile with Inserted Weir.
Depth of water above weir crest, H, is measured 4H upstream of weir edge.

Channel flow can be predicted for these two weir types as follows:

Rectangular or Cippoletti

$$Q(gpm\ or\ L/s) = \beta\ LH^{1.5} \qquad (3.16^{a})$$

In the case that there is significant kinetic energy, h, in the approach velocity at the weir, then use the following equation:

$$Q = \beta\ L((H+h)^{1.5} - h^{1.5}) \qquad (3.17^{b})$$

---

[a] Equations 3.16 and 3.17 are for a properly operating rectangular weir with suppressed end contractions, i.e. a weir with a crest equal to the channel width. End contractions are corrected by assuming effective width of crest length is reduced by 0.1H at each end or $L_{effective} = L - 0.2H$.

[b] Equation 3.17 shows a correction term for approach velocity, h. This was first introduced in 1852 and became known as the Francis forumla. It is based upon data that was for channel widths from 8–10 ft (2.4 to 3.0 m) and for hydraulic heads (H) of 0.6–1.6 ft (0.2–0.8 m) and kinetic energies of 0.2–1.0 ft (0.06–0.3 m). Published in the Handbook of Hydraulics by E.F. Brater and H.W. King, 1918.

$$h = \frac{V^2}{2g} \tag{3.18}$$

where Q is in gpm (L/s), H is in ft (m), V is in ft/s (m/s), g is in units of ft/s^2 (m/s^2), and β is:

- 1,511 for Cippoletti weir US units (1,857 SI Units)
- 1,495 for rectangular weir US units (1,837 SI Units)

Example 6

Calculate the flow rate from a Cippoletti weir if the height above the weir crest (4H upstre*am) is 10 inches (0.254 m) and the weir length is 1.0 ft (0.305 m) and the approach velocity is negligible.

IP[a] Units

$$Q(gpm) = 1{,}512 \cdot 1.00\,ft \cdot (10/12\,ft)^{1.5} = 1{,}150\,gpm$$

SI Units

$$Q(L/s) = 1{,}857 \cdot 0.305m \cdot (0.254m)^{1.5} = 72.54\ L/s$$

Vertical Pipe

Predicting the flow from an open ended round pipe pointed "straight up", Fig. 3.7, can be done using Eqs. 3.20 and 3.21 (more in depth treatment of this method is provided by Lawson, 1995):

IP units

$$Q(gpm) = 5.39 \cdot D^2 H^{0.47} \tag{3.19}$$

SI units

$$Q(L/s) = 0.034 \cdot D^2 H^{0.47} \tag{3.20}$$

Graphic representations of Eq. 3.20 is represented in Figs. 3.8 and 3.9.

[a] IP is English or Inch-Pound system.

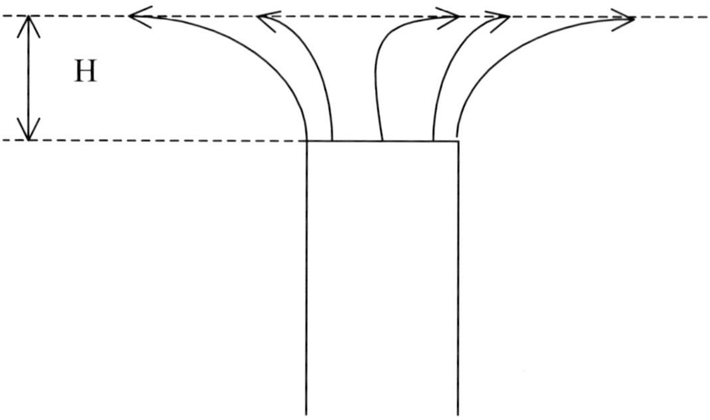

Figure 3.7 Flow From an Open Ended Round Pipe Pointed "Straight Up" Illustrating the Head Created "H".

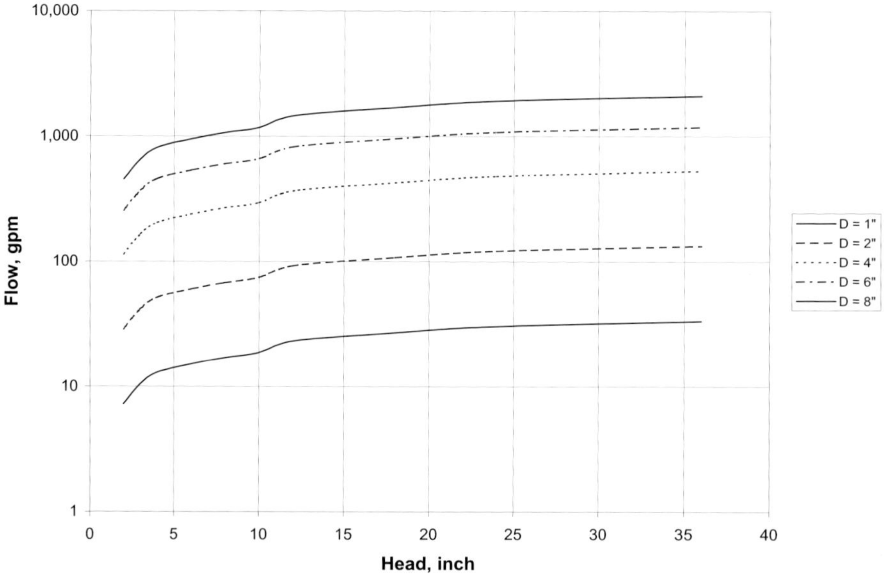

Figure 3.8 Flow Estimate for a Vertically Oriented Pipe (from Eq. 3.19).

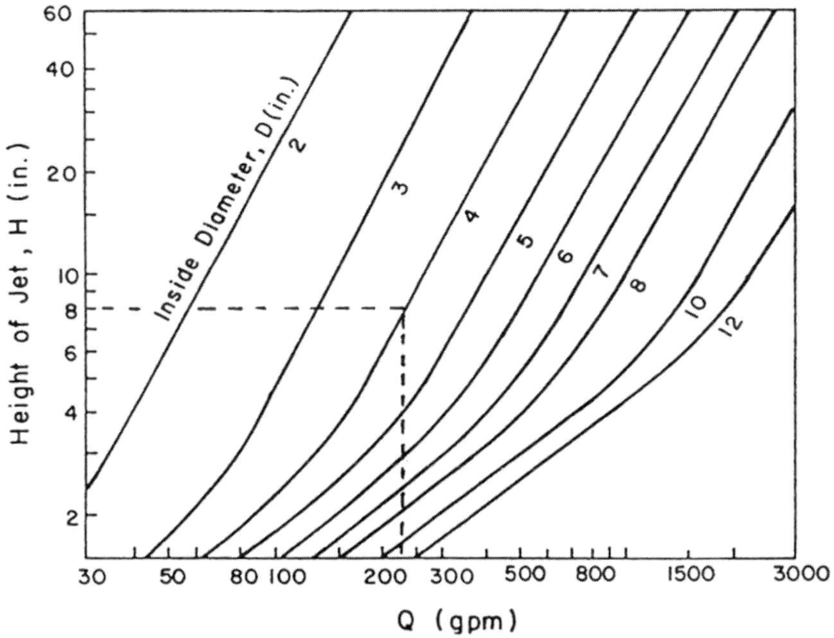

Figure 3.9 Flow Estimate for a Vertically Oriented Pipe (from Eq. 3.19) (from US Department of the Interior, 1967).

Example 7

Calculate the flow rate from a round pipe oriented vertically that is 4 inch (10.2 cm) in diameter and the height of the jet emitted from the pipe is 8 inch (20.3 cm).

IP Units

$$Q(gpm) = 5.39 \cdot (4\ inch)^2 \cdot (8\ inch)^{0.47} = 229\ gpm$$

SI Units

$$Q(L/s) = 0.034 \cdot (10.2\ cm)^2 \cdot (20.3\ cm)^{0.47} = 14.5\ L/s$$

3.4 Pumps and Pumping

All RAS employ some kind of pump to move water to a higher elevation or to increase the overall system pressure for filtration, aeration, degassing, etc. Pumps are fairly efficient mechanisms for transferring energy to water. However, to function efficiently and reliably, they must be carefully selected and integrated into the total system. The key requirement in pump selection is to match the aquaculture system requirements with the maximum operating efficiency of the pump. In pump design, pressure and discharge are inversely related, i.e., high pressure usually is with low discharge and high discharge rates are with low pressure. Poor pump selection can significantly increase operating and maintenance costs and could result in system failure. There are countless varieties of pumps commercially available, and each is usually engineered for a specific application. In aquaculture, two types are commonly encountered: centrifugal impeller pumps and axial flow propeller pumps. Other types, such as airlift pumps and diaphragm pumps, are also found and can be reasonably efficient in specific applications. Airlift pumps, for example, are often used where the pumping head requirements are low and partial re-aeration of the water is required. However, the pumping capacity of airlift pumps is generally not sufficient to drive the water treatment requirements of high-density recirculation systems.

Centrifugal pumps account for the majority of pumps used in the aquaculture industry. They are sometimes called radial flow pumps, because water enters the pump at the eye, or center of the impeller and is driven outwards by centrifugal force at right angles to the axis of rotation. The centrifugal pump characteristics are determined to a large degree by the impeller design. The impeller generally has the form of a disk with a series of curved, raised vanes radiating from the center. The spacing between the impeller and the volute together with the impeller geometry, limits the size of solids that can pass through the pump. Totally enclosed impellers have faceplates that completely enclose the waterways between the vanes on both sides. In open and semi-enclosed impeller pumps, the clearances between and around the vanes are large, allowing large solids to pass through the pump. Open and semi-enclosed impellers are used to pump liquids with high solids content, but at the cost of lower overall efficiencies. Totally enclosed impeller pumps have the highest efficiencies, but are not well suited for pumping liquids with suspended solids, due to increased wear on the impeller. To operate properly, a centrifugal pump must have a positive pressure at the inlet or be directly plumbed with a water filled, airtight pipe to the water source.

Centrifugal pumps are not self-priming, unless specifically so stated in the manufacturers specifications. Self-priming pumps are designed with a special cavity in the volute that retains water. When it is turned on, it is able to expel the air in both the volute and inlet pipe and re-prime itself. These self-priming pumps should not be confused with a simple filter basket or priming pot attached to many small centrifugal pumps. A pump with a filter basket and a foot valve will retain its prime, as long as no air enters the system. However, if any air enters the systems, due to a leaky foot valve for example, the impeller will spin in an empty volute once the water in the filter basket has been emptied out.

The best way to prevent loss of prime is to locate the pump below the surface of the water level being pumped. This is commonly referred to as a flooded suction, since when the pump is off, it's still flooded with water. If the pump must be located higher than the surface of the water, then a foot valve should be installed at the bottom of the suction pipe, alone with an inlet strainer. Some form of priming access, either a priming pot, filter basket or TEE fitting should be placed at the inlet to the pump, to allow the suction pipe to be completely flooded with water. Once flooded and sealed at the top and bottom, the pump should not lose prime when turned off. It is important not to restrict the flow of water into a centrifugal pump with a valve or extensive plumbing. This can result in a condition called cavitation, where some or all of the fluid has turned into water vapor. Cavitation can damage the shaft or impeller and significantly reduces or eliminates all flow from the pump. Burnt out pump motors can result.

Axial flow pumps are designed to pump water efficiently at low head and high discharge. The pumping element of an axial flow propeller pump consists of a revolving propeller, similar in appearance and function to a boat propeller. The propeller is located inside of a pipe or shaft that serves as a conduit to convey the water. The propeller is driven by either a submersible motor or a remote-mounted motor linked by a drive shaft. The water flows in essentially a straight line along the pump axis keeping friction and turbulence to a minimum. The propeller of an axial flow pump must be submerged in the water. Axial flow pumps can be extremely efficient at moving large volumes of water to modest head levels, e.g., 15 to 30 feet (4.6 to 9.2 m). In addition, they are able to tolerate some small debris and small solids.

Selecting a pump requires knowledge of at least the following factors:

- Total head required
- Discharge flow required
- Suction lift required
- Liquid to be pumped and its characteristics (freshwater, saltwater, solids)

- Continuous or intermittent pumping
- Power source available (single or three phase power, diesel or gas engine)
- Space, weight, and similar limitations
- Special requirements

The term "head" is commonly used to express the pressure at some point in terms of elevation (feet or meters). The following expressions of "head" are often found in the industry to describe performance and design parameters for systems:

- **Static Head:** the hydraulic pressure at a point in the water when the water is at rest

- **Friction Head**: the loss in pressure or energy due to frictional losses in flow

- **Velocity Head:** the energy in a fluid due to its velocity

- **Pressure Head:** a pressure measured in equivalent head units

- **Discharge Head:** the output pressure of a pump in operation

- **Total Dynamic Head:** the total pressure difference between the inlet and outlet of a pump in operation

- **Suction Head:** the vertical distance from the static water surface to the centerline of the pump intake, negative when the pump inlet is below the water surface

- **Suction Lift:** the vertical distance from the static water surface to the centerline of the pump intake, positive when the pump inlet is above the water surface

The total suction head or suction lift should also include the friction losses, minor losses, and velocity head on the suction side of the pump, but these are usually small in relation to the suction lift and can often be ignored. The water in the suction line of a pump is being pulled into the pump by the difference in the atmospheric pressure on the free water surface and the vacuum created in the pump inlet, just like sucking water through a straw. If the pressure drops below the vapor pressure of water in the pump inlet, the water will vaporize into small bubbles. This phenomenon also creates cavitation. As the bubbles form and collapse, they can stress the metal pump housings and impellers, and pitting can occur on their surfaces. Cavitation is easy to recognize, since it sounds like the pump is full of gravel or has a rattling sound. Cavitation reduces the pumps overall efficiency and can severally damage a pump and impeller if not corrected.

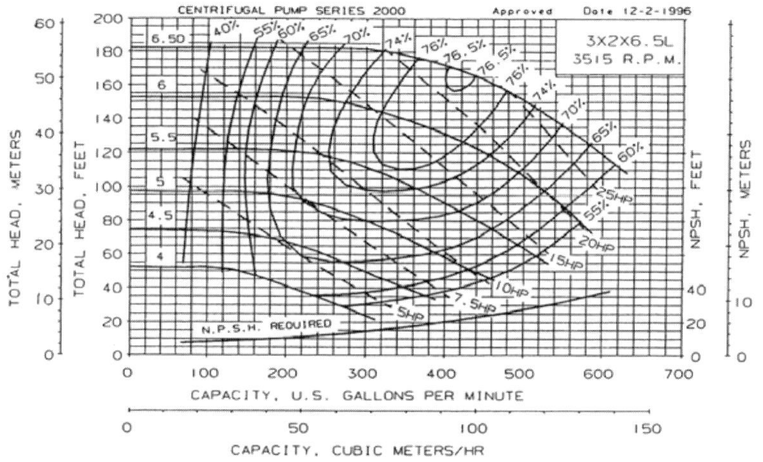

Impeller Code	A	B	C	D	E	F
Size	6.5	6	5.5	5	4.5	4
Standard HP	25	20	15	10	7 ½	5

Figure 3.10 Performance Curve of a Centrifugal Pump (Courtesy of Goulds Pumps).

Theoretically, the greatest such lift possible is 34 feet at sea level, but practical limits reduce this to between 15 and 20 feet for reasonably efficient operations. The possible lift for different elevations is shown in Table 3.7. The net positive suction head required ($NPSH_{required}$) to prevent cavitation is a function of the pump design and speed, and is usually specified by the manufacturer. A typical performance curve is shown in Fig. 3.10. This figure shows the total head that the pumps operates against (ordinate axis), the flow rate of the pump is on the abscissa axis; the $NPSH_{required}$ is shown on the right hand side of the performance curves on an ordinate axis. Efficiencies are shown on the graph for the different operating points of the pump.

The $NPSH_{required}$ must be less than the $NPSH_{available}$ or:

$$NPSH_{available} \geq NPSH_{required} \tag{3.21}$$

Where $NPSH_{available}$ is defined as the atmospheric pressure less friction losses on the suction side of the pump, the suction lift, and the vapor pressure of the water. The atmospheric pressure is given in Table 3.7 and the vapor pressure is given in Table 2.1.

Table 3.7 Standard U.S. Atmospheric Pressure at Different Altitudes

Altitude		Atmospheric Pressure				
M	ft	mm Hg	kPa abs	psia	m H_2O	ft H_2O
-200	-656	778	103.7	15.0	10.6	34.8
0	0	760	101.3	14.7	10.3	33.8
200	656	742	98.9	14.3	10.1	33.1
400	1,312	725	96.6	14.0	9.9	32.5
600	1,968	707	94.3	13.7	9.6	31.5
800	2,625	690	92.0	13.3	9.4	30.8
1,000	3,281	674	89.8	13.0	9.2	30.2
4,000	13,123	462	61.6	8.9	6.3	20.7

The efficiency of a pump is defined as the ratio of work output or water horsepower (WHP) and the electrical energy input used to operate the pump:

$$WHP = \frac{Q \cdot TDH \cdot \Omega_{sg}}{\beta} \tag{3.22}$$

$$E_{pump}(\%) = 100\frac{WHP}{BHP} = \frac{output}{input} \tag{3.23}$$

β =	6,116	for WHP in kW and Q in L/min and TDH in m
	0.102	for WHP in kW and Q in m^3/s and TDH in m
	3,960	for WHP in HP and Q in gpm and TDH in ft
	8.81	for WHP in HP and Q in cfs (ft^3/s) and TDH in ft

The Total Dynamic Head (TDH) is the sum of the static lifts from the water source to the discharge point, and all the frictional losses from the inlet suction side to the discharge side. The velocity head is often negligible but should be considered when extremely accurate values are needed. The frictional losses include piping, fittings, and processing equipment (for example UV, heating, oxygenation systems) on both the suction and the discharge sides of the pump. Since the frictional losses are a function of the flow rate, the TDH must be calculated over the range of anticipated flows. Considerations should be made of anticipated changes in operations and management, biofouling, or system configurations.

Once the system is initially designed and the TDH is estimated at the desired flow rate, then a pump must be selected that minimizes pumping costs by operating at or near its maximum efficiency to deliver the desired flow. This is accomplished by examining the performance data for the various pumps under consideration. Performance data is usually presented in graphical form, although for small pumps, tables of head versus discharge are usually provided. Each pump has its own unique set of characteristic curves, determined by the engineering design of the impellers, casing and to some extent by its age and wear of internal parts. A typical performance curve for a centrifugal pump was previously shown in Fig. 3.10 and performance data is given for a small submersible centrifugal pump in Table 3.8.

The flow of water is directly related to the TDH with maximum water flow occurring at the lower TDH values. Some pumps will have relatively flat responses in pumping rate to changes in TDH while other pumps are extremely dependent upon the TDH that the pump is operating against.

Each aquaculture system will have its own system operating curve. The system operating curve is constructed by computing the total dynamic heads required in the proposed system to deliver a range of specific flow rates. Once a system operating curve is constructed, then the manufacturer's pump characteristic curves can be imposed on the system curve and where the two curves intersect defines the system's operating point. Ideally, you try to chose a pump such that the intersection of the operating and system curves are as close as possible to the maximum operating efficiency of the pump (see Fig. 3.10). Note in this figure, that for a 15 HP pump, the maximum efficiency will occur at an operating point of approximately 310 gpm (70 m^3/hr) at a TDH of 110 ft (33.6 m). However, if you operated this same pump at a TDH of 60 ft (18.3 m), you would obtain a flow rate of 520 gpm (118 m^3/hr). Now, do not be totally misled by thinking that the only objective is to maximize pumping efficiency. In the previous example, it may be advantageous for you to operate the pump at a lower TDH and obtain the higher flow rate, particularly if your system design indicated that you only required approximately 50 to 60 feet (15.2 to 18.3 m) of TDH. The principles of system operating curve and pump operating characteristics is demonstrated in Fig. 3.11.

Table 3.8 Performance Data for a Submersible Sump Pump

			GPM @ Total Feet of Head			
HP	5 ft	10 ft	15 ft	20 ft	25 ft	30 ft
½	58	37	18	0	0	0
¾	65	47	31	15	0	0
1		58	43	29	15	0
1.50		81	63	48	32	16
1.75		94	78	62	47	32

Data for an ABS Stainless Steel Submersible pump

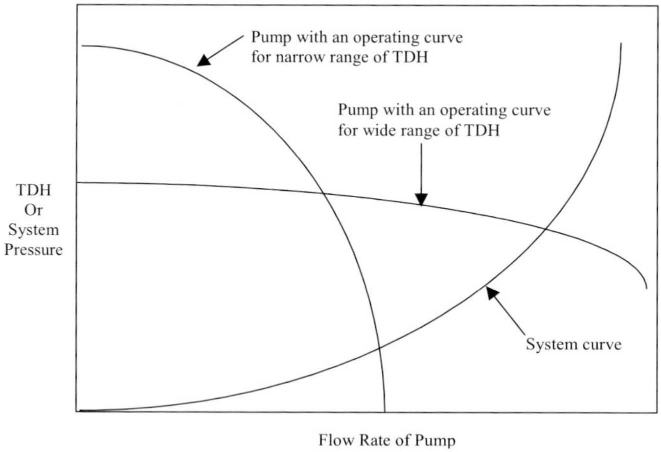

Figure 3.11 Pump Characteristic Curve and a System Operating Curve.

If a range of discharges were required, the pump selected would be capable of providing the maximum discharge. The pump would have the flattest possible efficiency curve over the necessary operating ranges, with the maximum efficiency at the operating point used the greatest amount of the time. Moreover, for large, commercial operations, this would be the design engineer's method of selection, plus some practical experience with other systems and pumps

In the real world, selecting a pump for small systems is usually done by estimating the overall system's total dynamic head and flow requirements. Then, the characteristic curves for several readily available pumps that can provide the flow rate at the required head are reviewed. In some cases, efficiency data for pumps is also provided, but normally not. So a guess is made as to the maximum efficiency, usually just to the right of the midpoint on the characteristic curve. A pump is chosen after considering other factors such as capital cost, operating cost, size, color, discounts, and availability.

Example 8

Calculate the brake horsepower required to operate a pump that had a pumping efficiency of 55%, operates against a TDH of 40 ft and pumps at a flowrate of 1,000 gpm.

IP

$$BHP = \frac{WHP}{E_{pump}} = \frac{1{,}000\,gpm \cdot 40\,ft \cdot 1.00}{3{,}960 \cdot \left(\frac{55}{100}\right)} = 18.4\ HP$$

SI

$$BHP = \frac{WHP}{E_{pump}} = \frac{3{,}785\,Lpm \cdot 12.2m \cdot 1.00}{6.116 \cdot \left(\frac{55}{100}\right)} = 13.7\ kW$$

In this example, you would need a pump that provides at least a power rating of 18.4 HP meaning you will probably use a 20 HP pump. In some cases, for larger pumps, the pump manufacturer can design the pump for you to operate specifically at a pre-selected operating condition and then the pump will be designed to achieve maximum operating efficiency for these specific conditions. This is particularly relevant to aquaculture applications where generally we need high flow rates against fairly low operating heads. In general, such pumps for large applications have not been available. The closest companion industries are irrigation pumps.

3.5 AIRLIFT PUMPS

In principal and theory, airlift pumps will by far move the most water per unit of energy supplied. However, airlift pumps have the following disadvantages:

- Limited application to create water lift or elevation change, e.g., 10 to 15 cm.
- Reductions in flow rates due to fouling of the air distribution mechanism, particularly for conventional sintered glass air stones
- Complete loss of water pumping when water elevation changes cause excessive requirements for water lift, e.g., water level dropped in tank where the unit is placed
- Reductions in water flow are not "obvious" to casual observer
- Energy efficiency is highly related to very closely matched requirements between centrifugal blower and the airlift requirements

In principle, all of the five disadvantages listed above can be eliminated with effective management, maintenance, and initial design of the overall system. Practically speaking, airlifts are problematic. In particular, the authors became discouraged with their application because airlifts are constrained to low water head differentials. We would suggest that airlifts are most appropriate in low density fish applications. Figure 3.12 demonstrates the important variables to consider when designing an airlift pump. Chapter 17 describes AIRPUMP, a software program that predicts the operating performance of such pumps.

The airlift pump has been in practical use as a pumping device for many decades. A gas, usually air, is injected at the base of a submerged riser tube. As a result of the gas bubbles suspended in the fluid, the average density of the two-phase mixture in the tube is less than that of the surrounding fluid. The resulting buoyant force causes a pumping action.

Complicating the prediction of the hydrodynamics of the airlift is the fact that there are three flow patterns possible, Fig. 3.13. When the initial bubble size is much smaller than the tube diameter and the gas void ratio is below about 25%, the bubble flow pattern results. Small bubbles are distributed over the pipe cross section. Bubbles remain close to their initial size, and there is little interaction between single bubbles. When gas void fraction exceeds 25%, coalescence occurs and large gas bubbles or "slugs' form.

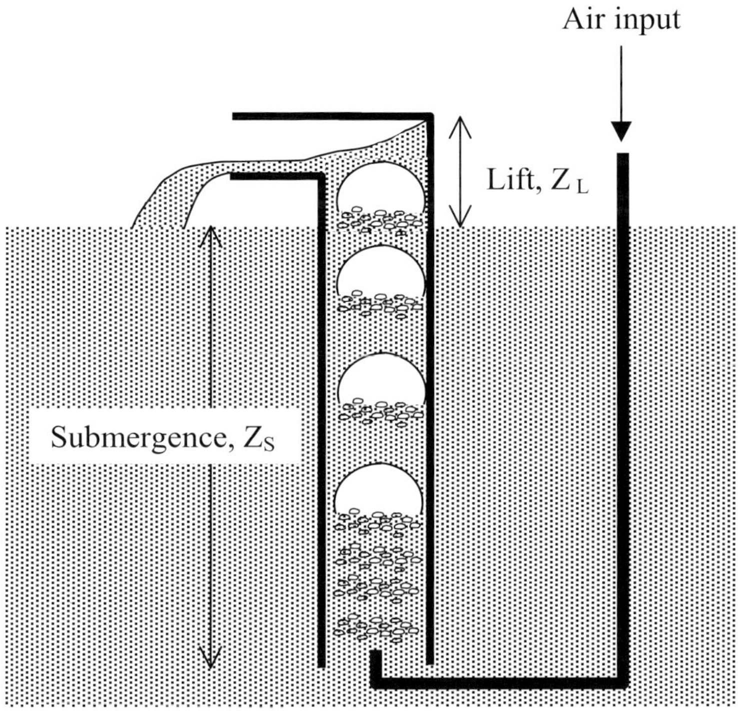

Figure 3.12 Typical Airlift Pump Application.

In slug flow, the gas phase is contained in large bubbles that nearly span the tube and range in length from the tube diameter to several times this value. These are referred to as gas slugs or Taylor bubbles. The liquid filling the space between the Taylor bubbles is referred to as the liquid slug. The liquid between the Taylor bubbles and the tube wall is referred to as the liquid film. The slug flow regime has been found to occur only when the riser tube diameter is below 25 mm (1 inch).

The bubbly-slug flow regime is most commonly encountered in airlift operation. In the bubbly-slug flow regime, small bubbles are found suspended in the liquid slug between the Taylor bubbles. The presence of these bubbles is due to the region of extreme turbulence encountered at the tail of the Taylor bubble. Small bubbles are broken off of the tail of the Taylor bubble and dispersed in the liquid slug. The bubbly-slug regime occurs in tubes with diameter greater than 25 mm (1 inch) when the gas void ratio is above about 25% regardless of the initial bubble size.

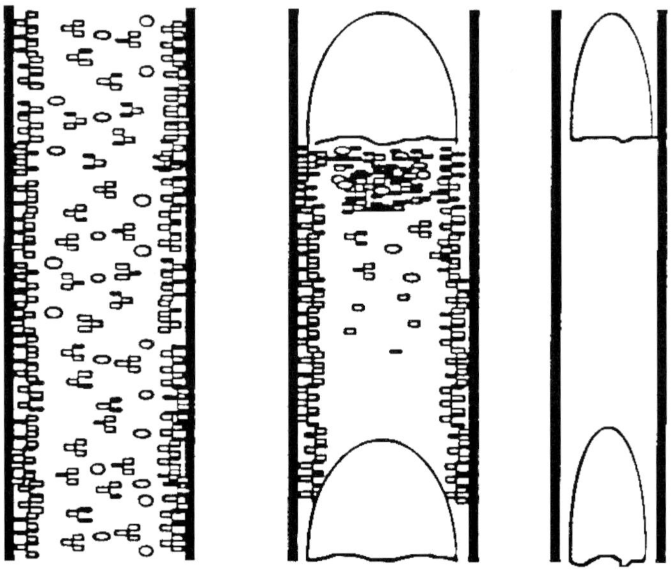

Figure 3.13 Three Flow Patterns Encountered in Airlift Pump Operation.

The gas concentration in the tube is a function of the submergence ratio and the amount of air being introduced into the pump. The submergence ratio is defined as the fraction of the tube that is submerged below the level of the water being pumped (see Fig. 3.12). Airlift pumps must be submerged at least 50% for any pumping action to take place. The pumping efficiency is defined as the useful work done raising the water divided by the work required to compress the air. Maximum pumping efficiencies are obtained for submergence ratios of approximately 80% depending on tube size and flow pattern. Maximum pumping rates are obtained with 100% submergence. Although there is no useful pumping work done when the pump is 100% submerged (since there is no lift), this configuration can be used to mix and aerate tanks and ponds.

The length to diameter ratio (Z:D) of the riser tube can have a significant effect on airlift pump performance (note the figure depicts Z_s

and Z_L where $Z=Z_s + Z_L$). The length to diameter ratio is defined as the length of the riser tube divided by its diameter, with both measurements in the same units, e.g., a 10 meter riser tube with a diameter of 10 centimeters has a length to diameter ratio of 100. When Z:D is below 50 the two phase flow patterns will not have sufficient time to develop fully. This results in increasing "slip" of the water past the rising gas bubbles and decreased pumping volume and efficiency. It is therefore advantageous to keep the ratio of Z:D above 50.

Airlift pumps have the added advantage of aerating the water in the process of providing the pumping function, since air is used to drive the pump. The gas transfer is affected by the flow rate of water and air through the riser tubes. The critical factors influencing the rate of gas transfer are the Reynolds number of the flow and the length of the tube. The Reynolds number is a dimensionless measure of flow velocity and the degree of turbulence in the flow. Gas transfer is not significantly affected by flow pattern, initial bubble size, or the presence of dilute organic wastes. Some caution must be exercised in using airlift pumps, since the water and thus gases entering the base of the airlift column may be above saturation due to hydrostatic pressure; this may cause gas bubble disease in some fish species. The degree of gas supersaturation can be controlled by limiting the submergence of the airlift riser. This will also limit the amount of lift available for a given tube length. An in depth treatment of the theoretical and experimental basis for the AIRPUMP program is in Reinemann and Timmons (1989) and Reinemann et al. (1990).

LIST OF SYMBOLS

A	Cross-sectional area, ft^2 (m^2)
BHP	Brake horsepower, kW (hp)
C	Coefficient relating pressure to velocity, dimensionless
$C_{H\text{-}W}$	Roughness constant used in Hazen-Williams Eq. 3.7
D	Diameter of pipe, ft (m)
D_{inch}	Diameter of pipe, inch
D_1	Diameter of upstream section, cm (in)
D_2	Diameter of throat or downstream section, cm (in)
E_{pump}	Pump efficiency, %
f	Friction factor (a function of the Reynolds number)
f_T	Frictional loss coefficient that depends upon the size of the fitting
$FITTING_{constant}$	Frictional loss coefficient that is size independent of the fitting
g	Acceleration of gravity, 32.2 ft/s^2 (9.81 m/s^2)
h	Correction for channel kinetic energy, ft (m)
H_L	Lost energy head due to friction, ft (m)
K	Resistance coefficient for particular style of fitting
L	Length of pipe, ft (m)
P	Pressure of the fluid, psig (kPa)
P_1	Upstream pressure, psi (kPa)
P_2	Throat pressure, psi (kPa)
Q	Massflow, lb/s (kg/s)
TDH	Total dynamic head that pump operates against, ft (m)
V	Fluid velocity, ft/s (m/s)
V_1	Velocity in upstream section, ft/s (cm/s)
V_2	Velocity in throat of venturi, ft/s (cm/s)
Z	Vertical distance from the reference plane, ft (m)
γ	Specific weight of the fluid, lbs/ft^3, [a]
ν	Fluid kinematic viscosity, ft^2/s (m^2/s) (see Table 2.7)
ρ	Density of the fluid, lb/ft^3 (kg/m^3) (see Table 2.7)
Ω_{sg}	Specific gravity fluid, 1.00 for water, dimensionless

[a] In SI units, the (P/γ) term is actually P, pressure, divided by (γ/g) and $g = 9.81$ m/s^2. Also, remember that SI pressure units are kPa (kilo pascals) and a Pa = Newton/m^2 and a Newton, N = kg m/s^2).

CHAPTER 4

MASS BALANCES, LOADING RATES, AND FISH GROWTH

4.0 INTRODUCTION

Water flow is the mechanism by which oxygen is transported into a fish culture vessel and the waste products being generated within are removed. The design of a recirculating aquaculture system (RAS) should insure that the important parameters affecting water quality and fish productivity, e.g., oxygen, ammonia, carbon dioxide, and suspended solids are properly balanced. This requires calculating the value of each of these parameters independently to determine the thresholds for each. Then, having done the necessary calculations, the system must be operated at the highest flow rate possible while still maintaining a particular parameter at or below it's maximum tolerable or design value, e.g., ammonia. Obviously, the maximum flow rate possible while maintaining one particular parameter may be too high for maintaining another. The same mass balance approach can be utilized on any variable affecting water quality. It simply comes down to balancing the transport in, the production of a particular parameter within the culture tank, and the transport out. In word equation form, we like to say:

*Transport **in** of "x" + **production** of "x" =*
*transport **out** of "x"* (4.1)

The production term can be the production of oxygen, ammonia, suspended solids, or CO_2. Note that the production term can be negative, meaning consumption of a certain component, e.g., oxygen.

We can depict a mass balance for the general case where part of the flow is recirculated and part of the flow is flow-through as:

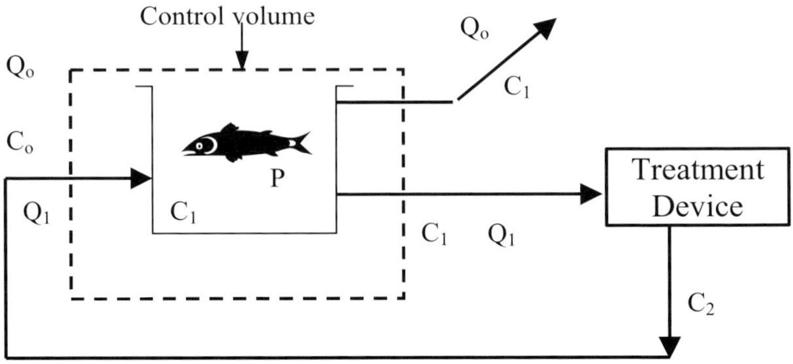

Figure 4.1 General Mass Balance on a Fish Culture Tank. Treatment occurs exterior to the tank.

We are assuming a completely well mixed tank and that the tank has reached a non-changing condition with respect to time or steady-state conditions. The box outside the fish tank represents some treatment device or process that changes the concentration of the noted parameter "x". (Note: there could be several treatment devices, each treating a different water quality variable.)

Returning to our word equation, Eq. 4.1 and Fig. 4.1, we can perform the mass balance assuming steady state conditions:

$$Q_1\,C_2 + Q_0\,C_0 + P = Q_0\,C_1 + C_1\,Q_1 \qquad (4.2^a)$$

To obtain accurate and reliable results from these equations, it is essential that each of the terms or products of terms in Eq. 4.2 is represented by the same unit value, e.g.,

$$\frac{kg_{oxygen}}{day}$$

For example, the unit balance example for a transport of oxygen flowing into the tank would be:

$$QC = \frac{kg_{water}}{day}\;\frac{kg_{oxygen}}{kg_{water}} = \frac{kg_{oxygen}}{day} \qquad (4.3)$$

---

[a] Symbols are defined at end of chapter.

If we were doing a mass balance using Eq. 4.2 for oxygen, then all terms or products would need to have the same units of kg oxygen per unit time. It is convenient to use "day" as the time unit, since growth rates and feeding rates are generally measured on a per-day basis. BE CAREFUL to be consistent with unit designations.

Transport is the key term in these calculations, and it is defined as the product of flow and concentration. For example, the remainder of oxygen transport into the tank minus the allowable minimum level of oxygen departing the tank defines the oxygen available for fish growth. Flow is measured as volume per time or mass per time, and will be usually defined in terms of:

- gallons per minute, gpm
- liters per minute, Lpm
- kg per sec, kg/s
- m^3/s

Typically, most water quality parameters are expressed in terms of:

$$\frac{mg}{Liter} \quad or \quad \frac{mg}{L}$$

The usage of mg/L is often called or referred to as:

ppm or parts per million.

These values are the same.

This usage comes from:

$$\frac{mg}{L} \cdot \frac{L}{1000g} \cdot \frac{g}{1000mg} = \frac{1}{10^6} \quad or \quad \frac{1\ part}{million} \text{ or } 1\ ppm$$

Thus:

$$10\frac{mg}{L}\ oxygen \rightleftarrows 10\,ppm\ \ oxygen$$

Now, to reinforce your understanding, let's calculate the available oxygen to support fish growth assuming a mass flow of water of 100 gpm of saturated inlet water at 60°F at an elevation 800 feet above sea level.

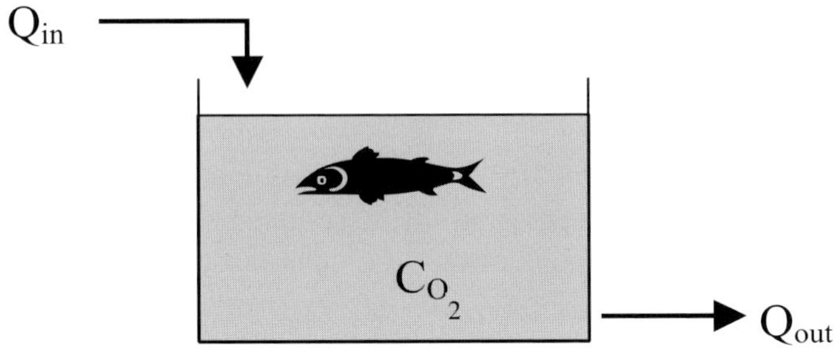

$Q_{in}\,C_{in}$ = O_2 transported in
$Q_{out}\,C_{tank}$ = O_2 transported out

Equation 4.4 provides an approximation that can be used to estimate oxygen concentration in non-saline waters (Chapter 8 provides a more detailed approach to determine gas solubility including oxygen, carbon dioxide, nitrogen, and argon):

$$C_{saturation}\frac{(mg)}{L}=\frac{132}{(T^{\circ}F)^{0.625}}\cdot\frac{760}{760+\dfrac{altitude(ft)}{32.8}} \tag{4.4}$$

Using Eq. 4.4:

Altitude of 800 ft.
Temperature 60°F

$$C_{sat}=10.21\cdot 0.97=9.89\frac{mg}{L}$$

For the mentioned flow rate of 100 gpm of water and with the incoming water having a concentration of 9.89 mg/L, the mass of oxygen transported into the tank on a daily basis is ("gpm" units are used since they are common terminology in US):

$$Q_{in}\,C_{in,\,oxygen}=100\;\frac{gal}{min}\cdot 9.89\;\frac{mg}{L}$$

Now, make the units consistent:

$$= 100\,\frac{gal}{min}\cdot 3.785\,\frac{L}{gal}\cdot 9.89\,\frac{mg}{L}\cdot 1440\,\frac{min}{day} = \frac{mg}{day}O_2$$

$$= 5{,}390{,}445.\frac{mg}{day}\cdot\frac{kg}{10^6\,mg} = 5.39\frac{kg\,O_2}{day}$$

This then is the oxygen available to the culture vessel on a daily basis to support fish growth and bacteria action. However, the water leaving the tank must still be at the minimum level necessary to support fish growth, e.g., 5 mg/L, so only a portion of the total available oxygen can be used.

The available oxygen is then (same as previous example except now a C_{out} for the discharge water is defined):

$Q\,(C_{in} - C_{tank}) =$

$$= 100\frac{gal}{min}\cdot 3.785\frac{L}{gal}\cdot(9.89-5)\frac{mg}{L}\cdot\frac{kg}{10^6\,mg}\cdot 1440\frac{min}{day}$$

$$= 2.66\frac{\text{kg}}{\text{day}}\text{O}_2$$

Returning to the general mass balance Eq. 4.2, the simplest RAS case is where all water flow is recirculated and there is no discharge. In this situation, the Q_0 terms in Eq. 4.2 drop out $Q_{in} = Q_{out}$ have:

$$Q_1\,C_2 + P = +\,Q_1\,C_1 \qquad (4.5)$$
$$Q_1\,(C_2 - C_1) = -\,P$$

Solving the general mass balance equation with the "magical" treatment box, we need to determine the concentration of each particular water quality parameter leaving the treatment device, so that the mass flow of the water quality parameter into the fish culture tank can be determined and the mass balance solved (C_2). Looking now at only the treatment device depicted in Fig. 4.1, the treatment box could be a biofilter, a CO_2 stripper, or a solids settling chamber. Each will have its own treatment efficiency for the particular water quality parameter it is designed to treat. We just use the box as a symbolic depiction of this treatment device.

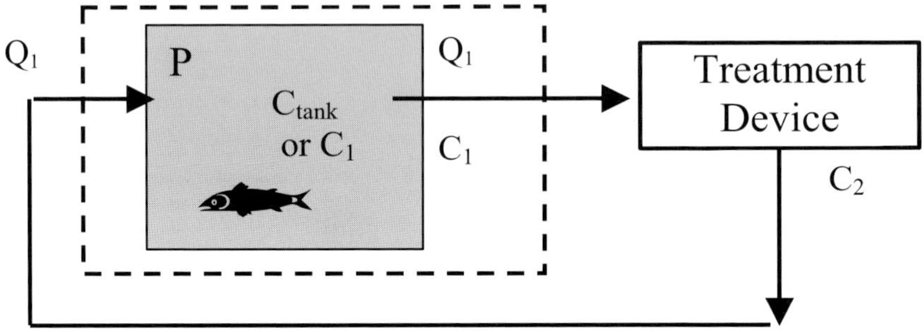

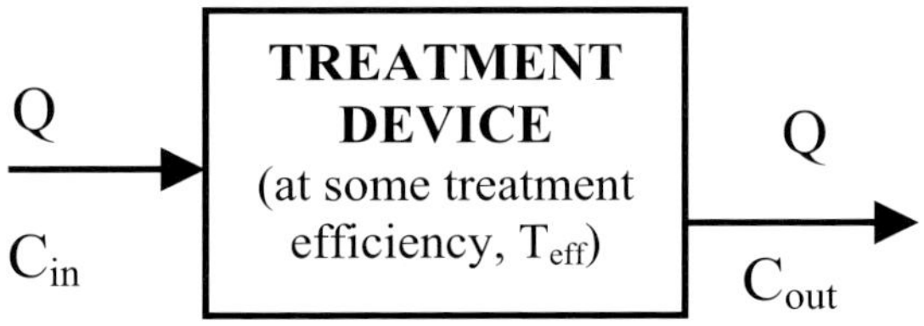

Looking at this treatment box and doing a mass balance on the control volume, you can solve for the leaving water quality concentration C_{out}. Since the water flow in equals the water flow out, the C_{out} term can be solved for directly:

$$C_{out} = C_{in} + \frac{T}{100}\left(C_{best} - C_{in}\right) \tag{4.6}$$

where, C_{best} is the absolute best result obtainable by a treatment system, e.g., zero ammonia or saturated oxygen, zero suspended solids.

We have adapted the term "C_{best}" to represent what the treatment device is trying to do. If you had the perfect treatment device, the device is still limited by basic physical laws as to what it can achieve. Thus, the reference to *best*. Note that if the device is an oxygen addition unit, the C_{best} term can be increased above atmospheric concentration values for oxygen by increasing the partial pressure above atmospheric oxygen partial pressure in the device. For example, a pure oxygen device will have a C_{best} value of roughly five times the C_{best} value that is available if normal air used at atmospheric pressure, e.g., trickling tower. C_{best} for most other parameters should be fairly obvious to the reader, e.g., ammonia and TSS are zero, but CO_2 will be around 0.5 mg/L since there is some CO_2 in the air.

Loading Rates

The relationships of production capacity to flow exchanges and space are important to aquacultural engineers. The terminology "loading" (L) is used to describe the fish mass that can be maintained per unit of flowing water, kg fish per liter per minute flow (kg/Lpm). Fish density (D_{fish}), defined as kg fish per cubic meter of space (kg/m^3), combined with the number of water exchanges per hour (R) through the rearing unit produces the loading rate (L):

$$L = 0.06 \frac{D_{fish}}{R} \qquad (4.7)$$

The constant 0.06 in Eq. 4.7 converts Lpm to m^3/hr (1.0 Lpm · 60 min/h = 60 L/h or 0.06 m^3/h, there are 1,000 L in a m^3). The loading capacity depends primarily on water quality, fish size and species. Using Eq. 4.7 and assuming that fish metabolism requires 250 grams of oxygen for each kg of feed fed (no allowance for nitrification), the allowable loading (kg of fish per Lpm of flow) due to oxygen constraints is:

$$L_{oxygen} = \frac{144\ \Delta O_2}{250 F_{\%}} \qquad (4.8)$$

The ΔO_2 term in Eq. 4.8 represents the allowable drop in oxygen from inlet to outlet. For example, if the inlet DO is 11.0 mg/L and the outlet is designed to maintain oxygen above 5 mg/L, then the ΔO_2 is 6.0 mg/L. Eq. 4.8 is demonstrated in Fig. 4.2 for various feeding rates and allowable changes in oxygen concentration.

As fish densities increase, minimum oxygen levels are better maintained by using pure oxygen systems, which permit inlet oxygen concentrations to be maintained at multiples of dissolved atmospheric concentration levels (see Chapter 8). However, other water quality parameters will begin to limit the allowable carrying capacity due to degradation in accumulated ammonia or carbon dioxide or suspended solids. The term for this condition is the cumulative oxygen consumption (COC) level. From basic stoichiometry and the production values for ammonia, solids, and carbon dioxide associated with feeding, then:

- 10 mg/L of oxygen consumed will produce up to 1.4 mg/L of ammonia,
- 14 mg/L of carbon dioxide, and
- 10 to 20 mg/L of suspended solids.

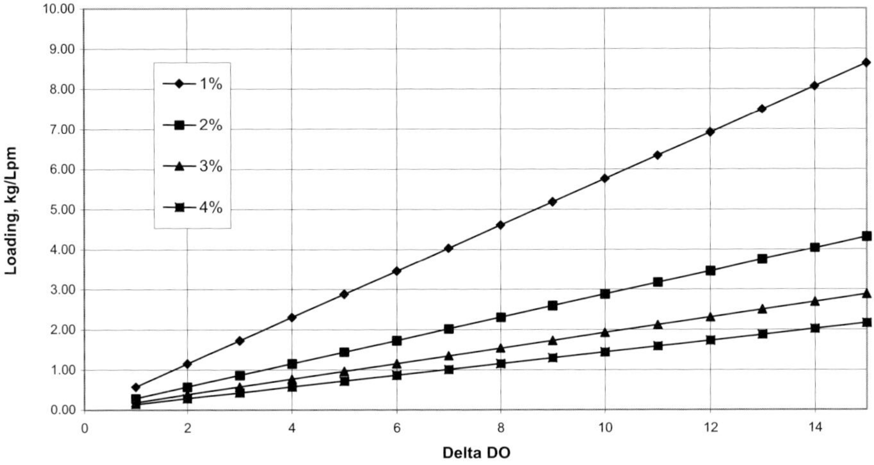

Figure 4.2 Allowable Fish Loading in kg per Lpm. Based upon the change in dissolved oxygen from inlet to outlet (Delta DO) and the percent body weight per day for the fish feeding rate (to convert kg/Lpm to lb per gpm, multiply by 8.33).

This relationship must be kept in mind when designing the fish culture vessel.

Maximum allowable or safe densities (D_{fish}) are much more difficult to ascertain than are maximum loadings. It still seems to be a very subjective process and has much controversy. Density is primarily a function of fish size, species, and the characteristics of the rearing environment and management skill. New growers tend to overestimate their own safe loading densities and assume they can establish and sustain densities from the very beginning that in fact require expert management skills. Do NOT fall into this trap. You will kill fish. New growers should target about 1/2 the densities recommended in this book for expert growers. This subject is discussed in detail in Chapter 6, Culture Tank Design.

4.1. PRODUCTION TERMS

The "P" term in Eq. 4.2 represents the production of some pollutant or consumption term. These terms in an RAS system can all be proportionately related to the fish feeding rate. In principle, if you are not feeding the system, there is no pollution. This generalization is valid, because even ammonia production rates reduce 10 fold within a day or so, once feeding activity has ceased. Oxygen consumption is also reduced by approximately 50% and fecal production goes to zero. However, the idea in any production system is to grow the animals, and to grow, they must be fed, so we relate the production terms exclusively to feeding rates as follows:

P_{Oxygen} = - 0.25 kg per kg feed consumed by fish (a negative production term)
- 0.12 kg per kg feed consumed by nitrifying bacteria
- 0.13 heterotrophic bacteria (estimate, can be as high as 0.5)
= - 0.50, kg (0.25 + ·12 + ·13) per kg feed for system including nitrifying and heterotrophic bacteria

P_{CO2} = 1.375 grams produced for each gram O_2 consumed (both fish and bacteria)

$P_{TAN} = F \cdot PC \cdot .092$

$P_{Solids,\,TSS} = 0.25 \cdot$ kg feed fed (dry matter basis)
(literature gives values from 20% to 40% of feed fed on dry basis)

4.2 WATER QUALITY DESIGN TARGETS

A difficult concept to grasp for most is that in performing the mass balances, the designer/manager must choose design or target operating conditions. These are the "C" values in Eq. 4.2 shown in the culture tank. These design numbers are species dependent and are continually being refined for RAS applications. Our recommendations are given in Table 4.1 for a common warm (tilapia) and cool (trout) water species.

Table 4.1 Water Quality Parameters for a Cool and Warm Water Species

Parameter	Tilapia	Trout
Temperature, °F (°C)	75 to 85 (24 to 30)	50 to 65 (10 to 18)
Oxygen, mg/L	4 to 6	6 to 8
Oxygen partial pressure, mm Hg	90	90
CO_2, mg/L	40 to 50	20 to 30
Total suspended studies, mg/L	<15	<10
Total ammonia - N, mg/L	<3	<1
NH_3-N	<0.6	<0.02
Nitrite-N	<1	<0.1
Chloride, mg/L	>200	>200

Selecting Target Values for Water Quality

Calculating the minimum flows required to maintain targeted values for water quality (and then using the largest minimum value found for all the different water quality variables) will show how sensitive the calculated flow rates are to the value selected for the design value. A typical scenario is to select a value, do the calculations, realize that there is no way you could afford to supply such a high flow rate, and then start to make adjustments in the targeted values, e.g., 4 mg/L oxygen is probably OK instead of the 6 mg/L you originally chose, etc, etc. In the end, one must choose realistic values and then stay with these choices and the ramifications of the resulting flows required to maintain the mass balances. Do NOT ever compromise on the required flow rates. You will be sorry if you do.

Oxygen

The major reason most fish die is from lack of oxygen due to a loss of water flow. This is because oxygen is consumed at a fairly high rate (fish metabolism) and oxygen is transported by water flow. Due to low inherent concentrations of oxygen, "high" flows are required to transport the required oxygen. Flows required to maintain a satisfactory oxygen level are generally the controlling flow rate parameter when solving the series of mass balance equations to determine the most restrictive parameter. Even a partial loss of flow will generally result in insufficient oxygen for the fish, resulting in death. And since this is a text on RAS, remember this rule about why fish die:

> **"RULE OF THUMB"**
>
> Loss of Water = Loss of Fish
> Loss of Flow = Loss of Fish

Effective methods of monitoring and control are discussed in Chapter 9 in detail, but remember the above rule; always sense and monitor these two activities in at least two independent ways. If you don't, we guarantee that you will lose fish and eventually you will lose fish catastrophically.

Oxygen concentrations versus temperature and salinity are provided in the Appendix. For salmonids, as a group, the rearing unit effluent should contain from 6.0 to 8.0 mg/L dissolved oxygen (DO). For catfish and tilapia, allowable minimum levels are much lower than for salmonids, e.g., 2 or 3 mg/L, while it is certainly recommended to stay much closer to 5 or 6 mg/L. This variation in what the allowable minimums are has to do with the fact that partial oxygen pressure ($p0_2$) appears to be a more valid way to determine the lower limits. A $p0_2$ of 90 mm Hg seems to be a reasonable target for salmonids (Downey and Klontz, 1981). The atmosphere contains 21% oxygen, and at standard pressure of 760 mm Hg, this represents a $p0_2$ of 0.21 • 760 or 160 mm Hg or 56% of saturation. Where dissolved oxygen saturation represents 9.0 mg/L, 90 mm Hg represents 5.1 mg/L DO, and where saturation represents 12.5 mg/L DO, 7.0 mg/L represents a pO_2 of 90 mm Hg. These ranges represent temperatures from 20°C down to 5°C at sea level elevation (760 mm Hg). As noticed, the warmer the water the lower the effluent DO in mg/L can go, while still representing a pO_2 of 90 mm Hg. Applying 90mm pO_2 @ 30°C for tilapia would set a target value of 4.2 mg/L for culture tank oxygen levels (consistent with Table 4.1).

For salmonids and using 90 mm Hg as the minimum target, the available oxygen at 5°C is 5.5 mg/L (12.5–7.0 mg/L), at 20°C it is 3.9 mg/L (9.0–5.1 mg/L). Thus, considerably less oxygen is available at higher temperatures when metabolic rates, too, are higher, significantly impacting production capacity on the basis of limited available oxygen. Note the dilemma between higher temperatures supporting higher fish growth and metabolism, but the increasing unavailability of oxygen in the water column to support growth.

Beyond the general rule that each unit of feed will require 0.25 units of oxygen for fish metabolism, usage rates will depend mostly upon the type of fish being considered. Westers (1979) uses 200 to 250 g per kg of feed fed, while Pecor (1978) recommends using 110 g per kg of feed fed for esocids, such as the tiger muskellunge, a coolwater, non-active fish. Huisman (1975) uses 230 g for the common carp, as a warmwater representative. These values are consistent with our general recommendation of 250g O_2 per kg of feed fed for fish oxygen needs.

Finally, remember that not only do the fish require oxygen, but also the biological filter is just as critically dependent upon adequate oxygen levels to support bacteria metabolism. The DO concentration within the filter must be maintained at or above 2.0 mg/L to insure that the rate of nitrification in the filter does not become limited because of oxygen depletion (Kumar, 1984; Manthe et al. 1988). Always measure the DO coming off the biofilters and if the concentration starts to approach 2.0, then take corrective action, e.g., increase flow rate through the biofilter by increasing the hydraulic loading rate on the filter.

AMMONIA

There is considerable confusion about ammonia. Definitive values for the toxic levels of ammonia and the differentiation between the toxic NH_3 form and the supposed non-toxic NH_4^+ have never been determined. Meade (1985) reviewed the published literature on the effects of ammonia on fish and concluded:

> A truly safe, maximum acceptable concentration of un-ionized, or of total ammonia, for fish culture systems is not known.

The apparent toxicity of ammonia is extremely variable and depends on more than the mean or maximum concentration of ammonia.

The European Inland Fishery Advisory Commission (EIFAC) of FAO has set 0.025 mg/L as the maximum allowable un-ionized ammonia (NH_3 or $A_{NH3\text{-}N}$). Note, this means that tank ammonia levels (TAN) can exceed 10 mg/L if pH is maintained below 7.0. We would be very uncomfortable using such a high design value. What if the pH rose to 7.3 and doubled your NH_3? As seen in Table 4.1, we use rule of thumb values of 1 mg/L TAN for cool water and 2 or 3 for warm water fish. You should always check your TAN target value selection by assuming some pH that you intend to maintain and see if your NH_3 concentrations will exceed the 0.025 mg/L value using the ammonia-ammonium pH Temperature tables from the Appendix. If so, rethink your design carefully before proceeding.

There seems to always be a disproportionate amount of attention applied to nitrification and ammonification. All these calculations depend upon the estimate of an ammonia generation load, which is based upon the fish feeding rate:

$$P_{TAN} = \frac{F \cdot PC \cdot 0.092}{t = 1\ day} \tag{4.9}$$

The constant in the ammonia equation is based upon a series of approximations and estimates that when multiplied together result in the 0.092.

$0.092 = .16 \cdot .80 \times .80 \cdot .90$

- 16% (protein is 16% nitrogen)
- 80% nitrogen is assimilated
- 80% assimilated nitrogen is excreted
- 90% of nitrogen excreted as TAN + 10% an urea (fresh water fish only)
- all TAN is excreted during time period "t"
- non assimilated nitrogen in feces is removed quickly
- (no additional mineralization of nitrogenous compounds)

The individual assumptions about digestion and ultimate production of ammonia that is diffused across the gill and excreted directly via the feces all lack crispness in their assignment. Thus, we tend to see the rate of ammonia generation as being a "soft" number. For simplicity, one could simply assume 10% of the protein in the feed becomes the ammonia-N generation rate.

Equation 4.9 represents a conservatively high estimate of the P_{TAN} production rate. We use the time period in Eq. 4.9 as one day, while others will use the time period between feedings. In RAS, feed can be fed uniformly over a 24 hour period, thus distributing the ammonia load uniformly over the entire day as well. If a uniform 24 hour feeding is not used, then the equation should be adjusted and the time period should be the time between feedings or if a single feeding per day is used, then use 4 hours as the time period as an estimate of the time for the ammonia to be excreted from a feeding event. The assumption that all of the TAN is excreted in a finite period of time (t) between feedings is founded in evidence that metabolic activity increases during the hours following feedings (Page and Andrews, 1974; Ruane et al. 1977). Although the value of t is dependent on many biological variables, experience has indicated that fish metabolic activity peaks from 1 to 4 hours following feeding. One quickly will conclude that many smaller feedings evenly spaced during the day would serve to minimize high values of P_{TAN}. In fact, this is a strategy employed in the production of fish in tanks through the use of automatic feeders or demand type feeders.

CARBON DIOXIDE, CO_2

Carbon dioxide is an important, but largely overlooked water quality limiting parameter. This is probably because until recently, most systems were generally low density (less than 40 kg/m^3) and relied on aeration as the main means of supplying oxygen. This type of management also kept CO_2 values at low levels, e.g., less than 20 mg/L. However, loading rates have increased in recent years, and it became necessary to inject pure oxygen into these systems, instead of using aeration. As result, the natural stripping of CO_2 that occurs when using aeration systems was no longer taking place. We now need to apply other means for CO_2 control.

The generation of CO_2 is based upon chemical stoichiometry by relating CO_2 production to oxygen consumption (1.375 is the ratio of molecular weights of the two gases, 44/32):

CO_2 = 1.375 grams produced for each gram O_2 consumed (both fish and bacteria)

Please refer to Chapter 8 for more extensive discussion on carbon dioxide control. Additionally, please note that a computer software program to design CO_2 removal systems is provided in Chapter 17. The relationship between CO_2 and alkalinity is shown in Fig. 4.3 for an alkalinity value of 100 mg/L. CO_2 concentration is proportional to alkalinity within normal ranges of pH encountered in aquaculture applications, e.g., pH of 6 to 8.5.

The available data on acceptable levels of CO_2 are difficult to interpret, at best. The authors have maintained tilapia systems in excess of 100 mg/L on a sustained basis, but this seems to be the exception. Colt and Watten (1988) recommend that the maximum level of CO_2 should not exceed 20 mg/L, while Needham (1988) recommends the maximum not to exceed 10.0 mg/L. Alabaster et al. (1957) report that in well aerated water, CO_2 levels don't become toxic for rainbow trout under 100 mg/L. On the other hand, 10 mg/L caused mortalities at a pH level of 4.5, and mortalities occurred at 20 mg/L CO_2 when the pH was 5.7 (Lloyd and Jordan, 1964). In other research, Piper et al. (1982) state that 40 mg/L CO_2 had little affect on juvenile coho salmon. However, they also mention that CO_2 in excess of 20 mg/L may be harmful to fish, and where DO levels drop to 3–5 mg/L, lower concentrations may be detrimental and long term exposure of one year or more should not exceed 12 mg/L. Smart (1981) argues that fish are able to acclimate to elevated levels of CO_2. High CO_2 levels may lead to nephrocalcinosis, the presence of white calcareous deposits in the kidney. The severity of this condition appears to vary greatly according to diet and environmental factors as well and in particular the form of alkalinity replacement used. Chen et al. (2001) reported that the usage of agricultural grade limestone in place of sodium bicarbonate in an intensive tilapia system led to increased levels of nephrocalcinosis.

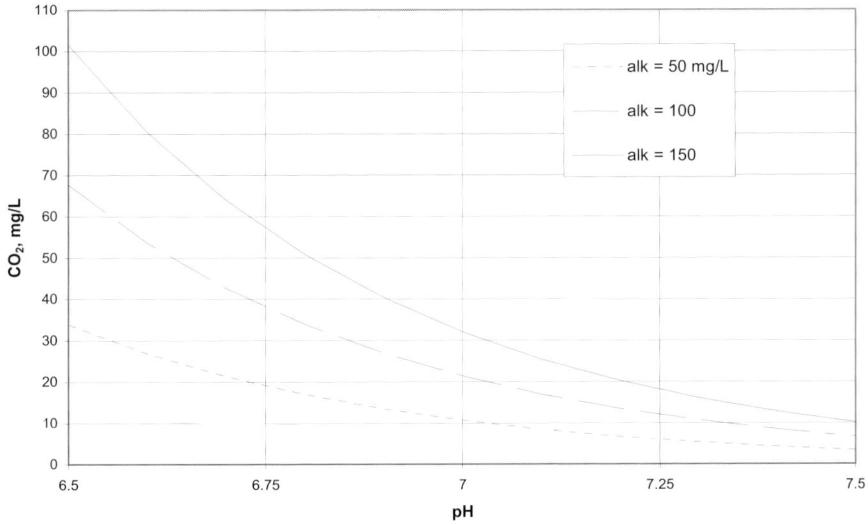

Figure 4.3 Concentration of CO_2 Versus pH at Alkalinities of 50, 100 and 150 mg/L.

A key point in the above discussion is that fish successfully acclimate to slowly changing water quality conditions, such as increases in the level of CO_2, but are adversely affected by sudden changes of water condition. Using CO_2 as an example, fish acclimated to CO_2 levels of 20 mg/L may die if exposed to a sudden spike to 80 mg/L CO_2 while other fish may exhibit good growth and feed conversion ratios at a gradually created but sustained level of 80 mg/L CO_2. An advantage of RAS is that system water quality can be maintained at fairly uniform values in a normally functioning system, but failures or malfunctions of the system that cause certain water quality parameters to fluctuate widely may cause fish to suffer or die.

SUSPENDED SOLIDS

The effective control of solids generated in RAS is probably the most important task that must be accomplished to ensure long-term successful operation of an RAS. This aspect of RAS is discussed in more detail in Chapter 6, but is also presented here as the suspended solids are a component of water quality. The quantity of suspended solids or Total Suspended Solids, TSS, generated per unit of feed being fed is estimated as:

TSS = 0.25 · kg feed fed (dry matter basis)
(from 20% to 40% of feed fed dry basis)

TSS is treated as a dilute waste. TSS design concentrations in RAS will be in the 10 to 30 mg/Liter range. Even after concentrating TSS with some type of treatment process, a certain volume of water will still contain only around 0.5 to 1% solids on a dry matter basis. In comparison, cow manure is 20% solids. TSS captured in a settling basin has a fluffy consistency and will require substantial volumetric space depending upon frequency of cleaning. As a "rule of thumb", assume that each kg of dry feed fed will produce approximately 8 liters of liquid waste, i.e., one lb feed produces one gallon of manure.

NITRATE

Nitrate nitrogen (NO_3) is the end product of the nitrification process. In general, concentrations of nitrate are not extremely adverse to RAS water quality. We have maintained some salmonid systems at nitrate levels above 1,500 mg/L without impact on the fish. Nitrogen should be conserved throughout the nitrification process. Thus, if 1 kg per day of TAN is being produced, then 1 kg of nitrate-N is being produced. The equilibrium concentration of nitrate will therefore be directly dependent upon the overall water exchange rate through the system. An effective exercise is to calculate the steady state nitrate-N balance assuming some water exchange rate through the system (see Eq. 4.4). Nitrate-nitrogen is relatively non-toxic to fish and as such will not influence the controlling flow rates in the system. One can choose some value such as 500 mg/L if you want a number to work with.

4.3 Fish Growth

The premise of RAS design is that we are endeavoring to grow fish at some defined rate, and that rate then defines the required fish feeding rate. The fish feeding rate in turn then defines waste generation loads and oxygen consumption. A convenient way of defining fish growth is based upon a temperature unit approach and some defined number of temperature units to create a unit growth rate, e.g., one inch per month

$$Growth = \frac{T - T_{base}}{TU_{base}} \tag{4.10}$$

Equation 4.10 predicts growth based with units of inches per month. The T_{base} and TU_{base} terms are defined in Table 4.2 and based upon historical observation and analyzing hatchery records. The terms for trout are from Piper et al. (1982) and the tilapia and perch terms are from the author's unpublished data:

Table 4.2 Temperature Growth Units for Trout, Tilapia and Perch (°F)

	Trout	Tilapia	Perch
T_{base}	32	65	50
TU_{base}	28	15	25
T_{max}	72	85	75

Use Eq. 4.10 subject to the limitation that if T is greater than T_{max}, then calculate the growth at T_{max}. Note that excessive temperatures will compromise growth and/or feed conversion.

For example, if we use tilapia and a water temperature of 80°F (26.7°C):

$$Growth \frac{inches}{month} = \frac{80 - 65}{15} = 1.00 \frac{inch}{month}$$

Condition Factor and Fish Weight

The weight of fish can be mathematically related to their length by using a term called the condition factor (CF); the bigger the CF, the more weight per unit length. For a given length, the bigger the CF, then more the weight of a particular fish. Each fish species will have an associated CF value to describe expected or normal body condition (see Table 4.3).

Table 4.3 Condition factors for various fish for use in Eq. 4.11 (from Piper* et al. 1982 or author)

Species	CF
Tilapia	750–850
Tilapia < 1 gm	500
Rainbow and Brown Trout	400
Lake Trout	250
Perch	490
Muskellunge	150
Northern Pike	200
Largemouth Bass	450
Walleye	300

*Note: Piper, pg. 406, uses CF's that are 10 times the above values, e.g., Piper would report the trout CF as 4,000, #400 and would divide by 10,000,000 instead of 1,000,000 in Eq. 4.11.

Thus, knowing the CF value is a good tool to gauge current feeding protocols for a given tank of fish. A CF too high means you are over-feeding the fish and a CF too low means you are under-feeding the fish. Using the CF and the fish length, weight can be calculated:

$$\text{WT (lb)} = \frac{CF(L\ inches)^3}{10^6} \tag{4.11}$$

Example - Tilapia that are 7 inches and using a CF of 760,

$$WT\,(7'', lb) = \frac{760(7)^3}{10^6} = 0.26\ lb$$

Weight Gain

Now, we can calculate actual gain by calculating the weights associated with two specific fish lengths over the time period required to achieve this growth. In the previous example, we grew 7 inch tilapia for one month at 80°F, which produced one inch of growth to a size of 8 inches. The new weight is then:

$$WT_{new}\ (7+1)'' = \frac{760(8)^3}{10^6} = 0.39\ lbs$$

The relative efficiency of converting feed energy into animal flesh is described by a term called the feed to gain ratio, or FG. FG is the inverse or feed conversion efficiency. FG is used to calculate feeding rates required to achieve projected growth rates:

$$FG = \frac{feed}{gain} \tag{4.12}$$

$$Feed = FG \cdot gain \tag{4.13}$$

$$\frac{feed}{month} = (\mathrm{WT_{new}} - \mathrm{WT_{old}}) \cdot \mathrm{FG} \tag{4.14}$$

Returning to our example of tilapia increasing in length for 7 to 8 inches in one month, the feed per fish required per month is:

$$\frac{feed}{month} = (0.39\ \mathrm{lb} - 0.26\ \mathrm{lb}) \cdot \mathrm{FG}$$

In the above example, each tilapia will gain 0.13 lb (0.39–0.26 lb = 0.13 lb) over the one month period. Converting this to a daily rate and using 30.5 days per month, the daily gain is:

$$daily\ gain = \frac{0.13\ lb}{month} \cdot \frac{454\ g}{lb} \cdot \frac{month}{30.5\ day} = \frac{1.93\ g}{day} \tag{4.15}$$

The expression in Eq. 4.15 is calculating the average daily gain based upon the average length gain over the month. Since weight gain is cubically related to length gain, the gain for the "last-day" of the month can be calculated based upon the length incremental gain on the last-day of the month. This concept is demonstrated in the following example.

Example

Calculate the maximum daily feeding rate for a tank containing 10,000 tilapia at a harvest weight of 2.00 lb (CF=760; a tank temperature of 80°F; and a FG of 1.1).

Solution
First calculate the length of a 2.00 lb tilapia:

$$L\ (inch) = \left[\frac{1{,}000{,}000 \times 2.00}{760}\right]^{1/3}$$

$L\ (inch)$ = 13.81 (from Eq. 4.10)

$$Growth\frac{inches}{month} = \frac{80-65}{15} = 1.00\frac{inch}{month}$$

Daily Increment of length (inch) = 1.00/30.5 = 0.033 inch

$Wt\ (day - 1) = 760\ (13.81 - 0.033)^3/10^6$

$Wt\ gain = 2.000\ lb - 1.986$

Feed /day/fish = 0.014 lb gain (6.4 g)

$$\text{Tank feed/day} = 10{,}000\ \frac{fish}{tank} \cdot 0.014\ \frac{lb\ gain}{fish \cdot day} \cdot \text{FG}\ \frac{lb\ feed}{lb\ gain}$$

$= 140 \cdot 1.1 = 154$ lb feed per day per tank

Feeding charts can be constructed using the above approach. Obviously, these calculations are well suited for spreadsheet applications.

FG can be very erratic. Assuming good management and good quality feed (protein content of 38 to 42%), guidelines for FG are as follows for tilapia:

0.7 to 0.9 (tilapia < 100 gram)
1.2 to 1.3 (tilapia > 100 gram)

Similarly, given length growth, the weight gain for any particular fish species can be calculated by using the condition factor for the particular species (both the tilapia and perch growth constants were developed by the authors).

Remember, all the "P" terms in Eq. 4.2 are related to the feeding rate term. Essentially, all design aspects of RAS technology are related to or in fact are based upon the daily design feeding rate. The daily design feeding rate is the critical value needed in designing a complete RAS. It should make some intuitive sense then, that one must calculate the largest anticipated feeding rate in order to properly design the water conditioning components of an RAS. And when will this largest feeding rate occur? The answer being when the fish tank reaches its largest biomass and associated fish mass. This is also where many designs fail, since the largest fish mass and largest feeding rate is the most stressful condition on the RAS. Any of the individual water quality parameters can quickly become the defining variable for water quality control, meaning the water flow rate through a particular water conditioning design component is not sufficient to remove the generated pollutant load, e.g., CO_2 removal device. The water parameters always come to an equilibrium based upon a balance between production load, water flow rate and change in concentration for different water quality variables across the conditioning component (see Eq. 4.6). The challenge to the designer and manager is to make sure these equilibrium values at least as "good" as the design target values for these variables.

WET WEIGHT MEASUREMENTS

Both terms in the FG are wet basis, but while feed will typically be from 5 to 10% moisture, the gain term is more nearly 80 to 85% moisture. FG values for broilers, pigs, and beef are something like 2.0, 2.5 and 3.0, respectively, while in fish, the FG value may be less than one. Quite typically FG values will be less than one for fingerlings up to a 100 gram or larger in size. This sometimes confuses people as to how the FG ratio could be less than one, but it is because a comparison of wet weight to dry weight terms is being made.

4.4 DESIGN EXAMPLES

Calculate Growth Period

The following calculation is applicable for calculating the growth period of tilapia from 50 grams to 800 grams, at 82°F; assume a CF = 760.

Calculate length at 50 grams (0.11 lb)

$$L\ (inch\ at\ 50\ g) = \left[\frac{1{,}000{,}000 \cdot 0.11}{760} \right]^{1/3} = 5.25 \text{ inches}$$

$$L\ (inch\ at\ 800\ g) = \left[\ \frac{1{,}000{,}000 \cdot 1.76}{760}\ \right]^{1/3} = 13.24 \text{ inches}$$

Total change in length required from 50 to 800 grams:

(13.24 –5.25) = 7.99 inches.

Calculate growth rate at the specified rearing temperature of 82°F:

$$Growth \frac{inches}{month} = \frac{82-65}{15} = 1.13 \frac{inch}{month}$$

Total time to achieve the required length =

$$Growth\ period = \frac{7.99\ inch}{1.13 \dfrac{inch}{month}} = 7.07 \text{ months}$$

Note that this growth could be forced to occur in a single tank, or the growth could be managed to occur in three tanks that are managed sequentially. If a three tank scheme is used, then the fish would reside in each tank for 1/3 of the total growth time, or

Time in a particular tank = 7.07 months/3 size classes
= 2.36 months/size class
(or 10 weeks in each tank)

EXAMPLE: REQUIRED FLOW RATE DESIGN PROBLEM

Calculate the required design flow rate for a 100% recirculating flow for a design fish feeding rate of 100 kg feed/day @ 38% protein. Calculate the required flow rate for each water quality parameter and then identify the controlling parameter. Assume your location is 228.6 m (750 feet) above sea level. Use the following design target values for water quality:

T^o = 28°C
TSS = 10 mg/L
TAN = 2 mg/L
$Oxygen$ = 5.0 mg/L
CO_2 = 40 mg/L

Assume treatment efficiency for each of the system components as follows:

TSS = 75%
TAN = 35%
Oxygen = 70% but gas concentration is 50% oxygen
CO_2 = 70%

Solution

Given a feeding rate of 100 kg per day of feed with 38% protein, you were to compute the required steady state flow rate for maintaining the following water quality levels: 10 mg/L TSS, 2 mg/L TAN, 5 mg/L O_2, and 40 mg/L CO_2. You were to assume the following efficiencies for the treatment devices: 75% for TSS, 35% for TAN, 70% for O_2, and 70% for CO_2. Additionally, the tank water temperature was set to 28°C, and the gas composition in the O_2 treatment device maintained at 50% O_2.

General Mass Balance

$QC_2 + P = QC_1$ or $Q(C_2 - C_1) = -P$ (repeat of Eq. 4.5)
or $Q(C_1 - C_2) = P$

Oxygen

$C_2 = C_1 + TE(C_{sat} - C_1)$
$C_1 = 5\ mg/L$
$TE = 0.7$

Calculate C_{sat} (Equations from Chapter 8 to determine oxygen saturation)

$T = 28 + 273.15 = 301.15K$
$\beta = \exp[A_1 + A_2(100/T) + A_3 \ln(T/100)]$
$\beta = \exp[-58.3877 + 85.8079(100/301.15) + 23.8439(\ln(301.15/100))]$
$\beta = \exp[-58.3877 + 28.4934 + 26.2864]$
$\beta = \exp[-3.6079]$
$\beta = 0.02711$ L/L-atm
$BP = 10^{(2.88014 - h/19748.2)}$
$BP = 10^{(2.880814 - 228.6/19748.2)}$ *(Height above MSL is 750 ft (228.6m))*
$BP = 10^{(2.880814 - 0.011576)}$
$BP = 10^{(2.869238)}$
$BP = 740$ mmHg
$P_{wv} = 28.971$ *(From Eq. 8.6)*

K = 1.42903
X = 0.5 *(50% oxygen atmosphere in device)*
$C_{sat} = 1{,}000\ K\ \beta\ X\ ((BP - P_{wv})/760)$
$C_{sat} = 1{,}000\ (1.42903)\ (0.02711)\ (0.5)\ ((740\text{–}28.971)/760)$
$C_{sat} = 1{,}000\ (1.42903)\ (0.02711)\ (0.5)\ (0.9356)$
$C_{sat} = 18.1$ mg/L

(atmospheric C_{sat} would be 7.57 mg/L; using Eq. 4.4, obtain 8.13 which is high)

$C_2 = 5 + 0.7\ (18.1\text{–}5)$
$C_2 = 14.2$ mg/L

Oxygen production (-P) is the sum of fish and bacterial oxygen consumption:

$$\frac{0.25 kg O_2\ Consumed\ by Fish}{kg\ feed} + \frac{0.12 kg O_2\ Consumed\ by Bacteria}{kg\ feed} = \frac{0.37 kg Oxygen}{kg\ feed}$$

$$P = \left(\frac{100\ kg\ \ feed}{day}\right)\left(\frac{0.37\ kg\ Oxygen}{kg\ \ feed}\right)\left(\frac{10^6\ \ mg}{kg}\right) = \frac{37{,}000{,}000\ mg\ Oxygen}{day}$$

Q (14.2 mg/L) + (-37,000,000) mg/day = Q (5 mg/L) *(apply Eq. 4.5)*

$$Q = \frac{\dfrac{37{,}000{,}000\ mg\ Oxygen}{day}}{(14.2-5)\dfrac{mg\ Oxygen}{liter}} = \frac{4{,}021{,}739\ liter}{day}$$

$$Q = \left(\frac{4{,}021{,}739\ liter}{day}\right)\left(\frac{1\ gallon}{3.785\ liter}\right)\left(\frac{1\ day}{1{,}440\ minutes}\right)$$
$$= \underline{\underline{738\ gpm\ required\ \ for\ O_2\ control}}$$

Carbon Dioxide (CO_2)

$C_2 = C_1 + TE\ (C_{best} - C_1)$

$C_1 = 40$ mg/L
TE = 0.7
$C_{best} = 0.5$ mg/L

$C_2 = 40 + 0.7\ (0.5–40)$
$C_2 = 12.4$ mg/L

Carbon dioxide production (+P) is based on oxygen consumption:

$$\left(\frac{0.37\ kg\ O_2\ Oxygen\ Consumed}{kg\ feed}\right)\left(\frac{1.375\ kg\ CO_2\ Produced}{kg\ Oxygen\ Consumed}\right)$$
$$= \frac{0.50875\ kg\ CO_2}{kg\ feed}$$

$$P = \left(\frac{100\ kg\ feed}{day}\right)\left(\frac{0.50875\ kg\ CO_2}{kg\ feed}\right)\left(\frac{10^6\ mg}{kg}\right) = \frac{50{,}875{,}000\ mg\ CO_2}{day}$$

Q (12.4 mg/L) – 50,875,000 mg/day = Q (40 mg/L) *(apply Eq. 4.5)*

$$Q = \frac{\dfrac{50{,}875{,}000\ mg\ CO_2}{day}}{(40 - 12.4)\dfrac{mg\ CO_2}{liter}} = \frac{1{,}843{,}297\ liter}{day}$$

$$Q = \left(\frac{1{,}843{,}297\ liter}{day}\right)\left(\frac{1\ gallon}{3.785\ liter}\right)\left(\frac{1\ day}{1{,}440\ minutes}\right)$$
$$= \underline{\underline{338\ gpm\ required\ for\ CO_2\ control}}$$

Total Suspended Solids (TSS)

$C_2 = C_1 + TE\ (C_{best} – C_1)$
$C_1 = 10$ mg/L
TE = 0.75
$C_{best} = 0$ mg/L

$C_2 = 10 + 0.75\ (0–10)$
$C_2 = 2.5$ mg/L

TSS production (+P) is based on feeding rate (assume feed has 0% moisture content):

$$P = \left(\frac{100\ kg\ feed}{day}\right)\left(\frac{0.25\ kg\ TSS}{kg\ feed}\right)\left(\frac{10^6\ mg}{kg}\right) = \frac{25{,}000{,}000\ mg\ TSS}{day}$$

Q (2.5 mg/L) – 25,000,000 mg/day = Q (10 mg/L) *(apply Eq. 4.5)*

$$Q = \frac{\dfrac{25{,}000{,}000\ mg\ TSS}{day}}{(10 - 2.5)\dfrac{mg\ TSS}{liter}} = \frac{3{,}333{,}333\ liter}{day}$$

$$Q = \left(\frac{3{,}333{,}333\ liter}{day}\right)\left(\frac{1\ gallon}{3.785\ liter}\right)\left(\frac{1\ day}{1{,}440\ minutes}\right)$$
$$= \underline{\underline{612\ gpm\ required\ for\ TSS\ control}}$$

Total Ammonia Nitrogen (TAN)

$C_2 = C_1 + TE\ (C_{best} – C_1)$
$C_1 = 2$ mg/L
$TE = 0.35$
$C_{best} = 0$ mg/L

$C_2 = 2 + 0.35\ (0–2)$
$C_2 = 1.3$ mg/L

TAN production (+P) is based on feeding rate (feed has 38% protein content):

$$P = \left(\frac{100\ kg\ feed}{day}\right)\left(\frac{0.38\ kg\ Protein}{kg\ feed}\right)\left(\frac{0.092\ kg\ TAN}{kg\ Protein}\right)\left(\frac{10^6\ mg}{kg}\right)$$
$$= \frac{3{,}496{,}000\ mg\ TAN}{day}$$

Q (1.3 mg/L) – 3,496,000 mg/day = Q (2 mg/L) *(apply Eq. 4.5)*

$$Q = \frac{\dfrac{3{,}496{,}000\ mg\ TAN}{day}}{(2-1.3)\dfrac{mg\ TAN}{liter}} = \frac{4{,}994{,}286\ liter}{day}$$

$$Q = \left(\frac{4{,}994{,}286\ liter}{day}\right)\left(\frac{1\ gallon}{3.785\ liter}\right)\left(\frac{1\ day}{1{,}440\ minutes}\right)$$
$$= \underline{\underline{916\ gpm\ required\ for\ TAN\ control}}$$

Required Flow Rates

Water Quality Parameter	Required Flow rate (gpm)
TSS	612
TAN	916*
Oxygen	738
Carbon Dioxide	338

* controlling flow rate

The only footnote to the above is that you must also calculate the required hydraulic loading for the biofilter as in $m^3/hr/m^2$ (gpm/ft^2). This may become the controlling flow rate over and above the flowrates required to maintain water quality conditions. Finally, just remember that if such were the case, it simply means that the equilibrium concentrations for all the water quality parameters would be just a little *better* and the fish will never complain about that

List of Symbols

C_x	Concentration of a particular water quality parameter X, $\frac{kg\ of\ X}{10^6\ kg\ water}$
C_0, C_1 and C_2	Concentrations of parameter X crossing the control volume, mg/L
CF	Condition factor that relates fish weight and length, dimensionless
D_{fish}	Fish density in culture vessel, kg/m^3
F	Feeding rate, kg/day
$F_{\%}$	Feeding rate per day as a percentage of fish body mass, %
FG	Feed to gain ratio, dimensionless
L	Loading rate, $kg/m^3/hr$
L_{oxygen}	Allowable fish biomass loading due to oxygen availability, kg/ Lpm
PC	Protein concentration or feed, decimal
P_{Oxygen}	Production rate (negative as related to feed consumption) of oxygen, kg/day
P_{CO2}	Production rate of carbon dioxide, kg/day
P_{TAN}	Production rate of total ammonia nitrogen, kg/day
$P_{Solids,\ TSS}$	Production rate of total suspended solids, kg/day
Q_0	Flow rate passing through culture tank (discharge), m^3/day (as kg/day)
Q_1	Water that is recirculated, kg/day
R	Number of water exchanges per hour through the rearing unit, hr^{-1}
t	Time period for feeding daily ration, day
TAN	Total Ammonia Nitrogen, kg
TSS	Total Suspended Solids, kg
T_{base}	Equivalent to the temperature at which no growth occurs, °F (°C)
T_{eff}	Treatment efficiency (removal or addition), %
TU_{base}	Temperature units required to produce an increment of growth per unit time, °F (°C)

CHAPTER 5

CULTURE UNITS

5.0 INTRODUCTION

About the only thing that aquacultural engineers can unanimously agree upon is that a body of water is required to raise fish. How this body of water should be contained is open to considerable debate. Should the fish be reared in ponds, raceways, or tanks? If tanks are chosen, to what geometry should the tank conform? How deep should a tank be or more descriptively, what should the tank diameter to depth ratio be? First, this book is restricted to tank culture and Recirculating Aquaculture Systems (RAS) technology. Therefore, we are only addressing tanks and not open ponds. Within that chosen focus, the paramount issue in all tank culture design is to maximize the capability of the tank system for self-cleaning. Without exception, the dominant reason RAS's fail is their inability to self-clean. A system that does not effectively and quickly remove manure from the culture water will never produce fish economically. All individual components of the system will fail to perform efficiently. Thus, the first question that one should ask is whether or not the tank will effectively clean.

Secondly, the capability to effectively manage the fish within the tank vessel is the next question that must be addressed. These and other issues will be addressed in this chapter. And for the stubborn, raceway design will also be reviewed briefly as will a newly designed hybrid of raceway and round tank concepts. All these concepts can be generally applied to either RAS or flow-through systems.

5.1 STOCKING DENSITY

One of the first questions that must be answered in designing a RAS is the number of fish that can be carried in the tank. The number of fish and their mass will define the feeding rates from which the individual engineering components are designed. The number of fish that can be stocked per unit volume ($D_{density}$) will depend on both the fish species and the fish size. We have developed an approach that is based upon body length (L) to estimate the number of fish that can be carried per unit volume of tank:

$$D_{density} = \frac{L}{C_{density}} \qquad (5.1^{a})$$

where:

$C_{density}$ 1.5 for L in inches (0.24 for L in cm) for tilapia
2.0 (0.32) for trout
2.5 (0.40) for perch

These values were developed based upon experience originally with brook trout and were later adapted to tilapia. There is a tendency to over-stock smaller fish. Some of our experience indicates that tilapia can be stocked in much higher numbers while the animals are small. However, crowding is risky, as it can have severe consequences on small animals during their rapid growth stage, e.g., fish less than 5 inches (13 cm) in length. Visual observations can be misleading, since you will see only the active dominant fish. Sampling fish populations will be your best guide to whether or not your stocking densities are too high.

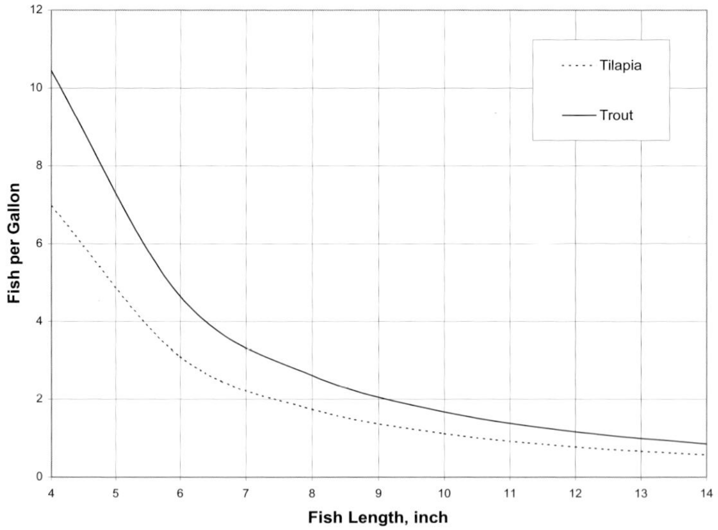

Figure 5.1 Fish per Gallon as a Function of Body Length for Trout (CF= 400) and Tilapia (CF= 800) Where CF = Condition Factor.

[a] Symbols are defined at the end of the Chapter.

Good management should produce a population that has a standard deviation that is less than 10% of the population mean. If this value increases, it is indicative of over-stocking, under feeding, or other management problems. The density relationship given above and expressed in Table 5.1 and Fig. 5.1 also show what might be considered excessive stocking densities for the larger fish. However, this may be more related to poor water quality and less than optimum effectiveness of feed distribution than the stocking density. For example, if you use a sinking feed you may be able to stock more heavily than when using a floating feed. Our experience is that we were able to maintain these high densities when water quality parameters were also maintained within the guidelines given in Table 4.1.

The relationship between required tankage volume, fish harvest size, and the harvests per year (turnovers) are shown in Fig. 5.2 for tilapia. The reader can develop similar charts for other fish. The percentage of rearing volume required for smaller fish of the final volume required for the market size fish (assuming the same number of animals) is demonstrated in Fig. 5.3. This figure can be used as a quick guideline on tank sizes needed for smaller fish. For example, if you intended to sell 1,000 gram tilapia, the tank volume required for 100 g tilapia would be 21% of the volume required for the 1,000 g market fish. In equation form, this relationship can be expressed as:

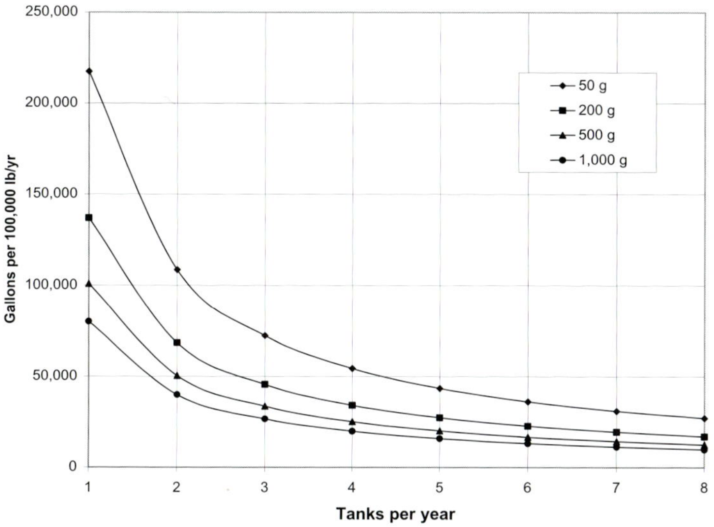

Figure 5.2 Required Volume of Tankage as Affected by Fish Size and Tank Turnovers per Year for Tilapia.

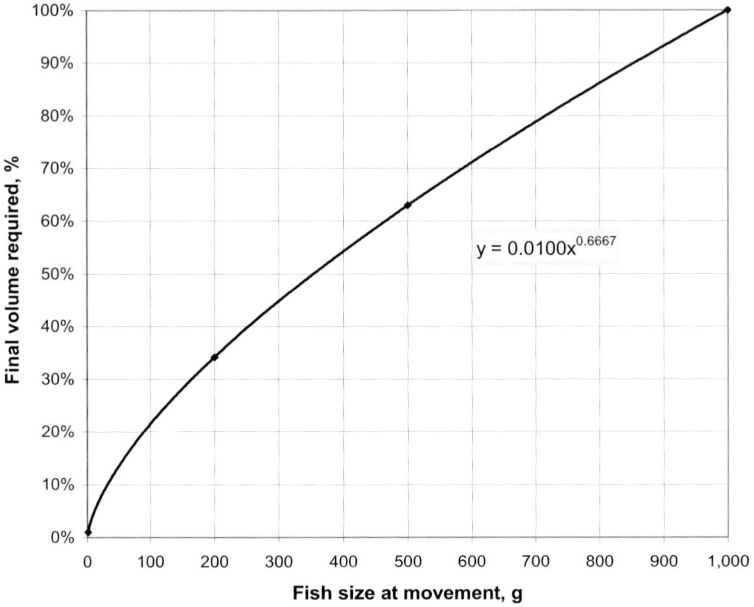

Figure 5.3 Percentage of Final Tank Volume Required (Y axis) as Affected by Size of Animal at Movement (X axis).

$$Y = 0.0100W^{0.6667} \tag{5.2}$$

Equation 5.2 is for a "standardized" fish of 1,000 weight units at maturity. For a particular growth scenario, substitute the particular size fish as a percentage of its mature size and 1,000 to standardize the size. For example, a 200 g fish on a 500 g target harvest is 40% or 400 weight units in Eq. 5.2.

The reader should also be aware at this point that the management methods used to maximize productivity from a given set of tanks can be quite inventive. For example, the simplest management method would be an "all-in and all-out" approach, where a tank is stocked with small fingerlings and then harvested once the fish reach target market size. Clearly, using this approach greatly under utilizes the rearing capacity of the tank, and for that matter, the water conditioning equipment associated with the system to maintain water quality at the highest feeding loads. Watten (1992) presented a mathematical analysis of the effects of sequential rearing on the potential production from the system. Generally, one would probably try to use at least three tank movements from fingerling to market size. Each time fish are moved, though, loss of growth and additional mortality due to stress will occur. It is this loss of productivity that actually limits the number of cohort movements from an economical standpoint, although the time and labor required to perform the movement are also important considerations. Other management methods to maximize the utilization of the RAS system would be to place several tanks on a common water treatment system where the fish are either left in the same tank they were stocked in or the fish are moved from one tank to the next as they increase in size. Any movement of fish is not a good thing and should be avoided by design whenever possible.

5.2 CULTURE TANK ENGINEERING

The production of food fish in large circular tanks has produced large cost savings in comparison to raising the same quantity of fish in more but smaller tanks. Larger circular culture tanks offer many advantages for food fish production. While just a few years ago, an 8 m diameter tank was considered large, now we are seeing 10 and 15 m and even larger diameter tanks being put into production. Substantial savings in both capital and labor costs can be realized by shifting production into fewer but larger culture tanks. Fundamentally, the time it takes to servicc a small tank or a large tank is similar. In fact, the capital costs associated with large versus small is not proportional to tank volume. Dramatic cost savings can be obtained by using larger and larger tanks. Circular tanks make good culture vessels for the following reasons:

- Improves the uniformity of the culture environment
- Allow a wide range of rotational velocities to optimize fish health and condition
- Rapid concentration and removal of settleable solids.

The flow inlet and outlet structures and fish grading and/or removal mechanisms should be engineered to reduce the labor requirements of handling fish and to obtain effective tank rotational characteristics, mixing, and solids flushing.

Table 5.1 Stocking Densities for Tilapia and Trout as a Function of Body Length

	Length (inch)	Weight (g)	Number/Volume (fish/gal)	Mass/Volume (lbs/gal)
Tilapia	1	0.4	111	0.09
	2	3	28	0.18
	4	23	7	0.36
	6	78	3.1	0.53
	8	186	1.7	0.71
	10	363	1.1	0.89
	12	628	0.8	1.07
	14	997	0.6	1.25
Trout	1	0.2	167	0.07
	2	1.5	41.8	0.13
	4	12	10.4	0.27
	6	39	4.6	0.40
	8	93	2.6	0.53
	10	182	1.7	0.67
	12	314	1.2	0.80
	14	498	0.9	0.94
		SI		
	(cm)	(g)	fish/m^3	kg/m^3
Tilapia	2	0.2	47,438	8.4
	4	1.4	11,860	16.8
	8	1.3	2,965	33.6
	12	38.3	1,318	50.5
	15	74.8	843	63.1
	25	346	304	105.1
	30	598	211	126.2
	35	950	155	147.2
Trout	2	0.1	71,157	6.3
	4	0.7	17,789	12.6
	8	6	4,447	25.2
	12	19	1,977	37.9
	15	37	1,265	47.3
	25	173	455	78.9
	30	299	316	94.6
	35	475	232	110.4

What is the downside of larger culture tank units? Tanks fail as units. Typically, you do not lose a part of a tank; instead you lose the entire tank of fish. Distributing a farm's population of fish among multiple tanks provides a measure of protection against entire farm failure. The more fish in a tank, the bigger the economic loss that will occur when a tank fails. However, as the experience of the management and design team increases, the risk of tank failure decreases, but should never be ignored. It would not be prudent to start out with large culture tank vessels, if this were your first RAS experience. Start small and build upon success. Other challenges of using large circular tanks include:

- distributing flow to obtain uniform mixing and rapid solids removal
- grading and harvesting fish
- removing mortalities
- isolating the biofilter while treating the fish with a chemotherapeutant.

Large tanks are more critically dependent upon tank hydraulic design than are small tanks, because in small tanks, $\leq 1\ m^3$, the overall rate of water exchange tends to be rapid. The rapid hydraulic exchange results in reasonably good water quality, because the high turnover rate (usually as rapid as 10 minutes per tank volume exchange, called hydraulic retention time, HRT) carries more oxygen into the tank and rapidly flushes wastes. Conversely, in large tanks, the HRT tends to be lower (around 30 minutes or greater) and, as a result, the inlet and outlet injection methods and flow rate become dominant factors affecting the uniformity of water conditions in the tank (aside from the feed loading rate). As reviewed in Chapter 4, the carrying capacity of a tank is influenced by water exchange, feeding rate, oxygen consumption, and waste production.

5.3 ROUND TANKS

Tanks are designed with considerations given to production cost, space utilization, water quality maintenance, and fish management. Tank designs used for intensive fish culture have been presented frequently and reviewed in the literature (Wheaton, 1977; Klapsis and Burley, 1984; Piper et al. 1982; Cripps and Poxton, 1992). There is a definite trend towards large circular culture tanks for food fish production. Tanks larger than 10 m in diameter, which used to be referred to as pools, are now reasonable choices for culture systems in intensive indoor operations.

Circular tanks are attractive for the following reasons:

- simple to maintain
- provide uniform water quality
- allow operating over a wide range of rotational velocities to optimize fish health/condition
- settleable solids can be rapidly flushed through the center drain
- designs that allow for visual or automatic observation of waste feed to enable satiation feeding are possible

Recommended tank diameter to depth ratios vary from 5:1 to 10:1 (Burrows and Chenoweth, 1955; Chenoweth et al. 1973; Larmoyeux et al. 1973); even so, many farms use tanks with diameter:depth ratios as low as 3:1. Selection of a tank diameter:depth ratio is also influenced by factors such as the cost of floor space, water head, fish stocking density, fish species, and fish feeding levels and methods. Choices of depth should also consider ease of workers handling fish within the tank and safety issues of working in waters that may be more than "chest" high.

In the early years of RAS, tanks that were actually deeper than their diameter were touted as a key design factor to economic success. None of these systems were successful, mostly for problems related to fish management. Even in more modest attempts to utilize deeper tanks, e.g., a 3:1 ratio of diameter:depth, not all fish will effectively distribute into the entire water column. An example of this species specific attribute includes tilapia, which utilize the entire water column, versus walleye or flounder, which are surface area animals and do not utilize the water column nearly as effectively as tilapia.

TANK WATER VELOCITIES

The ability to self-clean is a key advantage of circular tanks. To establish this ability, the water column must be in constant rotation within the tank. The rotational velocity in the culture tank should be as uniform as possible from the tank wall to the center and from the surface to the bottom, and it should be swift enough to make the tank self-cleaning. However, it should not be faster than that required to exercise the fish. Water velocities of 0.5–2.0 times fish body length per second are optimal for maintaining fish health, muscle tone, and respiration (Losordo and Westers, 1994). Velocities required to drive settleable solids to the tank's center drain should be greater than 15 to 30 cm/s (Burrows and Chenoweth, 1970; Mäkinen et al. 1988). For tilapia, Balarin and Haller (1982) reported an upper current speed of 20–30 cm/s.

For salmonids, Timmons and Youngs (1991) provided the following equation to predict safe non-fatiguing water velocities:

$$V_{safe} < \frac{5.25}{L^{0.37}} \tag{5.3}$$

In circular tanks, velocities are reduced somewhat away from the walls, which allow fish to select a more comfortable water velocity, as compared to raceway designs where velocities are uniform along the channel.

A phenomenon that has been observed by all culturists familiar with raceways is the fact that, more often than not, fish concentrate themselves in the upper one-third of the raceway, while sparsely occupying the lower two-thirds. This is, of course, a function of fish density, but unless densities are up to about 80 kg/m^3, raceway space is often poorly utilized. Higher densities require higher exchange rates or the injection of pure oxygen. Standard length raceways of 20 m to 40 m should be operated at six to four exchanges per hour respectively, but often are not. Even at these relatively high water exchange rates, R, the velocities are still below 5.0 cm/sec as Eq. 5.4 shows:

$$V_{raceway} = \frac{L_{raceway} \cdot R}{36} \tag{5.4}$$

The constant in Eq. 5.4 is the number of seconds per hour divided by 100 to convert meter (length or distance the water must travel) to cm, the unit used to express the velocity. Such velocities are well short of cleaning velocities of 15 to 30 cm/sec or recommended velocities for fish conditioning, which range from 0.5 to over 2.0 times their body length per second (Poston et al. 1969; Besner and Smith, 1983; Woodward and Smith, 1985; Leon, 1986; Totland et al. 1987; Needham, 1988; Josse et al. 1989; Timmons and Youngs, 1991).

Timmons and Youngs (1991) pointed out such deficiencies and concluded that, in practice, raceways are managed much closer to their design requirement for oxygen supply than for cleaning requirements. The means to overcome this deficiency is to either design very long raceways, or raceways having a very small cross-sectional area, neither of which is very practical. To at least partly overcome this drawback with standard length raceways, Boersen and Westers (1986) propose the use of baffles. These are evenly spaced throughout the raceway at distances approximately equal to the width of the raceway. The gap or space between the lower edge of the baffle and the bottom of the raceway dictates the velocity relative to the exchange rate. Their main function is to make the pond self-cleaning, removing solid waste as soon as it settles, thus preventing buildup and subsequent resuspension through fish activity. The relatively intact solids will settle quickly behind the fish retaining screen in a small section of the raceway dedicated to that function. This simple technique provides for an effective solids management approach. However, baffles increase velocities only over a small cross-sectional area of the pond. Fish will utilize this relatively high velocity zone but since the space is so limited, there is room only for a small percentage of the population. Yet it is desirable, especially for salmonids, to expose all fish to relatively high water velocities.

Using Eq. 5.3 for a 10.0 cm fish, the V_{safe} velocity should not exceed 2.2 times its body length per second (BL/sec); for a 20.0 cm fish it is 1.73 BL/sec. Totland et al. (1987) exercised large Atlantic salmon (56.3 cm and 2,038 g) at velocities of 0.40 to 0.45 BL/sec. They found improved survival rates over caged fish except during the initial two week adjustment period when losses were 1.2%, much greater than the reference group. Final losses were 4.4% versus 8.8%, and the weight gain was nearly 40% greater in the exercised fish. According to industry standards, quality was rated 9.2% higher. Based on Eq. 5.3, the recommended velocity for this size fish would be 1.2 BL/sec but favorable results were obtained at the lower velocities of 0.45 BL/sec, which represents a velocity of 25.3 cm/sec, far above raceway velocities.

Needham (1988) recommends velocities between 0.5 and 1.0 BL/sec for Atlantic salmon. Besner and Smith (1983) exposed Coho salmon to velocities of 0.2 (control) and 1.0, 1.5 and 2.0 BL/sec. Endurance improved over the control group. Woodward and Smith (1985) exercised rainbow trout at velocities of 1.5 BL/sec for 42 days. It improved fish quality in terms of better stress resistance. Indeed, it has been shown that sustained swimming speed improves disease resistance. Leon (1986) found this to be true along with improved growth rates and feed conversions when brook trout were reared in velocities of 1.5 to 2.0 BL/sec.

Josse et al. (1989) maintained rainbow trout at sustained velocities of 2.5 BL/sec with bursts of 3.8 BL/sec for a few minutes daily. This latter velocity was aimed at developing the white musculature. The continuous cruising speed served to develop the red musculature, but it was found to also stimulate the white musculature without exhausting the fish. Continuous swimming speed also had a positive effect on tail musculature development. The red musculature increased by 27% and the white by 9% over the control groups maintained in still water. The authors also concluded that the permanent rotary water movement (the direction of the water needed to be reversed from time to time to prevent uneven muscle development) ensured a perfect homogenization of the medium, encouraged optimal distribution of the fish, and inhibited their territorial behavior which, in turn, resulted in a 100% increase in rearing density over the control group (68.4 versus 36.0 kg/m^3). Mortalities during the early rearing phase were significantly less in the high velocity environment, and the experimental fish grew as well as the control group, despite the sustained swimming action. Much earlier studies by Poston et al. (1969) pointed out such benefits. Brook trout, exposed to velocities in excess of 2.0 BL/sec, showed increased stamina and a faster rate of replacement of muscle glycogen after exposure to strenuous exercise in a stamina tunnel, compared to unconditioned fish. The conditioned fish also showed a more efficient rate of food conversion.

Forced exercise, contrary to what one would expect, seems to result in reduced oxygen consumption. This has been attributed to physiological adaptations, such as increased white muscle activity (Nahhas et al. 1982), improvements in cardiac output, and enhanced oxygen carrying capability of the blood (Woodward and Smith, 1985). There can also be savings in "breathing" cost. Fish who can maintain their position in fast-flowing water need only to open their mouths to ventilate their gills. This is termed ram ventilation. Ram ventilation can contribute to saving energy in two ways. The fish does not have to pump water over its gills which, in turn, result in less turbulence, i.e., a more streamlined flow of water is maintained over the body that reduces the need to make continual positional adjustments. This hydrodynamic advantage results in small, but measurable, reductions in oxygen consumption (Randall and Dexboeck, 1984).

The cost of "breathing" in a dense medium, such as water, can be substantial. The costs, in terms of total oxygen uptake, can be from 10% (Jones and Randall, 1978) to 20% or more (Shelton, 1970) and, according to Schumann and Piper (1966) as much as 30%. Watten and Johnson (1990) offer yet another explanation. The elevated surface velocities in their cross-flow rearing tank, along with a homogeneous DO concentration, may accelerate the diffusion of oxygen at the air-water interface, thus making more oxygen available than the net decrease shows.

Round tanks, in contrast to raceways, do not have a distinct water quality gradient and frequently the rearing environment is considered to be homogeneous. Colt and Watten (1988) describe the ideal round tank as a continuous-flow stirred tank reactor, where the dissolved gas concentrations are well mixed and equal to the concentration in the effluent. However, Tvinnerein and Skybakmoen (1989) point out that in a complete mixed flow reactor, the maximum possible water exchange will be 63.2% during the theoretical mean retention time. Nevertheless, high concentrations of dissolved oxygen (DO) entering a round tank are rapidly diluted with lower DO water in the rearing unit, which is very different from raceways. Since the round tank has no gradient, ammonia and carbon dioxide are fully mixed into the rearing environment, resulting in the continuous presence of some level, in contrast with raceways. One must make sure that such continuous exposure to these potentially detrimental conditions is countered with healthy dissolved oxygen concentrations.

Water velocities in round tanks are, to a large degree, controllable independently of the water exchange rate, in contrast with raceways. The most critical factor, according to Tvinnereim and Skybakmoen (1989), is the design of the inlet and outlet arrangements. Properly designed inlet intake-outlet provisions also contribute to self-cleaning characteristics. Although round tanks are not as efficient in the utilization of floor space and are more difficult with respect to fish handling/harvesting (fish cannot be cornered) than raceways, they have a number of important advantages. They provide cheaper rearing space and they can be operated at low water exchange rates while still creating desirable velocities, hydraulic patterns and self-cleaning characteristics. Round tanks can also receive very high DO inputs without creating "hot spots" and they can be easily equipped with feeders, requiring only one, or at most a few feeding stations. With round tanks it is often possible to accomplish the maximum cumulative oxygen consumption (COC) within a single rearing unit. In other words, the water is "used up" after passing through one tank instead of a series of two or three. All units can, therefore, be placed on the same elevation level.

INLET FLOW STRUCTURES

Circular tanks can achieve relatively complete mixing, i.e., the concentration of a dissolved constituent in the water flowing into the tank changes instantaneously to the concentration that exists throughout the tank. Therefore, if adequate mixing can be achieved, all fish within the tank are exposed to the same water quality. Good water quality can be maintained throughout the circular culture tank by optimizing the design of the water inlet structure and by selecting a water exchange rate so that the limiting water quality parameter does not decrease production when the system reaches carrying capacity.

The water inlet and outlet structures and fish grading and/or removal mechanisms should be engineered to reduce the labor requirement for fish handling and to obtain uniform water quality, rotational velocities, and solids removal within the circular tank. There is nothing more irritating than pipes and other devices that are in the way, when it is time to harvest a tank. Such obstructions also may decrease the effectiveness of the self-cleaning attributes of round tank culture in the first place. It is useful to visually inspect a tank being operated without fish to ensure that the tank water is effectively rotating and the presence of any dead zones can be noted and eliminated. Also, inlet structures can be installed for ease of removal during harvesting procedures, when generally supplemental aeration devices are operated as well.

Circular tanks are operated by injecting water flow tangentially to the tank wall at the tank outer radius so that the water spins around the tank center, creating a primary rotating flow. The no-slip condition that exists between the primary flow and the tank's bottom and side walls creates a secondary flow that has an appreciable inward radial flow component at the tank bottom and an outward radial flow at the tank surface, Fig. 5.4. This inward radial flow along the bottom of the tank carries settleable solids to the center drain and thus creates the self-cleaning property so desired in circular tanks. Unfortunately, in a circular tank with such flow, an area near the center drain region will become an irrotational zone with lower velocities and poor mixing, resulting in solids settling onto the floor. The magnitude of the irrotational zone depends on the introduction of tangential flows near the walls, the diameter:depth ratio, and the overall rate of flow leaving the center bottom drain. Because this irrotational zone has lower water velocities and does not mix well, it can decrease the effective use of the culture tank by producing short circuiting of flow, by creating localized water quality gradients (especially of concern are reduced oxygen levels), and by providing a quiescent zone where solids can settle and collect.

The degree to which a circular tank can self-clean is in part related to the overall rate of flow leaving the center bottom drain. Further, effective solids removal also depends upon the fish re-suspending the settled materials through their movements. This explains in part why tanks with low fish biomass do not clean as well as tanks with higher biomass. In addition, because aquaculture solids have specific gravities that are relatively close to that of water (typically 1.05–1.20 vs. 1.00 for water; Chen et al. 1993; Potter, 1997), sloping the floor towards the center drain does not improve the self-cleaning attributes of a circular tank. Sloped floors are useful only when a circular tank is drained for maintenance purposes.

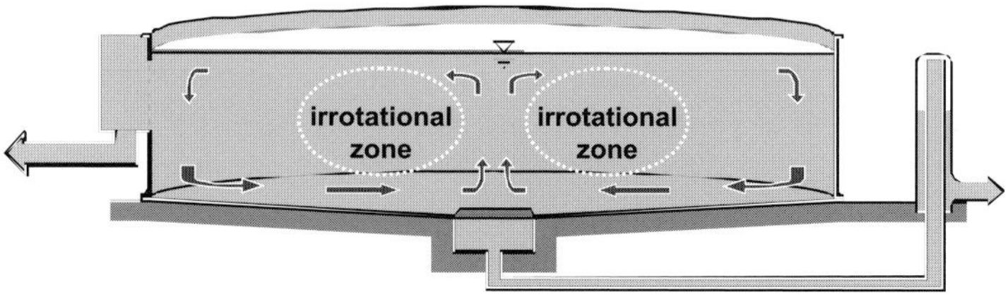

Figure 5.4 The Primary Rotating Flow (Not Shown, but Created by Injecting the Flow Tangential to the Tank Wall) Creates a Secondary Rotation That Flows Radially (Shown Here) and Carries Settleable Solids Towards the Tank's Bottom Center Drain in a Phenomenon Called the Tea-Cup Effect (Summerfelt et al. 2001).

Rotational velocity can be controlled by design of the water inlet structures, so water flow does not have to be increased beyond that required for the fish's culture environment. Tvinnereim and Skybakmoen (1989) reported that the current velocity in a tank can be largely controlled by varying the inlet impulse force (F_i), which is defined as:

$$F_i = \rho \cdot Q \cdot (V_{orif} - V_{rota}) \tag{5.5}$$

The inlet impulse energy largely dissipates as it creates turbulence and rotation in the rotational zone, Fig. 5.4. The impulse force, and thus the rotational velocity in the tank, can be regulated by adjusting either the inlet flow rate or the size and/or number of inlet openings. Paul et al. (1991) reported that the tank rotational velocity is roughly proportional to the velocity through the openings in the water inlet structure, especially near the tank wall, i.e.,

$$V_{rota} \approx \alpha \cdot V_{orif} \tag{5.6}$$

where the proportionality constant (α) is generally from 0.15–0.20, depending on the design of the inlet flow structure. The manner of flow injection influences:

- uniformity of the velocity profile through the tank;
- strength of the secondary radial flow along the tank bottom towards the center drain, i.e., the ability of the tank to move settleable solids to the center drain; and
- uniformity of water mixing.

Using Fig. 5.5, you can predict the velocity needed from the inlet orifices to create the desired tank water velocities. By installing a site tube immediately above the first orifice hole in the vertical inlet pipe, you can observe the available head to create velocity. Assuming the tank water velocity will be equal to 15 or 20% of the orifice velocity as noted above, you can adjust the number of holes or hole area in the inlet pipe to create the necessary velocities to achieve the desired tank water velocities. You should do this "on-paper" before you construct the inlet pipes. If the site tube is located immediately above the first orifice hole, the required dynamic heads, H, to create the velocity desired is 1.5 H ($H = V^2/2g$; V is velocity at the orifice and g is the acceleration due to gravity, 9.8 m^2/s; see Chapter 3 for more information on fluid mechanics).

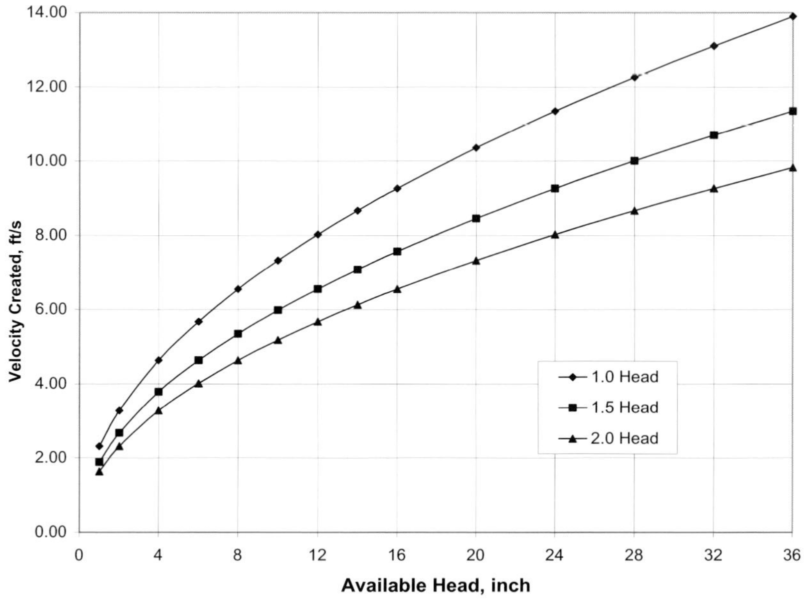

Figure 5.5 Resulting Orifice Velocities for a Specific Head Available (Typical Inlet Orifice Will Require 1.5 Dynamic Heads).

Skybakmoen (1989) and Tvinnereim and Skybakmoen (1989) compared the tank hydraulics that resulted from injecting the water flow tangentially at the outer radius of the circular tank with either:

- a traditional open-ended pipe point source;
- a short, horizontal, submerged, distribution pipe with its axis oriented towards the tank center and with evenly spaced openings along its length (directed at 30° below the water surface);
- a vertical submerged distribution pipe with evenly spaced openings along its length;
- and an inlet flow distribution pipe that combines both vertical and horizontal branches.

Open-ended pipe inlets create non-uniform velocity profiles in the tank, e.g., much higher velocity profiles along the tank wall; poor mixing in the irrotational zone, resuspension of solids throughout all tank depths; and poor flushing of solids from the bottom (Skybakmoen, 1989; Tvinnereim and Skybakmoen, 1989). Horizontal submerged pipe inlets improve water mixing effectively throughout the tank, but create a weaker and less stable bottom current (for solids cleaning). Vertical submerged inlet distribution pipes provide better self-cleaning than when injecting the water flow through an open-ended pipe or a horizontal distribution pipe, but the stronger bottom current (responsible for particle removal) also results in poorer mixing in the irrotational zone and short circuiting and therefore less efficient use of the flow exchange.

Maximum uniformity in tank water conditions can be obtained by using an inlet flow distribution pipe that combines both vertical and horizontal branches, Fig. 5.6. The inlet pipes must be placed somewhat away from the wall so that fish can swim between the pipe and wall. This design approach is an effective way to:

- achieve uniform mixing,
- prevent short circuiting of flow,
- produce uniform velocities along both the tank's depth and radius, and
- effectively transport waste solids to the tank bottom and out the center drain.

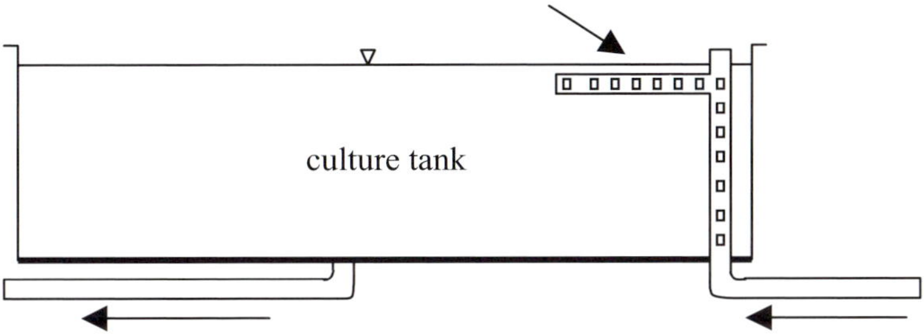

Figure 5.6 Water Injected into Culture Tanks through Evenly Spaced Openings (Holes Or Slots) in Both Horizontal and Vertical Injection Pipes Produces More Uniform Rotational Velocities Both Radially and Vertically, More Uniform Mixing, and Better Solids Flushing.

For large circular tanks, e.g., diameter > 6 m, placing multiple flow distribution pipes at different tank locations can improve solids removal, velocity uniformity, and water quality homogeneity. Again, perform visual observations prior to fish placement. Recent experience indicates that by using only vertical inlet pipes and then pointing the exhaust jets at approximately 45° from the tank wall is reasonably effective. This may produce a flow pattern in the tank effective enough to eliminate the need for any horizontal inlet pipes. The horizontal pipes will interfere more with harvesting and sampling activities than would just the vertical pipes near the tank wall. More elegant solutions to inlet distribution have been presented in the literature, e.g., Watten and Johnson, 1990; but economic considerations may eliminate these more elegant solutions from consideration.

The reader is referred to the extensive literature on circular tank design and inlet structures for further information: Burrows and Chenoweth, 1955; Goldsmith and Wang, 1993; Klapsis and Burley, 1984 and 1985; Larmoyeux et al. 1973; Paul et al. 1991; Skybakmoen, 1989; Tvinnereim and Skybakmoen, 1989; Wheaton, 1977.

Outlet Flow Structures

Circular fish culture tanks concentrate settleable solids, e.g., fecal matter, feed fines, and uneaten feed, at their bottom and center. The tank center is then the logical location for the bottom drain. The bottom center drain should be designed to continuously remove the concentrated settleable solids and for the intermittent removal of dead fish that are captured at the bottom center drain (discussed more fully in a later section). The bottom center drain structure is also used for water level control by connecting it to a weir, either on the inside, Fig. 5.7, or the outside of the tank, Fig. 5.8.

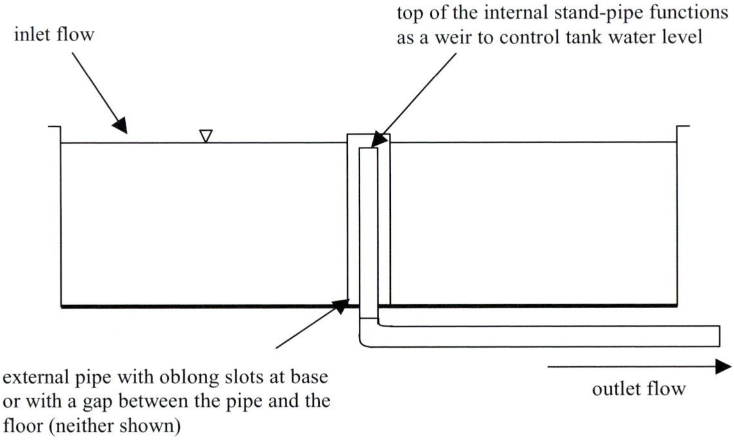

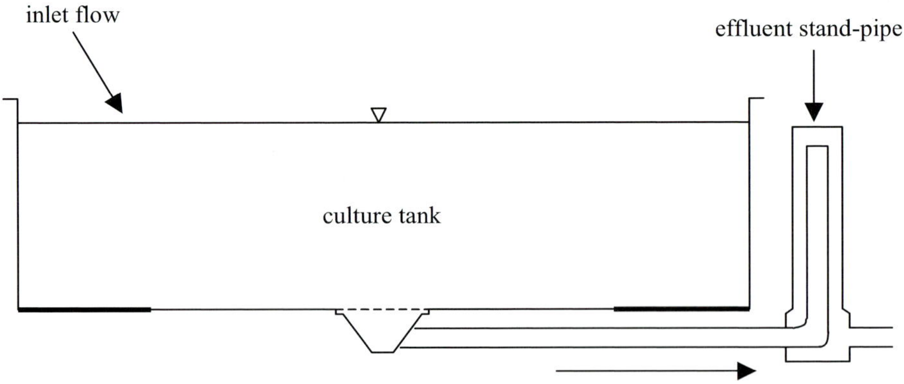

Figure 5.8 External Standpipe Arrangement. Excluding fish from the center drain with a vertical perforated plate or screen requires use of an external standpipe to control water depth, but makes the standpipe easily accessible and avoids obstructions in the center of the tank.

When a center standpipe is used on the inside of a tank, it can be designed so that it either captures and stores settleable solids near the drain (where they can be flushed at intervals, or it continuously withdraws settleable solids from the bottom of the culture tank. To continuously withdraw settleable solids through a center standpipe that also controls water depth requires use of two concentric pipes. Perforated slots at the base of the outer pipe or a gap at the base of the outer pipe forces flow to be pulled from the bottom of the tank (capturing settleable solids) and the inner pipe is used as a weir to set the water depth within the tank, Fig. 5.7.

As early as 1933, the use of self-cleaning center standpipe had been presented in the literature (Surber, 1933 & 1936). Figure 5.7 shows the recommended design for creating an adjustable gap between the bottom of the outer pipe and the tank bottom in order to increase suction while forcing the flow to leave at the tank bottom where settleable solids collect. The distance between the two pipes, i.e., the annular space should be selected to create a velocity large enough (0.3 to 1.0 m/s depending upon the size and density of the particles) to entrain solids up to the top of the inner pipe. A patented method that incorporates these principles uses an annular approach plate to enhance particle entrapment, Fig. 5.9.

Figure 5.9 A Solid Annular Approach Plate Fixed Above the Bottom Center Drain as Developed and Patented (U.S. Patent No. 5,636,595) by Lunde et al. (1997) to Exclude Fish and Enhance Particle Removal by Accelerating Particles into the Drain. This illustration of an Aqua Optima AS culture unit also shows a water injection mechanism (A), an external swirl separator attached to the bottom drain (B), a second drain (just above the bottom center drain) for withdrawing water relatively low in solids (C), and an external standpipe assembly (D) (reproduced by permission of Aqua Optima AS, Summerfelt et al. 2001).

When an external standpipe controls water depth, a vertical perforated plate or screen, Fig. 5.8, can be used to cover the bottom center drain; this allows solids to leave the tank but excludes fish. Corrosion-resistant screening material, such as perforated sheets of aluminum, stainless steel, fiberglass, or plastic are used to cover drain outlets. We recommend using perforated screening with horizontal oblong slots instead of holes, because the slots are easier to clean, provide greater open area, and do not clog as readily as round holes. Guidelines for slot size based upon fish size are given in Table 5.2. Ideally, openings through the screen covering the center drain should be small enough to exclude fish and yet large enough not to become clogged with feed pellets or fecal matter. Entrapment of fish on the outlet occurs when fish cannot escape the area in front of the drain because the water velocity in that area is too great. Fish impingement is minimized by providing a total open area through the outlet screen so that the water velocity through the screen is $\leq$ 30 cm/s. Depending upon the species and life stage, certain situations particularly with smaller fish require water velocities $\leq$ 15 cm/s, e.g., see Eq. 5.2. These velocities do not produce a significant pressure drop through the screen openings, thus minimizing fish impingement.

Table 5.2 The Horizontal Oblong Slot Size Depends Upon the Size of Fish to be Retained (Fish Species was not Specified, from Piper et al. 1982)

Slot Size (mm)	Fish size, g
1.6 x 3.2	fry to 0.45 g
3.2 x 6.4	0.45 to 2.3 g
6.4 x 12.7	2.3 to 15 g
12.7 x 19.1	15 g and larger

Not all aquatic species require screening to prevent their escape down the bottom center drain. Tanks (5 to 9 m diameter tanks) have been used to culture salmonid species with no fish exclusion screening over their center drain (S. Wilton, PRAqua, Nanaimo, British Columbia, Canada). This PRAqua design links the unscreened bottom drain to an external standpipe chamber, which contains a weir to control water depth, a screen to capture dead fish, and a drain, Fig. 5.10. PRAqua reported that fish escapement down the uncovered center drain was not a problem, because salmonids do not tend to swim down that kind of current. Depending upon the regulatory authority, more stringent escapement preventative methods may be required.

Further information can be obtained from the literature on outlet designs: Burrows and Chenoweth, 1955; Josse et al. 1989; Klapsis and Burley, 1985; Larmoyeux et al. 1973; Pankratz, 1995; Piper et al. 1982; Skybakmoen, 1989; Surber, 1933 and 1936; Tvinnereim and Skybakmoen, 1989.

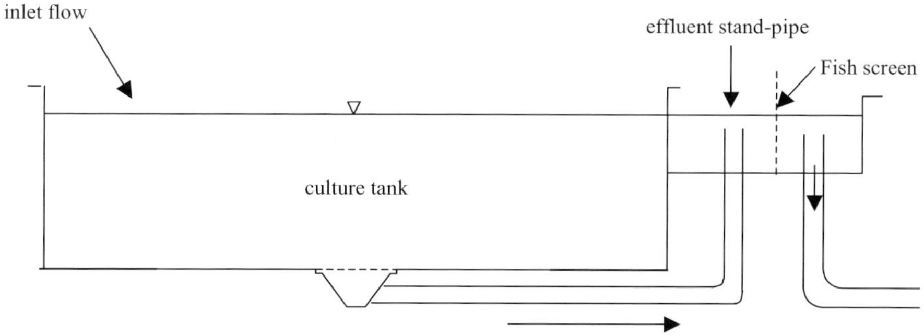

Figure 5.10 An Unscreened Bottom Drain Linked to an External Standpipe Chamber, Which Contains a Weir to Control Culture Tank Depth and a Bar Screen to Capture Dead Fish. (used by permission from PRAqua, Nanaimo, BC).

5.4 CORNELL DUAL-DRAIN DESIGN

Circular fish culture tanks can be managed as swirl settlers, i.e., settling basins with two effluents, because of their capability to concentrate solids at their bottom and center. Solids that concentrate at the bottom center can be removed in a small flow stream by using a bottom-drawing center drain, while the majority of flow is withdrawn at an elevated drain. Cobb and Titcomb (1930) and Surber (1936) were the first to report the use of a second bottom-drawing drain to remove solids that were settled in the center of circular culture tanks. More recently, settled solids have been reportedly concentrated in 5–20% of the total flow that leaves the bottom center drain of circular culture tanks when the remainder of the flow leaving the tank (roughly 80–95% of the total) was withdrawn through a fish-excluding port located above the bottom-drawing drain (Mäkinen et al. 1988; Eikebrokk and Ulgenes, 1993;

Lunde et al. 1997). There have been at least two patents filed on double-drain designs:

- Lunde et al. (1997), *Particle trap*, US Patent # 5636595.
- Van Toever (1997), *Water treatment system* US Patent # 5593574.

The location of the two tank drains have typically been at the center of the tank, which then takes advantage of both the 'tea-cup effect' and the strength of the overall flow when it drains through the tank center. This is the approach has been taken by the manufacturers of the particle trap (marketed by Aqua Optima, Norway), Van Toever (1997), and several others, where both the bottom drawing drain and elevated drain are placed at the center of the culture tank. The particle trap, Fig. 5.9, for example, endeavors to concentrate the majority of settleable solids into a reduced flow that passes through a gap created between the tank floor and an approach plate incorporated into the particle trap (Lunde et al. 1997).

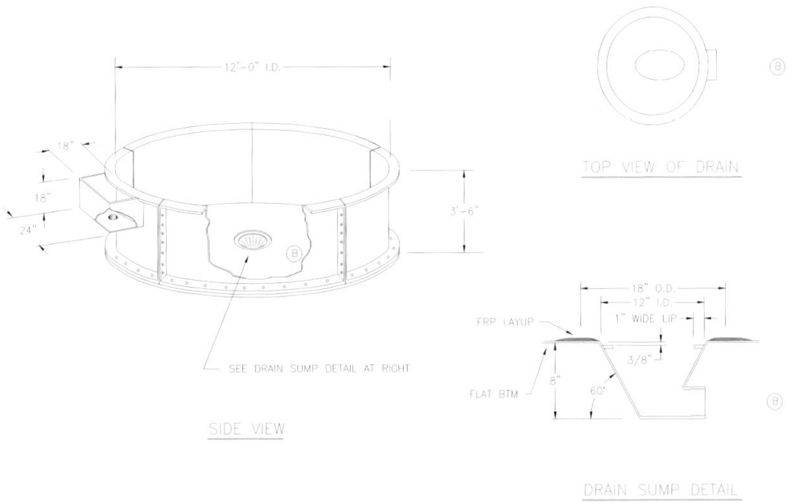

Figure 5.11 Cornell Dual-Drain Design Used at the Freshwater Institute (Shepherdstown, WV). Sidewall drain discharges 80 to 95% of the Total Flow (Drawing Courtesy of Red Ewald, Inc., Karnes City, Texas).

The dual-drain approach has significant economic implications. When using this approach, solids removal costs are controlled more by the volume of flow that is treated rather than the solids concentration of the effluent that is treated. By concentrating the majority of the solid wastes in only 5 to 20% of the flow leaving a tank via the center low-flow drain, treatment costs are proportionately reduced as well. The lower flow rate from the effluent pipe may also provide more effective results to applying ozone to a system to further enhance solids removal.

The Cornell-type dual-drain culture tank, Figs. 5.11 and 5.12, differs significantly from the other dual-drain designs, because it is the only design that removes the majority of flow through an elevated drain located on the tank's sidewall; all other double-drain approaches place both drains in the tank center.

Figure 5.12 Cornell Dual Drain System (28,000 Gallon Tank, Total Flow of 2,000 gpm, from Freshwater Institute)

Using the Cornell-dual drain design basically eliminates the ever present vortex usually created when both flows are extracted from the center of round tanks. The Cornell-dual drain also reduces the velocities near the center drain because the flow leaving the center drain is only 5 to 20% of the total flow from the tank. As a result, fish distribution is much more uniform in a Cornell dual-drain compared to conventional double drains. This is a key advantage because the fish effectively utilizes the entire tank volume. In a conventional dual-drain tank, it is not uncommon to see the center third of the tank being unoccupied by fish. The Freshwater Institute has adapted a dead fish removal system to the Cornell-dual-drain design where the center drain (low-flow) can periodically be flushed and in the process carry away any dead fish that have collected.

The Freshwater Institute (Shepherdstown, WV) and Cornell University recently completed a study of the Cornell dual-drain design.

The Freshwater Institute used rainbow trout at densities of 60 kg/m^3 and fish were fed 1% body weight per day; Cornell used 500 g tilapia on full feed at a tank density of 90 kg/m^3. Settleable solids fractionation and tank mixing were studied using a 3.66 m (12 ft) i.d. x 1.22 m (4 ft) tall circular Cornell-type dual-drain tank at the Freshwater Institute (FI) and a 2 m i.d. x 1 m (3.2 feet) tall (6.5 ft) tank at the Cornell site. A range of hydraulic exchange rates (1 to 7 tank volumes per hour) and percentage bottom drain flow (5, 10, and 20% of total flow) were used through the tank.

> **"RULE OF THUMB"**
> Choose the Center Drain Flow as the largest of:
> a) 0.15 gpm/ft^2 (6 Lpm/m^2) of floor,
> b) HRT center drain < 200 minutes, or
> c) 10 to 15% of total tank flow rate.

The solids removal capabilities of the Cornell-type dual-drain design have been determined in actual tests with and without fish present and at a variety of tank sizes. Sinking PVC cylindrical beads (3 mm in diameter, with a specific gravity of 1.05 times that of water) were added to the culture tank with fish present. Replicated bead flushing tests were used to determine the influence of hydraulic exchange rate (1 and 2 rearing tank volume exchanges per hour), diameter-to-depth ratio (12:1, 6:1, 3:1), and the percentage of flow leaving the bottom center drain (5, 10, and 20% of total flow) on solids flushing characteristics. The bead flushing data was analyzed using the solution to a non-steady-state mass balance and used to quantify the amount of solids flushing that occurs through the bottom center drain in comparison to that which was due to simple dilution versus solids transport produced by settling and radial flow dynamics.

Total suspended solids (TSS) concentrations were also measured in bottom and sidewall discharges and in the effluent from a commercial microscreen filter treating the bottom flow. Mean results from data collected at the Freshwater Institute indicated that the Cornell-type double-drain tank concentrated the majority of TSS in the bottom flow. The TSS concentration discharged through the bottom drain averaged 19.6 ± 3.6 mg/L (± standard error), while the TSS concentration discharged from the sidewall averaged only 1.5 ± 0.2 mg/L. This represents a 13 fold increase in solids concentration from the center drain. Practically speaking, the side wall drain discharges the best water quality in terms of suspended solids, since the solids are constantly moving towards the center drain and influent water also enters from the outside edge of a tank. The sidewall discharge, which represented 80–95% of the total flow, but only 1.5 mg/L TSS, would probably not require any further treatment and most likely could be discharged directly under most state and federal regulations. Further treatment of the bottom discharge across a commercial microscreen filter then captured 82% (± 4%) of the TSS in the bottom discharge, so that only an average of 3.5 ± 0.8 mg/L TSS would be discharged with the bottom flow.

When using the Cornell Double-Drain approach, the bottom flow is only 5–20% of the total tank discharge, so it would be possible and practical to further treat this flow with finer microscreen filters or by using settling tanks, wetlands, or sand filters, if required to meet stringent state or federal discharge regulations. Therefore, conventional flow-through systems that use circular culture tanks installed with the Cornell-type dual drain configuration can capture more of the waste solids and phosphorus from the water flow than can be captured within typical raceways operation that contain quiescent zones and off-line settling ponds. Additionally, when the Cornell-type dual drain tank is used within partial-recirculating systems, these systems are capable of supporting high production densities while providing uniformly healthy water quality, optimized water velocities, and efficient, rapid, and gentle solids removal (Freshwater Institute, unpublished data). This type of partial recirculating system would be able to capture approximately 80–95% of the total waste solids produced, which is significantly better waste capture than is typically achieved within raceways operated under serial water reuse, e.g., 50% solids capture (Mudrak, 1981). Also, the solids filter treating the discharge from the partial-reuse system could be relatively small and inexpensive in comparison to a similar unit scaled to treat the much larger flow discharged from a raceway operation of relatively similar annual fish production. These combinations of treatment methodology offer a powerful approach to effective and efficient waste stream control from intensive culture systems.

Theoretical basis for solids removal and practical application

The ideal theoretical mean residence time (MRT_{ideal}), meaning with perfect and uniform mixing, or hydraulic retention time (HRT) can be calculated using the water volume in the tank (V_{tank}) and the water flowrate through the tank (Q):

$$MRT_{ideal} = \frac{V_{tank}}{Q} \tag{5.7}$$

Solids will be removed from a tank based upon mass action or simply HRT and an enrichment factor (k) due to concentration of solids near the center drain. The relative importance of these two solids flushing mechanisms can be described as the enrichment factor as a percent (φ_{enrich}):

$$\varphi_{enrich} = \frac{k}{k + \frac{Q}{V_{tank}}} \cdot 100 \tag{5.8}$$

Solids removal in the absence of incoming, generation, or accumulation of solids can be described in an unsteady-state mass balance broken into a component representing simple mass action of flow to the bottom drain and a component representing solids enrichment at the bottom drain due to a combination of sedimentation and radial flow (assuming that the solids outflow at the tank's sidewall is negligible). This is done to distinguish between the two different mechanisms transporting solids to the bottom center drain:

$$Loss = -\ Outflow\ (mass\ action) - Outflow\ (enrichment) \tag{5.9}$$

Or more specifically,

$$\begin{aligned} V_{tank} \cdot \frac{dC}{dt} &= -Q_{out,b} \cdot C_{out,b} - k \cdot C_{out,b} \cdot V_{tank} \\ &= V_{tank} \cdot \left\{ -\left(\frac{Q}{V_{tank}} + k \right) \cdot C_{out,b} \right\} \end{aligned} \tag{5.10}$$

Integration provides an equation that can be used to model solids removal through the bottom drain in real time:

$$C_{out,b}(t) = C_{out,b}(t = 0) \cdot exp\left[-\left(\frac{Q}{V_{tank}} + k \right) \cdot t \right] \tag{5.11}$$

Note that the flushing of a homogeneously distributed pollutant is only due to mass action, i.e., the culture tank exchange rate Q/V:

$$C(t) = C(t=0)\,exp\left[-\left(\frac{Q}{V_{tank}}\right)t\right] \qquad (5.12)$$

Now, it can be seen in the above equations how the k-value (1st order enrichment constant) increases the rate of solids flushing relative to the culture tank exchange rate.

The k-values estimated from the pellet-tracer tests are an indication of the rate that solids are flushed from the bottom-center drain and also of the relative strength of the radial flow. Strong radial flows and rapid solids flushing were created using 2 ex/hr and diameter:depth ratios of 3.1:1 and 6:1 in the Freshwater/Cornell trials. However, a very important observation was that solids removal was not effective at low exchange rates. At 1 ex/hr, the settleable solids frequently deposited near but did not exit the tank's bottom-center drain. As the diameter:depth ratios became larger (more shallow tanks), the removal rates became less effective. A tank with a diameter:depth ratio of 12:1 did not effectively discharge solids even at 2 ex/hr. During these trials, the radial flow transported the settleable solids to the center portion of the tank, but water velocities were so low in the middle of the tank that a good portion of these solids settled within in a near region about the center drain. Accumulated settled solids are usually near to the center drain so that pulling an external stand-pipe regulating the bottom-center drain flow, even for an interval of < 1 min, will usually create sufficient flow to flush these accumulated solids. The presence of fish will improve solids removal in this area.

The radial flow mechanism played a much larger role than the mass transport mechanism in the transport of solids to the bottom-center drain. However, the inlet flow and hence the overall HRT for the tank also plays a role in that some minimum HRT must be maintained in order to produce a radial force significant enough to move solids to the center. As mentioned above, one exchange per hour was not effective in removing solids, although the deeper tanks were more successful at this rate than were shallower tanks (DDR < 6).

Quite clearly, as the tank depth increases (or DDR decreases), the enrichment factor is much higher for the "deeper" tanks. This also indicates that previous recommendations (Larmoyeux et al. 1973) to use DDR between 5 and 10 should be revisited according to these results. It appears that DDR less than 5 and perhaps as low as 2 may be preferred to shallower tanks based upon the increase in enrichment as characterized by the k value and $k/(Q/V_{tank} + k)$. This is very important because it definitely impacts the quantity of fish that can be maintained as a standing biomass per unit floor area. Fish behavioral characteristics must also be considered, e.g., flounder that need floor space versus tilapia that distribute well vertically in the water column.

5.5 RACEWAYS

Raceways have primarily been associated with the culture of salmonids. Continued attempts have been made to adapt raceway culture methods to RAS, with mixed results. Proponents of raceways will generally argue for raceway design due to their better utilization of floor space and easier handling and sorting of fish. Their disadvantage is their inability to be self-cleaning. Figure 5.13 shows the all too frequent task associated with maintaining raceways, that being multiple weekly or even daily cleaning of the settling areas that must be placed somewhere in the race. Also notice in this figure that the worker is cleaning the bars that keep the fish confined in the race. This accurately demonstrates that every solid surface screening materials in particular will need frequent cleaning maintenance. Avoid placing materials in fish tanks whenever possible.

Figure 5.13 Some of the Solids Swept From a Raceway's Fish Rearing Area Can Settle in Quiescent Zones. Captured solids must be manually removed (photo courtesy of Pennsylvania Fish And Boat Commission, Big Spring Fish Culture Station, Newville, PA).

Raceways are the most common rearing tank design being used in locations where aquaculture has tapped into huge groundwater resources, such as the Thousand Springs area in Idaho where large amounts of rainbow trout are produced. In Idaho, raceway dimensions are typically about 3–5.5 m wide, 24–46 m long, and 0.8–1.1 m deep (IDEQ, 1998).

Raceways usually have a length to width ratio of 1:10 and a depth < 1.0 m, and require a high water exchange rate, e.g., one tank volume exchange every 10 to 15 minutes.

Raceways depend upon this high turnover rate to maintain sufficient water quality for fish growth. Water enters the raceway at one end and flows through the raceway in a plug-flow manner, with minimal back mixing. As a result, the best water quality exists at the head of the tank where the water enters, and then deteriorates steadily along the axis of the raceway towards the outlet. It is not be uncommon to see fish crowded towards the inlet end of the raceway in order to obtain oxygen, or at the end of a raceway to avoid high concentrations of ammonia or some other accumulating contaminant. During feeding activities, fish will also crowd to the feeders, since it is also more difficult to distribute feed throughout raceways than it is in circular tanks.

Velocity through the raceway will generally be between 2–4 cm/s, which is insufficient to effectively remove settled solids from the rearing space. Thus, frequent maintenance for cleaning of raceways is required. A variety of mechanisms can be used to increase solids removal, such as placing periodic baffles across the raceway to artificially increase velocity (20 to 40 cm/s or 10 times the raceway average water velocity) and sweeping action. However, these baffles must be moved to work the fish, and also provide a substrate for biosolids growth in warmer temperatures. In practice, the management of raceways is based on their oxygen design requirements, rather than their cleaning requirements. The velocity required to flush solids from unbaffled raceways is much greater than the velocity required to supply the oxygen needs of the fish. In practical terms, raceways are incapable of producing the optimum water velocities recommended for fish health, muscle tone, and respiration.

Since the difference between flow rates necessary to achieve a cleaning velocity and required flows for oxygen supply are so different, designs are based upon minimizing cross-sectional area to promote maximum water velocity. As a result, many raceway systems are operated in series, with the discharge of the upstream raceway serving as the inflow water of the next one downstream. Water may be treated between race sections to restore a particular water quality parameter for subsequent downstream utilization or removal, e.g., oxygen addition, CO_2 removal. Low head oxygenators (LHO's) have become popular in such applications (Watten, 1989). LHO's vary in configuration, but all are fundamentally similar in operation. These units consist of a distribution plate positioned over multiple (5 to 10) rectangular chambers, Fig. 5.14.

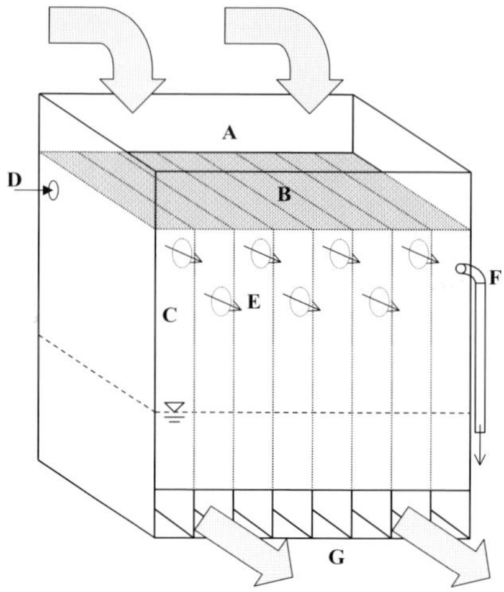

Figure 5.14 Low Head Oxygen (LHO) Units are Popular in Raceway Culture to Restore Oxygen in Serial Reuse Raceway. A typical LHO configuration and components are shown: water flowing into a collection trough or plate (A), through a perforated distribution plate (B), and is oxygenated in the chambers (C), as gas flows from inlet gas port (D), through holes between chamber to chamber (E), to the off gas port (F), where excess gas is bubbled off under water. Water exits at the bottom of the unit (G).

Water flows over the dam boards at the end of a raceway or is pumped upwards from an indoor fish tank, through the distribution plate, and then falls through the rectangular chambers. These chambers provide the gas-liquid interface needed for mixing and gas transfer. The streams of falling water impact a collection pool at the bottom of each chamber where the effluent water flows away from each chamber equally in parallel. Pure oxygen is introduced into the outer or first rectangular chamber, passes through the series of individual chambers, and finally is vented to the atmosphere at a much lower concentration. Each of the rectangular chambers is gas tight and the single holes between the chambers are properly sized and located to reduce back-mixing between chambers (Watten, 1989). These devices are completely described in Chapter 8 (Gas Transfer) and a computer program to predict their performance is given in Chapter 17.

Raceways can be constructed side-by-side, with common walls, for maximizing the utilization of floor space and reducing construction costs, Fig. 5.15. However, when constructed without common walls, because of their large aspect ratio (L:W), raceways require 1.5 to 2.0 times as much wall length as circular tanks. Circular tanks can also better structurally handle the weight of the confined water and can thus use thinner walls than rectangular tanks.

A quiescent zone devoid of fish is usually placed at the end of a raceway tank to collect the settleable solids that are swept out of the fish rearing area. At many large trout farms, these solids collection zones are the primary means for solids removal to meet discharge permit requirements. The maintenance of these settling zones can often be 25% of the total labor required for the entire fish operation.

Figure 5.15 Raceways are Often Constructed Side-By-Side, with Common Walls. The raceways is slightly sloped, so that each raceway can be broken into several sections and the water serially reused as it flows from section to section.

MIXED-CELL RACEWAY TANKS

It should be fairly evident to the reader at this point that we strongly favor circular tanks because of their advantages of elevated water velocities, uniform water quality, and good solids removal characteristics. Still, we acknowledge that linear raceways make better use of floor space and facilitate harvesting, grading, and flushing operations. The cross-flow tank is a recent hybrid design that incorporates the desirable characteristics of both circular tanks and linear raceways (Watten and Johnson, 1990). Water is distributed uniformly along one side of a cross-flow tank via a submerged manifold and is collected in a submerged perforated drain line running the length of the opposite side.

The basis for this design is that the influent is jetted perpendicular to the water surface with sufficient force to establish a rotary circulation about the longitudinal direction. As a result, the standard raceway section is modified to create a series of horizontal counter rotating mixed cells with cell length equal to vessel width, Fig. 5.16. Cells receive water from vertical pipe sections extending to the tank floor and positioned in the corners of the cells using jets that direct water into the cell tangentially to establish rotary circulation. The pipe sections can be swung up and out of the water during fish crowding or grading operations. Water exits each cell through a centrally located floor drain.

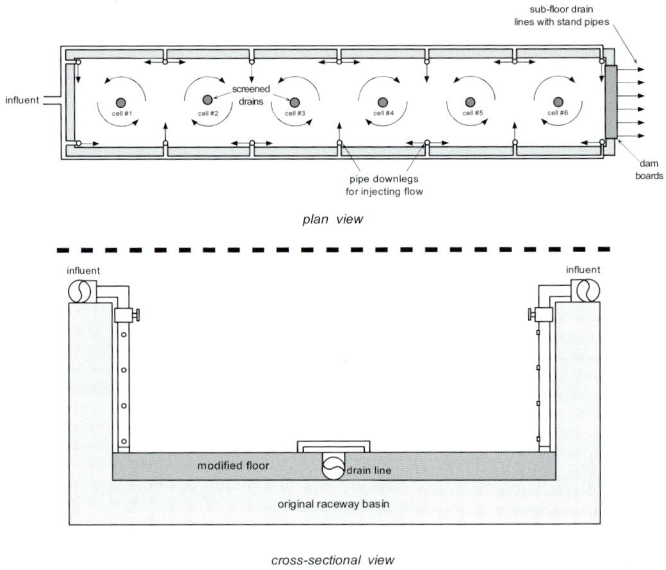

Figure 5.16 Illustration of a Mixed-cell Raceway Tank.

Hydraulic characteristics for a mixed cell raceway approximate that of a circular tank (mixed-flow reactor), both with and without fish present. Water velocities were reasonably uniform from top to bottom in the water column and were sufficient to scour and purge fecal solids. Cell interaction was significant with cell to cell exchange rates representing about 3 to 4 times the tank inflow rate. This characteristic contributed to the observed uniform distribution of fish throughout the vessel. The energy requirement of the design was 1.32 m of water head primarily for the inlet system and jets. Given that the tank center drain in each cell is similar to that of a circular tank, the Cornell-double drain could also be applied in this design. Before implementing this design, please recognize that there will be some additional construction costs associated with this raceway design, and performance over the long-term is yet to be established.

5.6 CARRYING CAPACITY ISSUES

Production efficiency in culture tanks can be boosted by increasing the culture tank's carrying capacity, which in simplistic terms is the maximum fish biomass that can be supported at a given feeding rate. Dissolved oxygen is usually the first water quality parameter to limit culture tank carrying capacity, because the fish consume dissolved oxygen during respiration. The amount of dissolved oxygen available in fish culture tanks is dependent upon the water flow rate multiplied by its concentration of available dissolved oxygen, i.e., the inlet dissolved oxygen concentration minus the minimum allowable dissolved oxygen concentration, assuming no in-tank aeration or significant photosynthesis. For example, assuming that water is not limiting and that its dissolved oxygen concentration is constant, doubling the water flow, i.e., hydraulic exchange rate, through a culture tank will double the carrying capacity of the tank. However, moving more water through the culture volume is expensive, as it requires large pumps, pipes, or water pressure requirements, which may not be feasible, and will increase the farm's fixed and variable costs. Alternatively, supersaturating the dissolved oxygen in the water flowing into the tanks has also been popular, and often is a more cost effective method to improve the profitability of tank-based fish farms (Colt et al. 1991). For example, assuming a minimum allowable outlet dissolved oxygen concentration of 7 mg/L, increasing the dissolved oxygen concentration entering a culture tank from 10 mg/L to 16 mg/L would triple the culture tank's available oxygen and thus triple its carrying capacity.

Supersaturating the flow with dissolved oxygen can be achieved cost effectively with many different oxygen transfer devices, even in low head applications (Boyd and Watten, 1989). The cost of adding oxygen at a commercial tank-based fish farm can be estimated from the following farm and species specific variables:

- oxygen consumption rate per unit feed consumed, e.g., 0.25 kg oxygen per kg feed

- feed conversion rate, e.g., 1.3 kg of feed is required to produce 1.0 kg of fish

- transfer efficiency of oxygen gas into water, e.g., 1.0 kg of oxygen feed-gas must be supplied to dissolve 0.6 kg oxygen (as an example)

- density of oxygen gas, e.g., 0.75 cubic meter oxygen gas (at 20°C and 1 atm pressure) per kg oxygen

- cost of oxygen gas, e.g., \$0.15 per cubic meter oxygen gas (at 20°C and 1 atm pressure)

As an example, estimates for each of the variables listed above are used to make a simplistic prediction of the cost of oxygenation, assuming all dissolved oxygen requirements are provided by the purified oxygen feed gas:

$$oxygen\ \ cost = \frac{1.3\,kg_{\ feed}}{1.0\,kg_{\ fish}} \cdot \frac{0.25\,kg_{\ O_2}}{kg_{\ feed}} \cdot \frac{1.0\,kg_{\ O_2}\ supplied}{0.6\,kg_{\ O_2}} \cdot \frac{0.75\,m^3\,O_{2\ gas}}{kg_{\ O_2}\ supplied} \cdot \frac{\$0.15}{m^3\,O_{2\ gas}}$$

$$= \frac{\$0.061}{kg_{\ fish}}$$

In this example, the cost of the oxygen feed gas required to produce 1.0 kg fish would amount to \$0.061. At this same feed conversion rate, fish feed alone might cost more than \$0.50 for every kg of fish produced. Therefore, the cost of the oxygen would be relatively low, much lower than the cost of the required fish feed. Thus, adding a pure oxygen transfer system to increase fish farm carrying capacity can be used to decrease production costs by increasing fish production in the same culture tanks while using relatively similar labor requirements.

Other fish metabolites such as dissolved carbon dioxide, ammonia, and suspended solids will limit culture tank carrying capacity, after oxygen is no longer limiting. In such cases, fish can produce roughly 1.0–1.4 mg/L total ammonia nitrogen, 13–14 mg/L dissolved carbon dioxide, and 10–20 mg/L of total suspended solids (TSS) for every 10 mg/L of dissolved oxygen that they consume. Without some form of ammonia or carbon dioxide control, dissolved carbon dioxide and un-ionized ammonia concentrations can rapidly accumulate to toxic levels when fish consume large concentrations of dissolved oxygen. Colt et al. (1991) developed a method to estimate the carrying capacity of water if dissolved oxygen were not limiting. The Colt et al. (1991) method uses mass balances and chemical equilibrium relationships to predict the amount of dissolved oxygen concentration that fish can consume before the dissolved carbon dioxide or ammonia concentrations become limiting to further oxygen consumption. Accordingly, intensive fish farms can use the water flow without worry of ammonia or carbon dioxide limitations (assuming no biofiltration or air-stripping) up to a cumulative dissolved oxygen consumption of about 10 to 22 mg/L, depending upon pH, alkalinity, temperature, and the fish type and life stage (Colt et al. 1991). After reaching this cumulative oxygen demand, the water flow cannot be used again until the dissolved carbon dioxide and ammonia concentrations are reduced.

5.7 Stock Management Issues

Fish farm production can also be increased (approximately doubled) through the use of a continuous production strategy, rather than a batch production strategy (Watten, 1992; Summerfelt et al. 1993; Hankins et al. 1995; Summerfelt and Summerfelt, 1996). The maximum economic productivity of the culture system can be obtained with year round fish stocking and harvesting because continuous production maintains the culture system at or just below its carrying capacity. Also, harvesting at a given size increases product value by providing more uniformly sized fish for the market. In a full-scale production experiment, Heinen et al. (1996A) showed that rainbow trout stocked every 8 weeks and harvested weekly could achicve a steady-state annual production (kg/yr) to maximum system biomass (kg) ratio (P:B) of 4.65:1 per year. Many commercial farms operate at P:B ratios of 3:1 per year or less. Adopting stock management strategies that increase the P:B ratio on a commercial farm would significantly reduce production costs.

Continuous stocking and harvesting strategies requires culturing several size groups at the same time, handling the fish frequently, and improving inventory accounting techniques. This can stress the fish and increase labor costs if automated equipment is not used. To keep the cost of handling and grading fish to a minimum, and the stress on the fish to a minimum, a convenient mechanism for size sorting the fish, counting them, and moving them to other locations should be incorporated into the culture tank and facility design. Simply netting the fish out of the tank, or using a net to crowd the fish for harvest or grading is an obvious solution. More sophisticated crowding and grading can be achieved using crowder and grader frames or gates that move down the length of a raceway or pivot around the center of a circular culture tank (Piper et al. 1982, Brannon, 1991; Karlsen, 1993). Fish sufficiently small in size can swim through the grader bars while the larger fish are retained behind the gate. Use of crowder and grader gates is thought to be less stressful to the fish, because their use does not require handling the fish or moving them out of the water. And once crowded, fish can also be induced to swim through channels, pipes, or raceways to another location with relatively little stress. Crowded fish can also be moved rapidly to other areas using more aggressive yet relatively safe fish pumps, or brail nets and cages when they are used with care.

Hand grading devices such as box graders are common at many hatcheries and have proven to work well on small farms, where the cost of automatic grading and counting equipment cannot be justified. At larger farms, however, labor and fish stress can be reduced with the wise use of automated grading and inventory tracking equipment. Commonly used automatic graders include mechanically driven belt graders and roller graders. These mechanical graders usually require removing the fish from water for a brief period as they pass through the sizing mechanism (see Chapter 11). Although mechanical graders produce some stress and trauma, many of the established commercial mechanical grading machines are considered safe, reliable, and fast methods to size sort and count large numbers of fish. However, as with all new technologies, before purchasing equipment of this type, it is best to check with other fish farmers who have used the equipment. Other interesting and new technologies that use ultrasonic, infrared light, and video systems are now commercially available to estimate fish size distributions within culture tanks and cages. This type of inventory tracking equipment can sometimes be used to track fish growth, feed conversion (if feed input and fish numbers are known), and estimate the fraction of fish reaching harvest size.

5.8 Scale Issues

The number and size of culture tanks is an important factor to be considered during the design of the fish farm and its stock management plan. It is now becoming a more common practice for fish farms to use fewer but relatively larger culture tanks to meet culture volume requirements. For example, Karlsen (1993) described how more recent land-based salmon-smolt farms in Norway have reduced the number of tanks to between six to eight units, much fewer in comparison to earlier farms. No matter where the fish farm is located or what species the farm is producing, it often becomes apparent that fewer (maybe six to ten) but relatively larger culture tanks can provide culture volume much more cost effectively than using many (perhaps 30 to 100) correspondingly smaller tanks. Also, the costs of miscellaneous equipment and labor decrease when a given culture volume can be achieved with a few larger culture tanks rather than with many smaller tanks. Use of fewer but larger tanks reduces the purchase and maintenance of feeders, dissolved oxygen probes, level switches, flow meters/switches, flow control valves, and effluent stand-pipe structures (Timmons et al. 1998). Use of fewer but larger tanks also reduces the time required to analyze water quality, distribute feed, and perform cleaning chores, i.e., the times are about the same for a larger tank as for a smaller tank, (Timmons et al. 1998). Also, the time and logistics of fish management in a large number of tanks can become quite costly.

However, the advantages achieved through the use of larger tanks must be balanced against the risk of larger economic loss if a tank fails due to mechanical or biological reasons. There are also difficulties that could arise in larger culture tanks when removing mortalities, grading and harvesting fish, and controlling flow hydraulics, e.g., water velocities, tank mixing, dead-spaces, and settling zones. Therefore, large culture tanks must be designed properly to allow for fish management and control over flow hydraulics, as discussed later in this article.

5.9 MECHANISMS TO REMOVE DEAD FISH

Daily removal of dead fish from the bottom center drain is important. Dead fish in the fish culture system negatively influence profits, fish health, water quality, solids removal, and the water level in the tank. Commercial fish farmers need a simple and reliable way to remove daily mortalities with minimal labor. There are varieties of approaches to address this task. When applicable, an uncovered bottom center drain makes the task of removing dead fish easy. When dead fish sink, they are carried in the radial flow to the bottom center drain where they are sucked through to the external standpipe chamber.

Methods to remove dead fish and/or waste solids from the bottoms of deep culture environments have been developed for large floating cages (Braaten, 1991; Skjervold, 1993), which when enclosed in bags rather than nets (Solaas et al. 1993) look very similar to circular culture tanks. These mortality and solids removal methods may be transferable to land-based circular culture tanks.

A dead fish collector can be incorporated into the dual-drain particle trap mechanism, Figs. 5.17 and 5.18. The illustration does not depict how fish are removed through the particle trap, but dead fish are flushed through the larger drain pipe to the external standpipe structure where they are removed.

Another method for removing dead fish incorporates the center drain outlet screen into the inner pipe of a two-pipe center post system. The outer pipe consists of a steel post secured in the tank floor such that large openings cut into the pipe just above the tank floor level allow dead fish to pass through the outer pipe when the inner pipe is lifted. The sizes of the outer and inner concentric pipes have been selected to make a close but unhindered fit. To conveniently flush dead fish captured at the bottom center drain, the inner pipe is raised inside of the fixed center post while the external standpipe over the mortality drain is removed; this produces a surge of flow that carries the dead fish out of the tank. The effectiveness of this mortality removing mechanism is still under review.

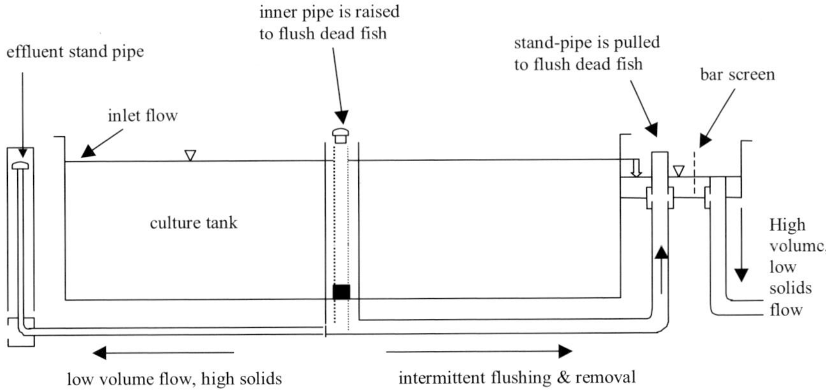

Figure 5.17 A Concentric Pipe System to Flush Solids and Remove Dead Fish From the Bottom Center Drain; Elevated Drain for Removing the High Volume, Low Solids Effluent.

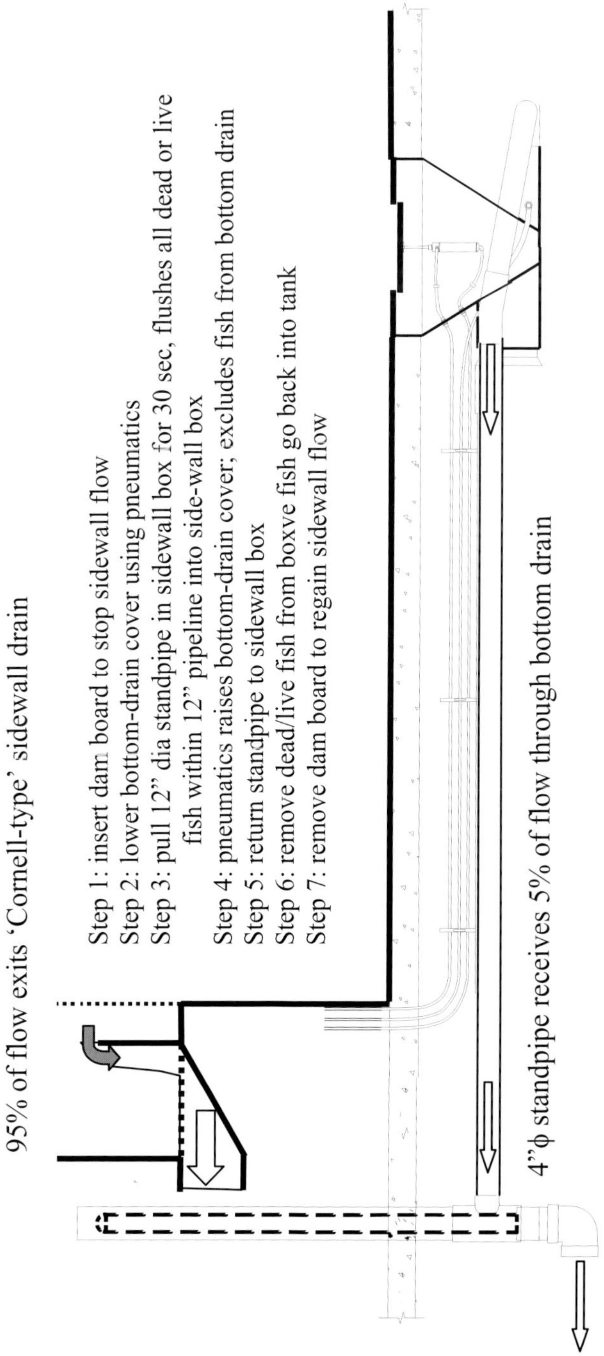

Figure 5.18 Dead Fish Remover

LIST OF SYMBOLS

BL	Fish body length, dimensionless
C	Concentration of solids in tank water, kg/m^3
$C_{density}$	1.5 for L in inches (0.24 for L in cm) for tilapia $C_{out,b}$ concentration of pellets leaving bottom center drain, kg/m^3
COC	Cumulative oxygen consumption, kg
$D_{density}$	Carrying capacity or biomass density, lb/gal (kg/m^3)
DO	Dissolved oxygen, mg/L
g	Acceleration due to gravity, m^2/s (ft^2/s)
H	Dynamic heads ($v^2/2g$), m (ft)
HRT	Hydraulic residence time, 1/h
k	1st order rate constant characterizing bead enrichment at the bottom-center drain
L	Fish length, inches (cm)
$L_{raceway}$	Length of the raceway in meters
MRT_{ideal}	Ideal mean residence time, 1/h
P:B	Ratio of yearly production to biomass carrying capacity
Q	Inlet flowrate, m^3/s
$Q_{out,b}$	Flowrate flushed through the bottom center drain, m^3/h
R	Water exchange rate in a raceway design, volumes/h
t	Time
V_{orif}	Velocity through the water inlet structure (orifices or slots), m/s
$V_{raceway}$	Raceway velocity, cm/sec
V_{rota}	Rotational velocity in the tank, m/s
V_{safe}	Fish body lengths per second, BL/s
V_{tank}	Water volume in the tank, m^3 or L
α	Proportionality constant to relate tank and orifice velocity
φ_{enrich}	Relative enrichment factor due to Cornell dual drain, %
ρ	Density of water, kg/m^3

CHAPTER 6

SOLIDS CAPTURE

6.0 OVERVIEW

Suspended solids adversely impact all aspects of a recirculating aquaculture system (RAS), so the first objective of any recirculating treatment scheme is the removal of solid wastes. Suspended solids are the result of feces, biofloc (dead and living bacteria), and uneaten food. These suspended particles will vary greatly in size from the cm size to the micron (μm) size. Aquacultural solids are characterized by size into classes, as shown in Fig. 6.1 (from Vinci et al. 2001). The term "fine solids" is used herein to identify the solid particulates that do not readily settle from the water column. As a prelude to understanding the descriptions of size distribution and contribution by weight of the various size classes, it is important to know that samples are usually pre-filtered to remove very large particles, i.e., >200 μm. The contribution of these large solids to TSS measurements are added to the solids concentration obtained from the filtered water when TSS is reported for a system's performance characterization. In practical applications, of course, these larger particles should always be removed first and must be a primary focus, since if they are not removed, they become "smaller" more difficult particles to remove.

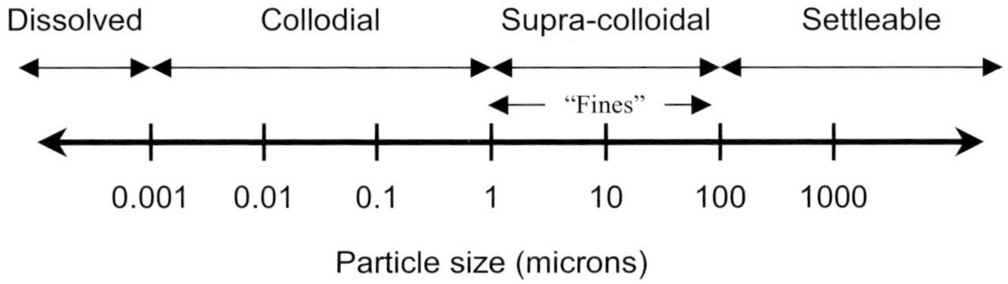

Figure 6.1 Solid Size Characterizations in Aquaculture Waters.

In RAS, the majority of particles by weight will be less than 100 μm in size; and in intensive RAS systems the majority of particles by weight will be 30 μm or less in size. In such cases, mechanical filtration will be ineffective. Chen et al. (1993) showed that 80% to 90% of the total weight of solids in a tank (prefiltered to remove particles >130 μm; see Fig. 6.2) was comprised of individual solid particles that were of 35 μm or less in diameter. All sizes of particulates must be addressed and managed by an appropriately selected treatment method for particles within each size range, e.g., sedimentation and screening for removal of large particles and foam fractionation or ozone treatment for fine solids removal. Granular media filters can control the widest range of solids. This type of filter is effective in removing solids down to about 20 μm, and is the favored choice of many designers for systems that have high reuse or high clarity demands. This and other solids removal techniques are reviewed in this chapter.

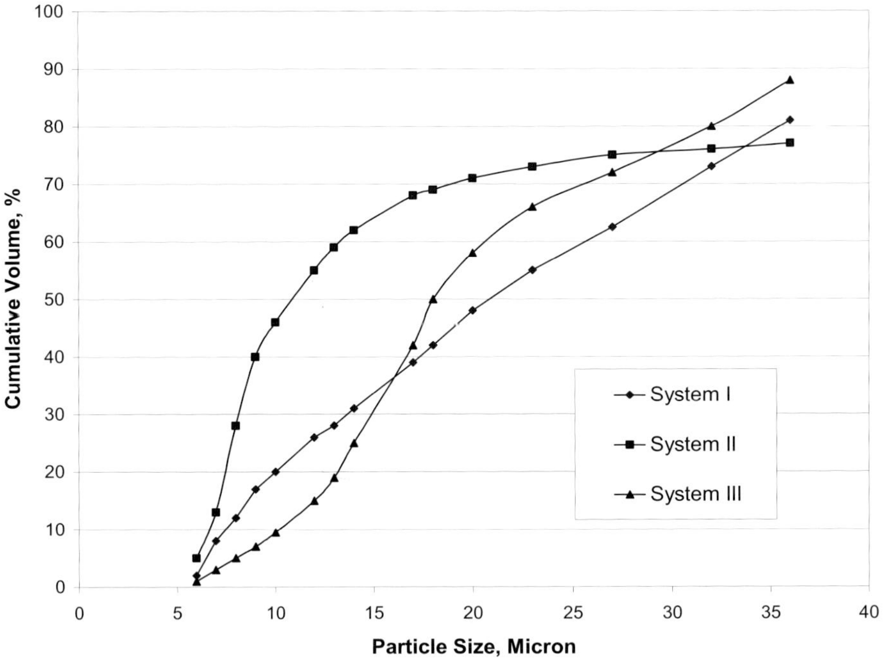

Figure 6.2 Cumulative Distribution of Particle Volume in Three Recirculating High Density Aquaculture Systems (Chen et al. 1993). (Systems I and III were rainbow trout and System II was brook trout. Feeding rates were approximately 0.75% BW per day. Water exchange was 5% per day. System III used ozone).

Total suspended solids (TSS[1]) concentration is defined as the mass of particles above 1 μm in diameter (APHA, 1989) occurring in a known volume of water. Suspended solids have both inorganic and organic components. The organic portion, known as volatile suspended solids (VSS), contributes to oxygen consumption and biofouling problems. The inorganic components contribute to formation of sludge deposits. Physically, suspended solids can be partitioned further into settleable solids, typically greater than 100 μm, and non-settleable suspended solids, which are less than 100 μm (EPA, 1975). The finer, non-settleable suspended solids are more difficult to control and cause most of the problems in recirculating systems.

Fine suspended solids are extremely detrimental to general fish health (Chapman et al. 1987; Magor, 1988; Alabaster and Lloyd, 1982). However, the experts in the field have not yet agreed on a set of definitive design values for acceptable TSS concentrations, which would then serve as a system design goal for the TSS removal efficiency of RAS designs. For example, according to Alabaster and Lloyd (1982), for inland fisheries, there is no evidence that concentrations of suspended solids up to 25 mg/L have any harmful effect on fish. The FIFAC (1980) suggests that TSS concentrations be maintained below 15 mg/L as a safe value in recirculating systems, while Muir (1982) recommends a limit of 20 to 40 mg/L for these same systems. The authors have grown tilapia in systems that had TSS in excess of 100 mg/L and still maintained good fish productivity, but this was in an absence of virtually all other stressors. Keep in mind that different fish species may have significantly different tolerance levels to solids concentrations and that other water quality parameters may impair a fish's ability to withstand high TSS concentrations.

Finally, during this initial introduction section, we cannot overemphasize the importance of rapid and complete solids removal from the culture vessel. All other unit processes will fail if this primary function is poorly performed. The case history failures described in Chapter 15 were primarily attributed to lack of effective solids removal from the culture vessels.

---

[1] All acronyms, abbreviations, and variable symbols are defined at the end of the chapter.

6.1 Solids Balance

Whenever possible, water flows should be managed to concentrate solids in a small portion of the total flow for the system. An effective way to do this is by using the Cornell Dual-Drain. In this system, 10% to 20% of the total flow exits the tank from a center bottom drain while the majority of the flow exits from the tank sidewall. (This concept is discussed in much more detail in Chapter 5: Culture Units). Use of the dual-drain approach greatly increases the concentration of solids being removed from the low flow bottom center drain. The concentration of solids in this low flow is typically 10 times or more higher than the concentration of solids that exit through the main flow drain, whether it is located in the tank sidewall or as an upper center drain.

As an example of the effectiveness of the dual-drain system, the author has measured the results achieved in an intensive tilapia system (daily feeding rates were approximately 80 kg/day; tank volume was 53 m^3). In this system, the concentration of suspended solids in the water exiting the side-wall drain was the same as the concentration of solids in the overall tank. Solids in both the incumbent tank water and the side-wall exit water were 6.4 mg/L (standard deviation of 3.6). This means that all solids removed from the tank were in fact removed by the center bottom drain flow, which exited the tank at a rate of 110 L/min. This center drain discharge flow of 110 L/min is in comparison to the total water flow of the system to the biofilters and water conditioning units of 3.6–5.5 m^3/min (Timmons, unpublished data). The net settleable solids that were removed through the center bottom drain were then subsequently captured by a mechanical screen filter and/or a settling tank (emptied daily; approximately 3 m^3). The authors have designed several systems that successfully use the Cornell dual-drain approach and the design has been widely adapted in the salmon smolt industry.

"Rule of Thumb"

Cornell Dual-Drain increases center drain TSS by 10-fold.

Research at the Freshwater Institute (Shepherdstown, WV) on solids removal within a "Cornell-type" dual-drain tank has shown that the mean total suspended solids (TSS) concentrations discharged through the elevated side-wall drain averaged only 1–2 mg/L, while the bottom drain discharge contained an average of 25 mg/L TSS (S. T. Summerfelt,

Freshwater Institute, unpublished data). The tank contained rainbow trout at a density of 60 kg/m^3 and were fed 1% body weight per day. Others (Lunde et al. 1997) have reported concentrating 91% of the fecal matter and 98% of uneaten feed within the bottom flow leaving dual-drain culture tanks.

The effectiveness of the dual-drain tank at concentrating solids within the bottom center drain discharge can be illustrated by a steady state solids balance written over the culture tank,

$$\begin{Bmatrix} TSS\ carried \\ in\ via\ influent \end{Bmatrix} + \begin{Bmatrix} TSS\ produced \\ from\ feeding \end{Bmatrix} = \begin{Bmatrix} TSS\ leaving \\ sidewall\ outlet \end{Bmatrix} + \begin{Bmatrix} TSS\ leaving \\ centers\ drain \end{Bmatrix} \tag{6.1}$$

or more explicitly,

$$\{Q \cdot TSS_{in}\} + \{P_{TSS}\} = \{Q_{out\,1} \cdot TSS_{out\,1}\} + \{Q_{out\,2} \cdot TSS_{out\,2}\} \tag{6.2}$$

P_{TSS} is determined by using the following equation:

$$P_{TSS} = a_{TSS} \cdot r_{feed} \cdot \rho_{fish} \cdot V_{tank} \tag{6.3}$$

The fraction of solids removed (f_{rem}) through the tank center drain can be determined from the following equation:

$$f_{rem} = \frac{Q_{out\,2} \cdot TSS_{out\,2}}{(Q_{out\,1} \cdot TSS_{out\,1}) + (Q_{out\,2} \cdot TSS_{out\,2})} \tag{6.4}$$

Equation 6.3 can be rearranged, substituted into Eq. 6.2 and then rearranged again to allow for TSS_{out2} to be calculated from f_{rem}, Q, Q_{out2}, P_{TSS}, and TSS_{in}:

$$TSS_{out\,2} = (\ Q \cdot TSS_{in} + P_{TSS}\ )\frac{f_{rem}}{Q_{out\,2}} \tag{6.5}$$

6.2 Basic Design Parameters for Round Tanks

Most of this design description material is presented in Chapter 5, but is repeated here for convenience. Round tanks will operate as self-cleaning vessels, if the diameter-to-depth ratios are maintained within a recommended range. It is for this reason that we highly recommend round tank culture vessels. Round tank vessels should be designed using the following criteria:

- use a tank diameter-to-depth ratio between 3 and 10 and preferably between 3 and 6. For example, if you are using a tank that is 4 feet deep, then the acceptable range of diameters is from 12 feet to 40 feet (the 3 to 10 range for diameter to depth ratio) or a 1 m deep tank could be from 3 to 10 m in diameter.

- employ the Cornell dual-drain design with the center drain sized to accommodate 5% to 20% of the total flow used to operate the tank; remove the remaining percentage of flow, i.e., 80% to 95%, from the upper half of the outside tank wall (see Chapter 5).

- maintain tank water velocities of at least 15 to 30 cm/s to promote the movement of solid wastes towards the center drain.

Once the larger solids have settled and been "flushed" from the culture tank, the next step is to remove the suspended solids from the water column before returning the water to the culture tank vessel. The various methods for removal of suspended solids are discussed in the following paragraphs.

6.3 Solids Generation

Virtually all the wastes generated within a recirculating system originate from the feed. This manifests in two ways: a) uneaten feed, and b) fish excrement in the form of solids, liquid, or gas. Of the feed that is eaten, 80% to 90% will eventually be excreted in some form (Hopkins and Manci, 1989). As a rule of thumb, use 25% of the quantity fed to the fish as the volume that will be produced as suspended solids (or total suspended solids, TSS) on a dry matter basis. TSS produced by fish is primarily in the form of feces. The mass production rate of feces is proportional to the feeding rate and is discussed in more detail in Chapter 4: Loading Rate and several literature references (Liao and

Mayo, 1974; Wimberly, 1990; Malone et al. 1990; Westers, 1989; Iwana, 1991; Page and Andrews, 1974).

> **"RULE OF THUMB"**
>
> TSS = 25% of Feed Fed
> (dry matter basis).

6.4 TSS PHYSICAL CHARACTERISTICS

From the perspective of solids control, the two most important physical characteristics of suspended solids in a recirculating system are:

- particle specific gravity
- particle size distribution

Specific gravity is determined by the source of the particles, while the size distribution is determined by a combination of factors, including the solids removal process, the source of particles, fish size, water temperature, and turbulence in the system.

The behavior of suspended particles in water is determined by the specific gravity of the particles. Specific gravity is defined as the ratio of the density of a wet particle to that of water (APHA, 1989). Based on the measurements of solids taken from two recirculating systems, Chen et al. (1993B) reported that the average specific gravity of the particulate matter was 1.19, somewhat greater than that of the water. Based on the author's findings, however, Chen's value is probably greater than should be used in filter design calculations, as the authors have found that a specific gravity of 1.05 is more representative of suspended particles originating as tilapia feces. This is perhaps due to the fecal casing that enshrouds tilapia feces, making it less dense. The point is that fish feces is not much "heavier" than water and therefore does not settle as rapidly as would aggregate material of the same size.

While feces is the source of most of the suspended solids, uneaten feed is also a significant source of TSS in fish culture water. TSS from feed typically has a different particle size distribution than the TSS originating as feces. Uneaten feed subsequently breaks down slightly in the water column, but even after several hours and repeated passage through pumps, over 97% of the feed particles will be greater than 60 μm in size and 73% will be larger than 500 μm (0.5 mm). The particles of suspended solids originating from these two sources (feces and uneaten feed) are notably different in size and specific gravity and therefore respond to control mechanisms in different ways. In RAS waters, fine particles (particles less than 30 μm) are the most prevalent and dominate the water column. In water reuse systems (the focus of this book), fine particles (particles less than 30 μm) will dominate the water column. Figure 6.2 shows that 90% of the total particle mass in RAS systems is from individual particles of less than 30 μm (Chen et al. 1993B). This is why it is impractical and almost useless to use 60 μm drum filter screens. Of course, large particles eventually become small particles, so rapid and efficient removal of the settleable (large >100 μm) particles is important, as it will minimize generation of the fine, more difficult to remove particles in the system. There have been several studies published on particle size distribution (Harman, 1978; Easter, 1992; Chen et al. 1993B; Chen et al. 1993C; Cripps, 1992).

Sedimentation techniques will not remove the fine particles from the water. This is because fine particles (<30 μm) have low settling velocities that make gravitational removal methods impractical. For example, fine particles need a retention time of several hours to settle a distance of 0.5 m. Sedimentation tanks are often regarded as being inefficient, but this opinion is usually caused by the settlement characteristics of fine particles, which require a lengthy retention time to settle (Chesness et al. 1975; Chiang and Lee, 1986), or because the sedimentation tank(s) were simply poorly designed to begin with.

6.5 REMOVAL MECHANISMS

There are three methods that are used to remove suspended solids from fish culture waters. These are:

- gravity separation
- filtration
- flotation

These classifications of methodology are based on the removal mechanisms used to effect the removal (flotation is sometimes considered as another kind of gravity separation, but it is a different principle of application so it is described separately). Large particles (larger than 100 μm) can be effectively removed by settling basins or mechanical screen filtration. However, fine particles cannot be removed effectively by either gravity separation or granular filtration methods.

Granular filters are effective only in the removal of particles larger than 20 μm (Task Committee, 1986).

Gravity Separation. Gravity separation works on the principle of sedimentation and settling velocities. Unit processes in this category include clarifiers (settling tanks), tube settlers, and hydrocyclones. The actual process of particle removal can be accomplished with either screen, granular media (GM), or porous media (PM) filters.

Filtration Removal. Particle removal from the water can be accomplished by one or more filtration processes. These are sedimentation, straining, Brownian diffusion, and interception. These processes are implemented in filtration systems by screen, granular media (GM), or porous media (PM) filters. In RAS applications, diffusion and interception have limited practical importance and will not be discussed (except as applied to foam fractionation, which is discussed later in this Chapter).

Flotation Process. In a flotation process, particles attach onto air bubbles and are separated from water. The flotation process involves all the transport mechanisms that occur in a filtration process with the exception of straining.

The filtration techniques of sedimentation and straining are the most practical and commonly used methods, and are emphasized in this Chapter. More detail on these methods is provided in the following sections.

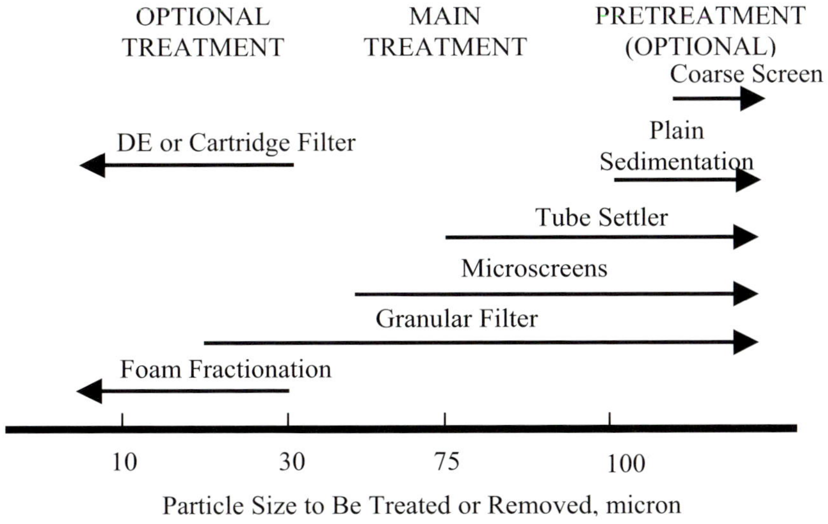

Figure 6.3 Solids Removal Processes and the Particle Size Range in micron Over which the Processes Are Most Effective.

Table 6.1 provides a summary of suspended solids removal techniques, their efficiency of removal and allowable hydraulic loading levels (Timmons and Losordo, 1994). Figure 6.3 categorizes technique appropriateness as related to TSS size characteristics.

Table 6.1 Characterizations of Solids Removal Techniques for RAS

Technique	Solids Size Removed (micron)	Head Loss (m)	Hydraulic Loading m^3/d per m^2	TSS Removed (%)	Reference
Sedimentation	>100		24–94	40–60	EPA 1975
Settling tank			24–61		Liao 1980
Tube Settler			30–90		Muir 1978
Granular Media	>20	0.1–3	175–430	20–60	Muir, 1982; EPA 1975
Rapid sand filter			94–351	67–91	EPA 1975
			285	70–90	Mayo 1976
Pressure sand filters		2–20	115–700	50–95	Muir 1982; EPA 1975
Floating bead filter		0.8–6	1935		Wimberly 1990
Screen	>75	negligible	100–2,200		
Porous Media	>0.1		40–130		Muir 1982
Dissolved Earth	>0.1		29–59	>90	Muir 1982; EPA 1975
Cartridge	1–10	~ 5	1–10 gpm		
Hydrocyclones	1–75	14–35			Huguenin & Colt 1989; Wheaton 1977
Foam Fractionation	<30		290–280		Chen 1991
Ozonation	<30		see Chpt 12		

When selecting a control process for TSS in a RAS, the following criteria should apply:

- hydraulic loading rate
- fine solids removal capability
- head loss
- water loss during filter backwash, and
- resistance to biofouling.

SEDIMENTATION

Sedimentation occurs due to the density difference between the solid particles and water. Assuming a particle to be heavier than water, under the force of gravity it will fall through the water with increasing speed until it reaches a terminal value for its settling velocity. Each discrete particle has an equilibrium settling velocity. Assuming spherical particles, the settling velocity can be calculated by the following equation (Montgomery, 1985):

$$= \sqrt{\frac{4g(\rho_p - \rho)D_p}{3C_D\rho}} \tag{6.6}$$

For a small particle having a low Reynolds number (less than one, see Chapter 3), Stoke's Law applies, and Eq. 6.6 can be rewritten as

$$V_s = \frac{g(\rho_p - \rho)D_p^2}{18\mu} \tag{6.7}$$

Both Eqs. 6.6 and 6.7 indicate that denser and larger particles will settle out of water faster than smaller, less dense particles. This is true for all types of removal processes and why you should do everything possible to maintain large particle sizes. The best technique for maintaining large particle sizes is to remove the particles as quickly as possible from the fish culture vessel and before any pumping has occurred. Also, you should try to minimize any turbulence/falling water situations prior to the primary TSS capture event.

REPORTED SETTLING VELOCITIES

Settling velocities for fish feces have been reported at 1.7–4.3 cm/s (Warren-Hansen, 1982). In tests we conducted, the authors measured settling velocities by using cylindrical beads (each 3 mm in length by 3 mm o.d.) with a specific gravity of 1.05, to simulate tilapia feces. The beads exhibited settling velocities of 3.8 cm/s. These results are similar to the settling velocity reported for fecal matter by Warren-Hansen. Certain species produce fecal matter with even lower specific gravities (Robertson, 1992), trout fecal matter is typically 1.005 specific gravity. This is reflected in slower settling velocities of these wastes. Wong and Piedrahita (2000) reported that the median settling velocity on a mass-basis for the settleable solids from rainbow trout is 1.7 cm/s.

These reported velocities are consistent with the results obtained using Eq. 6.6 and as shown in Fig. 6.4. This similarity in settling velocity also suggests that the specific weight of feces would be similar. We measured the settling velocity of feed pellets, and found them to have a much more rapid settling velocity, e.g., 14 cm/s, than the plastic beads. This finding is similar to that reported by Juell (1991) of 15–33 cm/s. However, finer and/or less dense particles can be produced in RAS, which may settle at only 0.01 cm/s (IDEQ, 1998). These particles would not effectively settle in a settling basin or concentrate at the bottom center of dual-drain tanks, thus staying in the water column until removed by some other process.

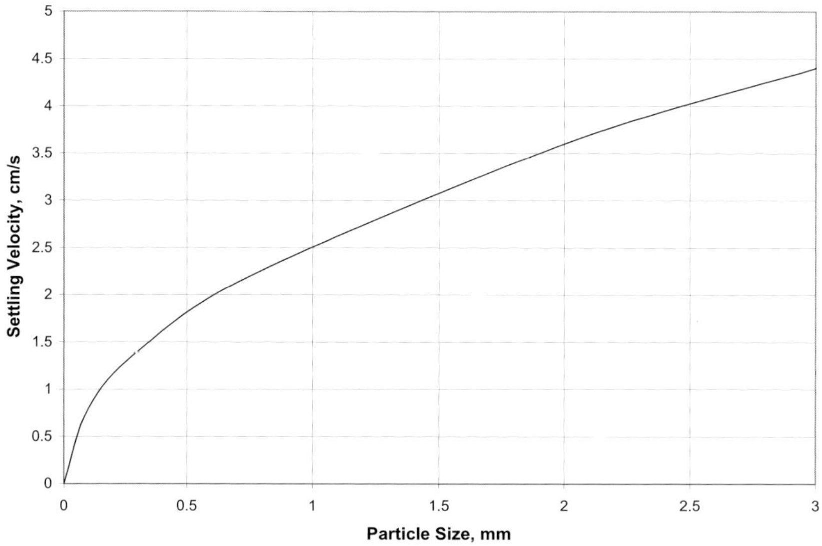

Figure 6.4 Predicted Settling Velocities Using Eq. 6.6 (Specific Gravity of Particle of 1.05).

SETTLING BASINS

Settling basins are very effective if properly configured and operated. Sedimentation, i.e., gravity separation, is one of the simplest of technologies available to control particulate solids in process water and wastewater. Sedimentation basins require little energy input, are relatively inexpensive to install and operate, require no specialized operational skills, and can be easily incorporated into both new and existing facilities.

The disadvantages of sedimentation are low hydraulic loading rates and poor removal efficiency of small suspended solids (<100 μm). Also, they require additional floor space for their incorporation in comparison to microscreen filters. Innovative uses of vertical space over the settling bed or placing the settling bed in less expensive space can reduce this cost considerably.

Another potentially serious disadvantage is that settled manure remains in the system until the settling basin is cleaned. This condition is one of the major concerns in their use. Dissolution of nutrients and the resuspension of solids that have settled and collected on the bottom of settling basins can markedly reduce the expected performance of these clarifiers (Cripps and Kelly, 1996). Henderson and Bromage (1988) estimated that settling ponds could capture an estimated 97% of their solids loading if resuspension of settled solids was not a factor. They suggest that settling basins are not effective in removing TSS when inlet concentrations are <10 mg/L or attaining effluent concentrations of <6 mg/L. Eliminating resuspension of TSS is difficult at best in most settling basins. Thus, settling basins will generally require further TSS treatment to meet the stringent removal criteria necessary to achieve mandated levels of TSS.

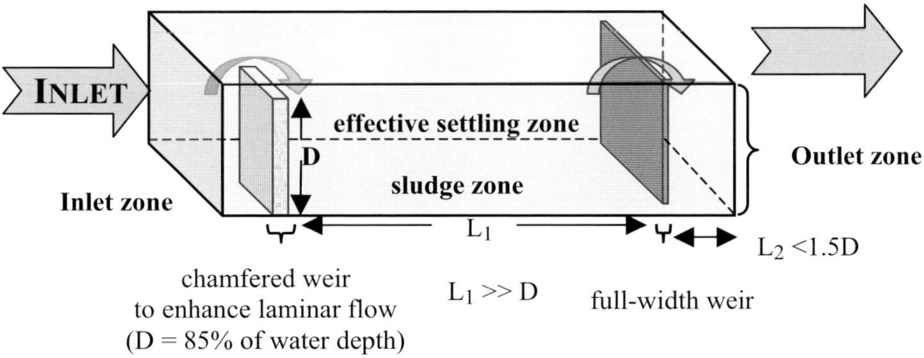

Figure 6.5 Four Principal Zones of a Rectangular Continuous Flow Sedimentation Basin.

All continuous flow settling basins are conceptually divided into four zones according to function, Fig. 6.5. The inlet zone serves to uniformly distribute the suspension over the entire cross-section of the basin.

Sedimentation occurs in the settling zone and, upon removal from the water column, the solids accumulate in the sludge zone. The clarified liquid is generally collected over the entire cross-section of the basin at the outlet zone and is discharged. Under ideal conditions (no mixing or turbulence), required retention time is the time required for a particle that starts at the top of the inlet zone and settles to the floor of the basin at or before the junction of the outlet zone. The key parameter for the design of settling basins is the volumetric flow of water per unit surface area of the basin or overflow rate (V_o):

$$V_o = Q/A_{sz} \tag{6.8}$$

"Rule of Thumb"

If you see any turbulence or currents in the settling basin, solids removal effectiveness will be very poor.

Any particle with a settling velocity (V_s) greater than the overflow rate (V_o) will settle out of suspension. Other particles, for which $V_s < V_o$, will be removed in the ratio V_s/V_o, depending upon their vertical position in the tank at the inlet (Hazen, 1904).

If one has access to TSS from a currently operating system similar to the one you intend to implement, then you can obtain a sample and test it to determine actual V_s for that particular waste; this is the ideal approach and has been described (Al-Layla et al. 1980). In doing these tests, use a settling column having an inside diameter greater than 13 cm. Results obtained from using narrower columns are not reliable, due to boundary layer resistance introduced by column walls. Use a test column that is the same depth as full-scale settling basin in the desired application. Without access to similar TSS, you can predict V_s using Eq. 6.6 or Fig. 6.4.

Settling basin designs were originally developed for municipal waste water treatment. Aquaculture waste is different, which is the reason for the erratic performance of settling basins when measured against the traditional approach of hydraulic retention time (τ) in the settling basin:

$$\tau = \frac{V_{basin}}{Q} \tag{6.9}$$

Hydraulic retention times of 15 to 30 minutes have been recommended for basin design (Liao and Mayo, 1974; Mudrak and Stark, 1981). Basin geometry is often used to maintain good settling conditions with recommended basin length to width ratios reported from 4:1 to 8:1 (Arceivala, 1983). It is also recommended that settling basins be designed so as to contain a minimum of 1 m (3.28 ft) of water depth (Liao and Mayo, 1974).

The overflow approach developed by Stechey and Trudell (1990) is an improvement over the hydraulic retention time approach. The overflow method provides much more effective TSS removal on a consistent basis. Stechey and Trudell (1990) recommend an overflow rate (V_o) for the design of settling basins in intensive salmonid aquaculture to be between 40–80 m^3/m^2 per day (982–1964 gpd/ft^2). These overflow rates translate to particle settling rates (V_s) equaling 0.046–0.092 cm/s.

Translating this into easy to understand language, for every gpm of water flowing through the settling basin, 0.73 to 1.47 square feet of surface area are required for settling; or, 1.0 gpm flow per square foot of settling zone area (40.7 Lpm/m^2). Mudrak (1981) reported on the performance of several settling basins used in intensive trout culture operations. He found that when the design overflow rate was at approximately 60 m^3/m^2 per day, the removal of settleable solids was 90% or greater, typically above 95%, although TSS removal was about 10% less. Also, there was no notable improvement in removal efficiencies when the loading rate was further reduced by as much as a factor of three, i.e., the conclusion is that a significant portion of TSS fine solids will not be removed by the settling basin. One design fundamental that must be kept in mind is that if you can see water currents in your settling basin, it will not efficiently remove the TSS except for the larger particles, e.g., >500 μm. Even if you double the settling basin floor area, this will not compensate effectively for a poorly designed settling basin where turbulence and mixing are present that are caused by ineffective inlet and outlet weir design.

The loading rates suggested above are applicable to off-line settling basins (the most widely used application in RAS). The Idaho Waste Management Guidelines for Aquaculture Operations (1998) suggests overflow rates for three typical settling basins used in aquaculture, i.e., full-flow, quiescent zone, and off-line (see Table 6.2).

Table 6.2 Surface Loading Rates for Settling Basin Design (Idaho Waste Management Guidelines for Aquaculture Operations, 1998)

	Surface Loading Rate	
	m^3/m^2 per hr (ft^3/ft^2 per s)	gpm per ft^2
Full-flow settling basin	14.3 (0.013)	5.9
Quiescent zone	34.0 (0.031)	13.9
Off-line settling basin	1.66 (0.00151)	0.7

The Freshwater Institute (Shepherdstown, WV) has successfully applied this approach to concentrating the effluent coming off of drum filters in thickening tanks. They used three off-line settling basins to capture and store solids from the intermittent backwash of three drum filters (see Fig. 6.6). The solids-laden backwash flow is introduced intermittently into the top and center of each tank. At the top of each tank, the flow is introduced within a cylinder with an open bottom that is centered within the tank. The cylinder improves the hydraulics of the tank's radial flow by directing the water to first flow down (underneath the cylinder and towards the cone of the tank) and then up as it travels radially towards the effluent collection launder about the top circumference of the tank. These thickening tanks have performed well, capturing 97% of the solids discharged from the microscreen filter backwash flows, Table 6.3.

However, the solids stored within the thickening tanks are degrading, as indicated by the >10-fold increase in total ammonia-nitrogen across the thickening tanks. The thickened manure, pulled from the cone's bottom on a monthly basis, contains about 10% total suspended solids and about 1000 mg/L of total ammonia nitrogen.

Inlet and Outlets for Settling Basins

The transition of water to and from each of the zones must be accomplished so as to keep turbulence and mixing to an absolute minimum, as the rate of solids settlement is affected by the motion of the water.

The settling basin design most frequently used today in aquaculture places a single pipe at each end of a basin, thereby placing all three zones in the same vessel. This is the simplest but unfortunately also the worst design for inlet/outlet structures and settling basins. This type of point-source inflow and outflow produces extremely poor flow, which does not permit effective sedimentation. Short-circuiting is intensified as the influent stream flows directly through the basin at relatively high velocity toward the discharge pipe. Consequently, a large portion of the fluid passes through the basin much quicker than the theoretical detention period or hydraulic retention time (τ) calculated based on the dimensions of the basin. To be effective, the design of a settling basin should address each of the zones as separate, identifiable functions of the basin. The design of the inlet and outlet structures is critical to the performance of the water being handled through the sedimentation process, and must address the specific function and constraints of each step within the process.

Figure 6.6 Off-Line Settling Basins Used for Solids Thickening and Storage at the Freshwater Institute (Photo by B. Vinci, Shepherdstown, WV).

Table 6.3 Change in TSS, TAN, NO_2-N, and NO_3-N Across Manure Thickening Tanks (Vinci et al. 2001)

	TSS mg/L	TAN mg/L	NO_2-N mg/L	NO_3-N mg/L
From the Drum Filter (Inflow to Thickening Tank)	5,147 ± 1,411	1.7 ± 0.1	0.19 ± 0.03	1.5 ± 0.1
Discharge from Thickening Tank	151 ± 24	19.5 ± 2.5	0.04 ± 0.01	0.7 ± 0.2

When designing inlet structures, the following factors must be considered (see Giles, 1943 for review):

- The influent stream should be introduced evenly across the entire cross-section of the settling zone.
- All flow through the settling zone should begin in even, horizontal path

- The influent velocity to the settling zone should be slow enough to prevent excessive turbulence and mixing.

Inlets. Inlets should consist of a submerged inlet weir that separates the settling zone from the inlet zone. The inlet weir should extend across the full width of the settling basin, and should be submerged approximately 15% of the basin depth. The weir crest should be about 20 to 30 cm wide (8 to 12 inches) and have rounded edges to smooth the flow as it enters the settling zone. For circular clarifiers, the inlet is generally at the center of the basin. A baffle surrounding the inlet pipe serves to reduce turbulence and distribute the flow in a radial pattern through the full depth of the basin.

Outlets. Rectangular settling systems are more efficient than are circular settling tanks, but they require considerably more floor space. The sub functions of rectangular systems are more easily recognized as functional zones. These are the inlet zone, the settling zone, and the outlet zone. The outlet weir divides the settling zone from the outlet zone, as it skims clear water from the surface of the settling zone. The outlet weir should be designed and constructed so that it distributes the water exiting the settling zone at a uniform depth and velocity across its width. This is necessary to avoid generating currents and the accompanying turbulence in the settling zone. The outlet zone area should be the same width as the settling zone, and the length not less than 1.5 times the depth of the settling zone. For example, if the settling zone is 8 feet (2.4 m) wide, 30 feet (9 m) long, and 4 feet (1.2 m) deep, the outlet zone should be 8 feet (2.4 m) wide, 4 feet (1.2 m) deep, and at least 6 feet (1.8 m) long.

If a circular system is necessary due to space or other constraints, the design of the system must recognize and accommodate the same functions as are more easily accomplished in rectangular systems. There must be an inlet zone, a settlement zone, and an outlet zone. However, the boundaries between these zones are not clearly identifiable as they are in rectangular systems. These zones are created within the circular tank by the placement of inlets, the hydraulic movement of the water within the tank, the physical properties of the solids within the water, and the placement of the drain and outlet pipes.

In these cases, the flow towards the outlet is managed to create minimal turbulence particularly near the outlet pipe in the center of the tank. A circular weir can be installed that surround the outlet pipe, mostly to prevent floating solids from escaping the settling tank.

It is critical that the weir edge be level to assure a uniform discharge rate across the entire weir length. The weir discharge rate (volume of water discharged per unit length of weir per unit time) governs the length of the outlet weir. For weirs that are long in relation to the flow, i.e., having a low weir rate, a saw-toothed or V-notch edge is necessary for uniform discharge along the weir length. Weir discharge rates should be 400 to 600 m^3/d per meter length of the outlet weir (22 to 33 gpm/ft). More information on outlet weirs is provided in the literature (ASCE, 1959; EPA, 1975; Mudrak, 1981). Weir length should be maximized to the extent possible.

"RULE OF THUMB"

Settling Basin Design

- basin floor area of 1 square foot per gpm of flow (41 Lpm/m^2)
- 20 to 33 gpm per foot width of weir for outflow (250 to 410 Lpm per m)
- submerge inlet weir 15% of basin water depth
- use 10 inch (25 cm) wide weirs and use rounded edges
- maximize length of settling chamber as much as possible

The total length of the settling basin is comprised of the actual settling zone plus the length (area) required for both the inlet and outlet zones. This total area requirement is often ignored and thus the settling basin will not perform as intended. In addition, remember that uncontrolled turbulence, i.e., mixing and stirring of the inlet waters with the incumbent waters, will decrease the effectiveness of the settling process. The solution for this problem is to lengthen the sedimentation basin. Do not compromise on the size of the settling basin.

Design Example

Design a settling basin to remove 90% of the TSS particles greater than 0.5 mm in diameter with a design flow of 0.38 m^3/hr (100 gpm).

Solution.

Using 0.05cm/s as a design settling velocity (based upon Stechey's findings):

$$V_o == 0.05\ cm/s$$

Converting to flow per unit area

$$0.05\frac{cm}{s}\cdot 60\frac{s}{min}\cdot\frac{1\ m}{100\ cm}\cdot\frac{m^2}{m^2}=0.030\frac{m^3}{m^2\ min}$$

$$(or\ 0.73\ gpm\ per\ ft^2)$$

Now determine the settling zone area:

$$A_{SZ}=\frac{Q}{V_o}=\frac{\left(100\frac{gal}{min}\right)}{\left(0.73\frac{gpm}{ft^2}\right)}=137\,ft^2\,(12.7m^2)$$

Similarly, if you wanted to capture 90% of 60 µm TSS particles, then you would have a V_s of 0.6 cm/s requiring a much larger settling basin (0.037 m^3/m^2 per min = 0.9 gpm per ft^2). This is equivalent to our recommendation of 2.4 m^3/m^2 per hr (1 gpm per ft^2) of settling basin floor area.

In summary, settling basins can be used very effectively and economically if properly designed, configured, and operated. The key factors to having a successful settling basin are given in the rule of thumb box.

Settling Deck in Settling Chambers

A major objection to the use of settling basins is that they require a large floor area, and square footage of floor area can be expensive. If this is a problem, the "footprint" of the settling basin can be decreased by adding obstructions inside of the settling basin to increase rates of settling. Tube settlers, also known as "settling decks", can be used to do this. The basic function of the settling deck (tube settlers) is shown in Fig. 6.7 where the incoming flow is brought into the settling basin under the settling deck and forced to upflow to exit the chamber. In the process, solids settle within the tubes.

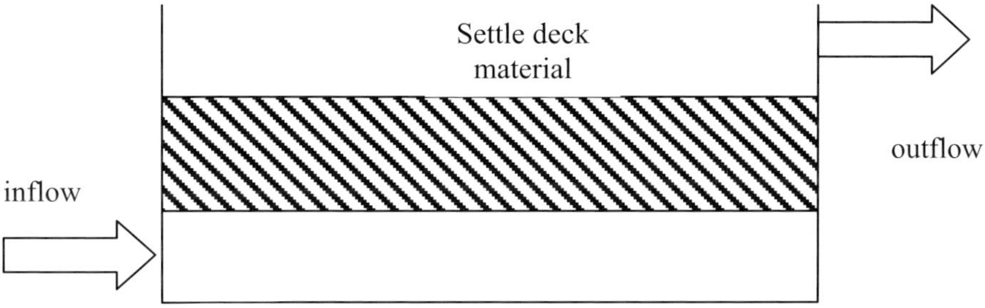

Figure 6.7 Settling Chamber with Settle Deck to Enhance Solids Removal Efficiency.

Separation distance between the inclined surfaces is typically 5 cm (2 inches) with a total inclined length of 0.9 to 1.8 m (3 to 6 ft) (Gregory and Zabel, 1990). Tube or plate plastic media are usually manufactured in structured bundles of tubes or stacks of parallel plates in a variety of opening shapes (square, rectangular, tubes, hexagonal, chevron). In operation, influent water flows into a tube or plate settler and then upward through the inclined tubes or plates as solids settle on the plastic surfaces. Generally the inclination of the tubes or plates is between 45°–60° above horizontal. This angle provides for the greatest degree of gravity self-cleaning of settled solids out of the media and into the basin bottom. A fairly broad range of hydraulic loading rates have been suggested, e.g., 1.5 m^3/m^2 per hr (McLaughlin, 1981), 7.4 m^3/m^2 per hr (Libey, 1993), and 67 m^3/m^2 per hr (Parker, 1981). The disadvantage to the use of tube settlers is that tube or plate settlers do not adequately self-clean, so they must be periodically cleaned by other means to prevent biofouling (Gregory and Zabel, 1990; Tchobanoglous and Burton, 1991).

The tube diameter determines the effectiveness of the capture and the allowable hydraulic loading rates (m^3/s per m^2 or gpm/ft^2). The physics of this technique are that a particle will settle and rest on the bottom of an individual tube prior to the transporting fluid emerging from the top of the tube. Thus, tube diameter and face hydraulic loading rate on the settling deck will determine the size of the settling deck area to treat the solids laden influent. The water velocity through the tube should be maintained in a laminar flow region to prevent turbulence and resuspending settled particles, i.e., Re <2,300 (see Chapter 3).

The effectiveness of tube settlers in removing fine solids can be impressive. Easter (1992) reported 80% efficiency in the removal of particles larger than 70 μm and 55% in the removal of all particles larger than 1.5 μm using a multi-tube clarifier. As mentioned above, the biggest disadvantage of the tube settling approach is that the tubes must be cleaned and this can be an extremely time consuming and unpleasant task. Larger tube diameters require less cleaning, but they also have a lower efficiency in capturing solids. If you are going to use tube settlers, use tube diameters of at least 4 to 5 cm (1.5 to 2 inches). Using smaller diameter tubes will make the cleaning task all but impossible. The authors used some 1.5 cm diameter tubes in a rainbow trout system. As expected, the tubes collected feces, but cleaning it required removing the settling deck from the settling chamber and using high pressure hoses to flush the solids from the decking. This was a dirty, disgusting task. (As a personal note: it was this particular experience that inspired Timmons to develop the Cornell dual-drain approach.)

Once the tubes begin to fill with fine solids settling out of the water flow, the water velocity rates through the tubes will increase due to reduced tube cross sectional area. As this happens and resistance to flow increases, water begins to seek a least resistance approach and will eventually simply bypass the tubes, thereby eliminating any solids capture at all. Therefore, periodic cleaning is necessary, but cleaning the settling deck is a dirty, nasty job that nobody likes to do and as a result is often neglected. In turn, neglect leads to poor performance of the settling device and subsequent deterioration in water quality. For this reason, we do not recommend tube settlers for use in highly loaded systems.

MICROSCREEN FILTERS

Microscreen filters for filtration are popular because they require minimal labor and floor space in comparison to settling basins. As in sedimentation processes, the head loss of a screen filter is small. Screen filters remove solids by virtue of physical restrictions (or straining) on a media when the mesh size of the screen is smaller than the particles in the wastewater. Microscreen filters are commercially available in a variety of different configurations. Typical microscreen filters used in aquaculture are the drum filter, Fig. 6.8, disk filter, Fig. 6.9, and inclined belt filter, Fig. 6.10.

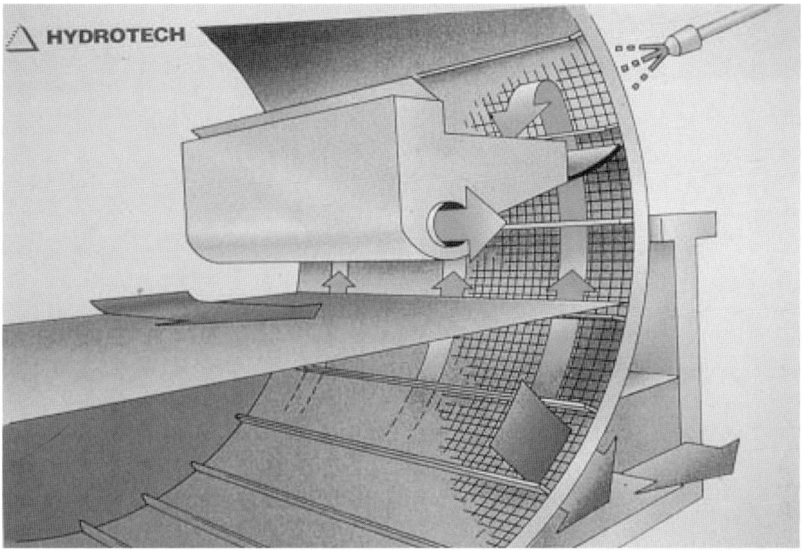

Figure 6.8 Microscreen Drum Filter Operation.
(Courtesy of Hydrotech AB, Vellinge Sweden & WMT, Inc., Baton Rouge, LA)

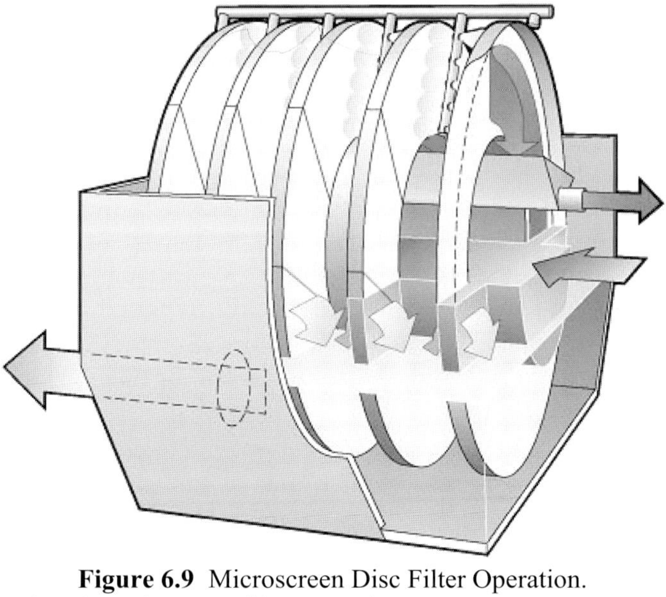

Figure 6.9 Microscreen Disc Filter Operation.
(Courtesy of Hydrotech AB, Vellinge Sweden & WMT, Inc., Baton Rouge, LA)

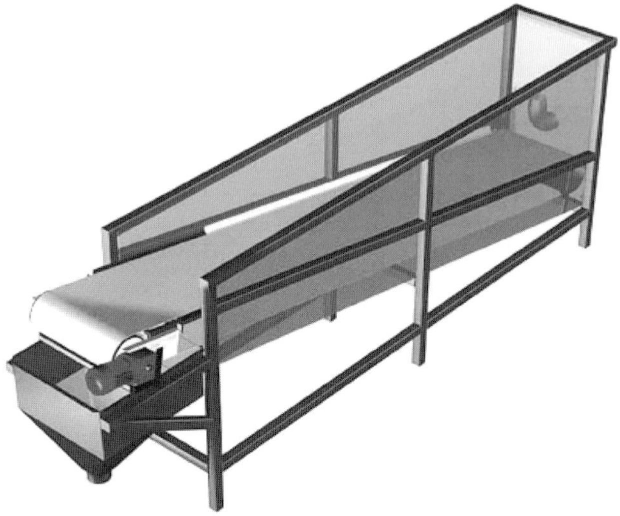

Figure 6.10 Inclined Belt Filter Used for Effluent Solid Removal. (Distributor: Sterner Aquatech AS; J. Ronhovde, pers. comm.)

Vinci et al. (2001) summarized the relative operational advantages, drawbacks, and costs of three types of waste water filters (see Table 6.4). All three types of microscreen filters are similar in that they have a separate solids waste stream that must be managed to result in a complete waste management system. This waste stream is a higher-solids, screen backwash flow. The backwash flow will vary in volume and solids content will vary based on several factors. These are the screen opening size, type of backwash control employed, frequency of backwash, and influent TSS load on the filter. Backwash flow is generally expressed as a percentage of the flow the filter treats, with reported backwash flows ranging from 0.2 to 1.5% of the treated flow (Summerfelt, 1999). This discharge is typically directed to a settling pond or other such device for final solids capture and storage.

Table 6.4 Operational Advantages, Drawbacks and Costs of 100 μm Mesh, 10 m^3/min Capacity Drum Filters, Disc Filters and Belt Filters: A=Sole Filter Unit; B=Filter Unit Equipped Backwash Water Processing for Dewatering (from Vinci et al. 2001)

Type of filter	Solid removal rate at 60–100 μm, %	Main advantages	Any drawbacks	Costs*, US$/unit		Source
				A	B	
Drum filter	SS inlet <5 mg/L: 31–67	Intermittent backwash		15,000	20,300	1
	SS inlet >50 mg/L: 68–94	flushing, reduced backwash volume				2
Disc filter	SS inlet <5 mg/L: 25–68	Lowest investment costs	High backwash flow volume	8,600	14,000	2,3
	SS inlet >50 mg/L: 74–92	(without basic investments)	Grounding/crushing of bigger particles			
Belt filter	SS inlet <5 mg/L: 0–62	Gently removing particles	Relatively high investment costs at low flow (<3–5 m^3/min)	18,000	—	4, 5
	SS inlet >40 mg/L: >89	Low maintenance work and costs				
		Cost effective at high flow				

*: purchase prices in Norway - no figures

1) J. Ronhovde, pers. comm.; 2) Ulgenes and Eikebrokk, 1992; 3) G. Indrevik, pers. comm.; 4) Bergheim and Bakkevig, 1998; 5) A. Naess, pers. comm.

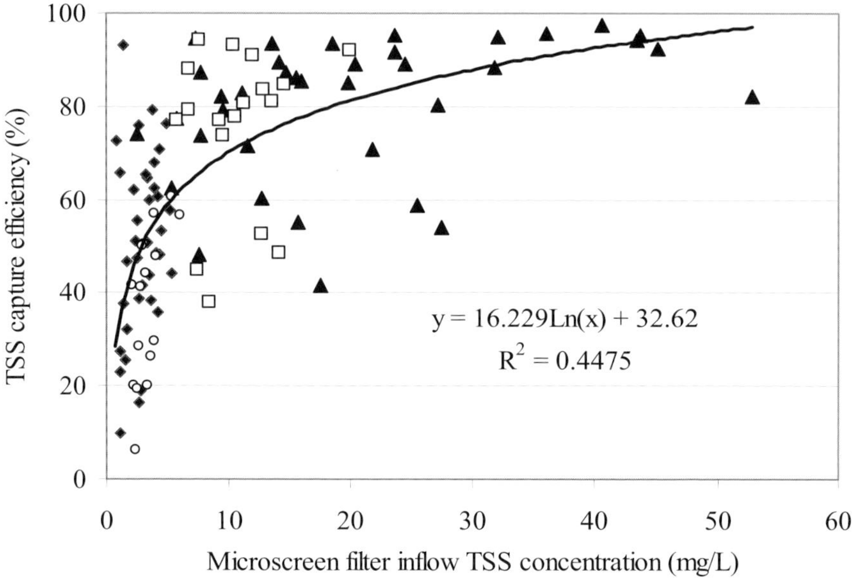

Figure 6.11 Removal Efficiencies versus Inlet TSS (Summerfelt et al. 2001).

Obviously, the size of the particles that can be removed by a screen filter is determined by the size of the screen opening. To a limited degree, even particles smaller than the nominal screen mesh size can be trapped if several smaller particles bridge together and subsequently become trapped by the screen. However, on a practical basis, simply assume that the screen size defines the smallest particles that will be removed. In addition to screen size, microscreen filter performance is dependent on the influent TSS concentration. Typical screen openings used in the treatment of aquacultural wastewater are 40–100 μm. In this screen opening size range, TSS removal can be between 30 and 80%. Summerfelt (2001) summarized a large amount of work that quantified the effectiveness of removing TSS using drum filters with screen sizes from 60 to 90 μm (see Fig. 6.11). This data was collected from four microscreen filters (60 to 90 μm) used at the Freshwater Institute (Shepherdstown, WV): one microscreen drum filter (◆) treated the flow discharged from the elevated drains of dual-drain culture tanks in a water reuse system; another microscreen Triangle™ filter (▵) sieved the bottom flow from a dual-drain culture tank in a single-pass system; another microscreen drum filter (□) treated the combined effluent from a reuse system and a single-pass system; and the other microscreen drum filter (o) sieved the effluent from a single-pass system.

The data shown in Fig. 6.11 shows a trend of increasing removal efficiency with increasing influent TSS concentration. In other words, the dirtier the water coming into the filter, the better the filter works. This data was used by Summerfelt (2001) to develop a predictive equation for the removal of TSS based upon influent TSS concentration:

$$TSS_{capture\ percent} = 16\ Ln\left(TSS_{in}\right) + 33 \qquad (6.10)$$

Smaller filter pore sizes increase the removal rate of TSS to a certain extent. Based on particle size distribution analysis of hatchery effluent (Cripps, 1993), an increased removal rate of solids is expected when using increasingly smaller filter pore size below 200 µm. There are limits to this increasing effectivity, however. In a follow-up study by Cripps (1995), they found that screen sizes below 60–100 µm no longer improved solids removal. This efficiency limit was confirmed by Kelly et al. (1997) who studied the particle removal in effluent water from two salmon hatcheries using test sieves with four mesh sizes between 30 and 200 µm, Fig. 6.12.

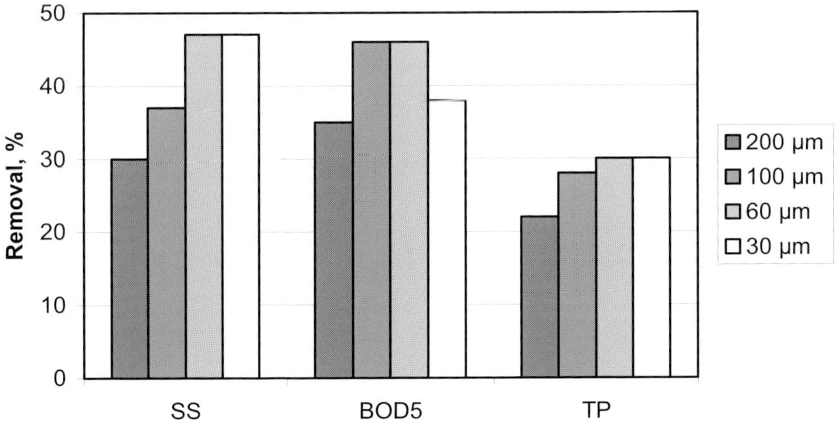

Figure 6.12 Daily Average Removal of Suspended Solids (SS), Biochemical Oxygen Demand (BOD_5) and Total Phosphorus (TP) after Passage through Four Apertures (200, 100, 60 and 30 µm) (Kelly et al. 1997).

Typically, the removal rate was significantly higher during tank cleaning operations (SS: 10–100 mg/L) compared with the removal during routine flow (SS <6 mg/L) throughout the day. The 60 μm aperture was estimated to reduce the main daily BOD_5, SS and TP outflow concentrations by 45, 46 and 30%, respectively. Consequently, the authors concluded that to achieve maximum effectiveness of solids filtration of waste waters from tank farms, the filtration system should use a screen with an aperture of 60 to 100 μm. This conclusion is supported by returning to Fig. 6.2 which shows that 80 to 90% of the solid particles smaller than 130 μm are in fact less than 30 μm in size, and using a screen that small is impractical due to pressure losses and excessive backwashing requirements. In slight contrast, Wedekind et al. (1995) did find improvement in solids removal when the screen size was reduced from 60 to 30 μm in an intensive rainbow trout production system. However, a markedly increased volume of backwash water (a bad thing) and sludge (a good thing) resulted when reducing that screen pore size.

Sizing and Managing Drum Filters

The use of the Cornell dual-drain system in conjunction with drum filters will result in a 10-fold higher concentration of TSS in the influent water to the drum filter. This means that the hydraulic capacity of the system will be reduced by approximately 30% when compared to non Cornell dual-drain influents. Consult your supplier carefully in this regard and be careful when comparing the drum filters of competing manufacturers so that the comparison is made using the same influent TSS concentrations. Table 6.5 demonstrates the relationship between unit selection and hydraulic loading, screen size, and influent TSS concentrations (Water Management Technologies, Baton Rouge LA, North America distributors of Hydrotech). Filter screens require regular maintenance. In a normally operating system, the drum filter should completely rotate in 2 or 3 minutes; more frequent rotations indicates that the screen needs maintenance, or it has an overload of suspended solids (this means you needed a bigger unit!). Increased backwashing frequency is a clear sign that screen maintenance is needed. In any case, pay careful attention to the manufacturer's recommendations for maintenance intervals and processes.

Disc Filters

Disc filters are an alternative to drum filters (see Fig. 6.9). Disc filters are less common in the USA, but are used in Europe especially on farms with high volumetric flows that require suspended solids (TSS) removal. Disc filters are generally less expensive than drum filters when treating high volumetric discharges, e.g., flows >2,000 gpm. A criticism of disc filters is that because of the vertical orientation of the filter screens, TSS can be retained longer before removal. This longer retention time than could cause TSS breakup into finer particles and perhaps additional nutrient leaching. Even if additional breakup of TSS occurs, the size range would still be much larger than 60 micron, which means practically the same TSS removal efficiencies should occur with a disc as a drum filter. A performance table is given for disc filters (see Table 6.6 on pg. 191, Water Management Technologies, Baton Rouge, LA, North American distributors for Hydrotech products).

HYDROCYCLONES

Hydrocyclones employ the principle of centrifugal sedimentation, i.e., the suspended solid particles are subjected to centrifugal acceleration, which makes them separate from the liquid more rapidly by effectively increasing their density (see Eq. 6.6). Hydrocyclones are also called swirl separators and tea-cup settlers. The Cornell dual-drain is effective because the tanks are round and are operated as swirl separators. The rotational flow of the inlet water imparts a centrifugal motion in the particles that causes the heavier particulate material to move to (or remain at) the outer portion of the vessel. Simultaneously, the particles are affected by gravity, which causes them to fall through the water, and move towards the bottom center drain. Here, at the bottom center drain, a small percentage of the total flow is removed, which is referred to as the underflow.

The underflow in a swirl separator should be about 5% to 15% of the total flow, which is the same percentage used for the Cornell dual-drain in the culture tank vessel. Downward spiral flow moves dominantly along the outside walls and creates an inward spiral flow in the center. Between these two spirals there is a layer, called the mantle, where zero vertical velocity exists. The entrance to the outlet pipe should be placed at the center of the mantle. This mantle plane is located at a distance equal to 1/2 to 2/3 D below the top surface of the water (where D is the diameter of the vessel). Strategic placement of the outlet takes advantage of the zero vertical velocity condition by minimizing TSS in the outlet flow. Ideally, you should test your own particular waste water prior to full design implementation to have any reasonable chance of getting it "right", or design the swirl separator so that adjustments in outlet placement can be made during operation.

Table 6.5 Example of Hydraulic Loadings, Screen Size, and Drum Filter Size as Part of an Overall Selection Process (Water Management Technologies, Baton Rouge LA, North America distributors of Hydrotech)

Hydrotech Operating Specifications

Filter Size	Screen Opening (um)	501	801	802	803	1201	1202	1203	1601	1602	1603	1604	1605	1606	1607	2005	2006	2007	2808	2810
		Maximum Flow Capacity (gpm)																		
	10	24	63	127	190	95	190	285	127	254	380	507	634	761	888	793	951	1,110	1,775	2,219
	18	71	190	380	571	285	571	856	380	761	1,141	1,522	1,902	2,283	2,663	2,378	2,853	3,329	5,326	6,658
Intake Water	23	84	222	444	666	333	666	999	444	888	1,332	1,775	2,219	2,663	3,107	2,774	3,329	3,884	6,214	7,768
	30	119	317	634	951	476	951	1,427	634	1,268	1,902	2,536	3,170	3,804	4,439	3,963	4,756	5,548	8,877	11,096
Max 10 mg/l SS	40	143	380	761	1,141	571	1,141	1,712	761	1,522	2,283	3,044	3,804	4,565	5,326	4,756	5,707	6,658	10,653	13,316
	60	174	476	951	1,427	713	1,427	2,140	951	1,902	2,853	3,804	4,756	5,707	6,658	5,945	7,133	8,322	13,316	16,645
	90	222	571	1,141	1,712	856	1,712	2,568	1,141	2,283	3,424	4,565	5,707	6,848	7,989	7,133	8,560	9,987	15,979	19,974
	30	71	190	380	571	285	571	856	380	761	1,141	1,522	1,902	2,283	2,663	2,378	2,853	3,329	5,326	6,658
Recycle System	40	95	254	507	761	380	761	1,141	507	1,015	1,522	2,029	2,536	3,044	3,551	3,170	3,804	4,439	7,102	8,877
Max 25 mg/l SS	60	132	349	697	1,046	523	1,046	1,569	697	1,395	2,092	2,790	3,487	4,185	4,882	4,359	5,231	6,103	9,765	12,206
	90	174	444	888	1,332	666	1,332	1,997	888	1,775	2,663	3,551	4,439	5,326	6,214	5,548	6,658	7,768	12,428	15,535
	30	108	285	571	856	428	856	1,284	571	1,141	1,712	2,283	2,853	3,424	3,995	3,567	4,280	4,993	7,989	9,987
Outlet Water	40	132	349	697	1,046	523	1,046	1,569	697	1,395	2,092	2,790	3,487	4,185	4,882	4,359	5,231	6,103	9,765	12,206
Flow Through System	60	174	444	888	1,332	666	1,332	1,997	888	1,775	2,663	3,551	4,439	5,326	6,214	5,548	6,658	7,768	12,428	15,535
Max 15 mg/l SS	90	206	539	1,078	1,617	808	1,617	2,425	1,078	2,156	3,234	4,312	5,390	6,468	7,546	6,737	8,085	9,432	15,091	18,864
	100	222	571	1,141	1,712	856	1,712	2,568	1,141	2,283	3,424	4,565	5,707	6,848	7,989	7,133	8,560	9,987	15,979	19,974
	500	317	824	1,649	2,473	1,236	2,473	3,709	1,649	3,297	4,946	6,594	8,243	9,892	11,540	10,304	12,365	14,425	23,081	28,851
Effluent from	6	16	32	48	79	32	79	111	48	95	143	190	238	285	333	301	365	412	N/A	N/A
Municipal WWTP	20	16	32	79	111	119	95	95	95	95	222	285	365	428	507	444	539	634	N/A	N/A
Max 40 mg/l SS	25	16	63	111	174	143	95	95	95	95	333	444	555	666	777	697	840	967	N/A	N/A

Technical Specifications

Filter Size	501	801	802	803	1201	1202	1203	1601	1602	1603	1604	1605	1606	1607	2005	2006	2007	2808	2810
Drive Motor (kw)	0.18	0.25	0.25	0.25	0.37	0.37	0.37	0.55	0.55	0.55	0.55	0.55	0.55	0.55	1.10	1.10	1.10	4.00	4.00
Rinse Water Flow @ 100 PSI (gpm)	3.20	4.75	9.50	14.25	4.75	9.50	14.25	4.75	9.50	14.25	19.00	23.75	28.50	33.25	23.75	28.50	33.25	38.00	47.50
Filter Area (sq. ft.)	4.84	10.33	20.67	31.00	15.50	31.00	46.50	20.67	41.34	62.01	82.67	103.34	124.01	144.68	129.18	155.02	180.85	289.36	361.70
Number of Filter Plates	2	2	4	6	3	6	9	4	8	12	16	20	24	28	25	30	35	56	70

Rinse Water Pressure	100 - 140 psi
Range of Filtration	10 - 3000 micron
Max. Water Temp.	140 F
Head Loss	8 - 11.81 inches

Liquid Level Control	On / Off Liquid Level Control - Operates the drum and booster pump on demand
High Pressure Rinse	Booster pump operated by choice of control listed to the right

Materials of Construction		Tank
	501	GRP Standard, 304SS & 316SS Available
	801 to 1203	GRP, 304SS or 316SS
	1601 to 2007	304SS or 316SS

Drum Structure
304SS or 316SS

Filter Elements
316SS or Polyester Supported by Polyethyelene Grid

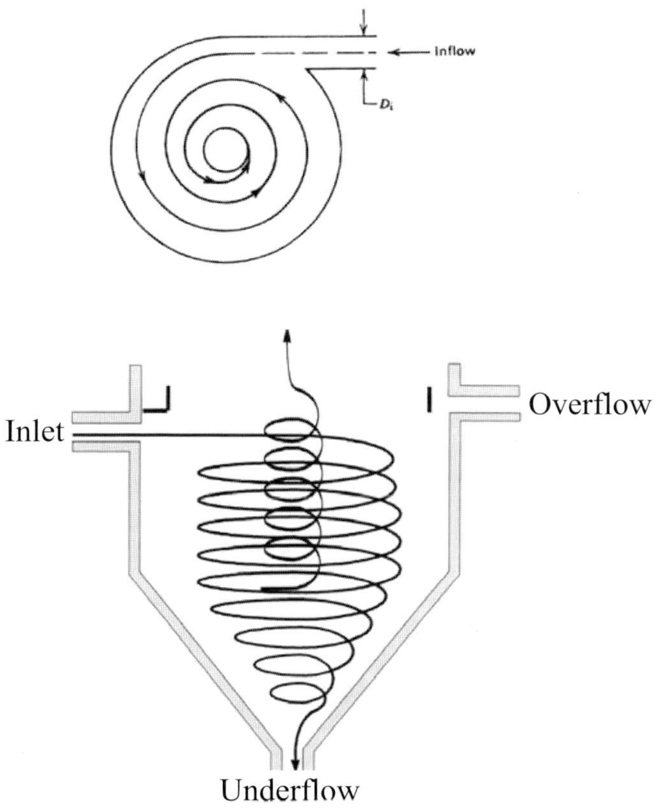

Figure 6.13 Design Concepts of a Swirl Separator or Hydrocyclones.

The inlet is typically placed about 1/3 diameter below the top of the water column with the flow directed tangentially to the vessel walls. Figure 6.13 illustrates the design concepts of a swirl separator. Wheaton (1977) provides more mathematical descriptions of the physics driving the operational characteristics of hydrocyclones; Paul et al. (1991) and Tchobanoglous and Burton (1991) provide additional design approaches and theory for vortex driven settling basins or hydrocyclones.

Loading rates for hydrocyclones are approximately four times greater than the recommended loading rates for conventional settling basins, or 10 m^3/m^2 per hr (4 gpm per ft^2). This means that the footprint area requirements for handling the same amount of water have been reduced by a factor of four. Some engineers will use even higher loading rates, but hydraulic retention times should be maintained at a minimum of 30 seconds. If loading rates are increased, then the volume of the vessel must be correspondingly increased so that the minimum 30 second hydraulic retention rate is preserved.

> **"RULE OF THUMB"**
> Swirl Separators
>
> - hydraulic loading of 3 to 5 gpm per ft^2 (122 to 204 Lpm/m^2)
> - underflow of 5 to 15% of total flow
> - tangential inlet flow
> - outlet submerged 1/3 vessel diameter
> - best used with Cornell dual-drain

Hydrocyclones are relatively expensive, and just as settling basins and drum filters, they are not effective at removing fine solids (diameter <50 µm). However, they can be quite effective in removing TSS; Scott and Allard (1984) reported that hydrocyclone prefilters removed over 87% of the particulate matter greater than 77 µm in diameter.

A currently popular approach is to use a commercially available swirl developed by SINTEF NHL (Trondheim, Norway) called the Eco-Trap™. Twarowska et al. (1997) reported that a particle trap (swirl separator) that took 5% of the total center drain flow (the underflow with concentrated solids) removed 80% ±16% of the solids. The surface loading rate on this unit, which accepted 16 L/min of flow, was approximately 5.6 m/h[a] (2.3 gpm/ft^2). Referring back to Table 6.2, this loading rate is less than a full-flow settling basin (14.3 m/h) but more than an off-line settling basin (1.66 m/h). The swirl settling feature has improved the effectiveness of the settling process by a factor of 3.4 which is consistent with the earlier note that surface loadings could be increased by a factor of four compared to conventional settling basins.

An interesting application would be to use a three times higher hydraulic loading level (30 m^3/m^2 per hr; see Table 6.2 equivalent to Full-flow settling basin) to remove large heavy solids as a pre-treatment step prior to the waste flow before being sent to a drum filter or other TSS removing area. This approach is being implemented in recently designed systems by taking the Cornell dual-drain flow and directing this into a swirl separator. Since the effluent from the center drain in a Cornell dual-drain is only 5 to 15% of the total flow, the size

[a] Note that m^3/m^2 per hour is equivalent to m/h or meter per hour.

requirements of the swirl separator can be reduced by a factor of 10 to 20. Also, the hydraulic loading rates may also be increased depending upon whether or not this treatment constitutes the total TSS removal process or is only a partial treatment, e.g., removal of easily removed large particle TSS.

Granular Media (GM) Filters

Granular media filtration involves passage of water through a bed of granular material (media) and deposition of solids onto the media. This type of filtration system is generally classified as packed-bed or depth filtration (Crites and Tchobanoglous, 1998). The major mechanisms that function to remove the particulates in a packed-bed filter are straining, sedimentation, impaction, interception, adhesion, flocculation, chemical adsorption, physical adsorption, and biological growth (Tchobanoglous and Burton, 1991). However, straining has been identified as the principal mechanism for the removal of suspended solids in the filtration of secondary effluent from biological treatment processes (Tchobanoglous and Eliassen, 1970).

Granular-medium filters are often used in combination with other treatment processes to increase the level of treatment. These filters provide increased removal of solids, phosphorus, algae, turbidity, and pathogens. In the case of phosphorus removal, the water is pre-treated by adding chemicals that precipitate phosphorus. Precipitated phosphorus is then subsequently removed in the particulate form by the granular-medium filter. It is important to note that the backwash from these filters will have high total phosphorus levels.

There are several types of granular-medium filters. These are: conventional monomedium downflow filter, conventional dual-medium downflow filter, conventional monomedium deep-bed downflow filter, continuous backwash deep-bed upflow filter, pulsed-bed filter, traveling-bridge filter, synthetic-medium filter, and pressure filter. The basis of operation of all these filter types is the same; they all use the previously stated removal mechanisms, and they all require a backwash or cleaning process to regenerate the particulate removal capacity of the filter (Crites and Tchobanoglous, 1998). For example, in the case of the continuous backwash deep-bed upflow filter, process water is evenly introduced at the filter bottom where it upflows through a sand bed. The filtered water flows out of the sand bed, into an overflow weir, and is discharged from the filter. Concurrent to this filtration, sand and solids trapped by the sand are drawn downward within the sand bed into the suction of an airlift chamber in the center of the filter. The airlift carries the sand and solids to the top of the filter into a reject compartment. Because the discharge weir from the reject compartment is lower than the filtrate weir, the reject compartment continuously receives a small flow of filtered water that carries solids out of the filter. This backwash flow also acts to clean sand that is settling back down to the top of the sand bed after having been airlifted to the reject chamber.

The most commonly used downflow pressurized sand filter is available in a variety of configurations because it is widely used in water treatment systems and can be readily purchased "off the shelf". Downflow sand filters (or swimming pool filters) are not appropriate for use in RAS that are even moderately loaded. There is simply so much TSS being generated in a RAS that a downflow sand filter will be constantly going into backwash mode. For the exceptionally stubborn, or those having very low loaded systems, design guidelines for large scale gravity or pressurized sand filtration units are of 12 to 30 m^3/hr per m^2 (Metcalf and Eddy, Inc., 1991). Bell (1962) gives design criteria for diatomite filters.

Upflowing sand filters used for biofiltration (see Chapter 7: Biofilters) will capture TSS depending upon their hydraulic loading rates. Upflow sand filters should not be used for the primary purpose of solids capture. If used in this manner, they will probably fail in their primary function of as biological filtration. However, they can be used to capture fine solids at low hydraulic loading rates. RAS applications have effectively used "bead" filters, which use a floating plastic media to reduce the water losses associated with backwashing of the media. Bead filters are operated to capture solids and provide biological filtration. Bead filters also require backwashing to remove solids, but their water loss compared to traditional downflow sand beds is very small. Figure 6.14 shows an industry supplied bead filter.

Bead filters can be designed to obtain most of the benefits of the sand filters without incurring the high water losses during the backwashing operation. These filters are multi-functional, providing both solids removal and biofiltration processes (Golz et al. 1999). Pressurized bead filters contain floating plastic beads that provide a medium to which nitrifying bacteria attach. The floating plastic beads commonly used in these filters are 3–5 mm (0.12–0.20 inch) diameter polyethylene spheres that have moderately high specific surface areas (1,145 m^2/m^3 (350 ft^2/ft^3) (Malone et al. 1993). In operation, water is typically pumped into the bead filter through a media retaining screen located at the bottom of the bead chamber, Fig. 6.14. Particles are captured within the filter as the water flows upward through the floating bead bed. The bead bed floats against a screen installed at the top of the pressure vessel, which is used to retain the beads while allowing the

Table 6.6 Performance Characteristics of a Disc Filter (Provided by Water Management Technologies, Baton Rouge, LA, North American distributors for Hydrotech products)

Flow Capacity for Hydrotech Discfilter HSF1704-2110

Flow capacity examples	Filter size	1704/3	1704	1706/5	1706	1708/7	1708	2102/1	2102	2104/3	2104	2106/5	2106	2108/7	2108	2110/9	2110
	Filter opening (micron)	Maximum flow capacity (l/s)															
Intake water from stream, lake or sea. Max 10 mg/l SS	6	29	39	48	58	68	77	16	31	47	62	78	93	109	124	140	155
	10	33	43	54	65	76	87	17	35	52	70	87	105	122	140	157	175
	15	87	116	145	174	203	232	47	93	140	186	233	279	326	372	419	466
	18	98	130	163	196	228	261	52	105	157	210	262	314	367	419	471	524
	20	105	140	175	210	245	280	56	120	169	240	281	360	394	480	506	600
	25	123	164	205	246	287	328	66	132	198	264	330	396	462	528	594	660
	30	163	217	272	326	380	435	87	175	262	350	437	525	611	700	786	875
	40	196	261	326	391	456	521	105	210	314	419	524	629	733	838	920	920
	60	244	326	407	489	570	652	115	230	345	460	575	690	805	920	920	920
	90	293	391	489	587	684	782	115	230	345	460	575	690	805	920	920	920
Recirculated fish farm. Max 25 mg/l SS	30	98	130	163	196	228	261	52	105	157	210	262	315	367	420	471	525
	40	130	174	217	261	304	348	70	140	210	280	349	420	489	560	629	700
	60	179	239	299	359	418	478	96	200	288	400	480	600	672	800	864	920
	90	228	304	380	456	532	608	115	230	345	460	575	690	805	920	920	920
Outlet water from flow through type fish farm. Max 15 mg/l SS	30	147	196	244	293	342	391	79	157	236	314	393	471	550	629	707	786
	40	179	239	299	359	418	478	96	192	288	384	480	576	672	768	864	920
	60	228	304	380	456	532	608	115	230	345	460	575	690	805	920	920	920
	90	277	369	462	554	646	739	115	230	345	460	575	690	805	920	920	920
Maximum hydraulic capacity for filter with standard tank.		180	180	180	180	180	180	230	230	230	230	230	230	230	230	230	230

filtered water flow to pass out through the top of the filter. Operating pressures, typically 0.34–1.02 atm gauge pressure (5–15 psig), increase as solids capture across the filter increases. To maintain adequate flow through the bead filter as solids build up, the filter is backwashed through the use of a motor-driven propeller mechanism or through injected bubbles. Backwashing occurs as the influent flow is stopped and the beads are turbulently disturbed, which breaks loose any collected solids. The solids are then removed from the filter through the media retaining screen located at the base of the filter, and normal flow is resumed (Malone and Beecher, 2000).

Solids captured by the filters are retained within the filter until the unit is backwashed. Typically, the plastic-bead filter unit is backwashed every 24 hours. During this retention period, 30–40% of the total retained solids will decay (Chen, Coffin and Malone, 1993). This is a very undesirable activity, because the entrapped solids begin to dissolve and mineralize, which adds soluble BOD and ammonia loads to the system. Removal of these solids is the only way to minimize this condition. There is a complex relationship between plastic-bead filter backwash intensity and frequency, solids degradation, and nitrification efficiency (Golz et al. 1999). The fact that the pressurized bead filter collects solids so effectively means that heterotrophic bacteria populations can grow and become a problem if the filter is not backwashed frequently. Heterotrophic activity in the filter depletes oxygen and generates ammonia as protein contained in the captured solids is metabolized. While backwashing will remove solids and unwanted bacterial growth, too frequent backwashing will also remove beneficial nitrifying bacteria. Obtaining optimum performance from the bead filter requires careful management of the intensity and frequency of backwashing these filters. (Golz et al. 1999).

Porous Media (PM) Filters

Porous media (PM) filters are often known as point-of-use devices and are used to treat low flow rates. They are very susceptible to clogging, even at low influent TSS concentrations, and experience high head losses. On a practical basis, the requirements for high recharging/reconditioning or replacement of the media as the filters become plugged or restrictive to flow prohibit the use of porous filters as the principle solids capture device in an intensive RAS. However, PM filters do remove fine solids and are thus effective polishing units. PM filters can be used to treat incoming water to an RAS or as an ultra polishing step in a particular RAS application, e.g., incubation system. Two examples of PM filters are diatomaceous earth filters (DE) and cartridge filters. Cartridge filters use disposable cartridges of different sizes. Like DE filters, cartridge filters remove particles by straining due to their thin media. Therefore, the removal of particles smaller than the pore size by cartridge filters is low. However, because of their fine pore size, both DE and cartridge filters can remove very small solid particles (<1 μm).

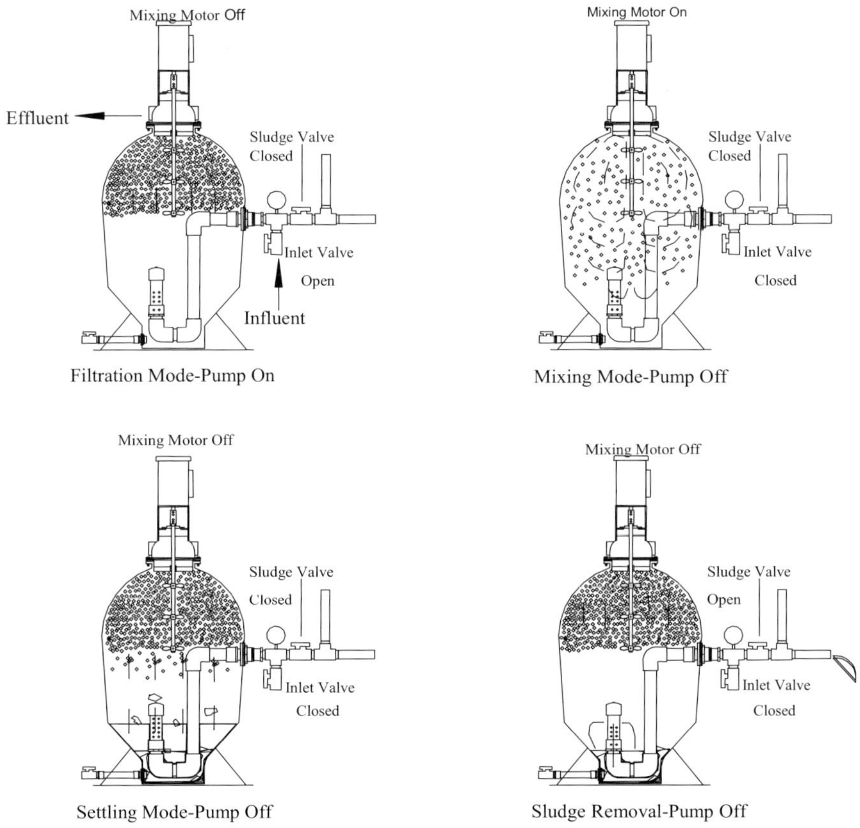

Figure 6.14 Bead Filter Operation.
(Courtesy of Aquaculture Systems Technologies, LLC, New Orleans, LA)

FOAM FRACTIONATION

Ugly foam in fish tanks is something we all detest seeing and something we all want to eliminate. However, as a practical bit of advice, if you can live with the foam, then don't try to get rid of it. Foam removal is typically a messy, unpredictable, and erratic process. But if you are convinced that you must remove the foam, this section will give you the basics of how to go about it. Remember that removal of wastes by foam fractionation is no substitute for effective primary TSS removal. Let's also include in this primary removal category the quick removal of dead fish, which consist of a high percentage of protein (assume 70% dry matter basis). Foaming is the result of the presence of surfactants, and surfactants are protein. Thus, a dead fish that is ground up through a pump or is left in the tank to decompose will provide a large amount of surfactant, which will in turn result in immediate and noticeable foaming.

If you ever walk in and see an unexpected level of foaming, start looking for a decomposing fish somewhere in your system. A simple example will illustrate the point. Assume that in a tank of 10,000 fish, that you will lose 5% of the animals over a 200 day growth period. This means you are losing an average of 2.5 fish each day. Using a carrying capacity of 120 kg/m^3, 2.5 fish at 1 kg each (20% dry matter) would be the equivalent of:

$$10{,}000\,fish \cdot \frac{1kg}{fish} \cdot \frac{m^3}{120kg} = 8.83m^3 = 83{,}000L$$

$$2.5\,fish \cdot \frac{1000g_{flesh}}{fish} \cdot \frac{0.2g_{DM}}{1.0g_{flesh}} \cdot \frac{0.7g_{protein}}{1.0g_{DM}} = 350g_{protein}$$

Now, the concentration of this amount of pure protein (a foaming agent) in the 83,000 L of water is:

$$\frac{350g \cdot 1000\frac{mg}{g}}{83{,}000L} = 4.2\frac{mg}{L}$$

This amount of protein in the water (say from a fish being ground up by a pump or left rotting on the bottom of a tank) will create enormous levels of foaming just by itself. Remove the dead fish!

Having now sensitized you to diligently remove dead fish and other organic solids as quickly as possible, we'll review the basics of foam fractionation. The process of foam fractionation is based on air bubbles in a water column that collect organic material as they rise to the surface. To a small degree, even some dissolved organics and proteins may precipitate out onto the bubble surface allowing these pollutants to be removed as well. The whole foam fractionation process is totally dependent on and functions because of the presence of surfactant materials in the water column.

Surfactant molecules are polar at one end. This end is hydrophilic (negative charge) and the other end is dipolar, thereby making the hydrophilic end of the molecule hydrophobic. The hydrophobic end (positive charge) thus "pokes" itself into the air bubble leaving the polar end in the water. The negatively charged polar end now sticking out into the water column then attracts positively charged materials in the water column. These positively charged materials in turn attract negatively charged particles that "stick" to them. Figure 6.15 illustrates the properties of a surfactant molecule attached to an air bubble. Charged surfaces of air bubble attract both negative and positive fine solids and bacteria.

The surfactant materials and other collected particles attached to the air bubbles will rise to the top of the water column as foam. The foam can then be removed from the water column. Foam fractionation is considered one of the few processes that is effective in removal of fine solids from a RAS. In fact, foam condensate consists mostly of dissolved organics and particles smaller than 30 μm. Removing and disposing of the foam is one of the problems encountered when using foam fractionators. Plan ahead to determine what you will do with the foam condensate, e.g., collection vats that must be emptied, or direct drainage to a floor drain. Foam fractionators have been well reviewed in the literature (Chen, 1991; Chen et al. 1992; Chen et al. 1994A, B; Lemlich, 1972). Typical foam fractionators are depicted in Fig. 6.16 (A = airlift type and B = venturi type).

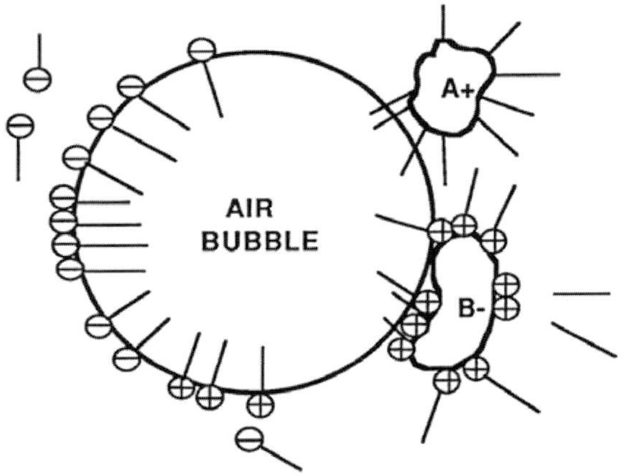

Figure 6.15 Surfactant Molecule Properties.

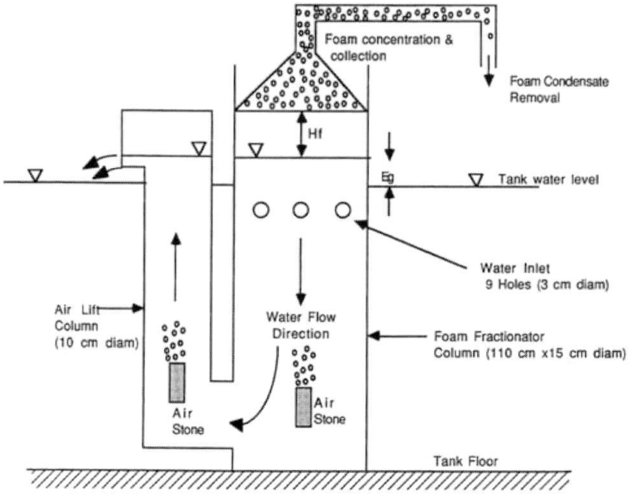

Figure 6.16A Airlift Type Foam Fractionator.

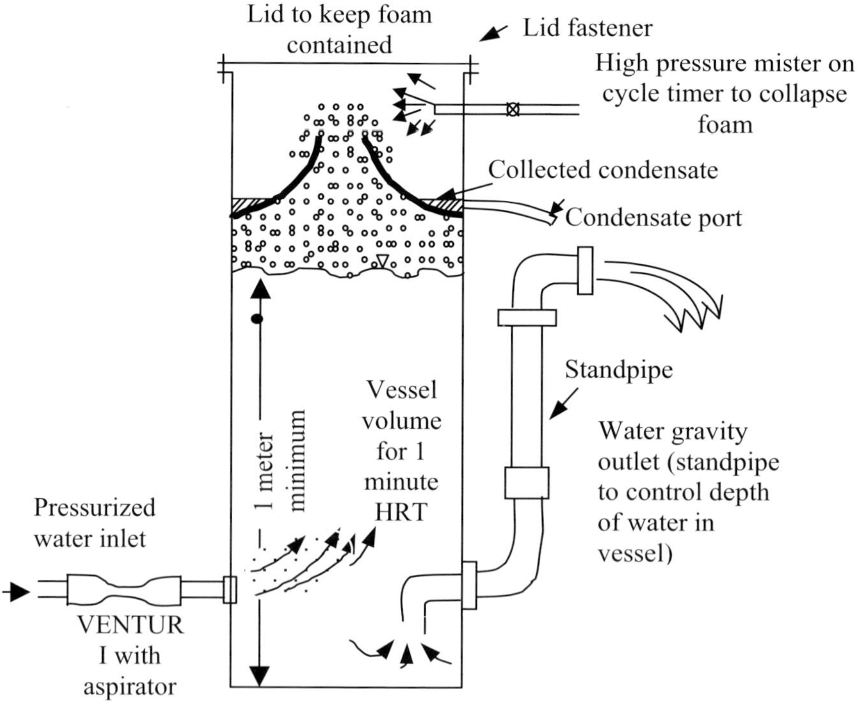

Figure 6.16B Venturi Type Foam Fractionators.

The use of foam fractionators requires a basic knowledge of the fundamental parameters used in their design and operation. They are as follows (see Timmons et al. 1995):

$$U_g = \frac{Q_{air}}{A} \tag{6.11}$$

$$E_g = 4.1 U_g^{0.83} \tag{6.12}$$

$$U_b = 0.21e^{-25.8U_g} \tag{6.13}$$

Predicting foam fractionator performance is problematic at best. For those interested, (see Chen et al. 1992; Timmons et al. 1995; Weeks et al. 1992; and Timmons and Losordo, 1994). While absolutely predicting the performance of a foam fractionator may not be possible, there are certain principles that will maximize foam creation, thereby maximizing waste removal, in a fractionator:

- Maximize the rising bubble's travel length
- Maximize the contact time between air bubbles and the water column
- Create as small a bubble as possible (decreases rising velocity)
- Operate fractionator columns at maximum gas holdup E_g of 25% (above this value, slug flow will occur; see Chapter 3, section 3.4)

Figure 6.17 shows the relationship between Eqs. 6.11 and 6.12, which must be calculated to determine the proper amount of air flow in a foam fractionator. Weeks et al. (1992) found that the removal rates of protein were independent of water flow but were simply proportional to airflow. (This approach will be demonstrated in the next section).

There are other pertinent facts about foam fractionation that must be addressed if this system is to be successfully used. As pH is increased, foam creation and corresponding protein removal is also increased, e.g., foam fractionators will remove twice the amount of protein concentrations from water having a pH of 8.3 in comparison to water with a pH of 5.3. This is demonstrated by seawater systems that always foam more readily than freshwater systems. And while use of smaller bubbles improves the overall fractionation process, generally gas bubbles will be between 2 and 3 mm as created by a wide variety of typical scintered glass air stones. It is difficult to create smaller bubbles. The authors have achieved some success in creating smaller bubbles by using venturis when a centrifugal pump is used to bring water to the foam fractionator. However, if you are using the type of fractionator that is based upon using air lift pumps (see Fig. 6.16A), using a venturi is really not a convenient option. Side-stream treatment where pumping is already in place is an ideal way to introduce the use of venturis to create small bubbles. In particular, consider using the type of venturis used with hot tubs and spa baths. These units are usually very economical as well.

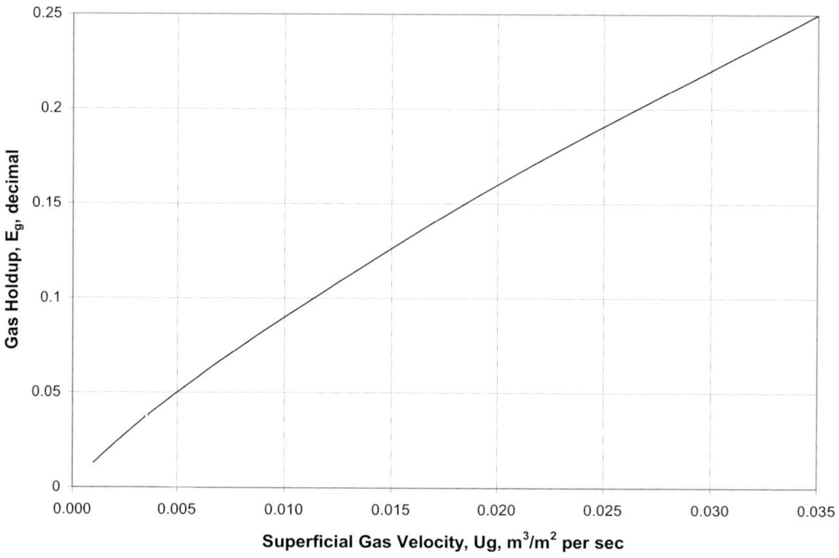

Figure 6.17 Gas holdup, E_g, Versus Superficial Gas Velocity, U_g.

Sizing Foam Fractionators

While it is not really possible to exactly predict foam fractionator performance, the findings of Weeks et al. (1992) show that the removal rates of protein were independent of water flow but were directly proportional to airflow. In other words, it is the quantity of air bubbles that was the important parameter. So, using some gross approximations, the volatile solids (VS) removal rate can be made proportional to the air flow rate used in the fractionator column (units do not balance):

$$R_{vs} = 0.4 \, x \, Q_{air,Lpm} \tag{6.14}$$

Equation 6.14 is an approximation to the Weeks et al. (1992) model and assumes that the foam fractionator is well designed and follows the general guidelines previously given for maximizing performance of a unit. It was developed from data for waste water that approximately had 300 ppm VS, 10 ppm TSS and 0.8 ppm TKN in the fish culture tank. Correlations and enrichment factors for volatile solids, TSS and Total-Kjeldahl-Nitrogen (E_{VS}, E_{TSS}, and E_{TKN}) were also developed for the foam condensate obtained from these waters using an airlift type fractionator:

$$TSS = 0.27\ VS \tag{6.15}$$

$$TKN = 0.038\ VS \tag{6.16}$$

$$E_{VS} = 2.7\ (307 \text{ increased to } 816\ ppm) \tag{6.17}$$

$$E_{TSS} = 25\ (10 \text{ increased to } 251\ ppm) \tag{6.18}$$

$$E_{TKN} = 44\ (0.8 \text{ increased to } 34.6\ ppm) \tag{6.19}$$

Example Problem

How much airflow is required for foam fractionation for a 10,000 fish @ 1 kg fed 1% of their body weight per day?

Solution.

TSS generated (Rule of thumb)

$$= 0.25 \frac{g_{TSS}}{g_{Feed}} \cdot \frac{0.01 g_{feed}}{g_{fish} \cdot day} \cdot \frac{1{,}000 g_{fish}}{fish} \cdot 10{,}000\, fish$$

$$= 25{,}000\ \frac{g_{TSS}}{day}$$

Assume 3% of the TSS is generated as fine solids (<30 μm) that are to be removed by the fractionators. Rearranging Eq. 6.14 to solve for the required airflow and assuming that all the TSS are VS:

$$Q_{air,Lpm} = \frac{R_{VS}}{0.4} = 0.03 \cdot \frac{\left(\dfrac{25{,}000\,g}{day}\right)}{0.4} = 1{,}875 \frac{Lpm}{cm^2}\ air\ flow$$

From Fig. 6.17, the superficial gas flow, U_g, for a fractionator column expanded 25% (for maximum effectiveness), shows a required U_g of 0.035 m^3/m^2 per sec or converting to Lpm per cm^2

$$U_g = 0.035 \cdot \frac{m^3/m^2}{s} \cdot 6 \frac{\dfrac{Lpm}{cm^2}\ air\ flow}{\dfrac{m^3/m^2}{s}} = 0.21 \frac{Lpm}{cm^2}\ air\ flow$$

Required cross sectional area of the foam fractionator units

$$A = \frac{1{,}875\ Lpm\ air\ flow}{0.21 \dfrac{Lpm}{cm^2}} = 8{,}930\ cm^2$$

These are the numbers for 100 kg of feed being fed, so we can express these requirements per kg of feed fed per day:

$$\frac{Q_{air,Lpm}}{\dfrac{kg\ feed}{day}} = \frac{19\ Lpm\ air\ flow}{\dfrac{kg\ feed}{day}} \cong 20\ Lpm\ air\ flow\ (roundup)$$

$$\frac{A}{\dfrac{kg\ feed}{day}} = 89\ cm^2 \cong 90\ cm^2\ (roundup)$$

And this gives us another rule of thumb (rounding the above numbers) for foam fractionators:

> **"RULE OF THUMB"**
> Foam Fractionators
>
> Per kg of feed fed per day (assumes 3% of TSS is quantity of fine solids to be removed by fractionation)
> - 20 Lpm of air flow
> - 90 cm^2 of column cross sectional area

If airlift type fractionators are used, these calculations would determine that a 15 cm diameter fractionator is required for each 2 kg of feed fed per day. This is consistent with the author's observations on an intensive trout operation that was discharging only 3% of system volume per day; the water was dark but very clear.

Finally, return to our earlier example of the amount of protein being generated from dead fish in a tank (350 g protein per day). Chen et al. (1994C) determined that only 1 mg of protein existed in 331 mg of VS collected as foam condensate. Doing the reverse math, the 350 g of protein (which effectively could become all surfactant material) would provide foaming potential for the following quantity of TSS or VS:

$$\theta_{TSS} = (350\ g_{protein}) \cdot 331 \frac{g_{VS}}{g_{protein}} = 115{,}850\ g_{VS}$$

Assuming that 100% of the VS is TSS type organic material, then this quantity TSS material would require about the following number of days for a foam fractionator process to remove these TSS (based upon the previous calculation):

$$\frac{115{,}850\ g_{TSS}}{25{,}000 \frac{g_{TSS}}{day}} = 4.6\ days\ \ supply$$

Thus, you might expect to see this tank under these conditions foam excessively for 2 to 3 days and it would be 4 or 5 days before the tank returned to normal foam levels.

LIST OF SYMBOLS

a_{TSS}	TSS produced as a proportion of feed fed (kg TSS per kg feed)
A	Cross sectional area of the column, m^2
A_{sz}	Settling basin floor area (settling zone only), m^2
BOD_5	Biological Oxygen Demand 5 day, mg/L
C_D	Drag coefficient, dimensionless
D_p	Diameter of particles, m
E_g	Gas holdup, decimal
E_{VS}	Enrichment factors for volatile solids
E_{TSS}	Enrichment factors for TSS
E_{TKN}	Enrichment factors for TKN
F	Settling zone safety factor
g	Gravitational acceleration, m/s^2
P_{TSS}	TSS production rate (kg TSS produced per day)
Q	Water flow, m^3/time
$Q_{air,Lpm}$	Air flowrate, Lpm
Q_{air}	Air flow rate through column, m^3/s
Q_{out1}	Flow rate leaving the sidewall drain (m^3 per day)
Q_{out2}	Flow rate leaving the bottom center drain (m^3 per day)
r_{feed}	Feeding rate (kg feed per kg fish per day)
R_{vs}	Removal rate of volatile solids, g/day
SS	Suspended solids, mg/L
TKN	Total-Kjeldahl-Nitrogen, mass
TP	Total phosphorus, mg/L
TSS	Total suspended solids, mg/L
$TSS_{capture\ \%}$	Percent removed of incoming TSS, %
TSS_{in}	TSS concentration entering unit (kg/m^3 or ppm)
TSS_{out1}	TSS concentration leaving the sidewall drain (kg/m^3)
TSS_{out2}	TSS concentration leaving the bottom center drain (kg/m^3)
VS	Volatile solids, mg/L
V_{basin}	Basin volume, m^3
V_{tank}	Volume of water contained within culture unit (m^3 culture volume)
V_s	Velocity of settling particles, m/s
U_b	Bubble rising velocity, m/s
U_g	Superficial gas velocity, m/s
ρ_{fish}	Density of fish in the culture tank (kg fish per m^3 culture volume)
ρ_p	Density of particles, kg/m^3

ρ	Density of water, kg/m^3
τ	Hydraulic retention time
μ	Dynamic viscosity, Pa·sec.
Θ_{TSS}	Mass of protein generated for foaming, g

CHAPTER 7

BIOFILTRATION

7.0 INTRODUCTION

Nitrogen is an essential nutrient for all living organisms and is found in proteins, nucleic acids, adenosine phosphates, pyridine nucleotides, and pigments. In the aquaculture environment, nitrogen is of primary concern as a component of the waste products generated by rearing fish. There are four primary sources of nitrogenous wastes: urea, uric acid, and amino acid excreted by the fish, organic debris from dead and dying organisms, uneaten feed, and feces, and nitrogen gas from the atmosphere. In particular, fish expel various nitrogenous waste products through gill diffusion, gill cation exchange, urine, and feces. The decomposition of these nitrogenous compounds is particularly important in intensive recirculating aquaculture systems (RAS) because of the toxicity of ammonia, nitrite, and to some extent, nitrate. The process of ammonia removal by a biological filter is called nitrification, and consists of the successive oxidation of ammonia to nitrite and finally to nitrate. The reverse process is called denitrification and is an anaerobic process where nitrate is converted to nitrogen gas. Although not normally employed in commercial aquaculture facilities today, the denitrification process is becoming increasingly important as stocking densities increase and water exchange rates are reduced, resulting in excessive levels of nitrate in the culture system.

7.1 NITRIFICATION

Ammonia is produced as the major end product of the metabolism of protein catabolism and is excreted by fish as unionized ammonia across their gills. Ammonia, nitrite, and nitrate are all highly soluble in water. Ammonia exists in two forms: un-ionized NH_3, and ionized NH_4^+. The relative concentration of each of these forms of ammonia in the water column is primarily a function of pH, temperature and salinity (Anthonisen et al. 1976). The sum of the two (NH_4^+ + NH_3) is called total ammonia or simply ammonia. It is common in chemistry to express inorganic nitrogen compounds in terms of the nitrogen they contain, i.e.,

NH_4^+–N (ionized ammonia nitrogen), NH_3–N (un-ionized ammonia nitrogen), NO_2–N (nitrite nitrogen) and NO_3–N (nitrate nitrogen). This allows for easier computation of total ammonia-nitrogen (TAN = NH_4^+–N + NH_3–N) and easy conversion between the various stages of nitrification.

Table 7.1 Concentrations of Nitrogenous Compounds when Normalized to the Molecular Weight of Nitrogen

Nitrogen based name	Nitrogen based concentration of 1 mg/L	Equivalent concentration of compound in mg/L
Ammonia-nitrogen	NH_3-N	1.21 NH_3
Ammonium-nitrogen	NH_4^+-N	1.29 NH_4^+
Total Ammonia-nitrogen (TAN)	TAN	1.21 NH_3 + 1.29 NH_4^+
Nitrite-nitrogen	NO_2^--N	3.29 NO_2^-
Nitrate-nitrogen	NO_3^--N	4.43 NO_3^-

An increase in pH or temperature increases the proportion of the un-ionized form of ammonia nitrogen, Table 2.5. For example, at 20°C and a pH of 7.0, the mole fraction of un-ionized ammonia is only 0.4%, but increases to 80% at a pH of 10. Un-ionized ammonia is toxic to fish at low concentrations, with 96-hour LC_{50}'s (LC is lethal concentration) varying widely by species from as low as 0.08 mg/L NH_3-N for pink salmon to 2.2 mg/L NH_3-N for common carp (Lawson, 1995). For long-term exposure, the concentrations of un-ionized ammonia is dependent upon the species and the culture temperature, but as a general rule of thumb should be kept below 0.1 to 0.05 mg/L NH_3 –N.

Nitrite is an intermediate product in the process of nitrification of ammonia to nitrate. Although it is usually converted to nitrate as quickly as it is produced, lack of biological oxidation of the nitrite will result in elevated nitrite levels that can be toxic to the fish. Nitrite is constantly produced as the intermediary step between ammonia and nitrate. High levels of nitrite are also indicative of biofilter impending failure and should always be addressed. The toxicity of nitrite is due to its effect on the oxygen carrying capacity of the blood hemoglobin. When it enters the bloodstream, nitrite oxidizes the iron in the hemoglobin molecule from the ferrous state to the ferric state. The resulting product is called methemoglobin, which has a characteristic brown color, yielding the common name brown-blood disease (Tomasso et al. 1979, Colt and

Tchobanoglous, 1976). The amount of nitrite entering the blood depends on the ratio of nitrite to chloride in the water. Chloride levels can be increased to lessen the effects of nitrite toxicity. At least a 20:1 ratio of chloride to nitrite-nitrogen (Cl: NO_2-N) is recommended for channel catfish in ponds, tilapia, and rainbow trout (Tucker and Robinson, 1990). Chloride levels can be increased by adding ordinary salt (sodium chloride) or calcium chloride.

Nitrate is the end product of nitrification and is the least toxic of the nitrogen compounds, with 96-h LC values usually exceeding 1000 mg NO_3-N /L (Colt and Tchobanoglous, 1976). In recirculating systems, nitrate levels are usually controlled by daily water exchanges. In systems with low water exchange or high hydraulic retention times, denitrification has become increasingly important as a control measure.

Biological filtration can be an effective means of controlling ammonia; as opposed to water flushing to control ammonia levels. There are two phylogenetically distinct groups of bacteria that collectively perform nitrification. These are generally categorized as chemosynthetic autotrophic bacteria because they derive their energy from inorganic compounds as opposed to heterotrophic bacteria that derive energy from organic compounds (Hagopian and Riley, 1998). Ammonia oxidizing bacteria obtain their energy by catabolizing un-ionized ammonia to nitrite and include bacteria of the genera *Nitrosomonas, Nitrosococcus, Nitrosospira, Nitrosolobus*, and *Nitrosovibrio*. Nitrite oxidizing bacteria oxidize nitrite to nitrate, and include bacteria of the genera *Nitrobacter, Nitrococcus, Nitrospira*, and *Nitrospina*. Nitrifying bacteria are primarily obligate autotrophs, which consume carbon dioxide as their primary carbon source, and obligate aerobes, which require oxygen to grow (Hagopian and Riley, 1998). In biofilters, the nitrifying bacteria usually coexist with heterotrophic microorganisms such as heterotrophic bacteria, protozoa, and micrometazoa, which metabolize biologically degradable organic compounds. Heterotrophic bacteria grow significantly faster than nitrifying bacteria and will prevail over nitrifying bacteria in competition for space and oxygen in biofilters, when concentrations of dissolved and particulate organic matter are high. For this reason, it is imperative that the source water for biofilters be as clean as possible with minimal concentration of total solids.

Nitrification is a two-step process, where ammonia is first oxidized to nitrite and then nitrite is oxidized to nitrate. The two steps in the reaction are normally carried out sequentially. Since the first step has a higher kinetic reaction rate than the second step, the overall kinetics is usually controlled by ammonia oxidation and as a result there is no appreciable amount of nitrite accumulation. Equations 7.1, 7.2 and 7.3

show the basic chemical conversions occurring during oxidation by *Nitrosomonas* and *Nitrobacter* and the overall oxidation reaction (U.S. EPA, 1975).

Typical startup characteristics for bringing a new biological filter system up to full capacity are shown in Fig. 7.1. Note that ammonia concentration peaks at 14 days followed by a nitrite peak at 28 days and nitrate accumulation begins after 21 days. Pre-seeding a biological filter with both ammonia and nitrite can accelerate this process. For safety, in a new system, you should observe a drop in nitrite before stocking fish as an indication that the biological filtration process is fully activated.

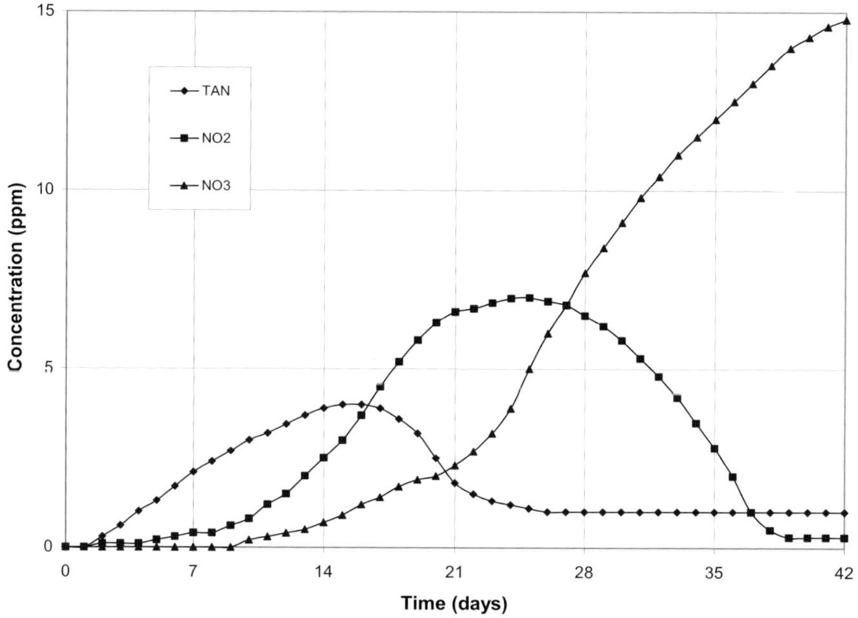

Figure 7.1 Typical Startup Curve for a Biological Filter.

Nitrosomonas

$$NH_4^+ + 1.5\ O_2 \rightarrow NO_2^- + 2\ H^+ + H_2O + 84\ \text{kcal/mole ammonia} \quad (7.1^a)$$

Nitrobacter

$$NO_2\text{-} + 0.5\ O_2 \rightarrow NO_3^- + 17.8\ \text{kcal/mole nitrite} \quad (7.2)$$

---

[a] Symbols are defined at the end of the Chapter.

Overall

$$NH_4^+ + 2\,O_2 \rightarrow NO_3^- + 2\,H^+ + H_2O + energy \qquad (7.3)$$

The overall reaction of nitrification and cell biomass formation can also be written as (Haug and McCarty, 1972):

Nitrosomonas

$$55\,NH_4^+ + 5\,CO_2 + 76\,O_2 \rightarrow C_5H_7NO_2 + 54\,NO_2^- + 52\,H_2O + 109\,H^+ \qquad (7.4)$$

Nitrobacter

$$400\,NO_2^- + 5\,CO_2 + NH_4^+ + 195\,O_2 + 2\,H_2O \rightarrow C_5H_7NO_2 + 400\,NO_3^- + H^+ \qquad (7.5)$$

Based on the above relationships, 4.57 g of O_2 and approximately 7.14 g of alkalinity, as $CaCO_3$ are needed for the complete oxidation of one gram of ammonia-nitrogen. Alkalinity should be maintained at 50 to 100 mg/L $CaCO_3$ through the addition of chemicals containing hydroxide, carbonate, or bicarbonate ions. Sodium bicarbonate (baking soda) is usually used since it is relatively safe, easy to obtain and dissolves rapidly and completely in water (Loyless and Malone, 1997). As a rule of thumb, for every kg of feed fed, approximately 0.25 kg of sodium bicarbonate is needed to replace the lost alkalinity consumed during nitrification.

Biological treatment processes employ bacteria that grow either attached to a surface (fixed films) or that grow suspended in the water column (Wheaton et al. 1991). Almost all recirculating systems use fixed-film bioreactors, where the nitrifying bacteria grow on either a wetted or submerged media surface. The ammonia removal capacity of biological filters is largely dependent upon the total surface area available for biological growth of the nitrifying bacteria. For maximum efficiency, the media used must balance a high specific surface area, i.e., surface per unit volume, with appreciable voids ratio (pore space) for adequate hydraulic performance of the system. The media used in the biofilters must be inert, non-compressible, and not biologically degradable. Typical media used in aquaculture biofilters are sand, crushed rock or river gravel, or some form of plastic or ceramic material shaped as small beads, or large spheres, rings, or saddles. Biofilters must be carefully designed to avoid oxygen limitation or excessive loading of solids, biochemical oxygen demand, or ammonia.

> **"RULE OF THUMB"**
>
> For each 1 gram of ammonia-N nitrified
> 4.57 g of Oxygen
> &
> 7.14 g as $CaCO_3$ or 0.143 equiv.
> of Alkalinity are required.
> *(50 g as $CaCO_3$= 1.00 eq. of Alkalinity)*

The following section will discuss in the major factors affecting the rate of nitrification, which include: pH, alkalinity, temperature, oxygen, ammonia, and salinity

pH

Both the rate of nitrification and the relationship between the ionized and unionized forms of ammonia-nitrogen will affect the rate of nitrification. The effect of pH on the nitrification rate for biofilters has been researched for more than sixty years, yet there is a wide range in reported pH optima. This suggests that the history and condition under which the bacteria are cultured may affect their response to pH (Kaiser & Wheaton, 1983). The most recent results suggest that the optimum range is for a pH from 7.2 to 7.8 (Loveless & Painter, 1968, Antoniou et al. 1990). Nitrifying biofilters have been operated over a much broader range from 6 to 9, due to the adaptation of the bacteria in a filter to actual operating conditions. It is probably a good idea to maintain pH near the lower end of the optimum pH for the nitrifying bacteria to minimize ammonia stress on the cultivated fish species. In addition, rapid changes in pH over a few minutes of more than 0.5 to 1.0 unit will stress the filter and require time for adaptation to the new environmental conditions.

Alkalinity

Alkalinity is a measure of the buffering capacity of an aquatic system. From the relationships above, it was determined that for every gram of ammonia-nitrogen reduced to nitrate-nitrogen, 7.14 grams of alkalinity is consumed. This loss of alkalinity is easily made up by the addition of sodium bicarbonate, referred to commonly as baking soda ($NaHCO_3$), or other bicarbonate supplement. A rule-of-thumb is 0.25 lbs of baking soda per pound of feed, 0.25 kg per 1 kg (Loyless and Malone, 1997). Nitrification is an acid-forming process, and if the biofilter system's water is poorly buffered, the system pH will decline and impact the biofilter performance.

Temperature

Temperature plays a significant role in the nitrification reaction rate as it does in all chemical and biological kinetic reactions. In general, nitrification reaction rates follow an Arrhenius relationship, i.e., a 10°C decrease in operating temperature, results in a 50% reduction in removal rate. There is a wide range of optimum temperatures reported for nitrification (Jones and Morita, 1985), suggesting that nitrifying bacteria are able to adapt to a wide range of environmental temperature, if acclimated slowly. In practical application, however, the temperature at which the biofilter operates is normally determined by the requirements of the species being cultured, not by the needs of the biofilter bacteria.

$$R = 140 + 8.5\,T^{\circ}C \tag{7.6}$$

Wortman and Wheaton (1991) developed the above equation to predict relative nitrification rates, R, over the range of their data (7 to 35°C). For example, the nitrification rates at 17°C would only be 77% of the rates obtained at 27°C, or a 23% reduction in rate. This is less severe than the 50% reduction that would be predicted based upon Q-10 (Arrhenius relationship) effect for a 10°C drop in water temperature.

Oxygen

Oxygen often becomes the rate-limiting factor in biofilters, because of the low levels in the influent and the competing demands of the heterotrophic bacteria. For every gram of ammonia-nitrogen oxidized to nitrate-nitrogen, 4.57 g of oxygen is required. Wheaton (1985) and Malone et al. (1998) state that biofilter effluent levels of at least 2 mg/L of oxygen are probably adequate to maintain maximum nitrification rates.

Ammonia

Ammonia concentration itself will affect the nitrification rate directly. In general, as ammonia concentration increases the biofilter performance increases proportionately over some limited range of concentrations. We assume this proportional relationship to exist from 0 mg/L to at least 3 mg/L. The proportional increase will at some point decrease, and then eventually not provide any additional nitrification at all. This phenomenon is shown graphically in Fig. 7.2 (Ebeling, 2000). There is some evidence in the literature that at extremely high concentrations of ammonia and nitrite and much above any expected concentrations that will be seen in aquaculture applications, accumulating ammonia will become inhibiting to nitrification (Anthonisen et al. 1976).

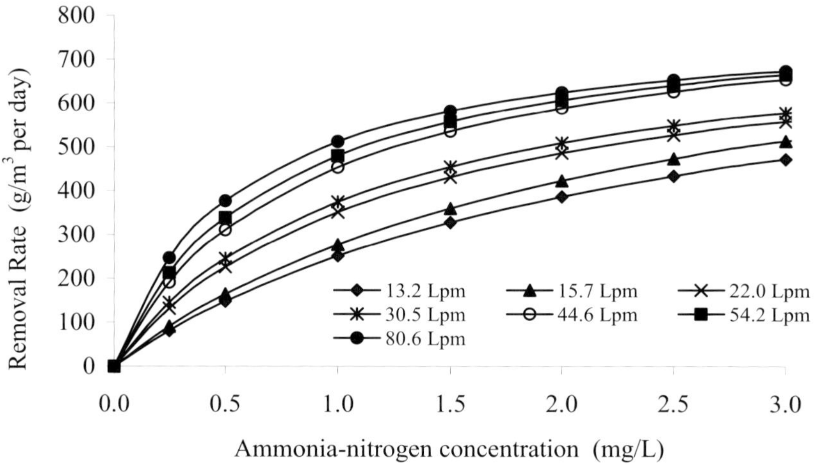

Figure 7.2 Rates of Nitrification as Affected by Water Column Concentration for a Bubble-Washed Bead Filter (Ebeling, 2000).

Salinity

Salinity is similar to both temperature and pH, in that nitrifying bacteria can acclimate to almost any salinity range, given sufficient time. Abrupt changes in salinity of greater than 5 g/L, will shock nitrifying bacteria and decrease the reaction rate for both ammonia-nitrogen and nitrite-nitrogen removal (Hochheimer, 1990).

7.2 BIOFILTERS

There is considerable debate as to the most appropriate biological filter technology for intensive aquaculture applications. An ideal biofilter would remove 100% of the inlet ammonia concentration, produce no nitrite, require a relatively small footprint, use inexpensive media, require no water pressure or maintenance to operate, and would not capture solids. Unfortunately, there is no one biofilter type that meets all of these ideals, each biofilter has its own strength and weaknesses and areas of best application. Large scale commercial recirculating systems have been moving towards using granular filters (expanded beds, fluidized beds and floating bead beds). However, there are many types of biofilters that are commonly used in intensive RAS: submerged biofilters, trickling biofilters, rotating biological contactors (RBC), floating bead biofilters, dynamic bead biofilters, and fluidized-bed biofilters.

It is helpful in any discussion of biofilter principals and advantages and disadvantages of the various choices to have a basic set of definitions and terminology. Generally, the following terms are used in the design and characterization of biofilters:

- **Void space** is the volume not occupied by biofilter media and void ratio is that volume divided by the total volume of the biofilter. High void ratios reduce clogging by having large open spaces that allow solids to pass easily through the filter.

- **Cross-sectional area** refers to the area of the filter bed looking in the direction of the water flow. Filter top area is usually one of the last parameters selected in the filter design, to yield a desired hydraulic loading rate.

- **Hydraulic loading rate** is the volume of water pumped through the biofilter per unit of cross-sectional area of the filter per unit of time, typically expressed as gpm/ft^2 or m^3/m^2 day. There is usually both a minimum and a maximum hydraulic loading rate for biofilters.

- **Specific surface area** is the surface area of the media per unit volume. The higher the specific surface area of a media, the more bacteria can grow on a unit volume and the greater the total ammonia removal per unit volume of filter. The media size, void ratio, and specific surface area are all interrelated, the smaller the size, the larger the specific surface ratio and the smaller the void ratio.

Table 7.2 Ammonia Assimilation Rates for Biofilters Based on Volumetric and Areal TAN Conversion Rates

Media Type	TAN Conversion Based On	TAN Conversion Rate (15 to 20 °C)	TAN Conversion Rate (25 to 30 °C)
Trickling or RBC (100–300 m^2/m^3)	Surface area of media	0.2 to 1.0 g/m^2 per day	1.0 to 2.0 g/m^2 per day
Granular (bead/sand) (>500 m^2/m^3)	Volume of media	0.6 to 0.7 kg/m^3 per day	1.0 to 1.5 kg/m^3 per day

Typical terms used to describe biofilter performance are based on either the volume of the media or its surface area. Although nitrification reaction rates are a surface area phenomenon, for fluidized sand beds and other granular media, the rate is easier expressed per unit volume rather than per unit surface area, due to the difficulty in measuring the actual media surface area. Table 7.2 provides a summary of design nitrification rates for general classifications of biofilters, e.g., warm or coolwater conditions and surface versus volumetric type medias.

- **Volumetric** TAN conversion rate is the grams of TAN per volume per day converted into nitrate (g TAN/ft^3 per day or kg TAN/m^3 per day).

- **Areal** TAN conversion rate is the grams of TAN per unit surface area per day converted into nitrate (g TAN/ft^2 per day or g TAN/m^2 per day).

Submerged Biofilters

The submerged biofilter is a volume of biofilter media upon which nitrifying bacteria grow. The wastewater flows in either an upflow or a downflow direction and thus the hydraulic retention time can be controlled by adjusting the water flow rate. Solids from the culture tank can accumulate within the submerged filter, along with cell mass from nitrifying and heterotrophic bacteria. This process can eventually block the void spaces, requiring some mechanism to flush solids from the filter for successful long term operation. To provide large void spaces to prevent clogging of the filters, the media used for submerged biofilters has been traditionally of large size, such as uniform crushed rock over 5 cm in diameter or plastic media over 2.5 cm in diameter. However, 5 cm diameter crushed rock would only have a specific surface area of 75 m^2/m^3 and a void fraction of only 40 to 50%. Random packed plastic media would also have a relatively low specific surface area of 100–200 m^2/m^3, but a much higher void fraction, greater than 95%. Drawbacks of this type of filter include problems of low dissolved oxygen and solids accumulation, resulting from heavy loading of organic matter and the difficulty of backflushing. Although this type of filter was promoted and used in aquaculture in the past, it has since been replaced in aquaculture due to the inherent high construction costs, biofouling problems, and operational expense.

A recent variation on the submerged biofilter, termed a moving-bed biofilm reactor or a dynamic bed biofilter, uses small slightly buoyant polyethylene tubular media (7 mm long and 10 mm in diameter), in a heavily aerated, submerged bed (Rusten et al. 1998). The tubular media has both internal and external ribs for enhanced surface area and a protected divided interior section to protect the biofilm from being completely stripped off during agitation in the moving bed. The heavy aeration keeps the bed in constant motion, which minimizes dissolved oxygen problems and solids accumulation. These biofilters report low total energy use and a high nitrification rate. The effective surface area for bacterial growth is around 350 m^2/m^3. One advantage of this type of biofilter is its low hydraulic head and aeration; its disadvantage is the large aeration requirement to maintain the bed in motion.

TRICKLING BIOFILTERS

Trickling biofilters operate in the same way as submerged biofilters, except the wastewater flows downward over the media and keeps the bacteria wet, but never completely submerged (Wheaton et al. 1991). Since the void spaces are filled with air rather than water, the bacteria never become oxygen-starved. Trickling filters have been widely used in aquaculture, because they are easy to construct and operate, are self-aerating and very effective at off gassing carbon dioxide, and have a moderate capital cost. In municipal waste water treatment systems, trickling filters were traditionally constructed of rocks, but today most filters use plastic media, because of its low weight, high specific surface area (100–300 m^2/m^3) and high void ratio (>90%). A range of trickling filter design criteria has been reported. Typical design values for warmwater systems are hydraulic loading rates of 100 to 250 m^3/day per m^2; media depth of 1–5 m; media specific surface area of 100–300

m^2/m^3; and TAN removal rates of 0.1 to 0.9 g/m^2 per day surface area. Trickling biofilters have not been used in large scale coldwater systems, probably due to the decrease in nitrification rates that occurs at the lower water temperatures and the relatively low specific surface area of the media. They have found a use in smaller hatchery systems where loads tend to be low and variable (Wilton, 2001).

Figure 7.3 A Gravel Biofilter with a Pipe Distribution Manifold on the Surface.

Figure 7.4 A Moving-bed Biofilm Reactor or a Dynamic Bed Biofilter.

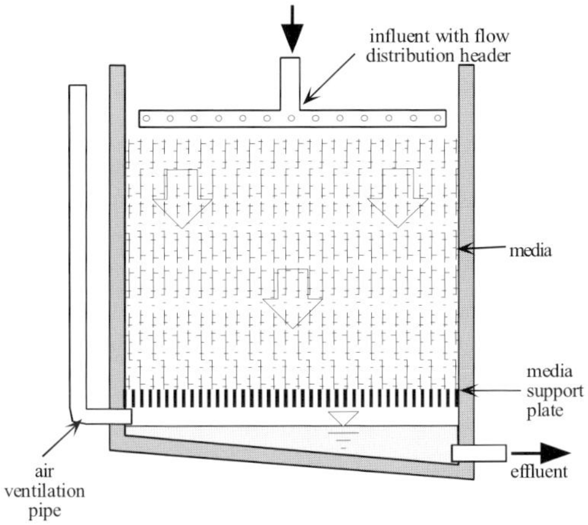

Figure 7.5 Schematic of a Trickling Filter With a Rotary Water Distribution Arm (From Summerfelt, 1999).

Figure 7.6 Two Trickling Filters With a Rotary Water Distribution Arm.

ROTATING BIOLOGICAL CONTACTORS (RBC)

Rotating biological contactors operate by rotating the biofilter media, consisting of disks or tubes, through a tank containing the wastewater, Fig. 7.6. Bacteria attached to the rotating media are exposed alternately to the wastewater and the atmosphere, which provides oxygen to the biofilm. Media is typically submerged 40% of the drum diameter and is rotated at speed of 1.5–2.0 rpm. Early RBC designs were fabricated from discs of corrugated fiberglass roofing material. Currently, media with a higher specific surface area are utilized in construction of RBC, which reduce the physical size and increase the ammonia and nitrite removal capacity. Rotating biological contactors have seen some use in fully recirculating systems, because they require little hydraulic head, have low operating costs, provide gas stripping, and can maintain a consistently aerobic treatment environment. In addition, they tend to be more self-cleaning than static trickling filters. The main disadvantages of these systems are the mechanical nature of its operation and the substantial weight gain due to biomass loading of the media and the resultant load on the shaft and bearings. Early efforts using RBC's often employed under-designed shafts and mechanical components, which resulted in mechanical failure, but a properly designed RBC is very functional and reliable. Figure 7.7 shows an RBC (manufactured by Fresh-Culture Systems, Inc,) which rotates using pumped air and/or water, eliminating the mechanical difficulties often reported with the gear motors, pillow blocks, chain drives and shafts used with most commercially available RBC's. These industrial-scale units each provide 930 m^2 of surface area, and have been operating successfully for over 7 years.

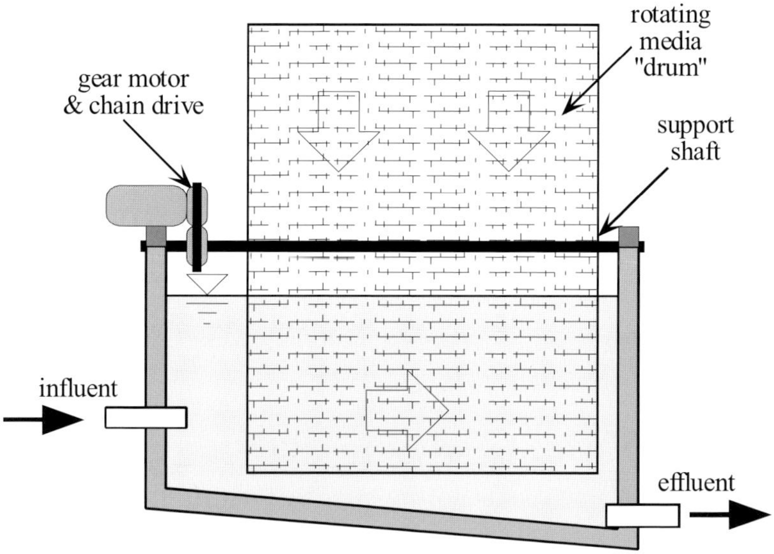

Figure 7.7 Schematic of a Rotating Biological Contactor, RBC (from Summerfelt, 1999).

Figure 7.8 Commercially Available Rotating Biological Contactor (RBC) Fresh-Culture Systems, Inc.

BEAD FILTERS

The floating bead filter has become a popular biofilter for the treatment of small or moderate flows, usually less than 1,000–2,000 L/min. The floating bead filters are expandable granular filters that display a bioclarification behavior similar to sand filters (Malone and Beecher, 2000). They function as a physical filtration device or clarifier by removing solids (Chen et al. 1993), while simultaneously encouraging the growth of desirable bacteria. They also remove dissolved wastes from the water through biofiltration (Malone et al. 1993). Floating bead filters are resistant to biofouling and generally require little water for backwash. The bead filter is typically either bubble-washed or propeller-washed during its backwashing procedure, which expands the bed and separates trapped solids from the beads. The beads used are food-grade polyethylene with a diameter of 3–5 mm and a specific gravity of 0.91, and a moderate specific surface area of 1150–1475 m^2/m^3 (Malone et al. 1993). Bead filters advantages include their modular and compact design, ease of installation, and operation. In addition, they can be used as a hybrid filter for both solids removal and nitrification.

One popular form of bead filter commonly used for small garden ponds and small aquaculture systems has an hourglass shaped internal geometry with a constricted washing throat. During continuous filtration, water from the production tank enters from the bottom through a slotted inlet pipe, flows upward through the bed of floating polyethylene beads and exits through a slotted discharge pipe at the top. The inlet pipe also serves as a sludge discharge line during backwashing. Backwashing consists of completely draining all the water from the filter, causing the beads to be sucked through the washing throat, where cavitation and bubbles from the air inlet valve vigorously scrub them. The solids laden water is then discharged and the filter refilled, and placed back into operation. Other methods of backflushing are used depending on the size of the filter include mechanical agitation with a motor driven propeller. Both methods are easily automated with a simple controller. Bead filters using propeller-washed backflushing have been built with bead volumes of up to 2.8 m^3. Most small-scale systems use the bubble-washed filters, typically less than 0.28 m^3.

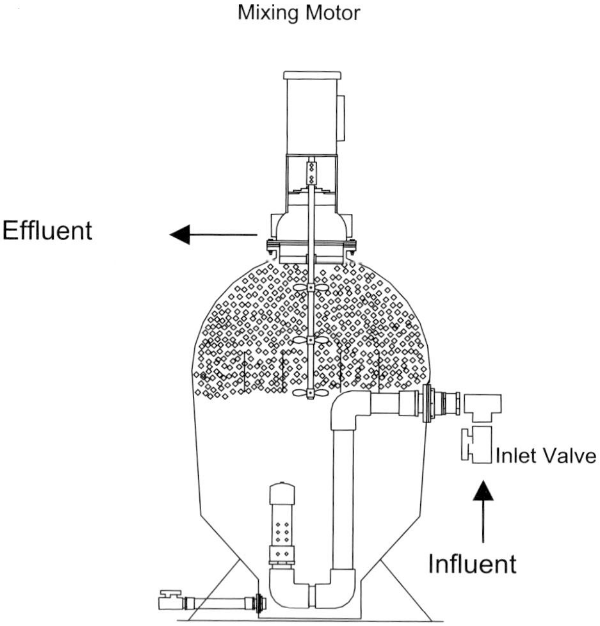

Figure 7.9 Schematic of a Pressurized Floating Bead Filter.

Figure 7.10 A Bubble-washed and a Commercial Size Propeller –washed Bead Filter.

FLUIDIZED-BED BIOFILTERS

Fluidized-bed biofilters have been used in several large-scale commercial aquaculture systems (15 m^3/min to 150 m^3/min or 400 to 4,000 gpm). Their chief advantage is the very high specific surface area of the media, usually graded sand or very small plastic beads. Specific surface areas range from 4,000 to 45,000 m^2/m^3 for sand versus 100 to 800 m^2/m^3 for trickling biofilter media and 1050 m^2/m^3 for bead filter media. The fluidized-bed biofilter can easily be scaled to large sizes, and are relatively inexpensive to construct per unit treatment capacity (Summerfelt and Wade, 1998, Timmons et al. 2000). Since the capital cost of the biofilter is roughly proportional to its surface area, fluidized-bed biofilters are very cost competitive and are relatively small in size compared to other types of biofilters (Summerfelt, 1999). Fluidized-bed biofilters are efficient at removing ammonia; typically removing 50–90% of the ammonia during each pass in cold- and cool-water aquaculture systems (Summerfelt et al. 2001). Nitrification rates for coldwater systems range from 0.2 to 0.4 kg TAN removal per day per cubic meter of expanded bed volume (Timmons and Summerfelt, 1998). In warmwater systems, TAN removal rates range from 0.6 to 1.0 kg per day per cubic meter expanded bed volume (Timmons and Summerfelt, 1998). The main disadvantages of fluidized-bed biofilters are the high cost of pumping water through the biofilter and that a fluidized-bed biofilter does not aerate the water, as do trickling towers and RBC's. Additional disadvantages are that they can be more difficult to operate and can have serious maintenance problems, usually due to poor suspended solids control and biofouling.

In fluidized-beds, water flows through the void spaces in the media, either upward or downward, depending upon the specific gravity of the media. The bed becomes fluidized when the velocity of the water through the bed is sufficiently large to suspend the media in the velocity stream, causing the bed to expand in volume. The resulting turbulent motion of the media provides excellent transport of dissolved oxygen, ammonia-nitrogen and nitrate-nitrogen to the biofilm and shears off excess biofilm. The result is high nitrification capacity in a relatively compact unit, but at the cost of the high energy required to fluidize the filter media.

The design of the flow distribution mechanism is absolutely critical for reliable operation of fluidized-bed biofilters (Summerfelt and Cleasby, 1996; Summerfelt, 1996). A variety of mechanisms has been employed to inject the water into the bottom of large fluidized-sand biofilters. Traditionally, some form of pipe manifold, starting at the top of the biofilter and running down through the inside of the reactor, has been used. This header and lateral system creates an additional operating pressure that the pumps must work against, generally on the order of ⅓ to ½ of an atmosphere. A recent innovation is the Cyclo-Bio™ that introduces the water flow into the bottom of the vessel through an outside annulus (incorporated into the bioreactor fiberglass wall) through a continuous slot that circumscribes the bioreactor wall (Timmons et al. 2000). This design saves energy by eliminating the conventional pipe-lateral manifold system and its high pipe friction losses. The Cyclo-Bio™ design also has very little pressure drop due to the lower water velocities produced across the slotted inlet.

The major advantage of fluidized-sand biofilters for both conventional units and the Cyclo-Bio™ is their ability to be scaled to capacities to assimilate ammonia production from standing fish biomasses on the order of 50,000 kg. In effect, the fluidized-sand biofilters can be made as large as they need to be to handle a specified fish biomass. Other considerations will dictate the actual fish load, with the primary one being risk associated with catastrophic failure.

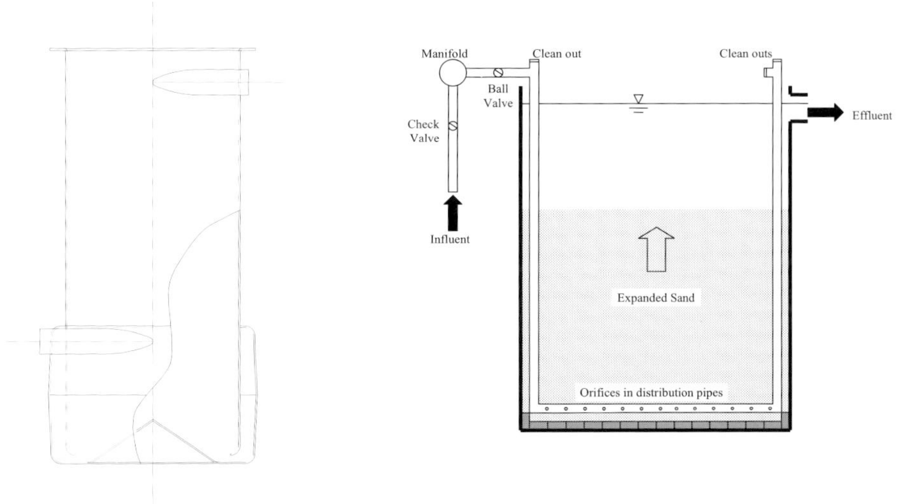

Figure 7.11 Schematic of a Fluidized Bed, Showing Pipe Manifold and the Cyclo-Bio™ (Courtesy of Marine Biotech, Beverly, MA).

Figure 7.12 A 4 m Diameter Fluidized-Sand Biofilter and Pipe Manifold Distribution (right). Photo is provided courtesy of Marine Harvest of their Big Tree Creek Hatchery (Campbell River, British Columbia, Canada). System engineering provided by PRAqua Technologies Ltd. (Nanaimo, British Columbia).

7.3 Generalized Engineering Considerations

All of the above biological filters are designed to perform the same function: oxidizing ammonia and nitrite to nitrate. Thus, the biological filter must be designed to fully oxidize the ammonia-nitrogen produced, with an additional safety margin to account for unforeseen events. From a practical perspective, the biofilter selection is less critical in small production systems, i.e., systems that feed at rates below 50 kg per day, than for larger production systems. In small systems, biofilters can be over-designed and the added cost is generally not of critical importance to the overall economic success of the venture. Smaller operations are not competing in the wholesale market were every penny counts. Usually, the smaller growers target the niche markets and gain sales by providing service or other product attributes that allows premium pricing to the seller/grower. On the other hand, for large production systems, i.e., those feeding 100 kg or more per day, economic designs for the biological filter are much more critical.

Each biofilter described above has advantages and disadvantages that need to be considered during the early design phase. One of the chief advantages of both the trickling biofilter and the RBC is that they both add oxygen to the water flow during normal operation. In addition, they provide some carbon dioxide stripping. In contrast, the submerged biofilters, bead filters, and fluidized-bed biofilters are all net oxygen consumers and rely completely on the oxygen in the influent flow to maintain aerobic conditions for the biofilm. If, for whatever reason, the influent flow is low in dissolved oxygen, anaerobic conditions are generated within the biofilters.

Both the trickling biofilters and the RBC filters have the distinct disadvantage of having low specific surface area media. Since the capital cost of the filter is proportional to its total surface area, the result is physically large and more costly filters. In contrast, bead filters and especially fluidized-bed filters use media with a high specific surface area. These results in reduced cost and space requirements in comparison to that required to achieve the same surface area in a trickling biofilter or RBC.

An additional disadvantage of the trickling biofilters and the RBC is that they readily biofoul, if suspended solids are not adequately controlled. Carbon eating heterotrophic bacteria grow 100 times faster than the autotrophic nitrifiers do. Their mass can double in an hour, while it takes nitrifiers days to double. This high growth rate and the associated oxygen demand consequently suffocate the nitrifiers buried deeper in the biofilms, resulting in death and sloughing of the biofilm from the bioreactor surfaces.

The advantages and disadvantages of several types of biofilters is summarized in Table 7.3 (Wilton, 2001).

Table 7.3 Advantages and Disadvantages of commonly used biofilters (Wilton, 2001).

Pro's	Con's
Trickling Biofilters	
Very simple design and construction requirements	Some biofilm sheared off is large enough to be problematic and many systems integrate post-biofiltration mechanical filtration for this reason
Currently a very popular method of biofiltration in the wastewater industry, which should improve material availability and cost	Filters using this media type tend to be very large in high feed load coldwater systems
Allows for passive aeration and CO_2 removal concurrent with biofiltration	Media itself can be costly due to low specific surface area
Media and design assistance is currently available from reputable commercial vendors facilitating the design effort	
Systems using these types of filters tend to be extremely stable	
Rotating Biological Contactors (RBC's)	
Low energy to move fluid across media	Can be expensive due to low specific surface area for large scale facilities
Provides passive aeration for nitrification process and limited CO_2 control	Mechanically more complex than most other biofilters
Can allow for efficient facility layout and combination of several processes (mechanical filtration, biofiltration, aeration and pumping in one common sump	Subject to rotational wear on bearing surfaces
Amenable to modularization, which can be useful for development of scalable facilities	

Pro's	Con's
Bead Filters	
Well developed product available from reputable commercial vendors. Can simplify system design and construction	Can be expensive due to relative low specific surface area for large scale facilities
Can be combined with other filter types in interesting hybrid systems as alternative design method	Relatively high head loss across filter can be an operational cost consideration
Can in some cases improve fine particle removal rates in well designed systems	Variable head loss across system can be problematic in systems without variable speed pumps
Amenable to modularization, which can be useful for development of scalable facilities	Has potential to leach nutrients into system or to fuel heterotrophic bacteria growth if not installed with pre-filtration systems or is backflushed infrequently
Fluidized-Bed Biofilters	
Very economical to build from commercially available materials	Can have problems with media carryover (initial fines) on system start-up
Large amount of design effort specific to coldwater systems using these types of filters	There are historical anecdotal reports of intermittent bed motility and system crashes
Raw filter media has very high specific surface area at low cost, which allows for very conservative design allowing for inherent capacity for expansion or load fluctuation	Can have problems with restarting if not designed to account for bed re-fluidization and distribution manifold/lateral flushing
Widest installed base of coldwater biofilters offers large operational and design experience base to draw from	Media density changes over time with biofilm accumulation in fine sand filters typical of coldwater systems, which necessitates a bed growth management strategy
Can be field built using a variety of proven methods or purchased from established and reputable vendors opening many design and construction options for facility designers or operators	Some systems can require relatively expensive plumbing to ensure that media is not back-siphoned on pump shut-down or power failure

7.4 Basic Design Concepts: Fluidized-Bed Sand Biofilter

All biological filters are designed to perform the same function of oxidizing ammonia and nitrite to the fully oxidized form of nitrate. From a practical perspective, biofilter selection is less critical in small production systems, but the right choice is essential for economical reasons in large operations. One primary benefit of the fluidized-bed biofilter is the ability to scale the biofilter to the system needs without paying large economic penalties as the fluidized-bed biofilter is scaled to an appropriate size. It is conceivable that a several hundred ton per year facility might be operated on one to three biofilters for the entire facility.

For simplicity, the production of total ammonia nitrogen (TAN) is approximately 3% by weight of the fish feed being fed per day, or more specifically 0.092% of the protein fed per day (see Eq. 4.9). Therefore, if 100 kg of feed are fed per day to a particular tank system, then there will be approximately 3 kg per day of TAN produced.

$$P_{TAN} = 0.03 \cdot R_{feed} \tag{7.7}$$

Biofilter TAN Removal Rates and Efficiencies

The biological filter must be designed to fully oxidize this TAN production; otherwise, the TAN concentrations in the tank will rise above design levels. Table 7.4 provides design values for TAN removal for both cool and warmwater systems and as affected by sand diameter size. This data is based upon replicated 10 cm (Cornell University) and 15 cm (Freshwater Institute) fluidized-bed biofilters. The sands were fluidized at fixed velocities that were set to achieve bed expansions of 50% with clean sand, e.g., a bed that is 1 meter in static depth would be 1.5 meters in depth when expanded. At Cornell University, inlet concentrations were 0.6 to 0.7 mg/L TAN, temperature 26°C, pH 7.3, 6 to 7 mg/L DO, and TSS <10 mg/L. At the Freshwater Institute, inlet concentrations were 0.5 to 0.6 mg/L TAN, temperature 15°C, pH 7.3 to 7.5, 10 to 11 mg/L DO, nitrite 0.04 to 0.06 mg/L, and TSS <10 mg/L (Tsukuda et al. 1997).

Note that the TAN removal rates in Table 7.4 are expressed on a unit volume basis, instead of on a surface area basis. As previously mentioned, low density medias such as RBC's and trickling towers provide nitrification rates proportionate to surface area provided by the media, but current research suggests that nitrification rates in granular medias are much more closely related to volume of media than surface area provided by the media. The large surface area provided by small sands provides no advantage in terms of nitrification rate. The fine sand biofilters, however, have demonstrated a much higher TAN removal efficiency (90% removal), when compared to the large sand biofilters (10–17%). The high removal efficiencies found with the finer sands are in part due to the low velocities required to fluidize fine sands. The lower velocities result in longer hydraulic retention times (HRT) in the reactor vessel and an associated higher percentage removal of ammonia. The negative aspect is that the reactor volume must be proportionately larger than a reactor that uses a lower HRT.

Table 7.4 Average TAN Removal Rates and Efficiencies Measured Across Coldwater (15°C) and Warmwater (26°C) Biofilters as a Function of Sand Size

Sand retained between sieve mesh sizes	40/70	20/40	18/30
Coldwater (15°C) Systems (Tsukuda et al. 1997)			
TAN removal rate, kg/d/m^3 clean static bed	1.5	0.51	0.51
TAN removal rate, kg/d/m^3 expanded bed	0.41	0.35	0.35
TAN removal efficiency, % each pass	90	10	10
Warmwater (25°C) Systems (M. B. Timmons, unpublished data)			
TAN removal rate, kg/d/m^3 clean static bed	NR	2.7	2.7
TAN removal rate, kg/d/m^3 expanded bed	NR	2.7	2.7
TAN removal efficiency, % each pass	NR	17	17

NR = not recommended due to excessive biofilm growth

The results, Table 7.4, show that a fluidized-bed biofilter used at warmer temperatures removed ammonia at a much higher rate than at cooler temperatures, i.e., 2.7 kg/d/m^3 @ 26°C versus 0.5–1.5 kg/d/m^3 @ 15°C. Based upon these numbers, it seems safe to use a nitrification rate of 1.0 kg TAN/m^3 per day of static bed as a design value for warmwater systems and a rate of 0.7 kg TAN/m^3 per day for clean static sand for coldwater systems (15°C).

The most important factors in the design of a biofilter are (1) the mass of TAN that it removes per day, i.e., the product of the flow rate across the biofilter and the change in concentration of ammonia across the biofilter; and, (2) the TAN removal efficiency (f_{rem}) of the biofilter. The mass of TAN removed per day can often be increased as the hydraulic loading rate is increased across a biofilter. However, increased hydraulic loading rate can decrease the TAN removal efficiency as increased hydraulic loading rates shorten the water retention time within the biofilter and increases the mass load on the filter.

The f_{rem} of TAN controls the accumulation of TAN in the recirculating system and thus the concentration of TAN discharged from a culture tank (TAN_{out}, mg/L). The TAN_{out} can be estimated using Eq. 7.8, which was first developed by Liao and Mayo (1972):

$$TAN_{out} = \left\{ \frac{1}{1 - R_{flow} + \left(R_{flow} \cdot f_{rem} \right)} \right\} \cdot \left\{ \frac{R_{TAN}}{Q} \cdot \frac{m^3}{1000\,L} \cdot \frac{10^3\,mg}{1\,g} \cdot \frac{\min}{60\,s} \cdot \frac{1\,day}{1{,}440\,\min} \right\} \tag{7.8}$$

where R_{TAN} (g TAN per day) is the rate at which the waste is produced, R_{flow} is the fraction of water flow that is reused, and Q (m^3/s) is the flowrate of water recirculated through the biofilter. This equation was derived from mass balances that assumes that no waste accumulation can occur in a culture tank, that the make-up water contains no TAN, and that the recirculating system is operating under steady-state conditions, i.e., water flow rates, waste production rates, and unit process treatment efficiencies are relatively constant.

The primary multiplicrs in Eq. 7.8 actually represent the "waste accumulation factor due to reuse" and the "concentration of waste that would be produced in a single pass through the tank." Therefore, Eq. 7.8 can be re-written as:

$$TAN_{out} = \left\{ \begin{matrix} accumulation\ factor \\ due\ to\ reuse \end{matrix} \right\} \cdot \left\{ \begin{matrix} concentration\ of\ waste \\ produced\ in\ a\ single\ pass \\ through\ the\ culture\ tank \end{matrix} \right\} \tag{7.9}$$

where,

$$\left\{ \begin{matrix} accumulation\ factor \\ due\ to\ reuse \end{matrix} \right\} = \left\{ \frac{1}{1 - R_{flow} + \left(R_{flow} \cdot f_{rem} \right)} \right\} \tag{7.10}$$

and

$$\begin{Bmatrix} \textit{concentration of waste} \\ \textit{produced in a single pass} \\ \textit{through the culture tank} \end{Bmatrix} = \left\{ \frac{R_{TAN}}{Q} \cdot \frac{m^3}{1000\,L} \cdot \frac{10^3\,mg}{1\,g} \cdot \frac{\min}{60\,s} \cdot \frac{1\,day}{1{,}440\min} \right\} \quad (7.11)$$

According to Eq. 7.8, the accumulation of waste within a recirculating system is dependent upon R_{flow} and the f_{rem} across the unit treatment process, i.e., the biofilter in the case of TAN. Most RAS operating in temperate climates reuse high fractions of their flow (to conserve heated water) and generally operate with only 5 to 100% of the total system water volume exchanged daily, equivalent to a fraction of water flow reused $R \geq 0.96$. In such recirculating systems, waste accumulation depends mainly upon the f_{rem} across the water treatment units and Eq. 7.8 simplifies to:

$$TAN_{out} \cong \left\{ \frac{1}{f_{rem}} \right\} \cdot \left\{ \frac{R_{TAN}}{Q} \cdot \frac{m^3}{1000\,L} \cdot \frac{10^3\,mg}{1\,g} \cdot \frac{\min}{60\,s} \cdot \frac{1\,day}{1{,}440\min} \right\} \quad (7.12)$$

Trout, char and salmon require relatively clean water and low levels of un-ionized ammonia, so high f_{rem} are required when designing systems to raise these species. For this reason, fluidized-sand biofilters containing fine sands are commonly used in coldwater recirculating systems because these biofilters will often achieve 70–90% TAN removal efficiencies. Fluidized-sand biofilters using fine sands are also capable of providing complete nitrification (due to their excess supply of surface area), which helps to maintain low nitrite-nitrogen concentrations within the recirculating system (generally <0.1–0.2 mg/L as nitrogen).

Species such as tilapia and African catfish do not require the high quality water required by salmonids. Therefore, the f_{rem} of TAN across the biofilter will not be as important as the mass of TAN that the biofilter removes each day.

Fine sands are not recommended for use in warmwater systems, due to probable difficulties in controlling excessive biosolids growth at warmer temperatures. Control of biosolids growth is still an issue when fine sands are used in coldwater applications, but biosolids growth is relatively easy to manage at 12–15°C.

Volume of Media Required

Given a defined rate of TAN production in g/d, and choosing an appropriate removal rate from Table 7.4, the volume of media required, V_{media}, can be determined:

$$V_{media} = \frac{P_{TAN}}{VTR} \tag{7.13}$$

As a general rule, design a fluidized sand bed with as much sand media volume as you can, i.e., minimize the ammonia loading on the bed per unit volume of sand media. This minimizes biological growth on the media and results in a less difficult bed to manage. The volume of sand required is independent of any of the other calculations, since the TAN removal rate is considered to be independent of the velocity through the bed.

> **"Rule of Thumb"**
> For fluidized sand bed,
> 1.0 kg TAN/d/m^3 warmwater
> 0.7 kg/TAN/d/m^3 coolwater
> (static sand bed volume)

Sand Bed Depth and Biofilter Cross Sectional Area

The selection of sand depth is primarily influenced by the physical constraints of the building, ceiling height, whether the sand filter can be partially submerged into the ground, and whether any additional elevation is required to gravity flow filtered water back to the culture tank through aeration and/or oxygenation unit processes. In any case, the design should result in requiring as much sand volume as practically possible. The larger the sand volume, the lower TAN load per unit volume of sand, which provides a factor of safety for the overall design. It also minimizes the biological film growth per unit particle of sand, which in turn will minimize sand bed growth and management problems related to changing expanded sand bed depths.

When fine sands are used, growth of biofilm can increase the expanded bed depth to the point that sand will be flushed from the reactor into the rest of the system. Therefore, fine sand biofilters are usually designed for eventual expansion of the biofilm-coated sand bed to achieve 200 to 300% of initial static sand depth, e.g., if a clean sand depth is 1 m before fluidization, after 50% fluidization the clean sand will be 1.5 m, once a thick biofilm grows in this bed the total sand expansion may be 200%, or around 3 m of total depth.

Based upon the above guidelines, then, choose and define your unexpanded bed depth, h_{sand}. Then, given the sand bed depth and the volume of media required for ammonia removal, the cross sectional area of the sand bed can be calculated:

$$A_{bed} = \frac{V_{media}}{h_{sand}} \tag{7.14}$$

Selecting Sand Size

The selection of the proper sand size is related to the previous discussion on sand depth. The overall design for the fish system will include calculating flow rates to maintain target levels for the various water quality parameters: primarily oxygen, ammonia, and carbon dioxide. Depending upon the type of oxygenation and CO_2 stripping units being used, large variations in required design flow rates can result. Thus, the design of the different components of the overall system is often manipulated until the flow rates for the different water quality parameters being controlled by the different system components are somewhat in balance.

Selecting the sand size and bed expansion determines the water velocity through the biofilter, V_{sand}. A finer sand will require lower water velocities than will a larger sand. Selecting smaller sands will result in larger reactor vessels and higher HRT's. The rate of TAN production by the fish will determine the pumping rate from the fish tank. This flow rate is the same flow rate that the biofilter vessel will see. Given a defined water flow rate, the selection of sand size will determine the allowable velocity through the biofilter vessel, V_{sand}, which can be expressed as gpm per ft^2 (m^3/day per m^2). The flow rate per unit area reduces to (unit balance):

$$\frac{\frac{m^3}{day}}{m^2} \overset{\circ}{=} \frac{m}{day} \cdot \frac{100cm}{m} \cdot \frac{day}{1440 \cdot 60s} \overset{\circ}{=} \frac{cm}{s} \tag{7.15}$$

Bed Expansion and the Associated Required Flow Rate

The failure of the fluidized-bed biofilter to properly expand can result in severe problems. Under-fluidization of the fluidized-bed biofilter will result in bed channeling with water leaving the biofilter untreated. Degrees of poor fluidization will result in the larger sands moving to the bottom of the bed and becoming static. Such areas then are apt to become anaerobic or anoxic resulting in denitrification and other undesirable water chemistry changes, e.g., sulfide gas production.

The author's recommendation is to choose an overall clean sand expansion of around 50% when designing the fluidized sand bed system. More generally, we would recommend expansions to be between 40 to 100%. Table 7.5 provides the required V_{sand} as affected by sand size and degree of expansion chosen. Larger expansions ensure greater assurance that all the sands are fluidized, but this comes at a higher flow rate or a reduced biofilter cross-sectional area. Choosing too low of an expansion will increase the likelihood that some of the largest sand particles will not fluidize and may create anaerobic pockets at the bottom of the sand bed column. Avoid this at all costs! Anaerobic pockets can lead to hydrogen sulfide gas (toxic at extremely low concentrations) and most certainly off-flavor conditions in the fish.

Sand expansion is a function of water temperature and of several sand characteristics. Water velocity requirements increase with increasing expansion and increasing sand size, Table 7.5. Estimates of expansion velocity requirements for a given sand can be erroneous due to the influence of water temperature on viscosity, as well as by variations in sand characteristics from different quarries. Therefore, the numbers in Table 7.5 should be used only for preliminary design estimates; a hydraulic test on a sample of the sand selected should be completed to determine the actual expansion velocities on a case-by-case basis.

Construct hydraulic testing columns with careful attention to how the flow is distributed and use at least 1 m sand depth. Experience with small test columns of 10 to 15 cm diameter have not been reliable in predicting full scale bed behavior. The observed fluidization and expansion in these smaller columns is very sensitive to the flow distribution mechanism. Try to use as large a diameter test column as possible and also try to mimic the flow distribution system that will be used on the biofilter units for the full-scale system.

The expansion water velocity values in Table 7.5 must be applied with caution because each particular sand source may have specific characteristics that cause it to expand more or less than the values predicted using Table 7.5. We recommend that you develop your own data for your own specific sands and use the data in the Table7.5 as an estimate or starting prediction point to design your sand filters.

Table 7.5 Upflow Velocity Requirements for Specific Sand Size and Degree of Expansion

Sieve Designation Number	D_{eq} (mm)	Specific Surface Area (m^2/m^3)	Velocity (cm/s) Estimated to Achieve Fluidization at:				
			0% exp (ε=0.45)	20% exp (ε=0.542)	50% exp (ε=0.633)	100% exp (ε=0.725)	150% exp (ε=0.780)
100	0.15	29,530	0.024	0.030	0.081	0.23	0.39
80	0.18	24,859	0.034	0.045	0.14	0.35	0.56
70	0.21	20,952	0.048	0.086	0.23	0.51	0.77
60	0.25	17,600	0.068	0.15	0.35	0.71	1.02
50	0.30	14,815	0.10	0.24	0.51	0.95	1.32
45	0.35	12,429	0.14	0.36	0.71	1.24	1.69
40	0.42	10,476	0.19	0.51	0.94	1.58	2.09
35	0.50	8,800	0.27	0.70	1.22	1.98	2.57
30	0.60	7,395	0.37	0.93	1.56	2.43	3.10
25	0.71	6,223	0.51	1.20	1.95	2.95	3.70
20	0.84	5,232	0.70	1.52	2.39	3.53	4.38
18	1.00	4,400	0.94	1.89	2.89	4.19	5.14
16	1.19	3,697	1.24	2.31	3.47	4.93	5.98

Sand size and degree of fluidization is based upon the water conditions expected for the biofilter, e.g., a warmwater system will generally use larger sands than cool water systems. Smaller sands will achieve higher removal efficiencies per pass and will approach 100% ammonia removal, but often require biosolids management at the top of the sand bed column. Larger sands require much higher upflow velocities and have very short hydraulic retention times in the biofilter reactor vessel, often as short as 2 or 3 minutes. TAN removal efficiency will be in the 10 to 30% per water pass. These type of sand bed systems will not collect biosolids and require minimal maintenance. However, these beds also constantly emit fine suspended solids.

From the continuity equation, Eq. 3.1, the chosen upflow velocity, V_{sand}, and the calculated cross sectional area of the sand bed, A_{bed}, the flow required for fluidization is defined:

$$Q_{sand} = V_{sand} \cdot A_{bed} \tag{7.16}$$

This flow rate must be compared to the other flow rate calculations that were performed to maintain fish water quality conditions, Q_{wq}, at some design or target level for the overall system, e.g. ammonia, oxygen, CO_2, or solids control. The design flow rate chosen for the tank system is the largest of all of these. If one of the Q_{wq}'s is larger than Q_{sand}, then one must determine if this larger Q will result in a sand bed expansion that is too large. If so, then the sand bed cross section must be increased. Generally, one should make the sand bed cross sectional area as large as possible, again for the reason of maximizing the sand bed volume and minimizing the nutrient loading per unit sand bed particle. This becomes an iterative approach and is often confusing to the first time designer.

Sand Specifications and Selection

Suppliers of graded filter sands usually report the effective size (D_{10}) and uniformity coefficient (UC) of their sand. Fish farmers purchase sand for fluidized-bed biofilter by specifying either an effective size or some range of sieve sizes, Table 7.6, in addition to an acceptable uniformity coefficient for the sand (where smaller UC's result in sands with less variation in particle diameter). Specifying sand that is sized through a 20/40 mesh means that the largest sand passes a 20 mesh sieve (D_{eq} = 0.841 mm) and the smallest sands are retained on a 40 mesh sieve and are larger than 0.42 mm. There is always some small percentage of sands that can be considered dust, but this should be only 1% to 3% of the total mass of sand. Upon start-up of a new fluidized-bed biofilter, the bed will flush the small sand from the system for several days. Several culture tank flushings are required to clear the fines from the system, which should be completed before fish are added to the system.

Table 7.6 Opening Sizes of U.S. Sieve Series Designation Number (Perry and Chilton, 1973)

Sieve Designation Number†	Size of Opening, (mm)	Sieve Designation Number†	Size of Opening (mm)
4	4.76	35	0.500
5	4.00	40	0.420
6	3.36	45	0.354
7	2.83	50	0.297
8	2.38	60	0.250
10	2.00	70	0.210
12	1.68	80	0.177
14	1.41	100	0.149
16	1.19	120	0.125
18	1.00	140	0.105
20	0.841	170	0.088
25	0.707	200	0.074
30	0.595	230	0.063

† Number of meshes per inch.

The ***effective size*** (D_{10}) is defined as the opening size, which will pass only the smallest 10%, by weight, of the granular sample. The D_{10} provides an estimate of the smallest sand in the sample and is the size used to estimate the maximum expansion at a given water velocity.

The ***uniformity coefficient*** (UC) is a quantitative measure of the variation in particle size of a given media and is defined as the ratio of D_{60} to D_{10}.

The D_{90} is the sieve size for which 90% of the grains by weight are smaller. The D_{90} provides an estimate of the largest sand in the sample and is the value used during design to calculate the water velocity required to fluidize even the largest sand to some minimal expansion, e.g., 20%. One should be sure to check that the design fluidization value meets this minimum requirement. The D_{90} can be estimated from the effective size (D_{10}) and the uniformity coefficient UC:

$$D_{90} = D_{10} \cdot \left(10^{1.67 \cdot \log(UC)}\right) \tag{7.17}$$

The ***mean size*** (D_{50}) is the sieve size for which approximately 50% of the grains by weight are smaller. The D_{50} provides an estimate of the average size of the sand in the sample and is the value used during design to estimate the average bed expansion at a given superficial velocity. Using the uniformity coefficient of the sand, you can approximate the mean sand size.

$$D_{50} = D_{10} \cdot \left(10^{0.83 \cdot \log(UC)}\right) \tag{7.18}$$

If desired, the specific surface area of the static sand bed (S_b) can be approximated from the static bed void fraction ($\varepsilon \approx 0.45$) and the ***sphericity*** of the sand ($\Psi \approx 0.75$):

$$S_b = \frac{6 \cdot (1 - \varepsilon)}{\Psi \cdot D_{50}} \tag{7.19}$$

BIOSOLIDS MANAGEMENT

The disadvantage of the smaller sand sizes is flocculation growth in the fluidized sand bed, especially as water temperature increases (Thomasson, 1991; Monoghan et al. 1996; Timmons, unpublished data). Biofilm growth on sand decreases individual particle effective density, which causes these particles to migrate toward the top of the biofilter column, and increases the total biofilter expansion. Heterotrophic growth continuously occurs and tends to trap sand particles in its growth. This biological growth plus the dead bacteria from biofilms also migrates to the top of the sand column. The smaller sands can be trapped in this flocculant material and remain at the top of the sand column. When the biosolid-coated sands remain at the top of the column, the shearing of biofilm material from the sand particles that occurs at the bottom of the bed near the orifice on the horizontal laterals does not occur.

If the fluidized-bed biofilter reactor vessel has translucent walls, one can easily observe the interface between the fluidized sand and the flocculant layer. When using small sands, D_{10} <0.42 mm (sieve size #40), in warmwater systems, the growth of the flocculant layer must be actively managed, or it can become uncontrollable. The fluidized-bed biofilters operator must have a regular routine of removing the flocculant layer or the whole sand bed could become engulfed. Removing the flocculant layer will also generally require replacement of sand, especially when fine sands are used. Replacement of sand has an inherent disadvantage beyond the obvious problems (sand cost and labor) in that the fine material introduced with the new sand will foul the water column. Depending upon the fish species and the rigor of the required water quality, pre-flushing of the new sand may be required.

Cornell University research experience when using sands that have D_{10} >0.42 under warmwater conditions has been that the beds have no appreciable collection of flocculant material at the top of the fluidized-bed biofilters. This is a large advantage in terms of simplifying management. The ***disadvantage*** is that the larger sands require higher fluidization velocities and this reduces the size of the biofilter.

> **"RULE OF THUMB"**
> Well managed FSB's will lose ¼ to a full tank volume of sand each year

For nearly 10 years, the Freshwater Institute has consistently managed fine sand (D_{10}= 0.20–0.25 mm) fluidized-bed biofilters in several recirculating rainbow trout culture systems. The fluidized-bed biofilters operated reliably during this period, with TAN removal efficiencies typically ranging from 70–90%. Biosolids growth in the fluidized-bed biofilters was usually controlled by siphoning biosolids from the top of fluidized-bed biofilters as their beds reach a maximum depth, and then replacing lost sand as needed. The Freshwater Institute has also controlled biofilm thickness by shearing the biofilm in-vessel, using a pump to transport the flocculant particles from the top of the biofilter to the bottom of the bed, where shear forces are greatest. This approach to controlling biofilm growth effectively maintained bed expansion at a fixed level without significant flushing of sand. All fluidized bed sand filters no matter how well managed will lose approximately ¼ to a full tank volume of sand per year. These sand losses may be insignificant on a daily basis, but over the course of several months to one year will generally require the replenishment of sand. Fortunately sand is inexpensive. In particular circumstances, pro-active capture of sand particles for recycling back into the vessel may be warranted.

In the Freshwater Institute's coldwater recycle system, it was found that biosolids did not accumulate within expanded beds using sands with effective sizes of (D_{10}) 0.60 and 0.80 mm. However, biosolids did sometimes collect in a distinct layer above the expanded sand layer. Siphoning the biosolids layer was simple with these larger sands, because the expansion depth of these sands remained fairly constant and the biosolids could be removed relatively free of sand. Because the biosolids layer is expanded, it is also fluid, which greatly reduces sand loss and the need to replace old sand with new sand (when larger sands are used). A siphon can be strategically placed at a height so that only biosolids are removed and not clean nitrifier-biofilm-covered sand. A distinction is being made here between biosolids that are sand particles encased within bioflock and heterotrophic bacteria and sand that is covered with biofilm from nitrifiers.

Design the Water Delivery System

Designing an effective fluidized-bed biofilter water delivery system requires appropriate sizing of the plenums, pipes, and orifices to promote equal fluidization of the sand bed. Generally, this involves using a collector plenum to receive water from the pumps and then routing a given number of laterals from this plenum to provide an effective lateral grid across the bed floor. Maximum spacing of laterals (center to center) is in the 15 to 30 cm (6 to 12 inch) range. The basic idea is that the cross sectional area of the pipe upstream of a group of laterals or laterals upstream of orifices should be 2 times the collective area of the downstream flow area of the laterals or downstream area of the orifices. This results in the pressure gradient being dominated by the loss across the reduction section so that each reduced section, e.g., lateral or orifice, will see approximately the same pressure regardless of its location along the lateral (orifice case) or lateral attached to a collection plenum/manifold. This approach ensures that the flow will be approximately the same to each of the laterals from the manifold/plenum or from a lateral to each orifice.

The following area ratios should be maintained between manifolds and laterals and between laterals and orifices:

$$1.5 < \frac{manifold\ area}{lateral\ area} < 3 \tag{7.20}$$

$$2.0 < \frac{lateral\ area}{orifice\ \ area} < 4 \tag{7.21}$$

As a rule of thumb, the flow distribution manifolds designed according to these criteria have a total pressure requirement (in meter or feet of water pressure) equal to approximately three times the sand depth plus the height that the water must be lifted from the floor sump to the top of the biofilter. Therefore, total pumping head requirements range from 5 to 11 m (16 to 36 ft), depending mostly on the depth of the biofilter column. For precise pressure estimates, the hydraulic grade line should be calculated across the piping from the pump sump to the water depth in the FSB column. Summerfelt (1996) and Summerfelt et al. (1996) give a more detailed discussion on how to estimate pressure losses through the FSB system and how to size laterals and plenums/manifolds. An example of how a spreadsheet can be used to size the pipe manifold, pipe laterals, and distribution orifices is provided, Table 7.7.

Sizing of Orifices

Orifice size is selected to fluidize the bed by assuming the pressure loss across the orifice is greater or equal to the static sand depth:

$$\frac{C_{orifice} V_{orifice}^2}{2g} > h_{sand} \tag{7.22}$$

Orifice coefficient values, $C_{orifice}$, are in the range of 0.6 to 0.8. While the choice of this orifice coefficient does not appear to be terribly significant, since the known value in the equation will be the flow rate, small changes in $C_{orifice}$ can significantly affect the estimate of the pressure loss term. The severe negative consequence of using a $C_{orifice}$ value that is too low is that the bed may not fluidize in all regions. A $C_{orifice}$ value of 0.7 has been used in past designs, or a $C_{orifice}$ value of 0.6 and an orifice size is selected to create a pressure loss 1.2–1.5 fold greater than the sand depth.

Plugging Concerns

A major concern in operating fluidized-bed biofilters is the plugging of the lateral systems with sand. This can happen if the check valves in line with the pumps and biofilter malfunction and do not close at the time pumps are shut down. Check valve failure should be rare, but if it ever does happen, water will siphon out of the biofilter to the pump sump (which is at a lower elevation) and carry sand into the pipe laterals until the pipes are plugged. To prevent this, premium quality swing check valves should be used to reduce the chance of failure. In the event of a failure, laterals can be unplugged, usually in a matter of hours, with the use of clean-outs on the pipe laterals at the top of the biofilter. One must anticipate that the pipe laterals will at some point become plugged, and know how to clean them effectively if the biofilter is to be restored to operating mode.

Table 7.7 Spreadsheet Example of Fluidized Sandbed (FSB) Design Process Spreadsheet is used to calculate the size of the FSB manifold, pipe laterals, and distribution orifices. Values that are in **BOLD** are user assigned numbers and (except for the expansion velocity) must be entered in an iterative procedure until the criteria in ***Italics*** are met.

Design Parameter		Comments
Expansion velocity, cm/s	**2.6**	
Expansion velocity, ft/min	5.1	
Flow Rate, gpm/ft^2	38.0	
Diameter of sand tank, inches	72	
Diameter of sand tank, ft	6.0	
Area of floor cross section, ft^2	28.3	
Total flow, gpm	1,074	
Lateral pipe diam, inches	**3.00**	
Lateral pipe area, sq inches	7.07	
Lateral Length, avg, feet	4.80	note: for round tank, avg L is 86% of vessel Diameter
Lateral spacing, inches	**12**	
Number of laterals (calc)	7.0	
Number of Laterals assigned	**6**	you must have a unit number
Orifice diameter, inches	**0.563**	
Orifice area, in^2	0.248	
Orifice spacing (='s lat space) inches	12	
Number of orifices/lateral	11.6	assigned in pairs
Total # of orifices assigned per lateral	**12**	
Orifice area per lateral, in^2	2.99	
Total Orifice area, ft^2	0.124	
Pipe area: orifice area	2.4	recommended between 2 to 4
Manifold area: lateral pipe area	**1.5**	recommended area ratio is 1.5 to 3.0
Min Req Manifold diameter, inch	9.0	using one header pipe for all the laterals
Min Req Manifold diameter, inch	6.4	using two header pipes for all the laterals
Orifice Coefficient (0.6 to 1.0)	0.8	
Velocity across orifice, ft/s	24.1	
Head loss across the orifices	9.0	This number needs to be > sand depth

Currently, this siphoning action has been successfully prevented by placing swing check valves above or below the pump(s). Heavy, brass or PVC swing check valves with a solid rubber swing flap or a brass swing flap and a well-machined seat ledge were used. These valves are still working effectively after several years of operation. Others have used siphon breaks to let air into the manifold above the biofilter when pumps are shut down, which allows the water in the pipe manifold to drain into the biofilter and prevents siphoning. The Freshwater Institute has tried a siphon break such as this and found it to be undesirable, because the air filling the laterals and manifold is forced through the sand bed when pumping is restored. The large slug of air passing through the biofilter has caused significant volumes of sand and biosolids to dump out of the biofilter. This phenomenon is probably much more of a problem with the finer sands and the associated mass of biosolids that accumulate in these biofilters. Similarly, any air leak into the pumping and lateral system will also promote sand loss from a bed and prevent proper operation, especially when smaller sands are used.

When using check valves, some allowance should be made to prevent air locks in the system. Simply examine the system, and if there is the potential for air lock to occur, make provisions to allow air to escape from the pump. This is a typical problem if the pumps are submerged and the collection sump is drained once a pump is turned off. When the sump area is refilled, the pipe above the pump will have trapped air and will prevent the pump from moving enough water to open the check valve.

Cost Basis of Sand Filters

As stated in the beginning of this chapter, the main argument for using fluidized-bed biofilters is that their cost per unit of TAN treated is low compared to competing technologies. This is true only for larger systems. Fingerlakes Aquaculture, LLC (Groton, NY) operated 1.83 m diameter by 4.6 m high fluidized-bed biofilter systems using a static depth of sand of 2.14 m. Volume of sand in these systems was 5.61 m^3. Using a TAN nitrification rate of 1 kg TAN/day/m^3 indicates a nitrification rate of 5.61 kg TAN per day per fluidized-bed biofilter, which would support a fish feeding load of 187 kg feed per day. More recently, Fingerlakes installed and operated 3.5 m diameter by 1.83 m high Cyclo-biofilters with good success. Fingerlakes Aquaculture uses a design value of 100 kg and 275 kg feed per day for these biofilter systems, respectively.

Costs provided by Fingerlakes Aquaculture for their 1.83 m diameter fluidized-bed biofilter tank systems are approximately \$2,200 for the fiberglass reactor vessel, $800 for a fiberglass plenum/manifold, plus the costs of pipe and valves. Fingerlakes Aquaculture estimates plumbing labor requirements to be approximately 50 man-hours to fully plumb a system to the fish tank unit.

Summerfelt and Wade (1997) report that two fluidized-bed biofilters built at two commercial fish farms sized to treat flows of 1.5 to 2.3 m^3/min cost around $6,000, including biofilter vessel, sand, piping, valves, shipping, and labor for installation. These prices were approximately five times less than the estimated cost for a installing a trickling filter of similar TAN removal capacity.

Summary of Steps

In summary, the following design steps should be followed when designing a fluidized sand bed:

1. Determine the TAN load
2. Determine the sand volume required to match TAN load
3. Select the design depth of sand bed
4. Select the sand size in relation to flow rate available or desired
5. Determine the required flow rate to fluidize the bed
6. Compare to the other required flow rates for maintaining water quality parameters in the fish tank; make adjustments in the sand bed cross section if one of the Q_{wq}'s is larger than Q_{sand}
7. Design the water delivery system

Design Example – Fluidized-Bed Sand Biofilter

Design a FSB for a warmwater system using 20/40 sized sand for a feeding rate of 100 kg feed/day.

Step 1. Determine the TAN load

$$TAN\ Load = 100\frac{kg_{feed}}{day}\cdot 0.03\frac{kg_{TAN}}{kg_{feed}}\cdot = 3{,}000\frac{kg_{TAN}}{day}$$

Step 2. Determine the sand volume required to match TAN load

$$Volume = \frac{\dfrac{3kg_{Tan}}{day}}{\dfrac{1\,kg_{TAN}}{day\cdot m^3_{sand}}} = 3m^3_{sand}\ required$$

Step 3. Select the design depth of sand bed

$$h_{sand} = 2\ m$$

This is the unexpanded sand bed height; placed on floor level. Selection based upon ceiling height and elevation required to permit gravity flow back to culture tank through aeration column and/or oxygenation column or other equipment devices.

Step 4. Cross sectional area of bed

$$A_{bed} = \frac{V_{media}}{h_{sand}} = \frac{3m^3}{2m} = 1.5m^2$$

Step 5. Fluidization velocity

For a 20/40 sand, Table 7.6 can be used to estimate D_{10} (40 mesh) = 0.42 mm. If the sand has a UC of 1.5, Eqs. 7.17 and 7.18 can be used to estimate D_{90} and D_{50}:

$$D_{90} = 0.42 \cdot 10^{1.67 x \log(1.5)} = 0.83mm$$

$$D_{50} = 0.42 \cdot 10^{0.83 x \log(1.5)} = 0.59mm$$

Using Table 7.5, the required velocity to expand the above 20/40 sand with a D_{eq} = 0.59 mm is 1.56 cm/s (Table 7.5 has a D_{eq} of 0.60 mm, so we use this value; or your could interpolate if more accuracy is needed for some particular application). To check if the largest sands will expand, the velocity required to expand a 0.83 mm sand 20% is 1.52 cm/s (again the table has a 0.84 mm D_{eq} value, so this is what we used for this example). Since the 1.52 cm/s is less than the bed design velocity used for the D_{50} sand at 50% expansion, the velocity selection is considered safe. If the D_{90} velocity criteria resulted in a velocity that exceeded the D_{50} value, then you should increase the overall velocity and go with a larger expansion than 50%. The design example will use 2.0 cm/s as a conservative safe estimate to ensure the large sands in the 20/40 sand used will fluidize. And going back to Table 7.5 and for the D_{50} value of 0.59 mm, the bed expansion is predicted as being around 75%. This expansion is considered a very safe value to ensure full mixing and is therefore consistent with our recommended conservative approach.

Step 6. Flow rate

Given the selection of an expansion velocity of 2.0 cm/s and the cross sectional area of the bed (1.5 m^2), the required flow rate is:

$$Q_{sand} = V_{sand} A_{bed} = 0.020\frac{m}{s} \cdot 1.5m^2 = 0.030\frac{m^3}{s} \cdot \frac{60s}{\min} = 1.8\frac{m^3}{\min}$$

Do a unit conversion from m^3/min to gpm:

$$1.8\frac{m^3}{\min} \cdot 264\frac{gpm}{m^3/\min} = 475gpm$$

From a mass balance on oxygen, the required flow rate (1.8 m^3/min) in this example would have to carry around 10 mg/L of available oxygen to meet the average respiration needs of the fish consuming 100 kg feed/day. Therefore, to achieve the available oxygen needs the flow would have to pass through an oxygenation unit to achieve on average a 15–17 mg/L of dissolved oxygen before entering the culture tank.

In a coldwater application, sands with D_{10} of around 0.25 mm are selected because of the high nitrification removal efficiency across the filter that can be achieved, due to the high hydraulic retention time of the vessel. These biofilters are usually oversized because the sand selection locks in a velocity of near 1.0 cm/s and the recirculating flowrate is controlled by the oxygen requirements of the fish. In this case, you solve for A_{bed} knowing Q_{wq} and V_{sand}:

$$A_{bed} = \frac{Q_{wq}}{V_{sand}}$$

7.5 BASIC DESIGN CONCEPTS – FLOATING BEAD BIOFILTER

The primary method for sizing the bead volume of floating bead filters is based on a volumetric organic loading rate (Malone and Beecher, 2000), assuming that the biofilter is used as a bioclarifier and the filter is well managed to sustain nitrification. Since the ultimate source of organics in a recirculating system is the feed, the sizing criterion, v_f is based on the feed load, Table 7.8. The volume of bead media required for a specific application, $Volume_{beads}$ (m^3), can be determined by the rates of the peak feeding, R_{feed} (kg/day), and the floating bead filter sizing criterion, v_f (kg feed/m^3 of media·day):

$$Volume_{beads} = R_{feed} \cdot v_f \tag{7.23}$$

Table 7.8 Performance Parameters for Pressurized Bead Filters (Malone and Beecher, 2000)

Performance Parameter	Units	Growout Systems	Fingerling Systems	Broodstock Systems
Feed loading, v_f	kg feed/m^3 media·day	≤16	≤8	≤4
Design TAN concentration	mg-N/L	1.0	0.5	0.3
Typical TAN concentration	mg-N/L	<0.5	<0.3	<0.1
Volumetric TAN conversion rate	g TAN/m^3 media·day	140–350	70–180	35–105
Volumetric oxygen consumption rate	kg O_2/m^3 media·day	2.5–3.0	1.4–2.5	0.7–2.5
Temperature	°C	20–30	20–30	20–30
Filter effluent dissolved oxygen	mg/L	>3.0	>3.0	>3.0
Alkalinity	mg/L as $CaCO_3$	>100	>80	>50
pH	pH units	7.0–8.0	6.8–8.0	6.5–8.0
Aggressive backwash interval	days	1–2	1–3	1–7
Gentle backwash interval	days	0.5–1	1–2	1–3

The feed load criterion, v_f, of 16 kg/m^3 day has been tested and proven to be stable in the local commercial sector (Beecher et al. 1997; DeLosReyes et al. 1997; Sastry et al. 1999). At this feeding level the floating bead filters can reliably provide solids capture, BOD reduction, and nitrification, while sustaining water quality conditions suitable for the growout of most food fish species, i.e., TAN and nitrite levels can be expected to remain well below 1 mg/L (Malone and Beecher, 2000). Reduction of the sizing criterion to 8 kg/m^3 allows the reliable maintenance of water quality conditions demanded by the fingerling category. Finally, a loading guideline of 4 kg/m^3 day is recommended for breeding and broodstock maintenance programs where the cost of maintaining pristine water quality conditions is justified by the value of the stock.

An alternative approach to sizing floating bead filters is based on the areal nitrification capacity (Malone et al. 1993). This criterion is based on the observation that for a wide range of floating bead filters, their areal conversion rates (VTR) are approximately 300 mg TAN/m^2 day, in systems with TAN and nitrite levels between 0.5 and 1.0 mg N/L (Malone and Beecher, 2000). In addition, observed VTR values tend to increase with the increasing TAN tolerances (0.3, 0.5 and 1.0) associated with the three categories in Table 7.8. These VTR values can then be used to estimate the size of the floating bead filter:

$$Volume_{beads} = (1.0 - I_s)\frac{R_{TAN}}{VTR} \tag{7.24}$$

The in situ nitrification fraction (I_s) recognizes the effect of nitrification occurring on the sidewalls of the tank and in the system piping configuration (Mia, 1996). This fraction is often conservatively estimated at 30%, although values in excess of 50% have been observed. In many designs, this fraction is ignored in favor of over designing the system to insure long term stability, even with the occasional overloading of the biofilter. Additional information and detailed design criteria can be found in Malone and Beecher (2000).

7.6 DESIGN EXAMPLE FLOATING BEAD FILTER

In addition to the procedure outlined for fluidized-sand biofilter, floating bead filters can be designed following the steps listed below, which includes oxygen as a design factor:

Step 1: Calculate the dissolved oxygen requirement (R_{DO}).

Step 2: Calculate water flow requirement (Q_{tank}) required for fish dissolved oxygen demand.

Step 3: Calculate TAN production by fish (P_{TAN}).

Step 4: Calculate the volume of media required to remove P_{TAN} from the volumetric TAN removal rate (VTR).

What follows is an example of the design process for a warmwater facility (250,000 lb/yr or 114,000 kg/yr tilapia system), utilizing a floating bead filter. The water flow requirements are based on the dissolved oxygen requirements and the sizing of the floating bead filter is based on the daily TAN production rate, which is directly dependent on the feeding rate. This design is based on a feed rate of 100 kg/day. Metric units will be used throughout.

The basic assumptions for this system are:

- Feed per day
  R_{feed} = 100 kg/feed per day (220 lbs/day)

- Feed rate
  r_{feed} = 1.5% bw/day

- Biomass
  M_{fish} = 6,700 kg (14,700 lbs)

- Fish rearing density
  $\rho = 80\ kg/m^3$ (0.67 lbs/gal)

- Volume of tank
  $V_{tank} = 83\ m^3$ (22,000 gal)

- TAN production
  a_{TAN} = 0.03 kg TAN/kg feed

- DO demand
  a_{DO} = 0.288 kg O_2/kg feed

- Temperature
  T = 20 to 30°C.

The first step in the design process is to calculate the dissolved oxygen requirement for both the fish and the biofilter. In this example, the oxygen demand has been estimated at 0.288 kg O_2/kg feed, based on design guidelines from Malone and Beecher, 2000[a]. This is often a difficult number to determine, and must either be based on existing system performance or estimated from literature data. As mentioned in Chapter 4, fish metabolism is often estimated at 250 g O_2/kg feed. Assuming that 4.57 g O_2 is required per gram of ammonia converted to nitrate, then 140 g O_2/kg feed is required for nitrification. This value does not include the oxygen requirement for the heterotrophic bacteria, which range from 140 to 500 g O_2/kg feed. The total oxygen demand is approximately 0.50 kg O_2/kg feed. Although this analysis would suggest that using 0.288 kg O_2 per kg feed is low, this value has been used in several commercial systems' designs.

Step 1: Calculate the dissolved oxygen requirement (R_{DO})

$$R_{DO} = a_{DO} \cdot r_{feed} \cdot \rho \cdot V_{tank}$$

$$= 0.288 \frac{kg\ O_2}{kg_{feed}} \cdot 0.015 \frac{kg_{feed}}{kg_{fish} \cdot day} \cdot 80 \frac{kg_{fish}}{m^3} \cdot 83 m^3$$

$$R_{DO} = 29 kg \frac{O_2}{day}$$

$$29 kg \frac{O_2}{day} \cdot 2.2 \frac{lb}{kg} = 63 lb \frac{O_2}{day}$$

Step 2: Calculate water flow requirement (Q_{tank}) required for fish DO demand

Assume: DO_{inlet} = 15 mg/L (pure oxygen aeration system)
DO_{tank} = 5 mg/L (warmwater: 25°C, Tilapia)

$$Q_{tank} = \frac{R_{DO}}{DO_{inlet} - DO_{tank}}$$

$$= 29 kg \frac{kg_{DO}}{day} \frac{1}{(15-5) \frac{mg_{DO}}{L}} \cdot \frac{1\ day}{1{,}440\ min} \cdot \frac{10^6\ mg}{1\ kg}$$

---

[a] Dr. Malone developed the bead filter and this example uses his numbers to be consistent with his published information. The authors would recommend using higher design values for oxygen demand in non-bead filter systems.

$$Q_{tank} = 2{,}000 \frac{L}{min}$$

$$2{,}014 \frac{L}{min} \cdot \frac{1 gal}{3.785 L} = 532 \frac{gal}{min}$$

Check to see if the number of tank exchanges per hour is sufficient for effective solids removal.

$$Tank\ exchange\ rate = \frac{V_{tank}}{Q_{tank}} = \frac{83 m^3 \cdot \dfrac{1{,}000 L}{m^3}}{2{,}000 \dfrac{L}{min}} = 41\ min$$

The 41 minute tank exchange rate is about 1.5 water exchanges per hour; this is adequate. Try to avoid tank exchanges that exceed one hour and strive for 3 tank exchanges per hour. Tank water velocities imposed on the fish become the limiting factor on increasing the water exchange rates through the culture vessels.

Step 3: Calculate TAN production using the approximation that the ammonia production by fish (P_{TAN}) is 3% of the feeding rate:

$$P_{TAN} = a_{TAN} \cdot R_{feed}$$

$$= 0.03 \frac{kg_{TAN}}{kg_{feed}} \cdot 100 \frac{kg_{feed}}{day}$$

$$= 3.0 \frac{kg_{TAN}}{day}$$

Step 4: Calculate the volume (V_{media}) required to remove P_{TAN} from the Volumetric TAN removal rate (VTR) (see Table 7.8). Assume a growout application, and a VTR at the maximum TAN value of about 1 mg/L, 350 g N/ m^3 day.

$$V_{media} = (1 - I_s)\frac{P_{TAN}}{VTR}$$

$$= (1.0 - 0.3)\cdot\frac{3.0\dfrac{kg_{TAN}}{day}\cdot\dfrac{1{,}000g}{kg}}{\dfrac{350g_{TAN}}{m^3 day}} = 6.0m^3$$

This design assumes an in situ nitrification fraction of 0.3, which is a conservative estimate, often ignored to create an even more conservative design. A 3 m^3 (100 ft^3) propeller washed bead filter is commercial available, constructed of heavy-duty food grade fiberglass. Design flow rates are from 1900–2250 Lpm (500–600 gpm). Two of these biofilters would be sufficient for this design, incorporating additional safety factors by having two separate filters and increasing the flow through the system and thus the overall oxygen carrying capacity of the system. This could be critical during periods of overloading of the system, increased oxygen demand due to unforeseen events, or if one biofilter is taken off-line.

7.7 DESIGN EXAMPLE TRICKLING TOWER

The design of a trickling tower follows a similar set of steps as the floating bead filter, except that the nitrification rate is based on surface area rather than volume and the oxygen requirement is limited to only fish metabolism.

Step 1: Calculate the dissolved oxygen requirement (R_{DO}).

Step 2: Calculate water flow requirement (Q_{tank}) required for fish dissolved oxygen demand.

Step 3: Calculate TAN production by fish (P_{TAN}).

Step 4: Calculate the surface area of media required to remove P_{TAN} from the areal TAN removal rate (ATR).

Step 5: Calculate the volume of media required based upon the required area (V_{media}) and the specific surface area (SSA) associated with the media being used.

Step 6: Calculate the biofilter cross-sectional area ($A_{biofilter}$).

Step 7: Calculate the biofilter depth from the biofilter cross-sectional area ($A_{biofilter}$) and volume (V_{media}).

The following example is the design process for a warmwater facility (250,000 lb/yr 113,630 kg/yr tilapia system), utilizing a trickling tower biofilter whose media has a SSA of 200 m^2/m^3 (61 ft^2/ft^3). The water flow requirements are based on the dissolved oxygen requirements of the fish and the sizing of the biofilter is based on the daily TAN production rate, which is directly dependent on the feeding rate. As in the floating bead filter design, it is based on a feed rate of 100 kg/day. Metric units will be used throughout this example.

The basic assumptions for this system are:

- Feed per day
  R_{feed} = 100 kg/feed per day (220 lbs/day)
- Feed rate
  r_{feed} = 1.5% bw/day
- Biomass
  M_{fish} = 6,700 kg (14,700 lbs)
- Fish rearing density
  ρ = 80 kg/m^3 (0.67 lbs/gal)
- Volume of tank
  V_{tank} = 83 m^3 (22,000 gal)
- TAN production
  a_{TAN} = 0.03 kg TAN/kg feed
- DO demand
  a_{DO} = 0.25 kg O_2/kg feed
- Temperature
  T = 20 to 30°C.

The first step in the design process is to calculate the dissolved oxygen requirement for both the fish and the biofilter. In this example, the oxygen demand has been estimated at 0.25 kg DO/kg feed. As previously mentioned, this is often a difficult number to determine, and must either be based on existing system performance or estimated from literature data. As discussed in Chapter 4, fish metabolism is often estimated at 250 g O_2/kg feed. For a trickling tower, the ambient atmosphere will supply the oxygen requirement for nitrification and for any heterotrophic bacteria. Thus, the overall oxygen demand should consist of only the fish metabolism of feed.

Step 1: Calculate the dissolved oxygen requirement (R_{DO})

$$R_{DO} = a_{DO} \cdot r_{feed} \cdot \rho \cdot V_{tank}$$

$$= 0.25 \frac{kg\ O_2}{kg_{feed}} \cdot 0.015 \frac{kg_{feed}}{kg_{fish} \cdot day} \cdot 80 \frac{kg_{fish}}{m^3} \cdot 83 m^3$$

$$R_{DO} = 25 kg \frac{O_2}{day}$$

$$25 kg \frac{O_2}{day} \cdot 2.2 \frac{lb}{kg} = 55 lb$$

Step 2: Calculate water flow requirement (Q_{tank}) required for fish DO demand

Assume: DO_{inlet} = 15 mg/L (pure oxygen aeration system)
DO_{tank} = 5 mg/L (warmwater: 25°C, Tilapia)

$$Q_{tank} = \frac{R_{DO}}{DO_{inlet} - DO_{tank}}$$

$$= 25 kg \frac{kg_{DO}}{day} \frac{1}{(15-5)\frac{mg_{DO}}{L}} \cdot \frac{1\ day}{1{,}440\ min} \cdot \frac{10^6\ mg}{1\ kg}$$

$$Q_{tank} = 1{,}736\,\frac{L}{min}$$

$$1{,}736\,\frac{L}{min}\cdot\frac{1\,gal}{3.785\,L} = 459\,\frac{gal}{min}$$

Check to see if the number of tank exchanges per hour is sufficient for effective solids removal.

$$Tank\ exchange\ rate = \frac{V_{tank}}{Q_{tank}} = \frac{83\,m^3\cdot\dfrac{1{,}000\,L}{m^3}}{1{,}736\,\dfrac{L}{min}} = 48\ min$$

The 48 minute tank exchange rate is about 1.5 water exchanges per hour; this is adequate.

Step 3: Calculate TAN production by fish (P_{TAN})

$$P_{TAN} = a_{TAN}\cdot R_{feed}$$

$$= 0.03\,\frac{kg_{TAN}}{kg_{feed}}\cdot 100\,\frac{kg_{feed}}{day}$$

$$= 3.0\,\frac{kg_{TAN}}{day}$$

Step 4: Calculate the surface area (A_{media}) required to remove P_{TAN} from the Areal TAN removal rate (ATR). Based on experience with the submerged trickling tower, the estimated Areal TAN removal rate was 0.45 g TAN/m^2 day.

$$A_{media} = \frac{P_{TAN}}{ATR}$$

$$= \frac{3.0\,\dfrac{kg_{TAN}}{day}\cdot\dfrac{1{,}000\,g}{kg}}{\dfrac{0.45\,g_{TAN}}{m^2\,day}} = 6{,}700\,m^2$$

$$6{,}700\,m^2\cdot\frac{10.76\,ft^2}{m^2} = 71{,}800\,ft^2$$

Note that for coldwater applications (12–15ºC) when TAN concentrations entering the trickling column are less than 1–2 mg/L, then the Areal TAN removal rate is only about 0.15–0.25 g TAN/m^2 day. Note also that the Areal TAN removal rate drops to only 0.1–0.2 g TAN/m^2 day for saltwater applications (@ 24ºC) when TAN concentrations entering the trickling filter are 1–2 mg/L.

Step 5: Calculate volume of media based on the specific surface area (SSA) associated with the media being used, for example BioBlock has a SSA of 200 m^2/m^3 (61 ft^2/ft^3).

$$V_{media} = \frac{A_{media}}{SSA}$$

$$= \frac{6{,}700 m^2}{200 \frac{m^2}{m^3}} = 33.5 m^3$$

Step 6: Calculate the biofilter cross-sectional area from required flow for the fish oxygen demand (Q_{tank}) and the hydraulic loading rate, HLR of 255 m^3/m^2 day (4.4 gpm/ft^2).

$$A_{media} = \frac{Q_{tank}}{HLR}$$

$$= 1{,}740 \frac{L}{min} \cdot \frac{1 m^3}{1{,}000 L} \cdot \frac{1}{\frac{255 m^3}{m^2}} \cdot \frac{1{,}440 \, min}{day}$$

$$= 9.8 \ m^2$$

The diameter of a single biofilter tank, $D_{biofilter}$, with this cross sectional area is:

$$D_{biofilter} = \sqrt{\frac{4 \cdot A_{media}}{\pi}} = \sqrt{\frac{4 \cdot 9.8 m^2}{3.14}} = 3.5 m$$

We recommend using two filters, so the cross sectional area of each of these biofilter units would be (9.8/2 = 4.9 m^2). The diameter would be:

$$D_{biofilter} = \sqrt{\frac{4 \cdot A_{media}}{\pi}} = \sqrt{\frac{4 \cdot 4.9m^2}{3.14}} = 2.5m$$

Step 7: Calculate the biofilter depth from the biofilter cross-sectional area (A_{media}) and volume (V_{media}).

$$Depth_{media} = \frac{V_{media}}{A_{media}} = \frac{33.5m^3}{9.8m^2} = 3.42m$$

LIST OF SYMBOLS

A	Area, m^2
A_{bed}	Cross sectional area of a fluidized sand bed, m^2
A_{media}	Surface area required to remove P_{TAN}, m^2
a_{TAN}	TAN production per unit of feed fed, kg TAN/kg feed
a_{DO}	DO demand per kg of feed fed, kg O_2/kg feed
ATR	Areal conversion rates, mg TAN/m^2 day
$C_{orifice}$	Orifice coefficient
D_{10}	Sand particle effective size, smallest 10% by weight of sample
D_{50}	Sand size for which 50% of the grains by weight is smaller, and also called mean size of the sand
D_{90}	Sand sizes for which 90% of the grains by weight are smaller
D_{eq}	Equivalent diameter of a sand particle assuming a spherical shape for the sand particle of equal mass
Depth	Depth of sand column unexpanded, m
DO_{inlet}	DO concentration at inlet to fish tank, mg/L
DO_{tank}	Dissolved oxygen concentration in the fish tank, mg/L
DO	Dissolved oxygen concentration, mg/L
f_{rem}	TAN removal efficiency across the biofilter as a fraction
g	Acceleration of gravity, m/s^2
h_{sand}	Height of unexpanded sand column, m
HLR	Hydraulic loading rate, gpm/ft^2 ($m^3/m^2 \cdot$ day)
HRT	Hydraulic retention time, minute
I_s	*In situ* nitrification fraction, unitless
M_{fish}	Biomass of fish in culture tank, kg
P_{TAN}	TAN production by fish, kg/day
Q	Flow rate, m^3/s
Q_{sand}	Flow rate required to fluidize the sand bed, m^3/s
Q_{tank}	Water flow requirement for fish DO demand, m^3/s
Q_{wq}	Flow rates required to maintain water quality design values in the fish rearing tank, m^3/s
R	Relative ammonia removal rate per unit time
R_{feed}	Feeding rates of the peak feeding, (kg/day)
r_{feed}	Feeding rate to fish on a % per day of bodyweight,%/day
R_{DO}	Dissolved oxygen requirement, kg/day
R_{flow}	Fraction of water that is reused
R_{TAN}	Ammonia nitrogen production rate, g TAN/day
RBC	Rotating biological contactor (acronym)
S_b	Specific surface area per unit particle volume, cm^{-1}
SSA	Specific surface area, ft^2/ft^3 (m^2/m^3)

T	Water temperature, °C
TAN	Total ammonia nitrogen
TAN_{out}	Concentration of TAN leaving fish culture tank, mg/L
UC	Uniformity coefficient (related to sand characteristics)
V_{sand}	Upflow velocity through sand bed, m/s or cm/s
v_f	Floating bead filter sizing criterion, kg feed/m^3 of media·day
V_{media}	Volume of media required to remove P_{TAN}, m^3
$V_{orifice}$	Velocity at orifice hole in a lateral, m/s
V_{tank}	Volume of tank, m^3 (gal)
$Volume_{beads}$	Volume of bead media required for a specific application, m^3
VTR	Volumetric nitrification capacity, g TAN/m^3 · day
ε	Static bed void fraction
ρ	Fish biomass density in culture tank, kg/m^3 (0.67 lbs/gal)
Ψ	Sphericity of the sand

CHAPTER 8

GAS TRANSFER

8.0 INTRODUCTION

The availability of dissolved oxygen (DO) is usually the first factor that limits increased carrying capacity and production in intensive recirculation systems. Using only aeration as a means of providing dissolved oxygen, a system can support only about 40 kg per m^3 (0.33 lb of fish per gallon) of water. However, by using pure oxygen and high efficient gas transfer devices to increase the amount of dissolved oxygen in the water column, stocking densities can easily be increased to over 120 kg per m^3 (1 lb of fish per gallon) of water. For example, by increasing the DO concentration at the inlet to a production tank from 10 mg/L (aeration alone) to 18 mg/L using pure oxygen, and assuming a DO concentration of 6 mg/L at the discharge, the carrying capacity of the system can be increased by a factor of three. Instead of only 4 mg/L DO (10 mg/L minus 6 mg/L) being available for respiration and metabolism by the fish, 12 mg/L becomes available (18mg/L minus 6 mg/L). Thus stocking densities can be increased from 0.33 lb/gal (40 kg/m^3) to 1.0 lb/gal (120 kg/m^3). An interesting by-product of this increased stocking density is that the concentration of the discharged solids is also increased, which means more efficient removal by, for example, microscreen filters.

In an aquaculture system operating at these high stocking densities, with low water exchange rates, oxygenation with little aeration and low pH, it is probable that dissolved carbon dioxide, a product of fish respiration, will accumulate to levels that create toxic conditions for both the fish and the biofiltration system. Other dissolved gases that also can be important to system operation are nitrogen, argon, and hydrogen sulfide, and in some special circumstances, methane, ammonia, and radon. In considering a treatment approach, it is important to understand that modification of the concentration of a particular gas, such as oxygen, may significantly change the concentration of other gases in solution. Thus, the design of a gas transfer system must take into consideration all the potential impacts of all the dissolved gases, as well as such water quality parameters as alkalinity, hardness, pH, and concentrations of iron and magnesium.

8.1 DISSOLVED GASES – FUNDAMENTALS

The atmosphere is composed of 20.946% oxygen (O_2), 78.084% nitrogen (N_2), 0.934% argon (Ar), 0.032%, carbon dioxide (CO_2) and other trace gases, Table 8.1. Barometric pressure (BP), or atmospheric pressure, is the sum of the partial pressures exerted by each of the individual gases in the atmosphere. The partial pressure of each gas is directly proportional to the mole fraction of that gas. Thus for oxygen at a standard temperature and pressure (STP) of 760 mm of Hg, its partial pressure is 760 mm times 0.20946 or about 160 mm Hg. Gauge pressure is equal to total pressure minus atmospheric pressure. In water, the Total Gas Pressure (TGP) is the sum of all partial gas pressures contributed by the individual gases, and can be greater than or less than barometric pressure. If the ratio of the Total Gas Pressure to barometric pressure is greater than one, the water is considered to be supersaturated.

Table 8.1 Dry Air Components

Species	% volume	% mass	Molecular Wt.
Nitrogen	78.084	75.600	28.0
Oxygen	20.946	23.200	32.0
CO_2	0.032	0.048	44.0
Argon	0.934	1.300	39.9
Air	100.000	100.000	29.0

The solubility of a gas (mg/L) in water depends on its temperature, salinity, gas composition, and total pressure. The solubility of the four major gases is presented in Table 8.2 as a function of gas composition.

From the Table 8.2 it can be seen that by increasing the mole fraction of oxygen from 0.20946 in air to 1.0000 with pure oxygen, the solubility or concentration of oxygen is increased from 10.08 mg/L to 48.14 mg/L (at 15°C).

Table 8.2 Solubility of Four Major Gases in Water

Gas Species	Solubility in Air* (mg/L)	Solubility of pure gas* (mg/L)
Oxygen	10.08	48.14
Nitrogen	16.36	20.95
Argon	0.62	65.94
Carbon dioxide	0.69	1992.00

*At 15 Deg. C.

The solubility of a gas is also proportional to its absolute pressure. Increasing the gas pressure will also increase gas solubility proportionately, i.e., doubling the gas pressure will double that gases solubility. Higher gas pressures occur in water injected into a pressurized system, or in water obtained from a deep well. One problem when using air as the oxygen source is that increasing the air pressure will also increase the solubility of nitrogen and argon above normal air saturation concentrations, which can result in potential gas supersaturation problems. Total gas pressure (TGP) for most systems should generally be kept below 105%, although the actual value is dependent on species and life stage.

Note that the percentage makeup of the air will not change with elevation (up to 90 km), but the total pressure will change as noted above. In addition, assuming 100% humidity conditions will change the above numbers; water vapor pressure will be approximately 10 to 15 mm Hg and is temperature dependent.

Gas Solubility Equations

Saturation levels for dissolved oxygen, nitrogen, and carbon dioxide ($C_{s,i}$) were calculated as a function of temperature and pressure based on Henry's Law (Colt, 1984):

$$C_{s,i} = 1000 K_i \beta_i X_i \frac{P_{BP} - P_{wv}}{760} \qquad (8.1^a)$$

[a] Symbols are defined at the end of the Chapter.

The Bunsen coefficients (β_i) and the water vapor pressure (P_{WV}) used in this determination were obtained from relationships developed by Weiss (1970, 1974) and ASHRAE (1972). The Bunsen coefficients are calculated separately for oxygen, nitrogen, and carbon dioxide as follows:

For Oxygen and Nitrogen

$$\beta_i = exp\,[A_1 + A_2(100/T) + A_3 ln(T/100)] \tag{8.2}$$

and for Carbon Dioxide

$$\beta_i = K_o\,(22.263) \tag{8.3}$$

$$K_o = exp\,[A_1 + A_2(100/T) + A_3\,ln(T/100)] \tag{8.4}$$

Temperatures (T) in Eq. 8.2 and 8.4 are in Kelvin (°C +273.15)

Table 8.3 provides the mole fractions for the different gases for standard air and the necessary constants to calculate individual gas saturation values in the above equations.

Table 8.3 Constants Used in Gas Solubility Equations and Mole Fraction for Standard Air

Gas Species	Mole Fraction	K_i	A_1	A_2	A_3	J_i
Oxygen	0.20946	1.42903	-58.3877	85.8079	23.8439	0.5318
Nitrogen	0.78084	1.25043	-59.6274	85.7661	24.3696	0.6078
Carbon Dioxide	0.00032	1.97681	-58.0931	90.5069	22.2940	0.3845

For example, assume water at 20°C, calculate the solubility of oxygen assuming a 100% oxygen environment:

$$C_s\,(20\,°C) = \frac{0.034 mL\,O_2}{mL\,H_2O} \cdot \frac{1{,}000 mL}{L} \cdot \frac{32 g\,O_2}{mol} \cdot \frac{mol}{22.4 L} \cdot \frac{L}{1{,}000 mL} \cdot \frac{1{,}000 mg}{g}$$

$$= 48.57 \ \frac{mg\ O_2}{L\ H_2O}$$

When gas concentrations are NOT 100% of a particular gas, then the dissolved gas concentration becomes proportional to the % gas composition present:

$$C_x\ (\%) = [C_{100\%}] \cdot [\%\ \textit{concentration of gas species } x]$$

For example, for standard air gas concentrations at 20°C (mg per liter)

Nitrogen: $100\% = 21.25 \frac{mg}{L}$ $78.084\% = 16.59 \frac{mg}{L}$

Oxygen: $100\% = 48.57 \frac{mg}{L}$ $20.946\% = 10.17 \frac{mg}{L}$

Carbon Dioxide: $100\% = 1886 \frac{mg}{L}$ $0.0314\% = 0.59 \frac{mg}{L}$

As mentioned above, equilibrium conditions can also be proportionately altered by changing the Total Pressure of the gas and maintaining the same % concentrations. For example, U-Tubes increase potential dissolved oxygen levels at the bottom of the U-Tube proportionate to the depth of the U-Tube. Similar changes in dissolved equilibrium concentrations can be caused by manipulating pressure or gas concentration.

Note: often $\frac{mg}{L}$ is expressed as parts per million, ppm

$$ppm = \frac{mg}{L} \cdot \frac{L}{1{,}000g} \cdot \frac{g}{1{,}000mg} = \frac{1}{1{,}000{,}000}$$

Note that ppm has no units or can be a ratio of the same units, e.g., lb per lb or g per g. There must be the same units in numerator and denominator so that the ratio is "unitless". There is some small error when temperature changes the actual weight of a liter of water from

1,000 grams, i.e., liter is a volume measurement not a weight measurement.

Observations

Problem #1: Weather pressure front comes through and oxygen stress in fish is observed.

Explanation

Low pressure weather front caused a 20 mm Hg drop in pressure.

Atmospheric pressure is now $\frac{(760-20)}{760} = 97.4\%$.

Oxygen concentration (assuming 20°C) is reduced to

$$10.17\ \frac{mg}{L} \text{ x } 0.974 = 9.91\ \frac{mg}{L}$$

Problem #2: Insufficient oxygen available in tank system. Gas conditioning device used to add oxygen is using pure air. Grower needs to double oxygen being added.

Solution

Increase the gaseous concentration of oxygen in the "device" to <u>2 times</u> atmospheric or approximately 42%. All other things remaining approximately constant, this should "double" oxygen being transferred into the fish tanks.

BAROMETRIC PRESSURE (P_{BP})

Barometric pressure can be obtained using the following equation

$$P_{BP} = 10^{a}$$

$$where \quad a = 2.880814 - \frac{h}{19{,}748.2} \tag{8.5}$$

h is the height above sea level in meters

WATER VAPOR PRESSURE (P_{WV})

Data given by Colt (1984) was used to regress water vapor pressure versus temperature between 10 and 30°C ($R^2 = 0.999$)

$$P_{WV} = A_0 e^{0.0645T} \tag{8.6}$$

where A_0 is 4.7603 or 1.65 x 10^{-7} for temperature in °C or K, respectively.

For example, the vapor pressures for water at 0, 10 and 30°C are 4.7, 9.1 and 33.0 mm Hg, respectively, using Eq. 8.6.

SUPER SATURATION OF GASES

The partial pressure of a gas (P_i^l , mm Hg) given a measured concentration level in the liquid phase ($C_{meas,i}$, mg/L) can be calculated as follows (mm Hg):

$$P_i^l = \left[\frac{C_{meas,i}}{\beta_i} \right] J_i \tag{8.7}$$

where values of J_i are given also in Table 8.3 and β_i is the Bunsen coefficient, calculated based upon the constants given in the same table.

Partial pressure of a gas in the gas phase (mm Hg) is proportional to mole fraction of gas:

$$P_i^g = X_i (P_{BP} - P_{WV}) \tag{8.8}$$

Users often prefer to view dissolved gas concentrations as its percent of saturation:

$$S_{\%} = \left[\frac{P^l}{P^g} \right] 100 \tag{8.9}$$

TOTAL GAS PRESSURE[a]

Finally, of particular importance to fish culture is the total gas pressure (P_{TG}) of the combined gases in solution:

$$P_{TG} = P^{l}_{oxygen} + P^{l}_{nitrogen} + P^{l}_{CO_2} + P_{wv} \qquad (8.10)$$

We often have fish culture recommendations based around an acceptable P_{TG} expressed as a percentage value:

$$P_{TG}(\%) = \frac{P_{TG}}{P_{BP}} \cdot 100 \qquad (8.11)$$

Example: Use of Gas Equations

The equations previously presented are a bit tedious to say the least. For sake of demonstration and to provide numbers for others to compare to once adapting these equations to spread sheet or computer type formats, some sample calculations are provided.

Assume: freshwater at 25°C and 228.6 m elevation

Barometric and Water Vapor Pressures

$P_{BP} = 10^{(2.880814 \text{ - } h/19748.2)}$

$P_{BP} = 10^{(2.880814 \text{ - } 228.6/19748.2)}$

$P_{BP} = 740$ mm Hg

$P_{wv} = 4.7603\ e^{(0.0645)(25)}$

$P_{wv} = 23.87$ mm Hg

Oxygen

K = 1.42903

$\beta = \exp[\ A_1 + A_2(100/T) + A_3\ln(T/100)\ ]$

$\beta = \exp[-58.3877 + 85.8079(100/298.15) + 23.8439\ln(298.15/100)]$

[a] total gas pressure is often noted as TGP

$\beta = 0.02844$ L/L-atm
$X = 0.20946$
$C = 1000KBX(P_{BP} - P_{wv})/760$
$C = 1000\ (1.42903)\ (0.02844)\ (0.20946)\ (740 - 23.87)/760$
$C = 8.022$ mg/L

Nitrogen

$K = 1.25043$

$\beta = \exp[\ A_1 + A_2(100/T) + A_3\ln(T/100)\ ]$
$\beta = \exp[-59.6274 + 85.7661(100/298.15) + 24.3696\ln\ (298.15/100)]$
$\beta = 0.01442$ L/L-atm

$X = 0.78084$

$C = 1000KBX(P_{BP} - P_{wv})/760$
$C = 1000\ (1.25043)\ (0.01442)\ (0.78084)\ (740 - 23.87)/760$
$C = 13.264$ mg/L

Carbon Dioxide

$K = 1.97681$

$\beta = K_o\ (22.263)$
$K_o = \exp[\ -58.0931 + 90.5069(100/298.15)$
$+ 22.2940\ln(298.15/100)\ ]$
$K_o = 0.03397$

$\beta = (0.03397)\ (22.263)$
$\beta = 0.7562$ L/L-atm

$X = 0.00032$
$C = 1000KBX(P_{BP} - P_{wv})/760$
$C = 1000\ (1.97681)\ (0.7563)\ (0.00032)\ (740 - 23.87)/760$
$C = 0.451$ mg/L

Now assume that we measured the following dissolved gas conditions for our site (25°C and 228.6 m elevation):

C_{oxygen} = 6.0 mg/L
$C_{nitrogen}$ = 13.3 mg/L

$C_{carbon\ dioxide} = 80.0$ mg/L

We can calculate individual gas concentrations in the liquid and then compare them to their saturation values:

$$P^l = \left[\frac{C_{meas}}{\beta}\right]J_i$$

Oxygen

$$P^l = \left[\frac{6.0}{0.02844}\right]0.5318$$

$P^l = 112.2$ mm Hg

$P^g = X(P_{BP} - P_{wv})$
$P^g = 0.20946(740 - 23.87)$
$P^g = 150.0$ mm Hg

$S_{\%} = (P^l/P^g)100$
$S_{\%} = (112.2/150)100$
$S_{\%} = 74.8\%$

Nitrogen

$$P^l = \left[\frac{13.3}{0.01442}\right]0.6078$$

$P^l = 560.6$ mm Hg

$P^g = X(P_{BP} - P_{wv})$
$P^g = 0.78084(740 - 23.87)$
$P^g = 559.2$ mm Hg

$S_{\%} = (P^l/P^g)100$
$S_{\%} = (560.6/559.3)100$
$S_{\%} = 100.2\%$

Carbon Dioxide

$$P^l = \left[\frac{80.0}{0.7563}\right]0.3845$$

$P^l = 40.67$ mm Hg

$P^g = X(P_{BP} - P_{wv})$
$P^g = 0.00032(740 - 23.87)$
$P^g = 0.23$ mm Hg
$S_{\%} = (P^l/P^g)100$
$S_{\%} = (40.67/0.23)100$
$S_{\%} = 17{,}682\%$

$P_{TG} = P^l_{oxygen} + P^l_{nitrogen} + P^l_{carbondioxide} + P_{wv}$
$P_{TG} = 112.2 + 560.6 + 40.67 + 23.87$
$P_{TG} = 737.34$ mm Hg

P_{TG} (%) = (737.23/740)100
P_{TG} (%) = 99.6%

Note that in the above example, argon has been omitted and would have contributed 6.82 mm Hg to the total gas pressure or 0.92% making the P_{TG} > atmospheric pressure. The reader can refer to Colt (1984) if these additional equations are required for their particular applications.

8.2 GAS TRANSFER

TRANSFER OF GASES

Gas Transfer is proportional to pressure difference for each particular gas. A species gas will move into a "higher" total pressure environment if the partial pressure of the species gas is higher than the species gas pressure in the "other" environment. Design and operational methods will influence and control *resistance* to gas transfer. Partial Pressure Differences determine *potential* for equilibrium dissolved gas conditions.

Principle: Gas transfer occurs due to *pressure differences*.

$$Q_i \frac{mass}{time} = \frac{1}{R_{pressure}} \cdot (P_{i,high} - P_{i,low}) \quad (8.12)$$

The rate of gas flow from a high pressure area to a lower pressure area is determined by the *resistance* to gas flow from the high to low pressure areas.

VALUES FOR PRESSURE

Atmospheric Pressure

At sea level 760 mm (29.91 inches) of mercury (Hg)
34 feet of water column
14.96 psi

Correcting for elevation:

$$P(h) = \frac{760}{[760 + (\frac{h}{32.8})]} \cdot 760 \tag{8.13}$$

where h is feet (not meters) of elevation above sea level

Example @ 1,000 ft,

$$P(1{,}000\,ft) = \frac{760}{[760 + \frac{1{,}000}{32.8})]} \cdot 760$$
$$= 0.96 \cdot 760 = 731\ mm\ Hg$$

TERMS USED TO DESCRIBE PRESSURE

Atmospheric Pressure is a sum of the ***partial pressures*** of contributing gases. ***Total gas pressure (TGP)*** is also sum of all the partial pressures (there may be non-typical gas concentrations for an individual gas component). Total Gas Pressure should not exceed 105% of atmospheric pressure (species dependent). Pure oxygen systems and green water systems can be subject to excessive total gas pressure problems. Incomplete degasification of “incoming” water, particularly for nitrogen, can lead to excessive total gas pressure (causes gas bubble disease in fish), since incoming water sources often are supersaturated with nitrogen gas.

Gauge pressure is:

$$Gauge\ Pressure = Total\ Pressure - Atmospheric\ Pressure \tag{8.14}$$

FUNDAMENTALS OF GAS TRANSFER

When air is in contact with water, dissolved gases in the water attempt to reach equilibrium with the partial pressures of the gases in the atmosphere. Two factors that directly impact the rate of gas transfer are first the area of gas-liquid interface and second the difference between the concentration (partial pressure) at saturation and the existing concentration of the gas in the water. For example, if the water is under saturated with the gas in question (a deficit), the gas will be transferred into solution and if the water is supersaturated, out of solution. In a simple trickling tower, it is possible to have supersaturated nitrogen gas being removed, while under saturated dissolved oxygen, increases in concentration. This overall gas transfer rate is dependent on the deficit (or surplus) of a dissolved gas and a proportionality constant, usually called the gas transfer coefficient. The overall gas transfer coefficient represents conditions in a specific gas transfer system. It is a composite term that includes such factors as the diffusion coefficient for the gas, the liquid-film thickness, and the area of gas-liquid interface. These factors also suggest ways to increase the overall rate of gas transfer. This could be accomplished, for example, by decreasing the liquid-film thickness by turbulence or mixing, increasing the gas-liquid interface by making the bubble size smaller, or increasing the concentration gradient.

In order to characterize the gas transfer, or more specifically the oxygen transfer ability of different system designs, a set of standard conditions and a standardized tests regime has been developed. The standard oxygen transfer rate (SOTR), defined as the amount of oxygen (kg O_2/hr) that a device will transfer to clean water at 0 mg/L DO and 20°C, is determined from these tests. (SOTR can be converted to pounds of oxygen per hour by multiplying by 2.205.) In addition, the amount of oxygen transferred by a device per unit of energy consumption (kg O_2/kWh) is defined as the standard aeration efficiency (SAE). Finally, the standard oxygen transfer efficiency (SOTE) or the absorption efficiency (AE) is defined as the oxygen transferred divided by the mass flow rate of oxygen supplied to the gas transfer device. Table 8.4 provides the correction factors for SOTR when initial conditions depart from the standard conditions of 0 mg/L initial DO and 20°C.

Overall gas mass transfer in a variety of devices can be represented by a G value as described originally by Colt and Bouck (1984), in an equation developed to predict gas transfer in packed columns:

$$G_{20} = \ln\left[\frac{C_s - C_{in}}{C_s - C_{out}}\right] \tag{8.15}$$

Table 8.4 SOTR[a] Correction Factors for Other Water Temperatures or Initial DO Levels

Dissolved Oxygen in Tank Water (ppm)	Tank Water Temperature in °F				
	50	60	70	80	90
0	.90	.91	.92	.93	.98
1	.82	.82	.82	.81	.95
2	.74	.73	.71	.69	.71
3	.66	.63	.61	.57	.57
4	.58	.54	.50	.45	.43
5	.49	.44	.39	.33	.29
6	.41	.35	.29	.21	.15
7	.33	.26	.18	.08	.02
8	.25	.16	.07	.00	.00
9	.17	.07	.00		
10	.09	.00			
11	.00				

[a] Ratings Obtained at Standard Conditions of 0 mg/L and 68° F (LSU, 1988)

The G value is very useful in evaluating the relative effectiveness of any particular gas transfer device. Once the G value is established for the device, it can then be used to predict gas transfer for any particular set of initial conditions of C_{in} and C_s. The effects of temperature (°C) can be adjusted for using a van't Hoff-Arrhenius relationship (APHA, 1995):

$$G_T = G_{20}\,\alpha\,(1.024)^{T-20} \tag{8.16}$$

Knowing influent conditions, Eqs. 8.14–16 can be used to calculate effluent conditions since effluent concentration is the only unknown in the equation:

$$C_{out} = C_s + (C_{in} - C_s)\,e^{-GT} \tag{8.17}$$

The α factor in (8.16) represents the increase in gas-liquid interfacial resistance to diffusion due to surface active compounds (Stenstrom and Gilbert, 1981). The α value is considered to be unity for clean water, but has been reported as 0.92 for surface waters from a reservoir (Ahmad and Boyd, 1988). Adjustments for G values for nitrogen and carbon dioxide are based upon the molecular diameters of the gas species relative to oxygen. Tsivoglou et al. (1965) applied Einstein's law of diffusion to estimate gas transfer for different gases is inversely proportional to the ratios of the molecular diameters. Applying this theory means that nitrogen gas transfer is 94% and carbon dioxide transfer is 90% of the rates occurring for oxygen.

Gas transfer efficiency values, E, are calculated based upon the change in DO across the system, expressed as a percentage of the initial dissolved gas deficit (Downing and Truesdale, 1955):

$$E = \left[\frac{DO_{out} - DO_{in}}{C_s - DO_{in}} \right] \cdot 100 \qquad (8.18)$$

Traditionally, gas transfer efficiency has been used to describe the effectiveness of various apparatus (see Table 8.5). Equation 8.19 relates E and G.

$$ln\left[\frac{DO_{out} - DO_{in}}{C_s - DO_{out}} \right] \cdot \frac{1}{E} = G \qquad (8.19)$$

Table 8.5 Aeration Effectiveness of Various Devices

Gas Transfer Device	Fall Distance cm	Aeration Efficiency (E), %
Simple Weir	23	6.2
	30	9.3
	67	12.4
Inclined Corrugated Sheet	30	25
	61	43
Splash Board	23	14
	30	24
	67	38
Cascade	25	23
	50	33
	100	52
Single Orifice	na	10
Packed Columns		see Eq. 8.20
Spray Towers		see Eq. 8.21

Oxygen absorption efficiency is calculated as the ratio of oxygen absorbed in a particular device to the amount of oxygen added and expressed as a percentage.

8.3 GAS TRANSFER OPTIONS

AERATION SYSTEMS – AIR STONES, PACKED TOWERS

Sources of Air

The standard sources of air in aquaculture are blowers, air pumps, or compressors. The primary differences between them are the pressure requirements and the volume of the discharge. Blowers supply high volumes of air at low pressure, while compressors supply small volumes of high pressure air. In specifying the type of air source required, two design parameters need to be determined: the required pressure and the required air volume. The operating pressure is determined by the requirements to overcome the water pressure at the diffuser's depth, the pipe friction losses, and the diffuser's resistance to air flows. For a typical application of air stones in a shallow (1 m or 3 ft) tank, this is about 2 to 3 psi (125 mm Hg). In deeper tanks or with diffusers requiring higher pressures, i.e., those with smaller bubbles or clogged pores, this could be considerably higher. The air volume required is determined by the pounds of oxygen required and the overall transfer efficiency of the system. For example, a 9 inch (23 cm) air stone operating in 3 feet (1 m) of water with 0.7 cfm (1.2 m^3/h) air supply transfers only 0.023 lbs/hr (0.25 kg/day) of oxygen.

Regenerative blowers are designed to provide large volumes of air at low pressure, typically less than 4 psi (190 mm Hg). They are most commonly used with either air stones or airlift systems. Advantages of regenerative blowers include their low noise levels, reliability, energy efficient motors, and lower comparative cost. Air pumps operate in the mid-range of performance, between blowers and compressors. Compressors are designed for high pressure operations, such as in very deep tanks or where long airlines are required.

Air stone

Air stones are very inefficient oxygen transfer devices (3–7%), but very inexpensive in terms of capital and operating costs. At low stocking densities and high exchange rates, they work very well at maintaining adequate oxygen levels. One disadvantage is the maintenance requirements due to clogging and biofouling, especially in very hard water.

Oxygen Transfer Systems

Sources of Oxygen

In aquaculture, three sources of oxygen are commonly used: high-pressure oxygen gas, liquid oxygen (LOX), and on-site oxygen generations. To insure availability and as backup, usually at least two sources are available at most facilities. High pressure oxygen gas is easily available in cylinders containing from 100 to 250 ft^3 (3 to 7 m^3) of gas at 2550 psi (170 atmospheres of pressure). A number of cylinders can be connected together using commercially available manifolds to increase the total capacity. Due to their cost and limited capacity, oxygen cylinders are normally used only as emergency backup systems.

In many areas, liquid oxygen is commercially available in bulk and can readily be transported and stored in on-site Dewar's type storage containers. At one atmosphere, liquid oxygen boils at -297.3°F (-182.96°C), thus special insulated cryogenic containers are required for storage. These containers range in size from 30 gal (0.11 m^3 liquid) to a much as 10,000 gal (38 m^3 liquid), and are usually rented or leased from the suppliers, although the smaller units can be purchased. One gallon of liquid oxygen is equal to 115 ft^3 (3.26 m^3) of gaseous oxygen. The maximum gas pressure in these containers is in the range of 150 to 200 psi (11.7 atmospheres). Prior to its use, the LOX is vaporized by directing it through heat exchanger coils. A liquid oxygen supply system will consist of a storage tank, vaporizer, filters, and pressure regulators. The economics of LOX use are dependent upon the transport cost, and the reduced capital and maintenance cost as compared to pressure swing adsorption (PSA) systems. In general, a LOX system is very reliable, operating even during power failures. Failures on farms using LOX systems as backup to power outage are caused by under-sizing the LOX system in the first place or unanticipated severe weather conditions that extend longer than predicted. Carefully consider your risks for such cases and size your LOX system with these potential dangers in mind. As a minimum, a LOX system should be able to maintain a facility with oxygen for 30 days. Remember that upon the first sign of major weather problems, it is probably prudent to take your fish off of feed, which will lower their oxygen demand dramatically over the next 24 hours.

Oxygen can also be generated on-site using either a pressure swing adsorption (PSA) or a vacuum swing adsorption (VSA) unit. In both cases, a molecular sieve material is used to selectively adsorb or absorb nitrogen from the air, producing an oxygen-enriched gas. Commercially available units can produce anywhere from 1 to 30 lbs (0.5 to 14 kg) of oxygen per hour at from 10 to 50 psi (0.7 to 3.3 atmospheres). A source of dry, filtered air at 90 to 150 psi (6.0 to 10.0 atmosphere) is required to produce an oxygen stream that is from 85–95% pure. PSA and VSA units operate on a demand basis and produce oxygen only when needed. They have proven to be very reliable and require little maintenance. However, they are both expensive in terms of capital and operationally expensive, due to the compressed air requirements. Also, since they require electrical power, some other source of oxygen is needed in the event of power failures or else the facility must be equipped with large backup generators and transfer switches.

U-tubes

The U-tube aerator operates by increasing the gas pressure, thus increasing the overall gas transfer rate. It consists of either two concentric pipes or two pipes in a vertical shaft 30 to 150 ft (9 to 45 m) deep. Oxygen is added at the upper end of the down-leg of the U-tube and as the water/gas moves downward through the contact loop, an increase in hydrostatic pressure increase the oxygen transfer rate. The overall oxygen transfer efficiency is a function of the depth of the U-tube, inlet gas flow rate, water velocity, diffuser depth and inlet DO concentration. Concentrations of dissolved oxygen ranging from 20–40 mg/L can be achieved, but the overall oxygen transfer efficiency if only 30–50%. Off-gas recycling can improve the absorption efficiency to 55–80%. Two advantages of the U-tube are the low hydraulic head requirements that allow operation with no external power if sufficient head is available, and that it can be used with water containing high levels of particulates or organics. Its chief disadvantages are that it does not vent off gasses such as nitrogen or carbon dioxide very efficiently and construction costs can be high, particularly if bedrock is present.

U-tubes are designed for flows where the downflow velocity is between 1.8 to 3.0 m/s. A particularly unique problem with U-tubes is that if too much oxygen is added a gas bubble blockage can occur that results in flow interruption. This will tend to happen if gas-liquid ratios exceed 25%. Be careful when adding oxygen.

For example, a U-Tube that was 40 feet (12 m) in depth would increased the potential for absorbing oxygen as follows (atmospheric pressure is equivalent to 34 feet, 10.4 m, of water):

$$C_{oxygen}(40\,ft) = C_{oxygen}(atmospheric) \cdot \left(1 + \frac{40\,ft}{34\,ft}\right)$$

$$C_{oxygen}(40\,ft) = 10.17 \cdot \left(1 + \frac{40\,ft}{34\,ft}\right) = 22.1\frac{mg}{L}$$

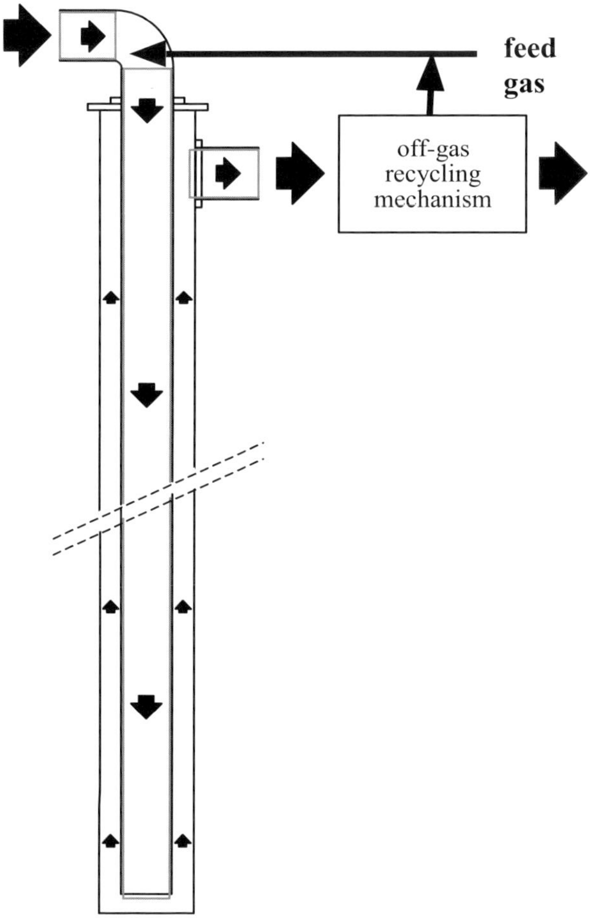

Figure 8.1 U-Tube Aeration System.

"RULE OF THUMB"
U-Tube Aerator

a) Design U-Tubes for a down flow velocity between 2 m/s to 3 m/s
b) Limit G/L ratio to <25%

Packed Columns

Packed columns consist of a vertical column filled with media having a high specific surface area. Water is uniformly distributed over the top of the media with a perforated plate or through a spray bar, and trickles down through the media. Oxygen is injected into the column and is transferred into the passing water through the large gas/liquid interface on the media. The column may be either open or closed at the top. Packed columns are efficient nitrogen strippers, but require a higher gas/liquid ratio (forced aeration) for carbon dioxide stripping. Packed columns are simple to build and easy to retrofit into existing facilities. Performance design characteristics include the water distribution method, media characteristics, media bed depth, gas/liquid loading rates, inlet DO concentration and operating pressure. Their main disadvantage is fouling due to the accumulation of organics and particulates on the media over time.

Packed columns have two additional advantages:

- Provide nitrification
- Provide CO_2 gas stripping

The above features allow to some degree that a packed tower can serve as the complete water conditioning system for lowly loaded systems. CO_2 stripping is discussed in more detail later in this Chapter.

Hydraulic loading rates should be at least 10 gpm/ft^2 (7 kg/m^{2}s), and can be as high as 70 gpm/ft^2 (48 kg/m^{2}s). Excessive hydraulic loadings will cause the packing to be flooded and must be avoided.

The gas transfer efficiency for packed columns depends upon the tower packing height (Z), the hydraulic loading (HL), and the characteristics of the packing material. Performance of packed columns can be predicted based upon the data developed by Watten (1990) and given in Table 8.6. The general form of the prediction equation is:

$$G_{20} = a + b \cdot Z \tag{8.20}$$

where the coefficients in Eq. 8.20 are given in Table 8.6

Closely related to a packed column and included within this section are spray towers. These units are also some type of tower or column but they are enclosed and then some type of spray nozzle is placed within them to provide the water gas breakup and gas transfer. Vinci et al. (1997) developed the following predictive equation using a 20 cm column diameter and a full cone spray nozzle (Spraying Systems Co., Wheaton, IL):

$$G = -0.05 + 0.3025 \cdot Z_{tower} + 0.000067 HL \tag{8.21}$$

Table 8.6 Coefficients for Predicting Packed Column Performance (Watten, 1990)

Packing Type	Geometric Surface Area, m^2/m^3	Coefficient a	Coefficient b
2.54 cm TriPack®	279		
HL[a]=32		0.327	1.655
HL=61		0.277	1.589
3.81 cm Nor-Pac®	144		
HL=32		0.324	1.555
HL=32		0.398	1.428
3.81 cm ACTIFIL®	139		
HL=34–73		0.357	1.349
5.08 cm Nor-Pac®	102		
HL=32		0.243	1.285
HL=32		0.162	1.855

[a]HL is the hydraulic loading on the packed column in kg/m^2s

LHO's

Low Head Oxygenators (LHO) are being used more frequently, particularly because of their adaptability to high flows using minimal hydraulic head, hence their name Low Head Oxygenator. The original LHO design was developed and patented by Watten (1989). LHO's vary in configuration, but all are fundamentally similar in operation. These units consist of a distribution plate positioned over multiple (5 to 10) rectangular chambers, Fig. 8.2. Water flows over the dam boards at the end of a raceway or is pumped upwards from an indoor fish tank, through the distribution plate, and then falls through the rectangular chambers. These chambers provide the gas-liquid interface needed for mixing and gas transfer. The streams of falling water impact a collection pool at the bottom of each chamber where the effluent water flows away from each chamber equally in parallel. All of the pure oxygen is introduced into the outer or first rectangular chamber. The mixture of gases in the first chamber, which now has a diluted oxygen concentration passes, sequentially through the remaining chambers. The gaseous mixture will decrease in oxygen concentration from chamber to chamber as the oxygen is continued to be absorbed. Finally the gaseous mixture will exit from the last chamber. This gas is referred to as off-gas. Each of the rectangular chambers is gas tight and the orifices between the chambers are properly sized and located to reduce back-mixing between chambers.

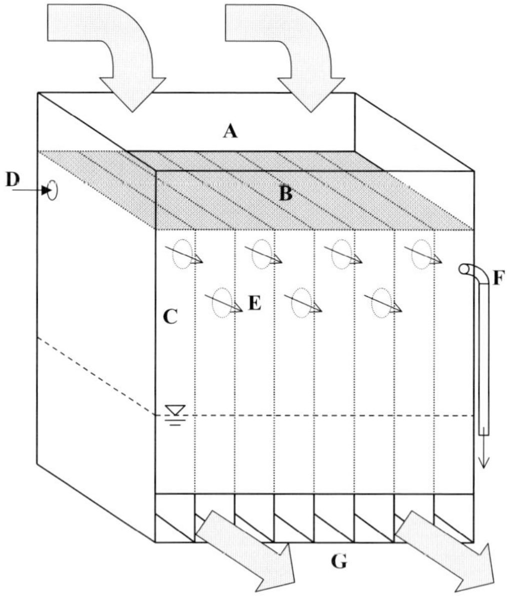

Figure 8.2 Low Head Oxygen (LHO) Units are Popular in Raceway Culture to Restore Oxygen in Serial Reuse Raceway. A typical LHO configuration and components are shown: water flowing into a collection trough or plate (A), through a perforated distribution plate (B), and is oxygenated in the chambers (C), as gas flows from inlet gas port (D), through holes between chamber to chamber (E), to the off gas port (F), where excess gas is bubbled off under water. Water exits at the bottom of the unit (G).

The operating performance of an LHO is primarily affected by the hydraulic head on the distribution plate (Y_1), the orifice hole size (Y_2), the depth of the receiving pool (Y_3), and the drop or fall distance from the orifice plate to the receiving pool of water (Y_4). Davenport et al. (2001) developed a regression model to predict the overall gas transfer coefficient G_{20} as a function of these geometric variables for a single chamber:

$$G_{20} = -0.0059(Y_2) + 0.017(Y_3) + 0.011(Y_4) - 0.00047(Y_3^2) - 0.000034(Y_4^2) + 0.00034(Y_2\,Y_3) - 0.000049(Y_2\,Y_4) + 0.000026(Y_3\,Y_4) \quad (8.22)$$

If (Y_3 >41 cm) Then $Y_3 = 41$
If (Y_2 >19 cm) Then $Y_2 = 19$

where hole size is in mm diameter and pool depth and drop height are in cm.

Timmons et al. (2001) developed an LHO model that is provided in the book's software package and is described further in Chapter 17. The Timmons LHO model calculates LHO performance as affected by hole size, hydraulic head over the flooded plate, hole orifice discharge coefficient (C_d), percent active hole area of the plate, plate size, and the number of chambers involved:

$$Q = C_d A\sqrt{2gY_1} \quad (8.23)$$

The value for C_d is predicted using the following equation (Timmons et al. 2001)

$$C_d = 0.914 - 0.00308Y_2 - 0.0519Y_1 + 0.000228Y_2^2 + 0.00298Y_1^2 - 0.000660Y_2Y_1 \quad (8.24)$$

If (Y_1 >13) Then $Y_1 = 13$
If (Y_1 <2.5) Then $Y_1 = 2.5$

A graphical representation of Eq. 8.24 is shown in Fig. 8.3. Classical approaches can be used to calculate C_d as well (Streeter, 1966) based upon jet Reynolds number and ratio of upstream cross sectional area to hole area.

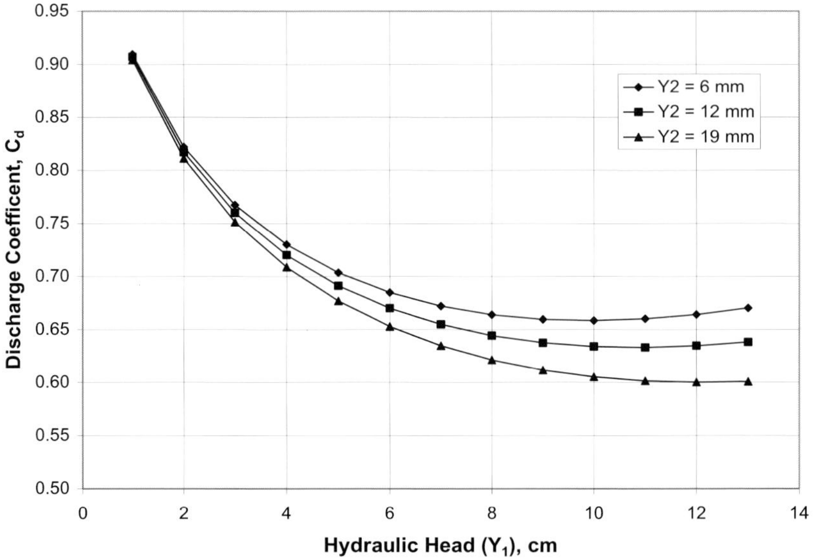

Figure 8.3 Discharge Coefficient Related to Hole Diameter (Y_2) and Hydraulic Head (Y_1).

Hole Size

The proper size of the holes to be used in the distribution plate can be determined by knowing the type of system the LHO will serve. Smaller holes generally have greater potential efficiencies. Smaller holes are also an advantage because the momentum of the water jets impacting the collection pool is less and the resulting gas bubbles that are formed will not travel downwards as far. This is advantageous because bubbles that become entrained in the effluent flow and do not rise back into the LHO chamber will be wasted to the atmosphere increasing operational cost. The trade off is that smaller holes will more readily clog from biological fouling or particulates. Some consideration needs to be given at this stage as to the nature of the system the LHO is serving. Potential sources of clogging particulates will be fish feces and scales. In general, the holes should be sized as small as possible but not so small as to incur frequent plugging.

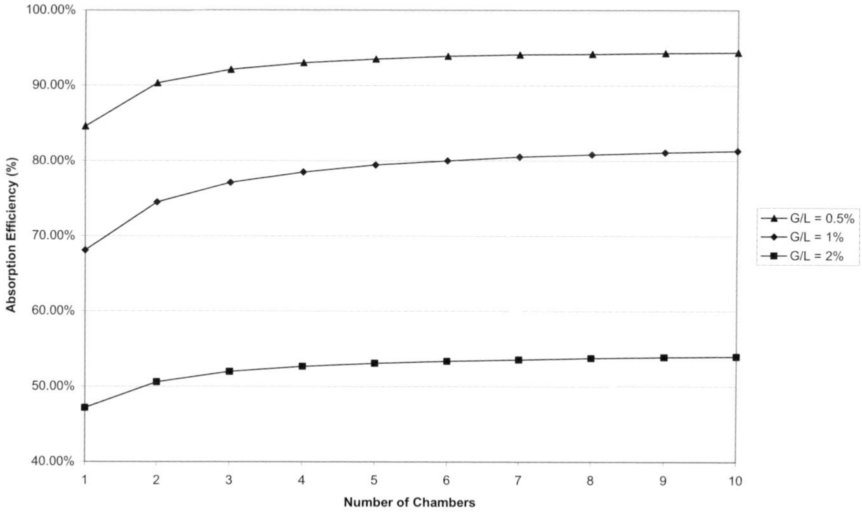

Figure 8.4 Absorption Efficiency as Affected by the Number of LHO Chambers and the Gas Liquid G/L Ratio (Arbitrary Set of Model Inputs were: Y_2 = 9.5 mm; Y_3 = 13 cm; Y_4 = 61 cm; Y_1 = 7.5 cm; Temp = 20.0°C; Top Area = 0.1 m^2; Active Hole Area = 10.0%; Chambers = Varies; G/L = Varies; DO_{in} = 6.0 mg/L; DN_{in} = 14.0 mg/L; DCO_2= 0.0; Pressure = 760.0 mm Hg; Oxygen Fraction in Inlet Gas = 0.99).

Predictions of LHO Performance

The LHO model is used to demonstrate the effects of LHO chamber number on gas absorption efficiency and effluent DO for various G/L ratios, Figs. 8.4 and 8.5, for an arbitrary LHO configuration (Y_1 = 7.5 cm; Y_2 = 9.5 mm; Y_3 = 13 cm; Y_4 = 61 cm; Temp = 20.0 °C; Top Area = 0.1 m^2; Active Hole Area = 10.0%; DO_{in} = 6.0 mg/L; DN_{in} = 14.0 mg/L; $DCO_{2,\ in}$= 0.0; Atmospheric Pressure = 760.0 mm Hg; Oxygen Fraction in Inlet Gas = 0.99). As can be seen in these two figures, an LHO should have at least 4 or 5 chambers to obtain high gas transfer efficiency. This is reflected in current commercial units that typically have 7 chambers. It is also quite evident from Fig. 8.4 that gas transfer efficiency is severely degraded at a G/L ratio of 2% (slightly over 50%). Thus, increasing G/L ratios to obtain higher effluent DO to meet biological fish demand is not an economical choice. In fact, the producer would probably be economically ahead to reduce fish density c than to try to maintain the higher densities by using elevated G/L flow rates.

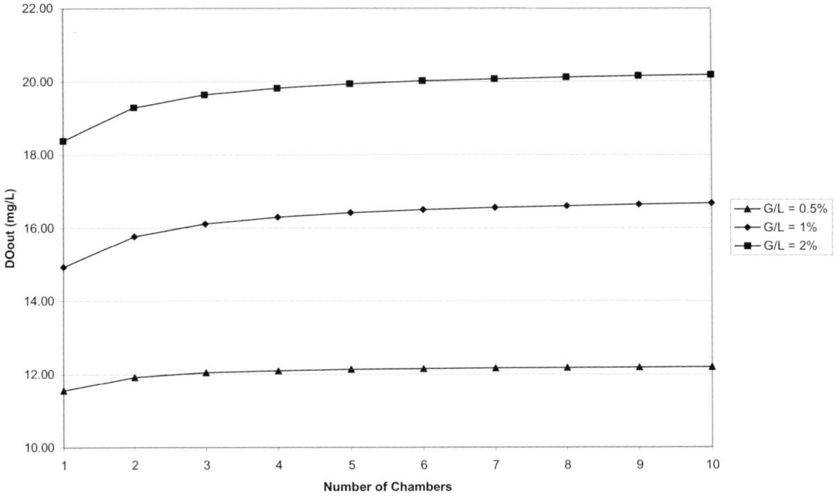

Figure 8.5 Effluent DO as Affected by the Number of LHO Chambers and the Gas Liquid G/L Ratio (Arbitrary Set of Model Inputs Were: Y_2 = 9.5 mm; Y_3 = 13 cm; Y_4 = 61 cm; Y_1 = 7.5 cm; Temp = 20.0°C; Top Area = 0.1 m^2; Active Hole Area = 10.0%; Chambers = Varies; G/L = Varies; DO_{in} = 6.0 mg/L; DN_{in} = 14.0 mg/L; DCO_2= 0.0; Pressure = 760.0 mm Hg; Oxygen Fraction in Inlet Gas = 0.99).

The effect of G/L ratio on absorption and effluent DO is demonstrated explicitly in Fig. 8.6 for the standard LHO unit selected as noted above. This graph indicates that a 1.4% G/L ratio is the largest gas flow that could be used if one were trying to achieve a minimum oxygen absorption efficiency of 70%; this would correspond to an increase in the effluent DO by 12 mg/L over the influent DO value of 6 mg/L. Rule of thumbs then emerge from this that delta DO's of 10 to 12 are target values for operating LHO units. The rapid drop in absorption efficiency as G/L ratios are increased is also a clear warning to the aquaculturalist that LHO gas usage should be closely monitored to avoid the easy solution of simply increasing G/L to increase effluent DO. The sensitivity of effluent DO and gas transfer efficiency to G_{20} is demonstrated in Fig. 8.7. This graph was created by assigning G_{20} values to the computer model instead of calculating them using Eq. 21. See

Chapter 17 for sample output screens and user instructions for employing the LHO model.

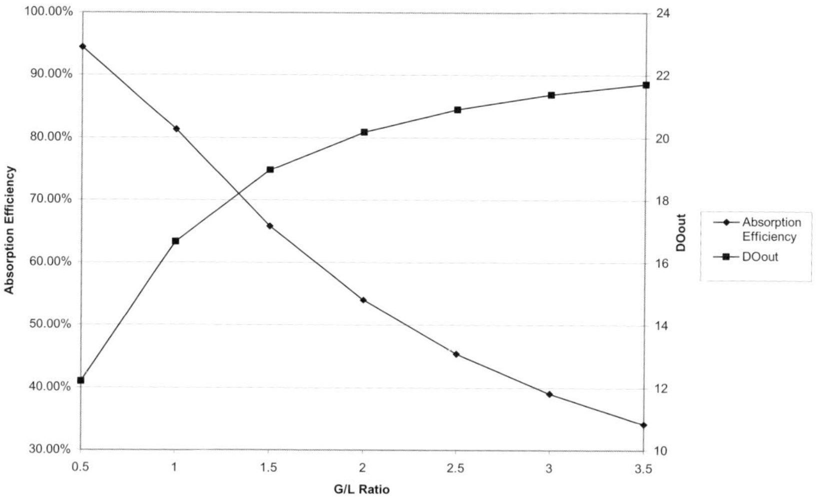

Figure 8.6 Absorption Efficiency and Effluent DO as Affected by Gas-Liquid ratio G/L (Arbitrary Set of Model Inputs Were: Y_2 = 9.5 mm; Y_3 = 13 cm; Y_4 = 61 cm; Y_1 = 7.5 cm; Temp = 20.0°C; Top Area = 0.1 m^2; Active Hole Area = 10.0%; Chambers = 10; G/L Varies; DO_{in} = 6.0 mg/L; DN_{in} = 14.0 mg/L; DCO_2= 0.0; Pressure = 760.0 mm Hg; Oxygen Fraction in Inlet Gas = 0.99).

Aeration Cone Or Down-Flow Bubble Contactors

The aeration cone, bicone, or downflow bubble-contact aerator consists of a cone-shaped cylinder or a series of pipes with reducing diameters. Water and oxygen enter at the top of the cone, flow downward, and out. As the cone's diameter increases, the water velocity decreases, until the downward velocity of the water equals the upward buoyant velocity of the bubbles. Thus, the bubbles are held in suspension, until they dissolve into the water. The performance of aeration cones is determined by gas and water flow rates, influent DO concentration, cone geometry and operating pressure. Absorption efficiency range from 95–100% with effluent concentrations from 30 to 90 mg/L. Commercial units are available that transfer from 0.4 to 10.8 lbs of oxygen per hour (0.2 to 4.9 kg/hr) at 25 mg/L, at flow rates from 45 to 600 gpm (170 to 2,300 Lpm) (see Figs. 8.8 and 8.9).

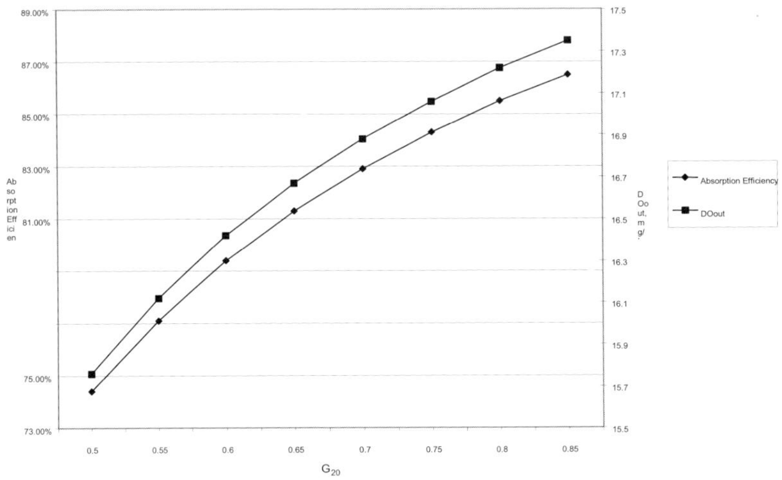

Figure 8.7 Absorption Efficiency and Effluent DO as Affected by Gas Transfer Coefficient, G_{20} (Arbitrary Set of Model Inputs Were: Y_2 = 9.5 mm; Y_3 = 13 cm; Y_4 = 61 cm; Y_1 = 7.5 cm; Temp = 20.0°C; Top Area = 0.1 m^2; Active Hole Area = 10.0%; Chambers = 10; G/L = 0.01; DO_{in} = 6.0 mg/L; DN_{in} = 14.0 mg/L; DCO_2= 0.0; Pressure = 760.0 mm Hg; Oxygen Fraction in Inlet Gas = 0.99).

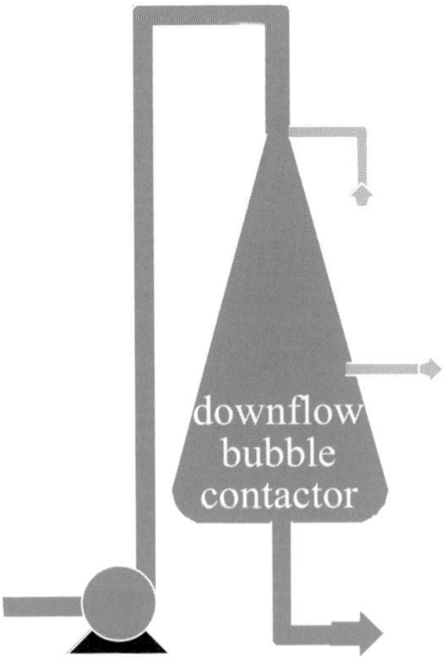

Figure 8.8 Schematic of a Downflow Bubble Contactor.

Figure 8.9 Downflow Bubble Contactor.

Diffused Aeration (Air Stones)

Due to their low absorption efficiency, the use of diffusers or air stones have been limited mainly to emergency oxygenation and fish live-haul systems. Although some of the recent fine-bubble diffusers (bubbles 100 to 500 microns) do perform well in deep tanks (50% oxygen transfer efficiency), they require a high pressure source of oxygen (25–50 psi) and are subject to both chemical and organic fouling.

Oxygen Injection

The most widely used form of oxygen injection takes advantage of the increased pressure available when pumping water. Oxygen is injected though a venturi nozzle or orifice, creating a fine bubble suspension in the pressurized line. Pressures of 30–235 psi (2 to 22 atmospheres) are needed to achieve satisfactory absorption, with contact times of 6–12 seconds. Absorption efficiency ranges from 15 to 70% with effluent DO concentration form 30–50 mg/L.

8.4 DEGASSING: CARBON DIOXIDE (NITROGEN)

Carbon dioxide is introduced into the water through respiration of the fish and bacteria. As stocking densities increase and water exchange rates decrease, dissolved carbon dioxide will become a limiting factor to production. When stocking densities were less than 30 to 60 kg/m^3, conventional aeration systems would generally provide sufficient removal of CO_2 in the process of transferring oxygen into the water with airstones and surface agitation or water falls. However, as carrying capacities have increased to 100 kg/m^3 and higher in order to be economically competitive, pro-active control of CO_2 removal must be included in any successful fish production system.

For every mole of oxygen that is consumed, one mole of carbon dioxide is produced. Alternatively, on a mass basis, for one gram of oxygen consumed; 1.38 g of carbon dioxide is produced. Calculating the concentration of dissolved carbon dioxide is complicated in that CO_2 is part of a chemical equilibrium system that includes carbon dioxide (CO_2), carbonic acid (H_2CO_3), bicarbonate (HCO_3^-) and carbonate (CO_3^-) ions. As a result, the concentration of any one ion is pH dependent, and to accurately determine their concentrations, values for pH, alkalinity, and temperature must be known. Figure 8.10 shows the effect of pH on the concentration of CO_2 as a function of pH for alkalinities of 50, 100 and 150 mg/L. Note that CO_2 concentration is proportional to alkalinity.

Carbon dioxide is toxic to fish because it reduces the capacity of the blood to transport oxygen. As the CO_2 concentration in the water increases, so does the blood CO_2 level. The presence of CO_2 in the blood reduces the ability of hemoglobin to bind oxygen molecules, a condition often referred to as the Bohr or Root effect. The safe operating levels of CO_2 depend on the species, development stage, and overall water quality. For tilapia and striped bass, concentrations of CO_2 as high as 60 mg/L have shown no adverse affect. For trout, values of 9–30 mg/L are generally assumed safe. Carbon dioxide may also affect the fish and the biological filter's bacteria by reducing the overall system's pH.

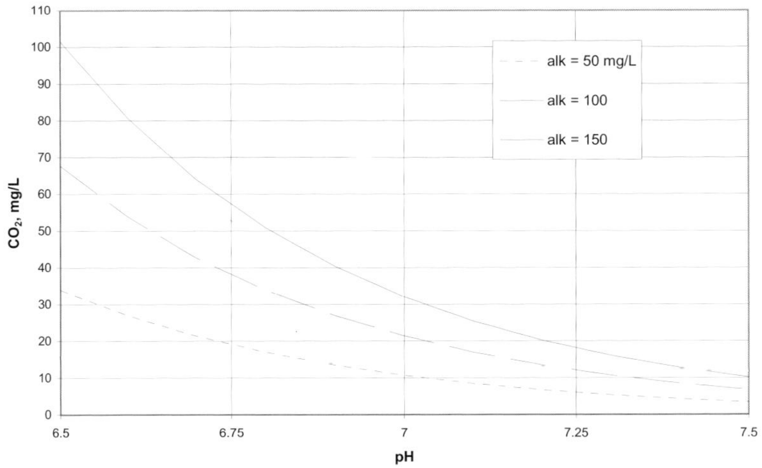

Figure 8.10 CO_2 Related to pH for Alkalinities of 50, 100, and 150 mg/L.

Removal of dissolved carbon dioxide is easily done through a gas exchange process, but accurately predicting the removal rate is very difficult. Unlike dissolved oxygen, carbon dioxide is part of a complex equilibrium system, where as CO_2 is removed, a shift in the carbonate carbon equilibrium occurs (see Chapter 2 for much more detail). For example, bicarbonate (HCO_3^-) acts as a reservoir of carbon dioxide, replenishing dissolved CO_2 as it is removed from solution. In addition, since the concentration of CO_2 in intensive systems is typically more than 20 to 100 times the ambient saturation concentration, the off gassing of CO_2 has a significant effect on the contacting gas. What this means in practice is that air bubbling through or passing over a gas exchange system is quickly saturated with CO_2.

Thus, large volumes of air per unit volume of water are required to strip off CO_2. For example, in normal aeration, gas to liquid ratios (G/L) are usually less than 3 to 1 (3 unit volumes of gas to every unit volume of water). In oxygenation, the G/L ratio is between 0.05:1 to 0.3:1. But for CO_2 stripping, the G/L ratio is between 5:1 to 10:1. Using a packed column and gravity flow of water to supply the packed column is one of the easiest ways to strip water of CO_2.

Figure 8.11 Two Types of Packing Media for Air-Stripping Columns.

Carbon dioxide removal as a separate water treatment process may be an issue when groundwater resources are used and has become especially important with the relatively recent intensification of aquaculture systems. While air-contact aeration processes remove dissolved carbon dioxide, their primary purpose is often considered the addition of oxygen. Air-contact aeration works well in low intensity aquaculture systems that do not use oxygenation and generally have sufficient water exchange and/or aeration to keep carbon dioxide from accumulating above safe levels. However, carbon dioxide problems have appeared in intensive systems using pure oxygen absorption because of the relatively low specific water exchange rates used (rate of make-up water flow per mass of fish in the system) and the fact that pure oxygen systems typically do not remove substantial quantities of carbon dioxide. Under intensive aquaculture conditions, the accumulated concentration of dissolved carbon dioxide in the culture tank will not be limiting (with no aeration or pH control) when the cumulative dissolved oxygen consumption is less than 10 to 22 mg/L, depending upon pH, alkalinity, temperature, and the species and life stage (Colt et al. 1991). Once this cumulative oxygen consumption level is reached, the water flow cannot be used again unless it is air-stripped of carbon dioxide, or a chemical is added to reduce carbon dioxide concentrations. Therefore, carbon dioxide control in water reuse systems may require aeration for stripping, the use of chemicals to adjust pH, or a combination of both methods.

The open packed column for carbon dioxide removal is a vertical tower filled with plastic packing similar to a pure oxygen packed column absorber. Packing media appropriate for this application is shown in Fig. 8.11. In operation influent water is introduced at the top of the column and is distributed uniformly over the plastic packing with a perforated plate, spray nozzle, or spray bar. Water flows downward through the packing and is broken up creating a large gas-liquid interfacial area for gas transfer. To overcome the potential for clogging or biofouling of plastic packing alternative water breakup approaches have also been utilized. It is common for the stripping column to have no packing or just a few splash screens to break the water up as it travels through the column (see Fig. 8.12). If plastic packing is used provisions are made to allow for particulates to pass through the media by using high porosity packing or packing that has an unobstructed downward path through the column.

Taking advantage of the water breakup created by the packing media and passing large volumes of air through the column achieve carbon dioxide removal. Fresh air has a low gas phase concentration of carbon dioxide (350 ppm) and is suitable for providing an adequate driving force for carbon dioxide gas transfer. Carbon dioxide is degassed from the water being treated and is carried out of the column with the fresh air blown through. In contrast with pure oxygen devices that use volumetric gas flow to water flow (G/L ratio) ratios of 0.3–5%, carbon dioxide stripping columns use G/L ratios of 500–2,000%. Low horsepower, high volume, ventilation-type air blowers provide the large amounts of airflow required for efficient carbon dioxide removal.

Stripping column design heights are generally limited to 1.0 to 1.5 m (3.3–4.9 ft) because of the diminishing returns on removal performance that correspond to packing heights greater than 1.5–2.0 m (4.9–6.6 ft). Hydraulic loading for carbon dioxide stripping columns are reported to range between 0.61–2.51 Lpm/m^2 (25–102 gpm/ft^2). However, values towards the lower end are recommended to reduce the back-pressure on the blowers when the column design calls for packing depths exceeding 1 m (3.3 ft) (Summerfelt, et al. 2000).

Figure 8.12 shows two types of counter-current air-stripping columns with media or if solids are a problem, splash screens.

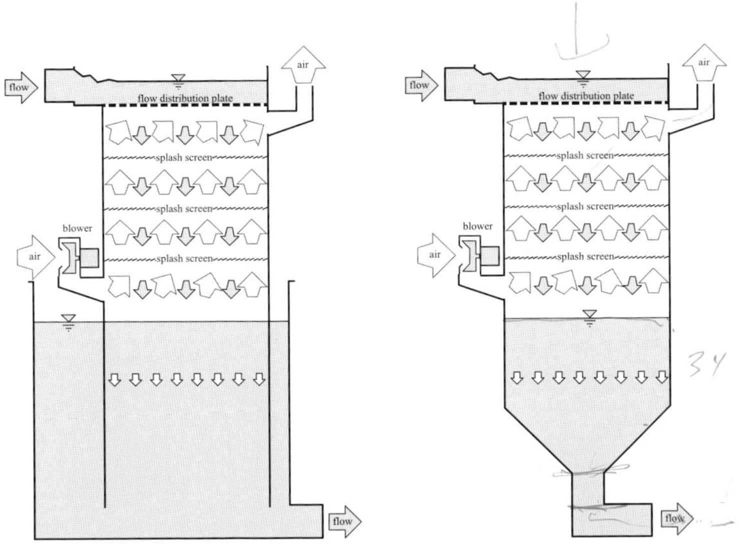

Figure 8.12 Carbon Dioxide Stripping Columns. Air is blown through stripping columns, often counter-current (shown here) to the water falling through splash screens or media. Air-stripping columns can also be constructed within cone-bottom cylindrical tanks (right diagram) to avoid a sump were settleable solids can accumulate (left diagram) (from Summerfelt et al. 2000).

When possible, try to blow the air upwards through the falling water streams to maximize stripping effectiveness. Design of counter current air-stripping columns requires determining the packing depth, hydraulic loading rate, volumetric air to water (G/L) ratio and estimated inlet CO_2 concentration. An interactive computer program to assist in the design of these systems is included in Chapter 17. In general, a CO_2 degasser is designed with a hydraulic fall of between 3 to 5 ft (1 to 1.7 m), a hydraulic loading of 25–35 gpm/ft^2 (17–24 kg/m^{2}s) a volumetric air-water ratio of from 5:1 to 10:1 and high porosity packing. Summerfelt et al. (2000) demonstrated the impact of increasing G/L ratios and column packing height on the percent of CO_2 removed per single water treatment pass (see Fig. 8.13).

Many times the primary interest of the design engineer is the impact of G/L ratio on the relative effectiveness of a particular device. For example, if a lower G/L ratio were used, what fraction of the maximum attainable removal would be attained for that particular device. The results shown in Fig. 8.13 were re-plotted in this fashion to show what percentage of the removal was attained as impacted by increasing G/L ratio (see Fig. 8.14).

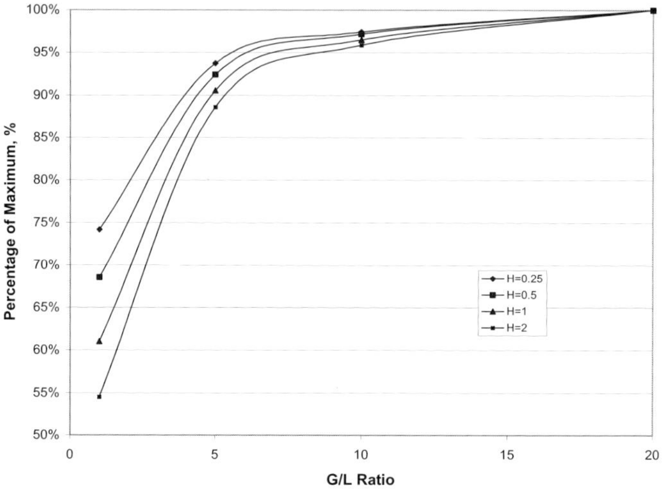

Figure 8.13 Packing Height and G/L Ratio Effect on Percent of Influent CO_2 Removed.

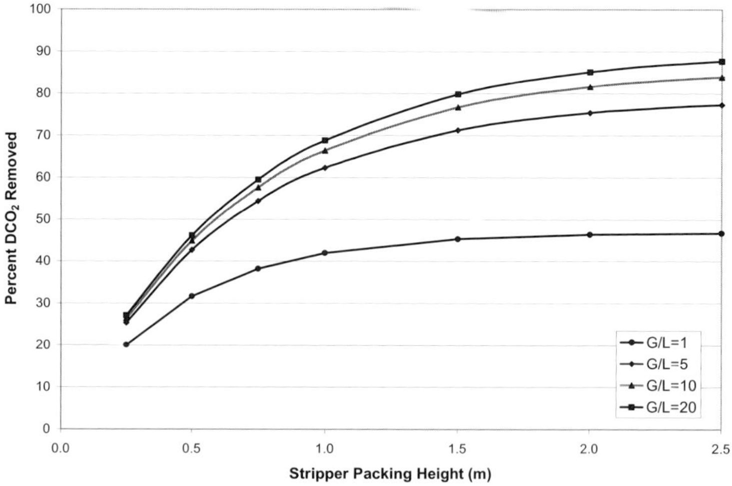

Figure 8.14 Effect of G/L Ratio on the Relative CO_2 Removed (G/L = 20 Being Defined as 100% of Maximum Achievable) for Packed Columns of Different Heights.

Where nutrient levels in the water may be high, stripping towers without packing media may be preferred to reduce maintenance time and variability in performance due to biofouling. The CO_2 removal percent as affected by fall height from a flooded plate to a receiving pool (similar to an LHO, except there is only a single chamber) is shown in Fig. 8.15. Note that the stripper column with the larger drop height requires a greater G/L ratio to remove the same amount of CO_2 as the shorter unit. A generic graph (see Fig. 8.16) is presented similar to Fig. 8.14 to show the relative impact of G/L ratios on removal fraction of the dissolved CO_2 for the two tower stripping heights. The LHO program given in Chapter 17 can be used to reconstruct these curves by setting the oxygen feed purity at 20.9%, i.e., air.

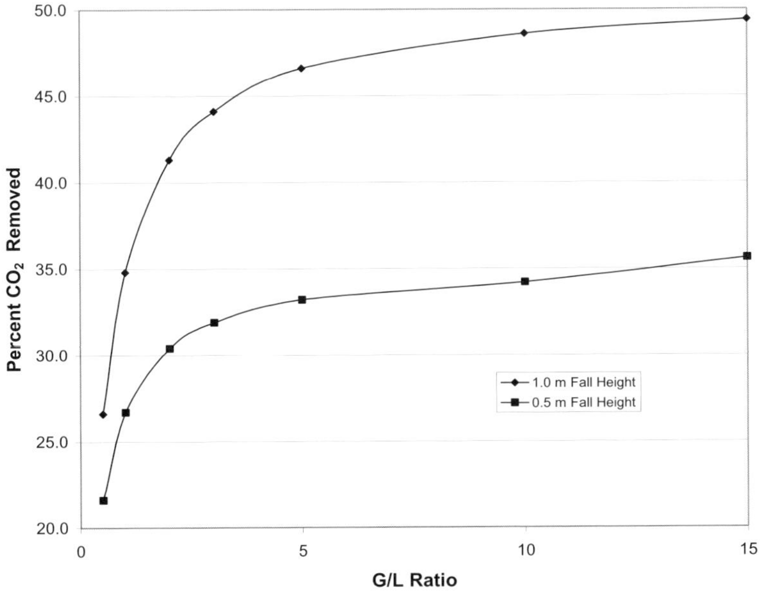

Figure 8.15 Effect of G/L Ratio on the % of CO_2 Removed as Affected by Height of Fall in a Stripping Chamber without Packing.

Air blown through the degassing column should be vented outside to prevent build-up within the production facility. Degassing of CO_2 should be done when it is at it highest level, which normally occurs after exiting the biofilter. Also, since the water leaving a degassing column is elevated to within 90% of saturation, additional oxygenation should occur after degassing and preferably just before the water enters the production tanks.

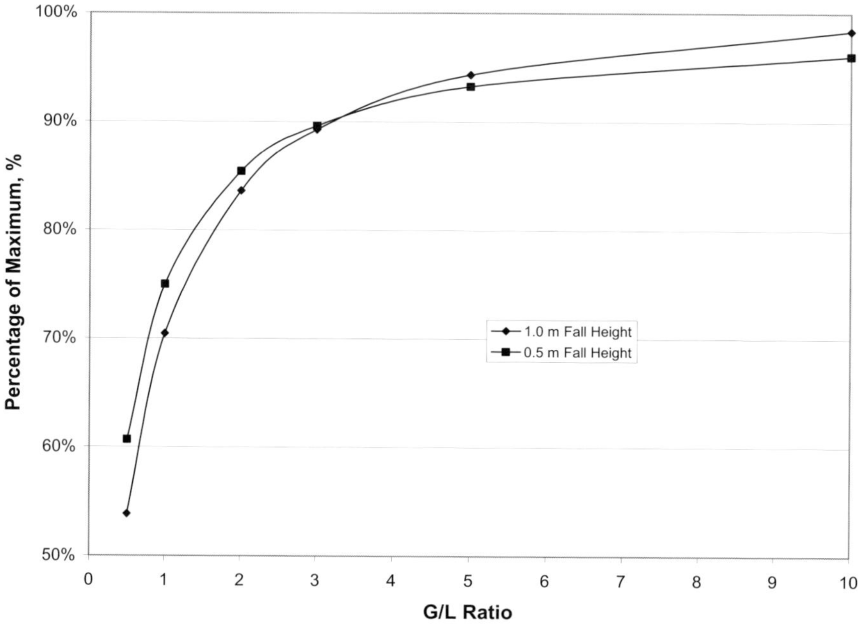

Figure 8.16 Effect of G/L Ratio on the Relative CO_2 Removed (G/L = 15 Being Defined as 100% of Maximum Achievable) for Stripping Columns with No Packing (Fall Height of 0.5 or 1.0 m).

Summary of Hydraulic Loading Rates

The many devices just discussed for gas transfer all have associated hydraulic loading rates that are considered appropriate for the particular device. Table 8.7 summarizes these for ease of reference. These values should be seen as guidelines and particular applications may warrant substantial deviation from the listed values. Exercise caution when doing so, though.

Table 8.7 Hydraulic Loading Rates for Various Gas Transfer Devices

Gas Transfer Device	Gas Species	Hydraulic Loading Rate kg/m^2s	gpm/ft^2
Packed Columns			
Sealed columns	oxygen	<166	<244
High Pressure Columns	oxygen	45–246	66–361
Pure Oxygen, atmosphere pressure	oxygen	73	107
Open to atmosphere	carbon dioxide	17–34	25–50
Spray Tower	oxygen	35–95	51–140
Low Head Oxygenator (LHO)	oxygen	34–68	50–100
Down Flow Bubble Contactors (Speece cone)			
at inlet	oxygen	1,800	2650
at outlet (to keep bubbles in)		150	220
U-Tubes	oxygen	2,000–3,000	2,940–4,410

INSTALLATION AND SAFETY CONCERNS

The handling and use of pure oxygen can be hazardous. Almost all substances will burn and in some cases explosively in an oxygen enriched environment. There have been rumors of used air stones exploding, when transferred to pure oxygen systems, when the organic residue oxidized. Special care and materials should be used in the construction and installation of pure oxygen systems. Stainless steel or copper tubing is recommended for oxygen lines, and both piping and fittings should be cleaned of grease and oil, which can present a fire hazard. Petroleum products should never be used on hardware that handles or contacts pure oxygen. Fittings and valves that have been specially cleaned and packaged for pure oxygen applications are available. Liquid oxygen is very, very cold and installation and servicing of storage containers should be left to the professionals.

LIST OF SYMBOLS

A	Total hole area over the entire LHO, m^2
A_O	Constant used for regression in water vapor equation, dimensionless
C_d	Discharge coefficient to predict velocity through an orifice, dimensionless
C_i	Concentration of gas species "i", mg/L
C_{in}	Influent dissolved concentration of a gas species, mg/L
$C_{meas,i}$	Measured concentration of gas "i" in water, mg/L
C_{out}	Effluent dissolved concentration of a gas species, mg/L
$C_{s,i}$	Dissolved gas saturation concentration for species "i", mg/L
DN	Dissolved nitrogen concentration, mg/L
DO	Dissolved oxygen concentration, mg/L
E	Gas transfer efficiency, %
F	Gas mean residence time for one flushing to occur for a chamber, hr
g	Acceleration due to gravity, 9.81 m/s^2
G_{20}	Overall mass transfer coefficient at 20°C, dimensionless
G_T	Overall mass transfer coefficient at a specific temperature T°C, dimensionless
G/L	Gas to liquid ratio, both being on a volumetric basis, dimensionless
h	height above sea level, m
HL	Hydraulic loading, kg/m^2 per s
J_i	Constants used to calculate gas partial pressure due to a specific gas "i" dissolved to some concentration in water, dimensionless
P_{BP}	Barometric pressure, mm Hg
P_i^l	Partial pressure of gas "i" in liquid form, mm Hg
P_i^g	Partial pressure of gas "i" in gas phase form, mm Hg
P_{TG}	Total gas pressure, mm Hg
P_{WV}	Water vapor pressure, mm Hg
Q	Rate of gas flow, kg/time
R	Gas flow rate into an LHO chamber, L/hr
$R_{pressure}$	Resistance to gas transfer
$S_{\%}$	Percent saturation of a particular gas, %
T	Temperature, °C
TGP	Total gas pressure, mm Hg
V	Volume of single mixing chamber in an LHO, L
Y_1	Hydraulic head over flooded plate, cm

Y_2	Hole diameter, mm
Y_3	Pool depth, cm
Y_4	Fall height of water from plate to receiving pool, cm
Z	Packing height of media, m
Z_{tower}	Height of an enclosed spray tower, m
α	Field water G_{20} / clean water G_{20}, dimensionless
β_i	Bunsen Coefficient for gas species "i", L/L-atm
K	Ratio of the molecular weight to volume (mg/mL)
X	Mole fraction of gas (dimensionless)

CHAPTER 9

SYSTEM MONITORING AND CONTROL

9.0 INTRODUCTION

Intensive recirculation systems have the potential for a significant increase in production per volume of water, but at the increased risk of catastrophic loss due to equipment or management failures. In addition, the managers of these intensive production facilities need accurate, real-time information on systems status and performance, in order to maximize their production potential. At production densities approaching and even exceeding a pound of fish per gallon of water (120 kg/m^3), failure of a circulation pump or aeration system can result in severe stress to the fish or even significant losses within minutes. Expensive and sophisticated monitoring and control systems and components from other industries, such as the wastewater and petroleum industries, have been successfully modified for use in aquaculture. However, only a small fraction of their processing power is usually employed, due to aquaculture's relatively simple monitoring and control demands, i.e., digital inputs/outputs. Today, with the rapid decrease in costs for computers, software, and off-the-shelf monitoring hardware, systems of this type are within the reach of even small producers and are mandatory for large-scale production facilities.

Before going any further, it should be emphasized that the most sophisticated (and normally, the most under appreciated and paid), monitoring and alarm system is an attentive human operator. Experienced staff can detect whether something is amiss the moment he or she steps into a facility, often just from the change in background noise. In the real world though, most facilities are not staffed 24 hours a day[a]. Moreover, the watery environment of aquaculture is totally different from what most "air-breathing" operators are accustomed too, requiring the need for automated monitoring of critical water quality parameters and system components.

---

[a] This choice depends upon production scale. Farms producing more than 500 tons per year will typically employ 24-hour coverage.

9.1 PARAMETERS TO MONITOR

Murphy's Law states simply states that: if anything can go wrong, it will (authors note: "and usually at 4:00 am on a Sunday morning"). Determining what can go wrong and generating a list of worst-case scenarios is a never-ending quest. From the authors' personal experiences, no matter how hard you try or how long your list is, there will always be a surprise in the near or far future and usually at the most inconvenient moment. Table 9.1 presents a short list of potential emergencies. It makes a good place to start, and then let your imaginations go wild, and assume that anything is possible, no matter how impossible it may seem!

Keep in mind though, that it is also important during this initial design process not to go overboard in terms of technological complexity or in the sheer number of monitoring points and alarms. Sophisticated alarm systems are of little use, if the part-time help disarms them due to their unreliability and frequent false alarms. Has such a thing happened? Yes!

When compiling a list of potential problem areas and water quality parameters, keep in mind the relative importance and the required response time each will require, Table 9.2. Life support priorities in aquaculture start with water, followed immediately by adequate levels of dissolved oxygen.

Table 9.1 A Short List of Potential "Emergencies" in Intensive Recirculating Systems

Type / System	Causes
Beyond your control	Flood, tornadoes, wind, snow, ice, storms electrical outages, vandalism/theft
Staff errors	Operator "errors", overlooked maintenance causing failure of back-up systems or systems components, alarms deactivated.
Tank water level	Drain valve opened, standpipe fallen or removed, leak in system, overflowing tank.
Water flow	Valve shut or opened too far, pump failure, loss of suction head, intake screen plugged, pipe plugged.
Water quality	Low dissolved oxygen, high CO_2, supersaturated water supply, high or low temperature, high ammonia, nitrite, or nitrate, low alkalinity.
Filters	Channeling/plugged filters, excessive head loss
Aeration system	Blower motor overheating because of excessive back-pressure, drive belt loose or broken, diffusers plugged or disconnected, leaks in supply lines.

Table 9.2 Life Support Priorities in Intensive Recirculating Systems

High (fast response time – minutes)	
	electrical power
	water level in tank
	dissolved oxygen –aeration system/ oxygen system
Medium (moderate response time – hours)	
	temperature
	carbon dioxide
	PH
Low (normally slowly changing – days)	
	alkalinity
	ammonia-nitrogen
	nitrite-nitrogen
	nitrate-nitrogen

Then come the other water quality parameters, correct temperature, pH, and alkalinity, and finally, acceptable concentrations of ammonia-nitrogen, nitrite, nitrate, and carbon dioxide. At high stocking densities (greater than 1/3 lb/gal, 40 kg/m^3), dissolved oxygen requires the most rapid response time. If either water flow or aeration is cut for any number of reasons, low oxygen and the resulting stress can lead to disease problems and/or mortality within minutes. Thus in the design of intensive systems, a simple audible alarm in the office may not be adequate, or even a pager if the manager lives 20 minutes away. Therefore, in addition to the monitoring, some form of backup aeration must be provided for and automatically engaged to insure survival of the fish. Except for dissolved oxygen, most of the other water quality parameters change relatively slowly and can take hours or days to reach levels of concern. This allows more time to discover and analyze the problem and take the necessary steps to correct them.

At low stocking densities (less than 0.33 lb/gal, 40 kg/m^3), basic parameters to be monitored include system electrical power, tank water level (high and low), aeration system pressure, and water flow through the filters and tank. All of these parameters can be monitored by simple digital sensors, i.e., either on or off. Analog sensors, such as dissolved oxygen levels and pH monitors, are more expensive to buy, install, and operate. Continuous dissolved oxygen monitoring is crucial at high stocking densities or whenever oxygen is used. For all these critical parameters, it is equally important *where* you monitor parameters, as *what* you monitor. Common sense should be the guide in this aspect of monitoring. For example, it is of little value to monitor the flow from a pump into a tank, if the tank drain line is left open, and all the water is flowing out as fast as it is flowing in. What is important to the fish is the tank water level. Similarly, there is no advantage in monitoring the power to a pump, if the discharge valve is shut or the motor thermal-overload switch has turned the pump off. The critical parameter to monitor is whether there is flow from the pump. Finally measuring air pressure next to the aerator is of little help, if there is a major leak at the far end of the distribution system, resulting in low air pressure for the last tanks on the line. The aeration pressure needs to be monitored at the farthest point in the system or at several different points. Table 9.3 lists some of the important systems and parameters that need to be monitored in intensive recirculation systems.

Table 9.3 Important Systems or Parameters to Monitor

Electrical power	Single and three phase supply, individual systems on GFIC's
Water level	Culture tank (high/low), supply sumps to pumps (high/low), chemical storage tanks, head tanks/ reservoirs (high/low), filters (high/low)
Aeration system	Air/oxygen pressure (high/low)
Water flow	Pumps, culture tanks, submerged filters, in-line heaters
Temperature	Culture tanks (high/low), heating/cooling systems (high/low)
Security	High temperature/smoke sensors, intruder alarm

Sensors in aquaculture can be roughly divided into two major types: digital (on/off signals) such as water level, aeration pressure and water flow switches and analog (continuous output) such as dissolved oxygen, temperature, pH, conductivity, and ammonia-nitrogen probes. In addition, most analog probes require some additional hardware or controllers to convert the probe output to a usable signal, provide a digital display, and allow for calibration and zeroing. Thus the higher cost for these types of measurements, compared to simple switch closures. Below is a short review of important monitoring parameters:

ELECTRIC POWER

Power failure is probably the most common emergency and the one most easily monitored. Monitoring power is especially important when systems such as filters or supply pumps are located some distance from the main building. Three-phase power can be especially confusing, because if only one phase is down, it is possible to lose power to some systems, but not all. Murphy's Law assures that the monitoring system will not be on the one phase that goes down. In addition, when power is lost on only one phase, severe damage can occur to three-phase motors and pumps, if not properly protected. One often overlooked result of power outage is the loss of lights, which means that either numerous flashlights need to be maintained in good working order or back-up emergency lighting provided. When the power does go out, back-up generators suddenly become worth their weight in gold, as long as they have been properly maintained, regularly tested, have sufficient fuel and come on line when needed.

Water Level

Water level is probably the easiest and most inexpensive parameter to measure, and should be monitored for both high and low levels in each production tank. High/low water level sensors will detect plugged drain lines, fallen standpipes or drain lines accidentally left open. Other locations to monitor include the intake side of pumps in wells or sumps. These should provide for automatic shutdown of the pumps to prevent their damage in the case of low water levels. Supply reservoirs or head tanks need to be monitored for both high and low levels. High levels can indicate unusual change in normal water demands, due to clogged pipes or valves accidentally turned off. Low levels can be caused by pump or water supply failure. If immersion heaters are used, low level monitoring should be designed to turn them off, to prevent overheating and burning out of the heaters and melting pipes. Alarm levels should be set so that normal operating transient do not activate an alarm. This can be accomplished either by setting the levels optimistically or by allowing some time delay before an alarm is activated after a sensor is triggered. Level sensors should be protected, so that active fish do not accidentally trigger them or even chew on them.

Aeration System Pressure

The aeration system is one of the most critical systems in any intensive recirculating aquaculture system. Response time to a detected failure is very short, and both monitoring and backup systems are important. Low pressure in the system may mean a ruptured airline, open or jammed pressure relief valve, disconnected diffusers or blower failure. Although not monitored as often, excessive high pressure could indicate blocked supply lines, valves turned off or clogged diffusers.

WATER FLOW

In some cases, the actual measurement of flow rate is important, such as when needed for the proper operation of chemical injection systems, for dechlorination, or for monitoring system performance. Normally however, simply monitoring whether water is actually flowing (flow/no-flow) with a digital sensor is adequate. One example of systems that these sensors protect is in-line heaters that require continuous water flow to prevent overheating and meltdown. Another example is submerged biological filters, where anaerobic condition due to pump failure can damage the nitrifying bacteria.

DISSOLVED OXYGEN

Dissolved oxygen (DO) is one of the most expensive and difficult parameter to monitor continuously, (Ebeling, 1998). Thus, the decision whether or not DO should be continuously monitored is dependent on the overall economics of the system, stocking density, and degree of risk a manager is willing to accept. Normally, the actual value of DO (mg/L) is not needed, just whether it is above or below a given set point. However, to provide a simple digital signal, both an expensive probe and a sophisticated hardware interface are required. The availability and costs of both oxygen probes and interface hardware has dramatically decreased in the last few years, but remains too high for many aquaculture operations.

TEMPERATURE

The continuous and precise monitoring of temperature in production tanks is important to optimize production, reduce stress, and minimize risk of disease. Systems should be monitored for both excessively high and low temperatures, keeping in mind the two extremes are not equal. While low temperatures may reduce growth, excessively high temperatures may yield a huge tank of fish soup and a new career path. Since most temperature controllers are cyclic in nature (either on or off), temperature alarm limits should not be set too close together, to prevent unnecessary alarms due to short term transients.

Other Water Quality Parameters

Other water quality parameters, pH, ammonia-nitrogen, nitrite, nitrate, alkalinity, and carbon dioxide change relatively slowly in comparison to dissolved oxygen. Although relatively expensive individual probes and automated systems are available to monitor these parameters, the most cost-effective method is daily or weekly manually monitoring with inexpensive off-the-shelf test kits.

Physical Plant Security

Intrusion alarms, smoke, and high temperature sensors (fire) are readily available and commonly used to protect against fire, theft, and vandalism. Often existing alarm systems can be connected to the proposed monitoring system.

9.2 Monitoring Sensors and Equipment Options

Over the past few years, the cost of computer hardware and software has dramatically decreased, while the processing power and computer programming sophistication has greatly increased, (Lee, 1995, 1998). Sensor technology has become more reliable and sophisticated with such innovations as embedded microchips in the sensors that provide signal processing and linearization. A large number of sensors and monitoring systems components have been adopted from the wastewater treatment and chemical and petroleum industry. In many cases, the sophistication and corresponding expense of these types of monitoring and control equipment is not necessarily required in aquaculture facilities. Nevertheless, until specifically designed equipment becomes available for aquaculture and for high valued products, the added costs of this equipment may be justified. Keep in mind, that for any monitoring system, its overall reliability is determined by the most unreliable part, i.e., the weakest link.

The following description of sensors represents the simplest and most cost-effective solutions to monitoring each individual parameter. These suggestions are not the only solutions available, however. For each of these parameters, there is a multitude of potential solutions, some of which are more expensive, more accurate, more reliable, more precise, with better interface capabilities, or simply more readily available. There is no simple right or wrong answer, and this is where the system engineering and design requirements come to play.

WATER LEVEL – FLOAT SWITCHES

Water level is probably the easiest and most inexpensive parameter to measure. The basic float switch is designed to monitor a single, discrete, preset liquid level, Fig. 9.1. Simple float switches are constructed with a float containing a small magnet, which moves with the water level and actuates a hermetically sealed reed switch within the stem or body of the float switch. The rugged construction of this design provides for long and trouble free service with minimum maintenance requirements. Several different designs are available for mounting either vertically or horizontally in the tank or sump. Two float switches can be wired in series to monitor both high and low levels in a tank. Although most float switches are designed to handle 110 VAC at small currents, they should be powered by low voltages, i.e., 24 VAC or DC, to minimize danger to personnel and fish. Float switches are simple, foolproof, and relatively inexpensive. Examples of two float switches that have been successfully used by the author are Aquatic Eco-Systems Liquid Level Switch ST-3M and Grainger Liquid Level Switch 2A554.

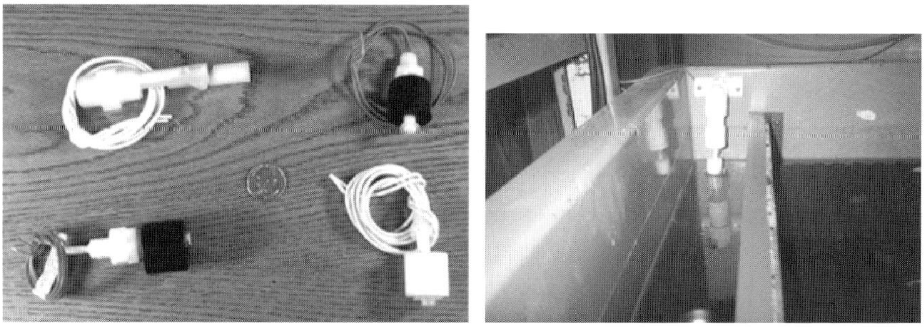

Figure 9.1 Level Sensors or Float Switches for Monitoring Tank Water Levels.

Other options for monitoring water level include optical liquid-sensing sensors that use an internal infrared circuit and the light refracting properties of water. Non-contacting ultrasonic level sensors measure the time required for the ultrasonic pulse to travel to the water surface and return. Conductivity level switches operate by detecting a small electric current between a single electrode probe and a grounded metal tank or between two electrodes. Finally, pressure-sensing systems use a pressure transducer to measure the pressure required to bubble air through an immersed pipe in the water column. Each of these has their application in specific design situations.

AERATION PRESSURE

Pressure is defined as a force per unit area, which is to be used to produce a deflection, distortion, or some other physical change in a sensor. A pressure control switch uses this deflection to trip an electrical switch at a preset pressure setting. Low and high pressure switches are available in a wide variety of configurations and price scales, from numerous manufacturers. One examples of a pressure switch that have been commonly used is Aquatic Eco-Systems Pressure Switch, Fig. 9.2.

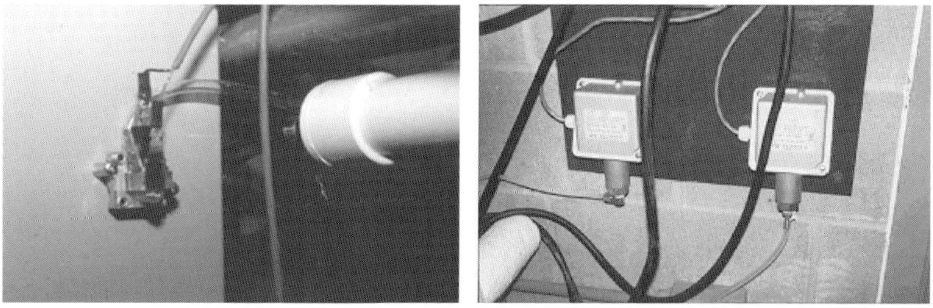

Figure 9.2 Pressure Switches for Monitoring Aeration or Water Pressure.

WATER FLOW - DRAG DISCS, PADDLE AND VANE FLOW SWITCHES

Drag discs, paddle and vane flow switches are all designed to monitor flow/no-flow or low flow conditions. Each operates on the drag force of the moving water against a small disk, paddle or vane in its path, which in turn controls a small micro-switch. They are available in a wide range of flow rates and pipe sizes. Normally drag discs and paddles are installed using a Tee fitting and vane types are installed in-line, Fig. 9.3. An example of a flow switch that has been commonly used is Aquatic Eco-Systems Flow Switch, ST-9.

Figure 9.3 In-line Paddle Flow Monitor and a Paddlewheel Flow Meter.

Other options for monitoring flow are rotameters. As the water flows through the rotameter, it raises a float in a tapered tube, increasing the area for passage of the fluid. The greater the flow, the higher the float raises in the tube, which is directly proportional to the flow rate. The float reaches a stable position, when the upward force exerted by the flowing water is equal to the downward force exerted by the weight of the float. These flow meters can be used to monitor flow by mounting a proximity switch externally, which is switched at a predetermined flow rate by a small magnet in the float. A second more expensive option is to use a turbine or paddlewheel flow meter. The flowing water turns a small turbine blade or paddlewheel, which generates an electrical pulse. This pulse is sent to the appropriate hardware, where the flow rate or the total flow can be displayed, and alarm conditions set and low/high flow alarm relays activated.

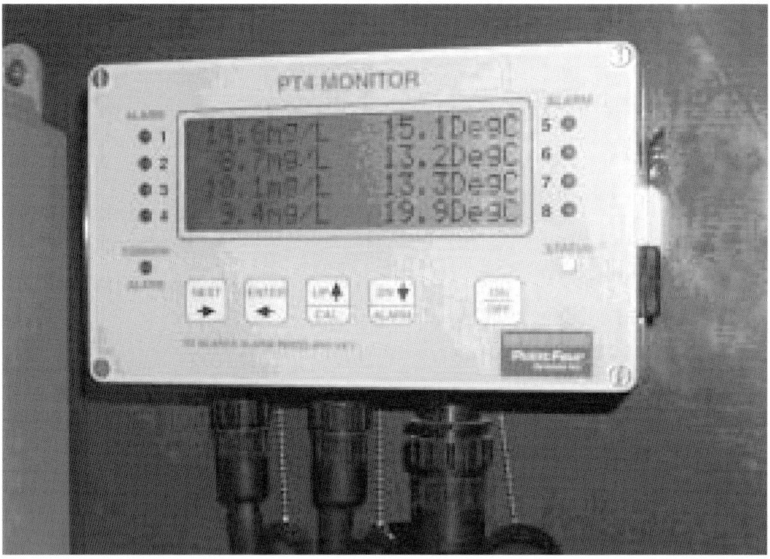

Figure 9.4 Dissolved Oxygen Meter From Point Four.

DISSOLVED OXYGEN

Over the past few years, a number of dissolved oxygen probes and analyzers designed specifically for the aquaculture industry have become available. Most of these are microprocessor-based instruments capable of measuring levels of dissolved oxygen up to 100 PPM, important for monitoring oxygen injection systems. Standard recorder outputs (0–5

VDC) are built-in and many include 4–20 mA current loop outputs. Several models also provide serial outputs (RS-232, or RS-485) for direct interfacing with microcomputers and local area networks (LAN's). These also include high/low set point control relays for automatically controlling external devices such as aerators, pumps, valves, or other alarm monitoring equipment. Although the initial investment in this equipment can be high, the cost must be weighed against the potential loss and poor growth due to low dissolved oxygen levels. Dissolved oxygen meters designed specifically for aquaculture are available from Point Four Systems, Inc., YSI, Inc. and Royce Instrument Corp., Fig. 9.4.

OTHER WATER QUALITY PARAMETERS

Temperature, pH, and other water quality parameters usually have a relatively long response time, when compared to dissolved oxygen or the effects of pump or aeration failures. Thus, these and other slowly changing parameters are probably best measured as part of a regular water quality monitoring program, using bench-top laboratory equipment. Although equipment exists for continuous monitoring of pH and temperature, Fig. 9.5, except for research purposes, it is not normally required in production aquaculture. On-line ammonia monitoring is possible, but in practice, it is very expensive and difficult to accomplish.

Figure 9.5 pH Meter and Temperature From GLI.

9.3 Automatic Phone Dialers

The final step in the development of the monitoring/control systems is to bring each of the potentially catastrophic alarms to the attention of the manager and staff, especially when they are home sleeping (Ebeling, 1994, 1995). A very inexpensive, simple, and versatile monitoring system can be constructed around readily available automatic telephone dialers/alarm systems. These units are readily available for a wide range of inputs, sophistication, and costs. One such unit, the Sensaphone (Phonetics, Inc., Aston, PA, http://www.sensaphone.com/) has been used in numerous research and production facilities with excellent results.

The Sensaphone unit automatically monitors the following conditions:

- AC electric power – power failure
- four digital alert inputs or 3 digital and 1 temperature
- temperature – reports actual temperature and checks for high or low limits
- high sound level – fire/smoke alarm, intruder alarms, unauthorized parties
- battery status – condition of its battery back-up

All monitoring is continuous and when an alarm condition occurs, the unit announces the alarm status locally for 30 seconds. If no response is received, it then sequentially dials up to four user-programmed telephone numbers (including pagers) with an alarm message. It will state in English the existing problem, disconnect, and wait for an acknowledging telephone call or coded response. It will continue dialing-out, up to sixteen attempts, until its message is properly acknowledged. In addition, it is also possible to call in, listen to a status report on the monitored conditions, and hear the background sounds through a built in microphone. For most small systems, this would provide all the necessary digital inputs for monitoring tank water level (high/low), aeration systems pressure, water flow, and sump water level or if desired, system water temperature.

The Basic System

The *basic monitoring system* was designed for low-density (less than 40 kg/m^3 or 0.33 lb/gal) recirculating systems, with aeration only and moderate feed rates, such as broodstock holding tanks, isolation/quarantine tanks, or educational systems. Basic system parameters (level, pressure, flow) are monitored by digital sensors, i.e., either on or off. Analog sensors, such as temperature and dissolved oxygen levels, are more difficult and expensive to utilize and are important only at much higher stocking densities or where pure oxygen aeration is required. Basic parameters to be monitored at this production level include electricity, tank water level (high and low), aeration system, and water flow. The actual number of subsystems monitored, depends on the specifics of the system design and the operating conditions. In most cases, only a few monitoring points should be necessary. Parameters monitored and sensors used include:

- **System electrical power**: monitored directly using the Sensaphone or indirectly due to loss of other subsystems (pump flow, aeration, etc).

- **Tank water level (high/low):** Aquatic Eco-Systems, Liquid Level Switch ST-3M, or Grainger Liquid Level Switch 2A554, wired in series.

- **Aeration system pressure**: Aquatic Eco-Systems, Pressure Switch B601.

- **Flow-sensing switch**: Aquatic Eco-Systems, Flow Switch ST-9.

- **Telephone dialer**: Aquatic Eco-Systems. Sensaphone Telephone Alarm System (see Fig. 9.6).

Figure 9.6 The Sensaphone for Continuous Monitoring and Phone Dialer.

Each of the sensors is wired directly into a Sensaphone input, with the two float switches wired in series to monitor both high and low water level. The fourth input on the Sensaphone could be used to monitor either temperature of the water or an additional alarm. With this system design, a single tank or perhaps several could easily be monitored for the basic system parameters: water level, flow, aeration, and electricity.

9.4 COMPUTER BASED SYSTEMS

Once it becomes necessary to monitor analog inputs, such as dissolved oxygen, pH, temperature or other output analog signals for control purposes, some form of computer-based system must be employed, (Tsukuda et al. 1998). With this added capability though, comes the cost of additional requirements for calibration of the probes and sensors and maintenance of the overall system. Until recently, the utilization of computer control and monitoring systems in aquaculture has been limited, with only a few custom-designed systems for research or large commercial operations. The vast majority of small producers has had neither the expertise nor the resources to custom-design and installs systems. However, in the past years, there has been a revolution in low-cost, high performance microcomputers and intuitive and relatively low cost process control software. With the rapid development of microcomputer technology, numerous 'user-friendly' software and hardware systems are becoming available for control and monitoring of industrial processes. These software packages use object-oriented programming that allows even inexperienced programmers to create customized programs. In addition, standardized data acquisition components and systems, software drivers and communication software are available for several different computer platforms and over a wide range of costs.

Supervisory control and data acquisition (SCADA) is the term used in industrial automation that refers to automated data collection usually by remote units and its display on a centrally located personal computer. The human machine interface (HMI) is the interface by which an operator interacts with the remote devices, i.e., pH, DO, temperature measurements. The HMI allow an operator to adjust set points, configure remote units, respond, and acknowledge alarms. In addition, system performance data can be recorded for future analysis and solutions to recurring and one-time problems. A data acquisition system has a variety of input/output (I/O) ports that provide connection between the sensors and the computer. Table 9.4 shows the four of the most commonly used: digital input (DI), digital output (DO), analog input (AI), and analog output (AO). Determining the type of I/O required by sensor is the first step to matching the sensor and the required I/O.

Table 9.4 I/O Ports Connecting the Computer to Sensors

I/O Type	Applicable sensors or control equipment
Analog input	DO, temperature, pH, flow sensors (voltage, current)
Analog output	Proportional control of chemical feeds, pump speeds
Digital input	Level switches, valve status, counters
Digital output	Activate relays controlling feed, oxygen, pumps

Computer control and monitoring systems in aquaculture can be separated into two design strategies:

1) stand-alone programmable logic controller (PLC) systems, or closed loop controllers,
2) centralized microcomputer based system.

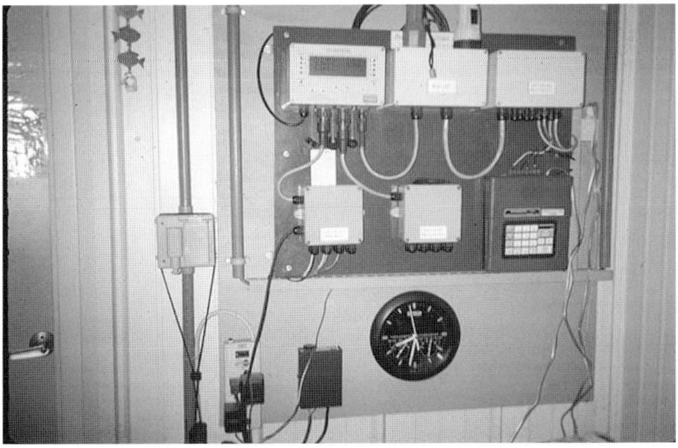

Figure 9.7 Stand-alone System Based on the Point Four PT4 Meter with Digital Input Monitoring.

The first design strategy can also be expanded to a distributed process control system, where each of the stand-alone monitor/control units relays data to a central supervisory microcomputer, Figs. 9.7 and 9.8. Examples of the stand-alone closed loop controllers are dissolved oxygen analyzers (YSI, Royce, Balanced *Aqua*Systems, and Point Four Systems, Inc.) and temperature controllers. Each of these closed-loop controllers is normally equipped with both control relays and high/low alarm relays. In general, these units have limited data display capabilities and normally do not store data, but are equipped to transmit the data to a central, supervisory microcomputer. With stand-alone systems, individual sensors are easy to service and calibrate, since each has its own hardware and display unit. In addition, monitoring and control is performed at the lowest system level, which provides a high degree of overall system stability and robustness. If a failure occurs in the supervisory microcomputer, the stand-alone units will continue to monitor and control critical processes. If a failure occurs in the stand-alone system, the supervisory microcomputer can by comparing to previous measurement or measurements from other sensors to detect the abnormal conditions and alert the operator.

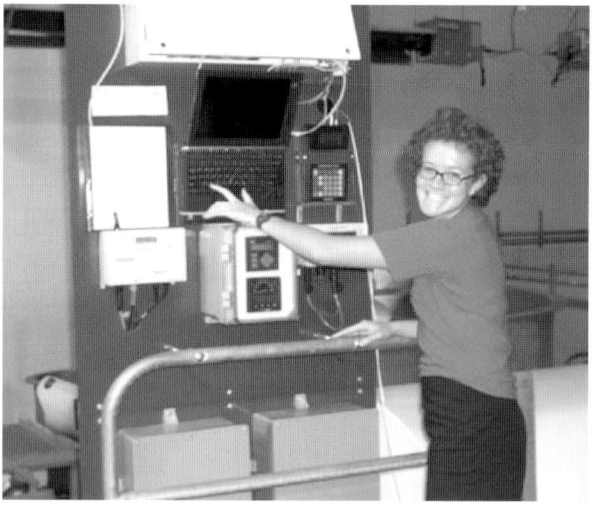

Figure 9.8 Control Monitoring System at Freshwater Institute. (Author's Daughter, Margaret Timmons, inspecting system).

The second design strategy utilizes commercially available data acquisition boards that are either located in existing expansion slots in the computer or communicated to it via a serial interface link. There is a wide selection of data acquisition boards available over a wide range of cost, performance and sophistication, including analog to digital (A/D) cards for monitoring voltages or currents from sensors, digital to analog (D/A) cards for outputting analog control voltages, and input/output cards (I/O) for monitoring and outputting digital control signals. They are easy to use, "just-plug-in", and come with a set of standard drivers and application software programs. Many types of sensors can be connected directly to these boards and most meters usually have some form of recorder output (0–5 V or 4–20 mA). In contrast to the distributed system, the microcomputer operates as the primary controller, monitoring, recording data, and controlling alarm functions. These systems are not as inherently reliable as the distributed systems, but overall systems cost is less, since they are based on fewer and less expensive components.

9.5 SYSTEM DESIGN AND MAINTENANCE

Listed below are some general suggestions about overall system design and maintenance:

System Design:

- Choose sensors carefully, use the fewest possible, label everything and include expansion capability in all components.

- Aquaculture facilities are now included under the National Electric Code; it may not be of concern to you, but it is to your insurance agent.

- Mount sensors and equipment where they are visible and easily accessible for service and calibration.

- Remember that water and electricity make for a fatal combination, so use low signal voltages (5 VDC, 12 VDC or 24 VDC or AC) to protect you and the fish.

- Clearly label the sensor's armed and unarmed modes preferably with LED's at each station to show sensor status.

System Maintenance:

- Have a well prepared maintenance manual accessible to the staff.

- Maintain a weekly/monthly/yearly maintenance scheduling plan and files of major service records and equipment manuals.

- Maintain daily/weekly/monthly instrument check lists.

- Perform regular (and some unannounced) system checks, including triggering of each sensor and checking operation of the automatic backup systems and phone dialer.

- Provide staff training to handle routine alarms.

- Ensure staff familiarization with the complete operating system, including water supply, aeration and emergency backup systems.

9.6 CONSTRUCTION HINTS

Probably the most important rule during design and construction is to *keep it simple silly*, known as the "KISS" principle. The other rule is always assume that someone else will have to repair it, thus complete design notes, wiring diagrams, and labels are important. If you change a configuration of a monitoring system, update the documentation, and date the update. In addition, system components should be readily available from local or reliable sources. A "one-of-its-kind" is just that, and will soon become extinct. While designing and constructing, plan for expansion and leave room for additional systems or more "bells & whistles".

Monitoring and Control equipment should be mounted and operated in a clean, dry, and safe environment. Provide adequate space around and in front of the equipment for easy access and future expansion. Do not place equipment were it will be subject to shock and vibration, dirt, dust or moisture. As much as possible, all materials used for system housing and hardware should be PVC, fiberglass or stainless steel to minimize corrosion. Water-resistant PVC junction boxes and fiberglass electrical cabinets, NEMA-4 enclosure, are corrosion resistant and easy to drill holes into and are ideal housing to protect the electrical components, Fig. 9.9. Include several vent holes to minimize heat build up in the cabinets. Do not install the system near motor starters, contactors or relays that switch inductive loads, i.e., motors! These devices generate large electromagnetic fields that can cause communications and system errors. If unavoidable, install the system in separate, grounded, steel enclosure to the shield from electrical interferences.

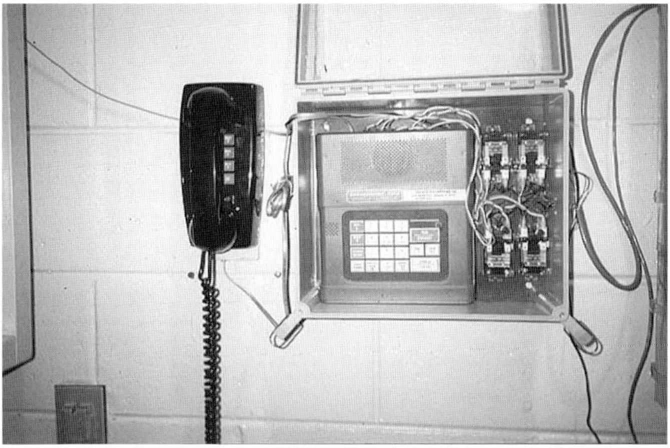

Figure 9.9 Fiberglass Electrical Cabinet to Protect Sensaphone and Relays.

All external sensors should be low voltage, i.e., 24 VAC or 24 VDC, ON/OFF, to minimize danger to operators and fish. Crimp style quick disconnect tab connectors on switches allow for easy construction and later modification. Solder joints should be covered with shrink-wrap tubing whenever possible. When buying individual components, look for extra options that may be useful in the future, such as extra alarm relays, voltage or current outputs and computer interfacing capabilities.

One simple trick to minimize the effects of aquaculture's harsh environment is to pressurize the control system housing using the aeration air supply. In this manner, relatively dry air is forced into the housing, preventing the high humidity and salt air from getting in. Alternatively, the step-down transformer in many of the systems provides a source of heat, thus preventing condensation from occurring.

WIRING CONSIDERATIONS

(adapted from Phonetics, Inc., Sensaphone SCADA 3000 manual Version 2.0)

Most sensors can be connected using inexpensive 2-conductor twisted-pair cable as small as #24 AWG (up to 700 ft, 210 m). For wiring distances up to 1500 ft (450 m) use #22 AWG and for distances up to 2500 ft (760 m), #20 AWG. If the senor is located far from the monitoring unit or if the cable is running in an electrically noisy environment, twisted pair shielded cable should be used. This will shield the signal from electrical interference and prevent false readings and possible damage to the monitoring system. When using shielded cable, you only need to connect the shield to the earth ground at the monitoring system, not at both ends. The guidelines below are designed to minimize electrical noise coupling between I/O lines, communications lines and control signals:

- Route the power supply, input and output wiring and communications cables in separate conduits, when ever possible.

- Segregate I/O wiring by signal type, i.e., dry contacts, thermistors, 4–20 ma, etc.

- Allow at least two inches between the monitoring system and I/O wiring ducts.

- Keep communications cable at least five feet (1.5 m) from any electric motors, transformers, rectifiers, generators, arc welders, induction furnaces or sources of microwave radiation.

- Keep communications cable at least 6 inches (15 cm) from AC power lines carrying less than 20 A, at least 1 foot (30 cm) from lines greater than 20 A, and 2 feet (60 cm) from lines greater than 100 KVA.

- If cable is run in a metallic wireway or conduit, keep the communications cable at least 3 inches (8 cm) from AC power lines carrying less than 20A, at least 6 inches (15 cm) from lines greater than 20 A, and 1 foot (30 cm) from lines greater than 100 KVA.

CHAPTER 10

WASTE MANAGEMENT & DISPOSAL

10.0 INTRODUCTION

As environmental regulations become more stringent, waste management and disposal will become increasingly important in an aquaculture operation. A proper waste management strategy is now considered critical for maintaining the legality, profitability, and sustainability of any aquaculture facility. The U.S. Environmental Protection Agency (US EPA) has initiated a major study to determine if aquacultural waste discharges should be regulated, see below. This is an ongoing assessment, which could have a major impact on the industry. One of the problems that EPA faces is the wide range of species, technologies, water sources, etc. that are encountered in the aquaculture industry. The Joint Subcommittee on Aquaculture is attempting to assist the EPA by providing information on the aquaculture industry. Recirculating systems may play a leading role in the future, due to their limited water usage and significantly smaller waste discharges.

An excellent review of federal and state policies affecting waste water discharge is given in NRAC Bulletin 300 "State Policies for Aquaculture Effluents and Solid Wastes in the Northeast Region (Ewart, Hankins, & Bullock, 1995; 24 p.) and NCRAC Technical Bulletin 101 "Aquaculture law in the North Central States: A digest of State statutes pertaining to the production and marketing of aquacultural products", 77 pp. These and other resource materials can be found at the Cornell aquaculture homepage: http://www.education.cornell.edu/aben/.

The EPA issues the National Pollution Discharge Elimination System (NPDES) permit to regulate various pollutants from point sources, including aquaculture. NPDES permits are required for fish hatcheries, fish farms, or any other facilities that raise aquatic animals under the following conditions:

1. Coldwater fish species or other coldwater aquatic animals in ponds, raceways, or similar structures that discharge at least 30 days per year, produce more than 9,090 kg (20,000 lbs) of aquatic animals per year, or receive more than 2,273 kg (5,000 lbs) of food during the month of maximum feeding.

2. Warmwater fish species or other warmwater aquatic animals in ponds, raceways, or similar structures that discharge at least 30 days per year. This does not include closed ponds which discharge only during periods of excess run-off or warmwater facilities, which produce less than 45,454 kg (100,000 lb) of aquatic animals per year.

3. Facilities determined on a case-by-case basis by the permitting authority to be significant contributors of pollution to waters of the United States. Discharge of pollutants to receiving waters from aquaculture production facilities, except as provided in the permit, is a violation of the Clean Water Act and may be subject to enforcement by EPA.

NPDES permits for aquaculture operations can set discharge limits on solids, nutrients, and chemical compounds used for water treatments. In addition, various state agencies also regulate the discharge of aquaculture effluents. These standards may be stricter than the current EPA NPDES discharge limits, to protect sensitive water bodies within their state jurisdiction.

10.1 Waste Management Issues

Background of EPA Actions

(Source: http://ag.ansc.purdue.edu/aquanic/jsa/effluents/Background.html)

On May 28, 1998, the EPA published its Notice of Proposed Effluent Guidelines Plan in the Federal Register (Volume 63, Number 102: pages 29203–29213). In this notice, Fish Hatcheries and Farms were identified as one of several candidate industries for effluent guidelines rulemaking projects. The EPA invited submissions of reports and other data on aquaculture discharges and requested public comment on its plan for development of effluent guidelines and preliminary studies. Comments were accepted until July 27, 1998.

On September 4, 1998, the EPA published in the Federal Register (Volume 63: pages 47285–47288) its announcement of Agency's fourth biennial for developing new and revised effluent guidelines, which regulate industrial discharges to surface waters and to publicly owned treatment works. This notice also discussed public comments on the proposed plan first released on May 28, 1998. This notice also reported the following:

1) Under Future Regulations, a court-order consent Decree requires the EPA to develop additional rules and the Agency may choose the next rulemaking projects from the following list of categories:

   Petroleum Refining
   Textile Mills
   Inorganic Chemicals
   Steam Electric Power Generating
   Photographic Processing
   Chemical Formulators and Packagers
   Urban Storm Water
   Airport Deicing
   Aquaculture (Fish Hatcheries and Farms)

2) Under Public Comments, the following was reported: "In the Proposed Plan, EPA asked for public comment on the need for effluent guidelines for aquaculture facilities, also called fish hatcheries and farms. The notice made reference to a recent report by the Environmental Defense Fund, which recommended promulgation of effluent guidelines for aquaculture (63 FR 29210). The Agency received comments arguing either that pollutant discharges are adequately addressed by National Pollutant Discharge Elimination System (NPDES) permits and that effluent guidelines are not needed, or that the permits issued by the various state programs are inconsistent and inadequate. Some comments also provided additional information such as explanations of how aquaculture facilities are regulated under existing permits, comparisons of state permitting policies, and technical papers describing the pollutants generated and treatment processes used in the industry. EPA needs additional information to determine whether the existing NPDES permit process is adequate to address these discharges. The Agency has not yet made a decision on whether national effluent guidelines will be developed for some or all parts of the industry. The Agency may conduct a preliminary study to obtain additional information and evaluate all of the information received."

In February 1999, the EPA's Office of Water announced its decision to conduct a preliminary study of the aquaculture industry to evaluate the current state of wastewater controls and the opportunity for environmental gain (EPA-821-F-99-006). This Aquaculture Study Announcement is posted at:

www.epa.gov/ost/guide/aquaculture.html

On April 20, 1999, the EPA released its draft Animal Aquaculture Study Objectives as a list of questions covering both fresh and saltwater aquaculture operations. It was noted that other items may be added as the issues clarify; therefore, this was considered to be a 'living' document. The major headings were:

- Wastewater Characterization
- Technology Performance, Operation Description, and Alternatives
- Cost & Benefits of Technologies
- Documented Problems; Ways to Avoid or Correct Them
- Description of Existing Regulations
- Description of Products
- Economic Status and Demographics of the Industry
- Industry Organization
- General Industry Trends
- Access to General Industry Source Data and Information

On January 21, 2000, the EPA released the following announcement: EPA Expands Focus on Nutrient Pollution. EPA's Office of Water is focusing new efforts to help reduce nutrient loadings from commercial agricultural and industrial operations nationwide. Among those efforts is a new activity to develop pollutant controls in the form of nationally applicable discharge standards (known as effluent limitations guidelines and standards) for commercial and public aquaculture operations.

In assessments of surface water quality, state regulatory agencies most frequently cite siltation, nutrients, and pathogens as the major cause of water quality impairment. Over the past two years, the EPA has directed resources of the Office of Water's Engineering and Analysis Division to address specific sources of these pollutants. Current activities addressing discharge problems in coal mining (remining operations and certain mine land reclamation activities in the arid west) and the construction and development industry are expected to result in significant reductions of soil and other solids reaching rivers, lakes and streams. Ongoing activities to control nutrients and pathogens from concentrated animal (pork, poultry, beef, and dairy) feeding operations will also improve water quality.

In 1974, the EPA issued a summary technical document for use as guidance in developing controls for wastewater discharges from fish hatcheries and farms. At that time, they decided not to issue final national effluent limitations guidelines and standards. Based on the 1997 agricultural census data, the aquaculture industry includes close to 5,000 land based and marine environment facilities. The aquaculture industry has facilities located in every state and territory, and is currently one of several growing segments of U.S. agriculture. Given the current growth of the aquaculture industry, and the inconsistent state regulatory oversight, the EPA has decided to examine technologies currently available for the control of pollutants, primarily nutrients from land based and marine environment aquaculture operations.

Although the aquaculture industry is currently subject to the NPDES permitting system, there are no national technology-based standards for aquaculture. New national standards for aquaculture will assist the 43 states that are delegated by EPA to administer the NPDES permitting program. Some aquaculture facilities contribute nutrients to the detriment of environmentally sensitive areas found in some estuaries, rivers, lakes, and streams throughout the country. Some aquaculture facilities have made successful improvements in wastewater treatment to reduce the nutrient pollutant load. It may be possible for more facilities to employ these types of successful technologies and management techniques to reduce pollutant discharge loadings to surface waters and, in some cases, water quality impairment in some locations of the U.S. By examining the cost and performance of pollution control technologies and practices, the EPA is committed to developing national effluent limitations guidelines and standards that are consistent with the principles of good environmental stewardship and support the long-term sustainability of the industry. Throughout this national regulatory effort, the EPA will work closely with USDA and other federal agencies, academia, industry trade associations, state and local governments, citizen groups, environmental groups, and other stakeholders. The EPA's efforts will build on the technical expertise of nationally recognized leaders, such as members and participants of the Federal Joint Subcommittee on Aquaculture (JSA) and it's newly created Aquaculture Effluents Task Force. The EPA will regularly provide information on its data needs and the status of their efforts throughout the regulatory development period to the JSA, the industry, and the public.

AQUACULTURE EFFLUENTS TASK FORCE

A proposal for an Aquaculture Effluents Study Task Force was presented and approved September 2, 1999 by the federal interagency Joint Subcommittee on Aquaculture (JSA) to assist the Environmental Protection Agency (EPA) with conducting a preliminary study on aquaculture effluents. The task force has since been renamed the Aquaculture Effluents Task Force subsequent to EPA's decision, announced January 21, 2000, to promulgate national effluent standards for aquaculture operations. More background on the task force and an update on activities can be found on the task force web page:

http://ag.ansc.purdue.edu/aquanic/jsa/effluents/index.html

MISSION:

To coordinate, facilitate, review, and provide input of science-based information via a broad-based stakeholder and technical expert process to assist the EPA in the development of national effluent limitations guidelines and standards for aquaculture facilities in the United States.

OBJECTIVES:

Create a nationally coordinated and systematic process for collection, analysis, and reporting of science-based data and information.

Integrate the best available and appropriate science, technology, data, and information into decision-making processes that best serve the nation.

Provide written scientific and technical reviews and analyses of data and information summaries for different production systems associated with major species in collaboration with EPA.

Prepare expert, peer-reviewed reporting as scientific and technical guidance for developing final national effluent limitations guidelines and standards.

FOR ADDITIONAL INFORMATION CONTACT:

Gary Jensen, Task Force Chair
USDA/CSREES
STOP 2220
1400 Independence Ave. SW
Washington, DC 20250-2220
Tel. 202-401-6802
Fax 202-401-5179
Email: gjensen@reeusda.gov

Edwin Rhodes, Task Force Vice-Chair
Aquaculture Coordinator
Dept. of Commerce/NOAA
National Marine Fisheries Service
1315 East-West Highway, Station 14601
Silver Spring, MD 20910
(301) 713-2334
Fax: 713-0596
Email: edwin.rhodes@noaa.gov

Maxwell (Max) Mayeaux, Task Force Associate
USDA/CSREES
STOP 2220
1400 Independence Ave. SW
Washington, DC 20250-2220
Tel. 202-401-3352
Fax 202-401-5179
Email: mmayeaux@reeusda.gov

10.2 WASTE MANAGEMENT

Handling wastewater is a major problem in all animal agriculture systems, but there are substantial differences between aquaculture wastewater and manure from hog or dairy systems. The latter are typically in the range of 5 to 15% suspended solids, while fish wastewater can be anywhere from 0.2 to 4.0% suspended solids (Kristiansen and Cripps, 1996; Twarowska et al. 1997; Bergheim et al. 1993). Typical suspended solids concentrations from drum filters used in intensive aquaculture operations are around 0.5%. Suspended solids are captured in a variety of ways, but capture techniques primarily rely on straining, settling, or a combination of both. These systems are reasonably effective for removing suspended solids from the culture water, but require periodic flushing of the device to maintain removal efficiency between flushings. Water lost in the flushing process must of course be replaced with clean or new makeup water.

The large flow rates involved in maintaining water quality and removing suspended solids can result in a significant cumulative waste load discharge from fish farms (Braaten, 1991; IDEQ, 1998; Summerfelt, 1999). Consistently meeting strict discharge standards can also be difficult because pipe, channel, and tank cleaning routines can produce fluctuations in discharge flowrates and in the consistencies and concentrations of waste material. The distribution of the nutrients and organic matter between the dissolved, suspended, and settleable fractions affects the choice of method used and the difficulty of effluent treatment.

The filterable or settleable solids contain most of the phosphorus discharged from tanks (50–85%), but relatively little of the total effluent nitrogen, about 15% (Braaten, 1991; Heinen et al. 1996). Most of the effluent nitrogen released (75–80%) is in the form of dissolved ammonia or nitrate when nitrification is promoted. The variability in the nutrient and organic material fractionation between dissolved and particulate matter is largely dependent on feed formulation and the opportunity for particulate matter to break apart, because the production of smaller particles increases the rate of nutrient and organic matter dissolution. Fecal matter, uneaten feed, and feed fines can be rapidly broken down into much finer and more soluble particles by water turbulence, fish motion, scouring along a tank/pipe bottom, and pumping. It is much more difficult to remove dissolved and fine particulate matter than larger particles. Therefore, culture tank designs and operating strategies that remove solids rapidly and with the least turbulence, mechanical shear, or opportunity for microbiological degradation are important to help the fish farm meet discharge limits (to be discussed in more detail below).

10.3 WASTE CHARACTERISTICS

A general mass balance on waste production from fish is shown in Fig. 10.1. Chen et al. (1998) summarized the waste production characteristics of the concentrated TSS coming from an RAS and compared it to domestic sludge characteristics (see Table 10.1). Compared with typical municipal sludge, aquacultural sludge has a relatively lower solid content and BOD5 concentration. Total ammonia nitrogen (TAN) is fairly low for fresh aquacultural sludge, but can increase drastically if the sludge is left undisturbed for a period of time and mineralization occurs under anaerobic conditions. Aquaculture sludge has a higher nitrogen and phosphorous content than domestic sludge. The average value of total phosphorous (TP) is 1.3% of the dry solid mass, while the typical domestic sludge contains only 0.7%.

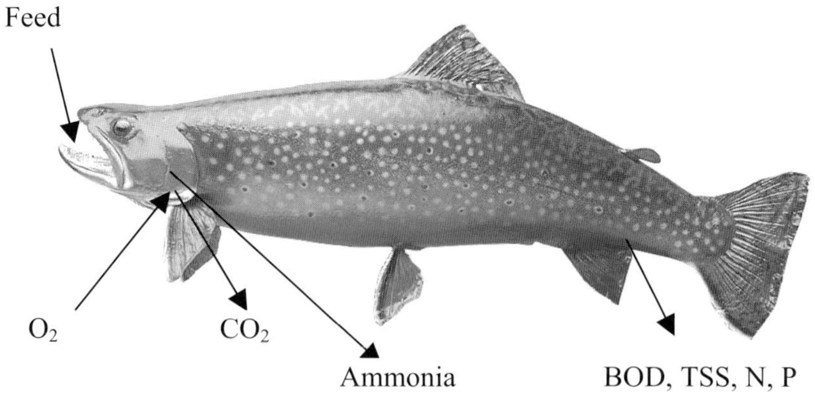

Figure 10.1 General Mass Balance on a Feeding Fish.

Table 10.1 Waste Production Characteristics of Sludge (Chen et al. 1998)

Parameter	Aquacultural Sludge			Domestic Sludge	
	Range	Mean	St Dev	Range	Typical
Total Solids (%)	1.4–2.6	1.8	0.35	2.0–8.0	5.0
TVS (% of TS)	74.6–86.6	82.2	4.1	50–80	65
BOD_5 (mg/L)	1,590–3,870	2,760	210	2,000–30,000	6,000
TAN (mg/L)	6.8–25.6	18.3	6.1	100–800	400
TP (% of TS)	0.6–2.6	1.3	0.7	0.4–1.2	0.7
pH	6.0–7.2	6.7	0.4	5.0–8.0	6.0
Alkalinity	284–415	334	71	500–1,500	600
BOD_{20} (mg/L)	3,250–7,670	5,510	1,210	—	—

Focusing on solid waste, the physical properties of different fish manures can be dramatically different. Tilapia manure is distinctly different than, say, trout feces in that tilapia produce a feces that is encapsulated in a mucous membrane resulting in sausage-like strands. The tilapia fecal strands will float to the surface about 30 minutes or so after being defecated due to gas bubbles being formed in the feces within the strand and then later sink again as gas bubbles diffuse through the membrane strands.

Carbohydrate makeup in tilapia feces will also have an impact on the fecal strand density. In general, higher carbohydrate diets and lower fat diets will result in fecal strands that float more readily than the fecal strands produced when the fish consume higher energy diets; the higher carbohydrate levels result in the fish creating a more definitive mucus membrane for the fecal strand. Digestibility also impacts fecal strands, with lower digestibility resulting in more floating fecal strands. Small changes in nutrient composition can have dramatic impacts on water quality. Be extremely cautious in switching feed formulations.

Waste generation is covered in Chapter 4 but as a quick means of reference, Table 10.2 provides a comparison of the amount of waste generated by the major animal groups. Adjustments can be made where you have a better estimate of feeding rates for a particular aquaculture application.

Table 10.2 Waste Generation (kg/day per 1,000 kg live weight) from a Catfish RAS and Other Commercial Animal Production Systems (Chen et al. 1993)

Animal	BOD_5	Solids	TKN	Sludge Volume (L)
Catfish RAS	0.8–1.3	4.2	0.20	70–420
Beef Cattle	1.6	9.5	0.32	30
Dairy Cow	1.4	7.9	0.51	51
Poultry	3.4	14.0	0.74	37
Swine	3.1	8.9	0.51	76

Note: Phosphorus generation is approximately 1% of the feed fed.

10.4 Waste Management Overview

The two most common methods used to recycle solid wastes from aquaculture facilities are land application and composting. Most states have guidelines or regulations that govern land application of manure and other organic wastes to fertilize agricultural crops by limiting the land application rates and the amount of associated pathogens, heavy metals, and other contaminants. The land application rates are based on the nutrient content, soil type, and plant nutrient uptake characteristics to prevent runoff or groundwater contamination. Odor problems can also limit land application in developed areas. Finally, the transportation from the point of generation to the application site can be a major factor in the costs of sludge management, since aquaculture sludge, even when thickened, is mostly water.

Before either of these disposal methods can be used, the sludge and effluent wastewater first needs to be transferred to some form of storage system for both sludge thickening and flow equalization. Sludge thickening will reduce the hydraulic loading on subsequent processes. In addition, sludge flow rates and concentrations will vary during cleaning and other maintenance activities. Therefore, some form of flow equalization is needed to equalize the flow on subsequent treatment processes. Sludge storage structures include anaerobic lagoons, earthen ponds, and above and below ground tanks. Sludge thickening can be accomplished by settling basins, lime stabilization, reed beds or wetlands, and sand beds.

10.5 Storage, Thickening, and Stabilization

Storage and Thickening

Thickening basins and separate storage structures are both utilized for storing aquacultural sludge. Thickening basins can be designed to accommodate the build-up of solids and provide temporary solids storage capacity. However, as sludge accumulates within these basins they will become less and less effective at solid-liquid separation due to impingement on the proper settling hydraulics, solids flotation due gas bubble production from fermentation, and dissolution of nutrients and organic matter.

In many cases, the thickened sludge from thickening basins is transferred to larger sludge storage structures capable of holding months worth of captured and thickened solids. Sludge storage structures utilized include earthen ponds, above ground tanks, and below ground tanks. Earthen ponds are generally rectangular basins with inside slopes (horizontal:vertical) of 1.5:1 to 3:1. Depending on site geology and hydrology, earthen ponds can have liners of concrete, geomembrane, or clay. Earthen pond design will include the capacity for storage of precipitation as well as a method for removing solids, Fig. 10.2. In the case where solids will be unloaded via pumping, the solids must be agitated to provide a uniform consistency. Pond agitation may be accomplished with hitch-type propeller agitators that are powered by tractors or by agitation pumps. Propeller agitators work well for large ponds, while chopper-agitator pumps work well for smaller ponds. Solids unloading may also be done with heavy equipment, in which case pond design should include ramp access (maximum slope of 8:1) and suitable load capacity in the unloading work area (NRCS, 1996; Wright et al. 1999).

Figure 10.2 Storage Lagoon for a Large Commercial Tilapia Producer (Fingerlakes Aquaculture, Groton, NY).

Sludge may also be stored in tank structures, above and below ground. Storage tanks are primarily constructed of reinforced concrete, metal, and wood. Reinforced concrete tanks including both walls, foundation, and floor slab-- may be cast-in-place or they may be constructed of pre-cast wall panels, bolted together, and set on a cast-in-place foundation and floor slab. Metal tanks are also widely used, with the majority being constructed of glass-fused steel panels that are bolted together. There are many manufactured, modular tanks commercially available in reinforced concrete and metal, as well as wood (NRCS, 1996).

Design of all structures, earthen or manufactured, should include considerations for internal and external hydrostatic pressure, flotation and drainage, live loads from equipment, and dead loads from covers and supports (NRCS, 1996).

SOLIDS THICKENING AND STABILIZATION

Captured solids from solids removal processes tend to be dilute, having less than 2% total solids content. These solids may be concentrated or thickened in settling basins up to 5–10% total solids content (Bergheim et al. 1993). Thickening basins operate according to the same discrete particle settling principles previously described. However, because thickening basins receive water with elevated solids content and are concentrating these solids by permitting particle settling, the solids are also subject to compression settling. Compression settling develops when a compressed layer of particles forms at the basin bottom. The particles in this region begin to form a structure of particle-particle contact and the slurry is concentrated further (Crites and Tchbanoglous, 1998). In general, the overflow rate for sludge thickening basins should be approximately 1.0 $m^3/m^2/hr$ (3.2 $ft^3/ft^2/hr$) with hydraulic retention times of between 20 to 100 minutes (Bergheim et al. 1993; Bergheim and Cripps, 1998).

Thickened sludge in the bottom of thickening basins may be further conditioned by adding lime. Lime addition is an effective method to kill sludge pathogens, Table 10.3, reduce odor problems, and improve solids thickening. It is recommended that 15–20 g (0.125–0.167 lbs) of unslaked lime (CaO) per gallon of sludge (10% solids content) be used to achieve a pH of 12 for lime stabilization, Fig. 10.3. In addition to stabilizing the sludge and improving its settling properties, lime stabilization also increases the removal of phosphorus (Bergheim and Cripps, 1998, Bergheim et al. 1998).

Table 10.3 Sludge Lime Treatment of Aeromonas Salmonicida and IPN-virus Sludge at Different pH Values and Temperatures of 15 to 20°C (from Bergheim et al. 1998)

Time	*Aeromonas salmonicida*				IPN-virus	
	pH	$\frac{CFU}{mL}$	pH	$\frac{CFU}{mL}$	pH	$TCID_{50}$ mL^{-1}
0	11.0	3.0×10^6	12.0	3.0×10^6	11.9	$10^{8.3}$
1 hr	11.0	2.1×10^3	12.2	$<1 \times 10^3$	11.9	$10^{7.6}$
1 day	—	—	—	—	11.6	$10^{6.1}$
3 days	10.5	$<1 \times 10^3$	11.4	$<1 \times 10^3$	—	—
7 days	—	—	—	—	11.4	$10^{5.1}$
10 days	10.6	$<1 \times 10^3$	11.4	$<1 \times 10^3$	—	—

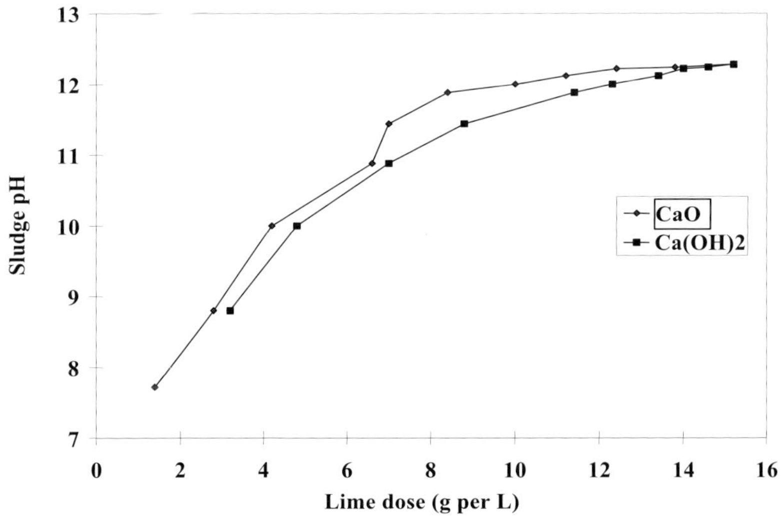

Figure 10.3 Effect of Lime Dose on Aquacultural Sludge pH for Two Types of Stabilization Amendments, CaO and $Ca(OH)_2$. Sludge dry matter content was 11.1% (from Bergheim et al. 1998).

Other sludge thickening methods include sand beds, wedgewire sieves, filter presses, centrifuges, vacuum filters, and reed beds (EPA, 1987; Tchobanoglous and Burton, 1991; Black and Veatch, Inc., 1995; Reed et al. 1995).

Reed drying beds

Depending on the location and local regulations, an aquaculture facility's only available options for sludge disposal may be limited and costly. However, if transportation costs make sludge disposal on cropland uneconomical, disposing of the sludge on-site within created wetlands may be an attractive alternative. A constructed reed drying bed can provide on-site treatment of a concentrated solids discharge with an uncomplicated, low-maintenance, plant-based system that could reduce solids disposal costs (Outwater, 1994).

Reed drying beds are vertical-flow wetland (VFW) systems that have been used over the past 20 years to treat thickened sludge (1–7% solids) produced in the clarifier underflow at wastewater treatment plants (Hofmann, 1990; Liénard et al. 1990; Nielson, 1990, 1993; Riggle, 1991; Outwater, 1994; Reed et al. 1995). When used for municipal treatment, these wetlands are loaded with 7 to 10 cm (2.76–3.94 inches) of 2% solids approximately once every 7–21 days (about 30–60 kg/m^2/yr. Other much higher hydraulic loading rates are also reported: 25 cm/day (Kadlec and Knight, 1966); 80 to 240 cm/day (Outwater, 1994); and 2,000 to 2,700 cm/day (Reed, 1990). Clearly, the specific application and the nature of the wastewater will have an impact on any sustained level of infiltration.

Reed drying beds are also beginning to be used in aquaculture (Summerfelt et al. 1999). Summerfelt et al. (1999) applied trout wastewater (7,800 mg/L total suspended solids) 6 times per day at a rate of 1.35 cm/day (30 kg dry solids per m^2 per year). During operation, a series of vegetated beds receives sequential batch applications of sludge. The sequential batch applications are such that the more recently flooded VFW cells are dewatering, while beds with older sludge applications are drying. Intervals between repeated sludge applications allow for dewatering and drying. Plants facilitate dewatering by conducting water along their stem and root paths through previous sludge layers and by removing water through evapo-transpiration. The plants also increase biological stabilization of the solids by transporting oxygen to their root zones. Reed bed treatment systems have been reported to have a useful lifetime of up to 10 years.

Rapid Sand Beds

The above section on reed beds is based upon the inherent hydraulic conductivity of the underlying sand bed. In the simplest form, one could design a dewatering bed that had no surface plants. In this case, you must design based upon the bed's hydraulic conductivity, or the coefficient of permeability, which is the standard measure of how quickly water flows through a soil (sand) column. Saturated hydraulic conductivity can be determined using Darcy's law (Ritzema, 1994). Sand beds may also incorporate marsh plants, such as phragmites, to sustain the hydraulic conductivity as described. In addition, removing such plants on a scheduled basis will also serve beneficially to remove nutrients from the wastewater (Ritzema, 1994; Kadlec and Knight 1996):

- Nitrogen ~225kg/ha/yr
- Phosphorus ~35kg/ha/yr

Sand bed drainage systems enhanced with plant cover crops have been used for a variety of applications including at least one commercial fish farm in Amherst, MA (Bioshelters, Mr. John Reid President, a 200,000 kg/year tilapia farm) and have been described in detail by others (Ritzema, 1994; Kadlec and Knight, 1996; Sanford et al. 1995).

Recent research by Palacios and Timmons (2001) developed predictive equations of infiltration including the effects of biosolids accumulation (the major resistance to infiltration). These equations are as follows:

$$\frac{Q}{A} = -K_s \frac{\Delta H}{L} \qquad (10.1^a)$$

$$\frac{\Delta H}{L} = \frac{Y/2}{N\,Y\,\beta} \qquad (10.2)$$

$$L = \frac{C_{TSS}}{16}\beta\,D \qquad (10.3)$$

where, K_S = 2.44 m/d and β is 0.05 m/m (based upon using tilapia wastewater with a TSS of 16 g/L)

These equations were developed from fish manure obtained from a commercial tilapia farm using RAS and employing approximately a 10% system volume exchange rate per day. The farm fed a 42% protein 12% fat feed; feed conversions at the farm were approximately 1 to 1. The TS concentration of the wastewater applied in these experiments was 2%, of which approximately 16 g/L was TSS (80% of the 2% was organic solids). Total manure depth (L) is the summation of solid layers accumulating from repetitive applications. The estimate of total L using Eq. 10.3 is used in Eq. 10.1 along with the K_S value of 2.44 m/day to predict infiltration rate. Note that the K_S value was specifically

---

[a] Symbols are defined in List of Symbols at the end of the Chapter.

developed from tilapia feces and thus may be different for other fish manures. The filter medium is not expected to have any significant impact on estimating infiltration rates, since the K_S for the sand type medium is >> than the K_S for the manure. As a conservative rule of thumb, assume a design infiltration rate after repetitive applications of 3.5 cm/day.

> **"RULE OF THUMB"**
>
> Sand beds will have an infiltration rate of 1.4 inches (3.6 cm) per day.

MANAGING SAND BEDS

Treatment wetlands should be divided into smaller beds that are easier to manage. Operate these beds in parallel to better handle variable discharge volumes by flooding only the necessary area, or to ensure that treatment is not completely interrupted if one of the beds needs servicing. Beds also could be operated in series to achieve more complete treatment of the wastewater.

Using Eqs. 10.1 and 10.3 to predict infiltration rate probably provides a conservative estimate, since the rate equation is for sand beds without any surface vegetation. Use of phragmites or other emergent type monocot herbs would provide some continual disruption of the sand column due to root and shoot growth and probably to the manure surface layer as well. All disturbances cause large increases in infiltration. Consideration should also be given as to how one can restore hydraulic conductivity after repeated applications of wastewater. Simply removing the accumulated biomass and skimming a top layer of the sand materials will restore much of the sand bed's initial conductivity.

10.6 UTILIZATION/DISPOSAL

LAND APPLICATION

The simplest and most useful method of sludge disposal is to take advantage of it as a fertilizer for direct land application. Processed aquaculture sludge dry matter contains high levels of nitrogen and phosphorus, but negligible amounts of potassium. Nitrogen is mainly organically bound and has to be decomposed by microorganisms in order to become available for plants (Bergheim et al. 1993). When the solids content is less than 1%, solids and slurries are easily pumped and distributed. Thus, there are a variety of land application systems that can be used, including conventional irrigation via sprinklers and surface flooding. Climate constraints on crops and vegetation may require winter storage of the wastewater. Other methods include rapid infiltration of the wastewater by intermittent flooding of shallow basins in relatively coarse textured soils of rapid permeability. Thickened sludge (>5% solids) can be used as a soil amendment or fertilizer and applied from tanker trucks, by surface spreading with or without incorporation into the soil, and by direct injection into the soil.

COMPOSTING

(adapted from Agricultural Waste Management Field Handbook, NRCS, 1996)

Composting is the aerobic biological decomposition of organic matter. It is a natural process that is enhanced and accelerated by the mixing of aquaculture wastes with other ingredients for optimum microbial growth. Composting converts the sludge and other waste products from aquaculture into a stable organic product by converting nitrogen from the unstable ammonia to a more stable organic form. The end result is a product that is safer to use than the original sludge and when applied to land, will improve the soil fertility, tilth, and water holding capacity. In addition, composting reduces the bulk of material that needs to be spread, improves its handling properties, reduces odor, fly and other vector problems and can destroy pathogens.

Composting methods include windrow, static pile, and in-vessel. The windrow method consists of piling the compost mix into long, narrow piles or windrows. To maintain aerobic conditions, this windrow needs to be periodically turned and mixed, exposing the decomposing material to air and to keep the temperatures from getting too high. The static pile method consists of mixing the compost materials and then stacking the mixture on plastic pipe or tubing through which air can be forced or drawn. Small compost piles may not need forced ventilation, if they are highly porous or with a mix that is stacked in layers with highly porous materials. The in-vessel method involves the mixing of the materials in a reactor, building, container, or vessel. Forced ventilation may be required. This process provides a high level of control over moisture, aeration, and temperature.

Composting of aquaculture wastes requires the mixing in of amendments or bulking agents in the proper proportions to promote aerobic microbial activity and growth and achieve optimum temperatures. This mixture provides a source of energy and nutrients, moisture, and oxygen for the bacteria. The composting amendment is added to the mixture to alter the moisture content, carbon to nitrogen (C:N) ratio, or pH. Many materials are suitable for use as composting amendments, including crop residues, leaves, grass, straw, hay, sawdust, wood chips or shredded paper, and cardboard. The bulking agent is used to improve the ability of the compost to be self-supporting and to increase porosity to allow internal air movement. Wood chips and shredded tires are two examples of bulking agents. Wood chips are an excellent bulking agent because they also alter the moisture content and C:N ratio. Figure 10.4 shows a compost bin using wood chips as both an amendment and bulking agent.

Figure 10.4 Compost Bin for Sludge.

Compost design requires knowledge of the characteristics of the aquaculture wastes and the amendments and bulking agents. The characteristics that are most important are moisture content, carbon content, nitrogen content and C:N ratio. The balance between the C:N ratio in the compost mixture is a critical factor for optimum microbial activity. These microbes multiply rapidly in the compost pile and consume carbon as a food source and nutrients to metabolize and build proteins. The C:N ratio of the compost mix should be maintained between 25 and 40 to 1. If the C:N ratio is low, a loss of nitrogen generally occurs through rapid decomposition and volatilization of the ammonia. If it is high, the composting time increases because the nitrogen becomes the limiting nutrient for growth. Moisture is required by the microbes to convert the carbon source to energy. Bacterial generally can tolerate moisture content as low as 12 to 15%. However at moisture contents of less than 40%, the rate of decomposition is slow. At greater than 60% moisture, the process turns from one that is aerobic to one that is anaerobic. Anaerobic composting decomposes more slowly and produces putrid odors. The determination of the compost mix design or retype is normally an iterative process of adjusting the C:N ratio and moisture content by the addition of amendments. If the C:N ratio is out of the acceptable range, then amendments are added to adjust it.

The composting of mortalities can be an economical and environmentally acceptable method. The process produces little odor and destroys harmful pathogens. As is shown in Fig. 10.4, compost bins are typically about 5 feet (1.5 m) high, 5 feet (1.5 m) deep, and 8 feet (2.4 m) cross the front. The width across the front should be sized to accommodate the equipment used to load and unload the facility. To prevent spontaneous combustion and to allow for ease of monitoring, a bin height of no more than 6 feet (1.8 m) is recommended. The depth should also be sized to accommodate the equipment used. Rapid composting of mortalities occurs when the C:N ratio is maintained between 10 and 20. This is considerably lower than what is normally recommended for more traditional composting, because much of the nitrogen in the dead animal mass is not exposed on the surface. A lower C:N ratio is necessary to ensure rapid composting with elevated temperature to destroy pathogens. The moisture content of the initial compost mixture should be between 45 and 55%, by weight, to facilitate rapid decomposition. Composting of mortalities should remain aerobic at all times throughout the process. This is easily accomplished by layering the mortalities and amendments in the mix. Layers of such high porosity material as straw, wood chips and bark allow lateral movement of air in the compost mix.

List of Symbols

A	cross sectional area, cm^2
D	cumulative depth of wastewater applied, m
K_S	saturated hydraulic conductivity, cm/d
L	depth of manure layer or material providing the resistance to flow, m
N	number of applications of wastewater
Q	volumetric flow rate, cm^3/d
Y	depth of wastewater applied per application event, m
β	constant used to estimate accumulation rate of manure per unit of waste water applied, m/m
ΔH	average hydraulic load imposed per application event (1/2 of treatment depth), m

CHAPTER 11

SYSTEM MANAGEMENT AND OPERATIONS[1]

11.0 INTRODUCTION

During almost all discussions of intensive recirculation aquaculture systems, the focus is on the culture tanks, filtration systems, aeration/oxygenation systems, and the species being cultured. Yet the system's support components, which are just as important and in some cases critical to commercial success, are rarely mentioned. Support components include all of the other parts of a recirculating system that are necessary for its profitable operation. How well these support systems are designed, integrated into the operation, and managed often determines whether a recirculating system survives commercially or not. Many of these supporting systems are common to all fish production facilities; some requirements are unique to recirculating systems. One common characteristic is that they are often ignored in estimating culture system construction costs. As a result, support systems are often installed late in the construction phase or as an afterthought subsequent to a disaster. When this happens, funding is limited and typically the least expensive support system is installed, regardless of its reliability or whether it is appropriate to the system. Such false economy usually compounds problems in an already marginal operation. The result of such activities is then just a matter of when, and not if, the system will fail.

The list of support components necessary for an intensive recirculating aquaculture system (RAS) is a reflection of the level of sophistication, the interplay of upfront capital investment versus daily operational costs, size of operation, number of employees, geographic location, and numerous other parameters. A short list includes such items as:

---

[1] Sections on Fish Product Handling and Transport were provided by Donald Webster, Extension Marine Science Educator, University of Maryland Cooperative Extension. Dr. Joe M. Regenstein, Professor of Food Science, Cornell University contributed Section 11.9 Post Harvest Handling.

- Backup Systems
- Laboratory Facility
- Quarantine Area
- Waste Management
- Storage – Feed, Chemicals, Product
- Handling of fish, both live and post slaughter
- Labor
- Access

The objective of this chapter is to identify some of the needed support systems with no attempt to be all inclusive or to prioritize. Each recirculating system will have it own special needs and the priority of needs for one system is often different from another system. Thus, the system design engineer and the manager must include the required support systems pertinent for any specific production unit.

11.1 Backup Systems

In science, a concept only becomes a law after being tested and observed countless times. Murphy's Law is an excellent example of this concept. It states simply: If anything can go wrong, IT WILL. This is the entire design rational behind backup systems and even backup systems for backup systems. During design, construction, and operation, it is critical to try to imagine the worst-case-scenario! Because if it can happen (and often when you think it cannot possibly happen), IT WILL! Anticipate! Plan! Train! Respond!

"Rule of Thumb"
Backup Systems
Anticipate! Plan! Train! Respond!

One of the primary systems requiring backup is the electrical power supply, which is required to operate pumps, aeration systems, instrumentation systems, and to perform a variety of other functions in a recirculating system. Failure of the electric supply can have devastating effects in a matter of minutes, especially in heavily loaded systems. Backup electricity can be provided by a generator that is fueled with gasoline, diesel, or natural or propane gas. Backup generators are a critical must for any commercial (for-profit) system. Several commercial manufacturers can supply turnkey systems over a wide range of power requirements, Fig. 11.1.

Figure 11.1 Backup Generators: Dedicated Diesel Unit and a Portable Standby Unit.

The cost of the generator is directly related to the generator size, i.e., the power in kilowatts that is generated. The generator size required for backup power is determined by the loads that are critical to the maintenance of good water quality or otherwise support the survival of the fish in the culture systems during power blackouts. Typically, this includes powering such items as the circulating pumps, aerators or blowers, the data acquisition system, and building emergency lightning. Usually, these loads are handled by a separate emergency circuit breaker panel. The capability of the backup generator in terms of voltage (120–240 V AC), current, and phase (single or three phase) will be determined by these essential support requirements.

One of the important design parameters in specifying a backup generator is whether it will or will not have automatic starting capabilities. Automatic starting systems monitor incoming power lines and when the external power supply goes down, the generator automatically starts, so power is always available to critical components. Automatic systems or cut-over switches are expensive, but are necessary when personnel are unavailable during certain periods of the day or are unable to get to the facility rapidly enough to carry out manual emergency measures. An example autotransfer switch with the emergency circuit breaker panel on the right is shown in Fig. 11.2.

An automatic transfer switch will cost roughly 35% of the total cost of the backup generator system, but is generally well worth the investment if continuous manual coverage is not possible. If you use an auto-transfer mechanism, make sure that protocols are in place to protect the fish if the auto-transfer fails. Sometimes, the alarm sensors will "sense" that power has been restored when in fact it has not. Three-phase systems are particularly problematic and tend to fail more often in the transfer operation than single phase systems. Remember, an auto-transfer is not nearly as reliable as 24-7 coverage by humans and a manual transfer switch.

Figure 11.2 Generator Automatic Transfer Switch and Emergency Circuit Breaker Panel.

Even people are not 100% reliable as a backup system. They must be properly trained to respond to emergency and to monitor continuously. It is a good practice for a night shift person to have duties that require them to remain active throughout their shift. Intentionally schedule activities that keep them busy, e.g., water chemistry measurements, and some hand feeding.

Regardless of the generator system chosen, an automatic or manual transfer switch will be required that disconnects the power company supply lines and connects the generator lines. These switches are required to prevent feedback of power into the power company grid from the system generator. Feedback into the power company grid is a safety hazard to power line workers repairing lines. In addition, if feedback occurs, shorts or loads on the power grid can overload the generator and burn it out. Whether manual or automatic, transfer switches are expensive and can cost as much as the basic generator. It is important to work with the local power supplier to determine exactly what their requirements are and in some cases, whether equipment can be leased from them.

> **"RULE OF THUMB"**
>
> There is NOTHING more reliable than "**24-7**" coverage by a person who can flip a manual transfer switch to backup power when you lose power!

Generator maintenance is an absolute must! If power outages are infrequent, the backup generator may sit for long periods of time between uses. Fossil fuel engines, whether diesel or gasoline, not started for prolonged periods tend to be difficult to start. Failure of the backup generator's engine to start due to dead batteries, low fuel levels, or other reasons will lead to catastrophic fish loss and perhaps the end of the business itself! Most commercial models are designed to be operated for a specified period every few days. Smaller generators should be periodically started and maintained in top running condition at all times.

Another style of backup generator is shown in Fig. 11.3. Note in the figure that the generator set has a roof over it (not shown) and that the generator set is well removed from the nearest wall. This provides fire protection, prevents heat buildup around the generator set, and provides a measure of sound reduction.

Figure 11.3 Backup Generator.

SIZING A BACKUP GENERATOR

It is first useful to have a basic understanding of the terms used to calculate loads and describe unit features. Here is the minimum information that you should understand:

- Standby Power is power available at variable load capacities for the duration of the main power network interruption.

- Prime versus stand-by power. According to ISO 3046 ratings, baseload power is defined as power available for continuous operation, with 10% overload available for 1 hour in every 12 hours of operation. Backup generators can be designated as a prime power unit (this is your primary source of power at this point) if the generator is selected for peak load operation at variable loads for an unlimited number of hours in place of the main power network.

- Transfer switches are the devices that switch the electrical feed from line power to the generator set. They can be automatic or manual.

- kVA is 1,000 volt-amps and is a power rating method used to describe the size of the generator. This number is used to size the generator and not the horsepower of the motor on the gen-set. The equations to determine the kVA of the generator are:

$$kVA_{1-phase} = \frac{Volts \cdot Amps \cdot PF}{1{,}000} \tag{11.1}$$

$$kVA_{3-phase} = \frac{Volts \cdot Amps \cdot PF \cdot 1.73}{1{,}000} \tag{11.2}$$

where PF is the power factor for the motor

Values for PF range from 0 to 1.0; typically use 1.0. PF accounts for the reduction in power output due to inductive reactance. When the device is all resistance, such as in an electric water heater, the PF is 1.0. When the electrical circuit contains a coil, which produces inductive reactance, the PF will be less than 1.0, but will be typically above 0.8. For exact information, you must contact the motor manufacturer.

Testing the Back-up Generator

Your back-up generator must work when the critical time comes. You will not have time to refine your protocols once you have lost power. The acid test is to go over to your main power panel and disconnect the incoming line power. Then check the following:

- Is all the critical equipment running (or was there insufficient power to start all the critical units at once). If not, consider installing delay timers on specific equipment so that the entire load is not activated at startup (startup amp requirements may be two to five times running load amps).

- Let the system run under full load for 6 hours at least twice a year to be sure that the generator set is adequate. This will allow any problems to be identified and corrected before a crisis occurs.

Final advice on this subject: make sure your backup generator set is installed and has been tested prior to your fish being placed into the facility.

LOSS OF OXYGEN

More fish probably die in recirculation systems due to a lack of oxygen than to any other single cause. Backup oxygen systems are a basic requirement for a culture system to be economically feasible. Because oxygen availability is so critical to heavily loaded culture systems, often a three tier emergency oxygen supply is prudent. There are many types of oxygen supply systems used in RAS and the type of emergency backup system needed varies with the primary oxygen supply system used. One of the simplest and most cost-effective emergency oxygen systems uses an oxygen tank, either liquid or gas, connected to micro-diffusers in the culture tank through a normally open, electrically operated solenoid valve (see Fig. 11.4). When electricity is applied to the solenoid valve, the valve is closed and oxygen is provided by the regular oxygen supply system. When the power goes out for some reason or the system is manually operated, the solenoid opens, allowing oxygen from the tank to flow into the culture tank. It is important for any backup system, but especially for oxygen systems, that they are automatic and engage at the first sign of a potential problem.

Figure 11.4 Solenoid Oxygen Backup System With Oxygen Flow Meters for Monitoring.

Have your oxygen back up systems in place. These units, whatever they are, need to be in or near the tank so that on a moments notice, you can activate them. Once you lose power (flow) you have only minutes to respond to protect all the tanks in the facility.

> **"Rule of Thumb"**
>
> You have **5 to 15 minutes** to restore oxygen to all of the tanks in the facility, once you've lost power. Can you do it?

So, do a dry run before the crisis occurs. Can you put all fish tanks on their backup oxygen systems in 5 minutes? If not, re-evaluate and re-plan until you can.

11.2 LABORATORY FACILITIES

The size and sophistication of the laboratory space needed for an aquaculture facility will vary with the size and complexity of the recirculation system. However, every recirculating aquaculture production system will require at least a minimum amount of space set aside for laboratory analysis. At a minimum the laboratory will include equipment for water quality analysis, a microscope for fish health management, a refrigerator for chemicals and samples, and a computer for data analysis and storage.

Water quality monitoring and control is a routine task in any aquaculture facility. The amount of laboratory space and equipment needed for this work will vary with the methods used, facility size, and the frequency of sampling. Smaller fish farms can often get by with commercial water quality test kits that are available from several manufacturers. They use pre-packaged chemicals, indicator strips, or color comparison techniques. They are convenient, relatively inexpensive per analysis, and provide sufficient accuracy for small production facilities with low stocking densities. More importantly, they require very little laboratory space.

As the number of samples and the corresponding risks to production increase, the sophistication of the analysis also needs to increase. Several important water quality parameters can be monitored using electrode sensors that provide an electrical output signal proportional to the water quality parameter being measured, such as temperature, pH, dissolved oxygen, conductivity, and oxygen reduction potentional (ORP). Portable meters do require that someone physically go to each sampling point and manually measure and record the required water quality parameter at whatever is the desired frequency. Labor costs for manual monitoring can be high for large system, but economical for small systems, in comparison to the equipment costs for dedicated systems.

In heavily stocked systems, dedicated systems can be used to both monitor and control critical water quality parameters, such as temperature, pH, and dissolved oxygen. Commercial systems are available that will monitor selected parameters, and also provide an alarm when a parameter deviates from the preset minimum and maximum. Combined with the necessary computer hardware and software, real time water quality data becomes available to the system's manager. This provides accurate information for management and historical data for system performance evaluation and analysis.

As the sampling protocols become more sophisticated, a higher quality laboratory will be needed, Fig. 11.5. Currently, for example, there are several commercially available laboratory spectrophotometers designed for water quality analysis. These dedicated pieces of equipment are able to analyze for a wide range of critical water quality parameters using pre-packed chemicals and simple laboratory procedures. At this point, the laboratory should be equipped with a sink, refrigerator, and computer, and have adequate room for equipment storage and personal work areas.

Figure 11.5 Laboratory Space for a Large Commercial and a Smaller Operation.

11.3 QUARANTINE FACILITIES

Many of the disease outbreaks occurring in recirculating systems come from diseases introduced on or in fish purchased from outside the system. Disease introduction can be minimized by quarantining incoming fish for one to a few weeks prior to introducing them into the recirculating system. During this period any disease problems can be treated without contaminating the recirculating production system. See Chapter 13 for more details on quarantining.

Quarantine areas require space and must be located such that the incoming fish are never in close contact with the growing systems. Personnel traffic must never be allowed to move from the quarantine area directly into the area containing the growing systems. Permanent physical barriers must make such movement impossible. Where possible, the quarantine area should be in a separate building from the main production systems. The size of the quarantine area is a function of the management techniques used in the facility. The quarantine area must be large enough to house all of the fish entering during one quarantine period, whether it is one week or several weeks. The water supply must be adequate for the volume of fish in the quarantine area and separate waste discharge should be provided. Further information on quarantine procedures and biosecurity issues are discussed in more detail in Chapter 13.

11.4 WASTE MANAGEMENT

Aquaculture waste discharge regulations are currently in a state of flux. However, it is clear that aquaculture producers will be required to meet discharge regulations and these regulations probably will become more stringent in the future. Thus, waste management facilities and management methodologies will be an important component of any aquaculture facility put on-line in the future. Because recirculating systems do not use as much water as many other types of aquaculture systems, their waste management problems may not be as severe, but there is waste that must be disposed of or put to use in an environmentally beneficial or at least benign way. See Chapter 10 for more details.

Waste disposal needs for an aquaculture facility depend on the system size, species cultured, feed used and other variables. The major problem is removing the wastes from the culture water. There are mechanical, biological, and chemical methods to accomplish this removal, and a variety of specific implementing systems. The best technique to use depends on the waste characteristics, concentrations, and form in which the waste is found. These filtration systems are covered extensively in Chapter 6 Solids Capture. Once the waste is removed from the culture water, it must be disposed of or used for some useful purpose. Solids from fish culture systems contain considerable nitrogen and phosphorous, which are useful nutrient elements in fertilizers. Thus, one disposal technique is to use the solids for organic fertilizers, by spreading it on agricultural land. Composting for later use as a soil amendment, mulch, or fertilizer can be used for both solid and liquid fish wastes. Holding facilities may be necessary to contain the wastes between periodic disposal cycles. Such facilities must contain the wastes in an acceptable manner and must prevent development of odors and other noxious nuisances.

Nearly all aquaculture facilities will suffer some mortality, although under normal operation the volume of dead fish will be low. Nevertheless, if mass mortalities occur due to oxygen stress, disease, or some other reason, the volume of dead fish that must be disposed of can become high. These fish must be disposed of in an environmentally acceptable manner, and they cannot create noxious odors or pose sanitation or health risks to either humans or other fish in the culture system. Acceptable disposal techniques will depend on the volume of dead fish, the land use near the facility, the depth of the groundwater table and other factors. In the event of a mass mortality, lack of planning for disposal of dead fish can be embarrassing at best and may result in a legal morass. If fish are processed on-site, processing waste must also be disposed of or used. Some of the options for large quantities of fish waste are composting, anaerobic digestion, burying, or moving to a landfill. The key is to be prepared ahead of time.

11.5 Storage—Feed and Chemicals

There are several obvious materials that must be temporarily stored at any aquaculture production facility including: 1) chemicals for disease treatment and for running water quality tests, 2) finished product, and 3) feed.

There are several potential uses of chemicals in recirculating systems, although only a limited number are available for disease treatment of fish. However, those chemicals that can be used must be stored in an appropriate manner. Some require refrigeration, while others require only dry storage conditions. All of these chemicals should be secured against accidental use and/or theft.

Chemicals used for water treatment are probably more common than disease treatment chemicals in recirculating systems. Typically, flocculants, disinfectants, and cleaning compounds are used in or around recirculating systems and space must be allocated for their storage. Normally such storage areas must be ventilated, dry, and relatively secure. Chemicals are necessary for calibration, titration, and other applications when measuring water quality. Most of these chemicals can be stored in a dry, secure area. Occasionally, a chemical is needed that is corrosive or volatile; these type of chemicals must be stored in an appropriate safety cabinet or refrigerated safety cabinet.

OSHA material safety data sheets (MSDS) describing the chemical and its effects must be available for all workers. EPA labeling and disposal requirements must be met. CVM (FDA) withdrawal times must be adhered to.

11.6 Fish Product Handling

Whatever the production facility, it will produce some final product, usually either live fish ready for sale, or dead fish in the round or processed to some extent (at least gutted, in most cases). Temporary storage must be available for the product. If live fish are to be sold, a holding tank and/or a depuration may be necessary. Often fish are graded (see below) from the production tanks just prior to sale. In this case, the fish to be sold are often collected and held in a separate holding tank that is readily accessible to the truck to be used for transportation.

Grading

Grading is generally accepted as a way to improve growth rates by eliminating negative interactions between fish of different sizes. Grading also allows for a more accurate feeding regimen, feeding the proper feed particle sizes, and it makes harvesting easier to plan and to carry out. The less variability in fish sizes across the cohort, the less grading that will have to be done. Generally, the less deviation in sizes, the more marketable the fish will be to a growout facility and to the processor, although some processors may want different sizes or a mix of sizes from time to time, to meet their market needs.

Some Norwegian salmon farmers use grading to cull the smallest, least efficient feeders and might remove as much as 50% of the population. Maintaining inefficient feeders can create a 30–50% increase in production costs; therefore, these fish should be removed as early as possible. Culling decisions will need to be made by the manager, based on experience with the facility's production. If it is not clear how well various size ranges of fish are converting feed, a small percentage of the smallest fish from a single tank can be removed. Feed conversion or other parameters can then be monitored to determine whether the initial cull improved production efficiencies.

There are many fish graders commercially available. Whatever the system used, facilities, and/or equipment are needed for grading. The type of grading equipment needed depends on the system design, species, management methods, and other factors. Grading procedures are discussed in detail later in the chapter.

Grading Method

The box grader is often used to grade fingerlings. It consists of a floating box that contains an adjustable or replaceable set of grader bars. In most cases, the appropriate bar width will be determined by trial and error. After grading a few times, grader size may be correlated with condition factors estimated from weight and length samples, and grading can be planned more accurately. The production notebook should contain records of fish size and grader bar width so that benchmarks relating fish size and weight or condition factor to grader width can be established.

Fish to be graded are placed in the top of the grader while it floats in the tank. The small fish pass through, and the larger fish that are trapped inside are emptied into a separate tank. Gently raising and lowering the grader up and down in the water can facilitate the grading process. Grading of fish into two size groups is most easily done using three tanks, but can be done with two. When three tanks are used, the source tank is gradually emptied as the other two tanks are filled with small or large fish.

When two tanks are used, fish are crowded to one side of the tank using a partition, such as a hinged screen. A portion of the fish are then netted and placed into a grader floating on the other side of the partition. The small fish fall through the grader into the empty half of the first tank. The grader is then lifted from the water and the large fish remaining in the grader are placed into the second tank. The empty grader is then returned to the first tank and the process continues until all fish are removed from the crowded side of the first tank. Ideally, all grading and sampling for growth estimates should be finished in one or two days so that fish growth during the days of grading does not negatively affect data quality.

> **"RULE OF THUMB"**
>
> Important: Grading is stressful. Remember to take fish off feed for 24 hours prior to grading.

11.7 TRANSPORTING LIVE FISH

Fish transport is life support for the animals. In order to have a successful business, you have to keep your fish alive during the transport event. In a business using RAS technology, this frequently means learning how to get the fish into the system as well as shipping to market in live form. It doesn't make sense to spend a great deal of time and effort to develop and operate an efficient system if you don't also learn how to move your animals without causing them harm. Many times fish that are transported live are stressed during shipment, which can lead to disease and death losses for several months.

Live shipment is involved in two aspects of the business:

- Fingerling transport from supplier to the culture facility
- Shipment of product to live markets

In the first instance, fingerlings must be transported from the supplier to the grow-out facility. They must arrive in good condition in order to minimize stress-related damage and mortality. We must emphasize that you should always deal with reputable suppliers who have a track record of producing quality fingerlings. Stressed or diseased fingerlings are no bargain at any price. They can lead to disease organisms entering the culture facility that can cause many problems for the operator and be extremely difficult to repress. Health certification should always be provided in accordance with state laws and regulations. Proper quarantine procedures should be followed upon arrival to the facility (see Chapter 13). Secondly, if the fish are marketed live at the end of their growth cycle, the product must be shipped so that the animals arrive at market in undamaged form, free from stress that can induce disease and subsequent death. Stress induced from poor hauling techniques can lead to quick mortality of the animals. This will result in a poor reputation for you as a supplier.

It is very important to minimize stress during transport. The fish will be stressed by: (1) the move from tank to transport truck and back to tank; (2) poor water quality in the transport tank; and, (3) high densities of fish in the tank. There are a variety of methods to reduce stress when transporting fish outside of the facility. Reducing water temperature, adding supplemental oxygen to maintain concentrations at or above saturation, adding salt (0.5%), taking the fish off feed 24 to 48 hours before transport, changing the water between multiple trips, and using anesthetics such as MS-222 are all ways to reduce stress during transport. Note that if MS-222 is used, the fish should probably be taken off feed for more than the minimum 24 hours prior to transport, as the fish will be more likely to defecate and reduce water quality following exposure.

Even if only a few fish are being transported, always fill the transport tank as high as practical to avoid sloshing. When sloshing occurs, fish can be damaged by hitting the sides and top of the live-haul tank. Sloshing also makes the transport truck difficult to drive.

A summary of important techniques used to reduce fish transport stress are:

- Replace the hauling tank water with fresh water on successive trips
- Maintain oxygen between 100 and 150% saturation
- Add a salt solution to each tank prior to adding fish (5 ppt)
- Consider using 15 mg/L MS-222 to sedate the fish
- Take fish off feed 24–48 hours prior to transport

Further discussion and elaboration is provided in the following sections.

EQUIPMENT

Tanks

Hauling tanks have changed considerably over the years. Years ago, tanks were constructed of wood, iron, and steel. They were heavy, often leaked quite a bit, and were hard to sanitize. Today, most modern hauling tanks are constructed of either fiberglass or aluminum. Properly constructed with foam insulation, these tanks can safely haul fish for long distances while minimizing temperature changes. These materials also lend themselves to proper sanitizing between uses to ensure that disease organisms are kept under control.

While tanks can often be built locally, the professional fish farmer would do well to consider commercial companies that have a good track record of building these units. Frequently, they have established designs and the experience to build tanks that can prevent many problems encountered by those who have had tanks built by people with little fish hauling experience. Tanks ranging is size from 50 gallons (190 L) to several thousand gallons in size are available, although the most common ones hold from 100 (380 L) to 300 gallons (1,136 L) of water. Multiple tank units are quite common, combining several individual tanks into one hauling system. This allows for partial loads, multiple sizes or species, or several points of delivery to be included in a single haul.

The key features of a good hauling tank are:

- Durable
- Easily repairable
- Made of non-toxic materials

- Easy to sanitize
- Insulated
- Watertight
- Proper features for hauling distances (dissolved gas control)
- Easy to load and unload fish

Durability is a key feature that is required in a tank since it will take a lot of abuse over its lifespan. Aluminum and fiberglass are both good materials to use, although aluminum will normally stand more abuse. The material should also be resistant to degradation by ultraviolet (UV) radiation coming from sunlight, which will make some materials brittle and subject to cracking within a year or two. Good hauling tanks should be constructed using foam as an intermediate insulation layer, which further helps to stabilize temperature in the hauling unit. While sturdily constructed units will have little need for repair, sometimes damage does occur. When it does, fiberglass can be more easily repaired without specialized equipment or skills than can aluminum, which requires special welding techniques. The use of non-toxic materials is very important not only because we do not want fish to die, but also that we are dealing in most cases with food and that should always be a priority factor in our culture operations.

Sanitizing the hauling equipment is very important and should be considered a normal part of the maintenance. Sanitizing agents that wipe out disease causing organisms and keep the transport units clean should always be used at the conclusion of a delivery and before the tanks are stored for the next run. Preferably, this process should occur "off-site", e.g., a car wash facility. Biosecurity in the operation should be paramount in order to prevent disease organisms from becoming established in the system or within the fish populations (see Chapter 13).

Tanks should be sized properly for the intended loads as well as the truck or other vehicle that will carry them, Fig. 11.6. Remember, the weight of water is approximately 8.34 pounds per gallon (1,000 kg/m^3). For a tank that will hold 500 gallons (1,890 L) of water then, you must figure the weight of the water (4,170 pounds or 1,890 kg) plus the weight of the tank and related gear in order to know what the load on the vehicle will be. In most cases, small pickup trucks will not be useful in live hauling any significant amount of fish because of the weight involved.

Figure 11.6 Two Sizes of Live Haul Tanks for a Small Pick-up and Commercial Carrier.

The shape of the hauling tank is another consideration. Most tanks are built to get the most volume from the hauling unit. Therefore, they are square or rectangular in shape. However, other shapes have been used and may be more efficient to work with. Some producers have used circular tanks. Other fish hauling units are elliptical, with partially rounded bottoms. This helps to provide mixing of the water for circulation, and the lack of sharp corners keeps fish from being injured. The shape tends to concentrate weight over the center of gravity of the hauling vehicle, which can help to stabilize the ride, and the elliptical areas on the sides can be filled with expanded foam to further insulate the tank from road heat. These shaped tanks are also easy to unload. The drawback to these units is that volume and area will be lost. A circular tank, for instance, may lose over 20 percent of the area that could be covered with rectangular or square tanks designed to fit in the same area. The fish farmer should consider all alternatives and investigate the available options before making a decision.

When constructing or purchasing tanks, thought needs to be given not only to the size of the tank but the fittings and options available that make it more efficient. Proper construction of the tank can help in getting fish loaded and unloaded quickly, thereby keeping them at lower stress levels. Large hatches in the top of the hauling unit should be hinged and provide space sufficient to place the fish into the hauling tanks. These hatches should be equipped with dogs or clamps that keep these doors sealed when the vehicle is underway. Gaskets that provide watertight seals are frequently added to ensure that no water is lost when moving. Most hauling tanks are constructed of several different chambers. These may be built either longitudinally or crossways into the tank. These chambers prevent the "free surface effect" of water moving from side to side and front to back in the normal operation of a vehicle.

It is possible for this motion to cause the vehicle to become unstable if not checked. This is why the smaller compartments are built to minimize this motion. Also, the higher the water level is, the less the free surface effect will be in the tank.

The tank should have unloading or dump gates that allow the tank to be emptied quickly. Fish frequently swim against the current, so that there will be few fish in the beginning of offloading while many fish will still remain in small puddles of water near the end of the dump. Some tanks are built with sloping bottoms so that they funnel the fish down and out when unloading, however this configuration also uses some of the volume of the tank and may not be efficient when maximum capacity is desired. If round drains with plugs are used in the tank, there should be a gate inside the tank so that a controlled discharge can be carried out when the entire tank is not being dumped at one time. Also, the drain should be flush with the bottom of the tank so that all of the water and fish can be released. If a lip exists on the drain, the unit will be hard to completely empty and fish will very likely end up remaining in the tank, and disinfection of the tank will be difficult as well. Make sure that the ability to completely drain a tank compartment exists, even if it is a small additional floor drain that is only used for water drainage and not fish removal. Sanitation is paramount.

DISSOLVED GAS CONTROL

Delivery of dissolved oxygen and removal of carbon dioxide (CO_2) must be carried out through the hauling process. Oxygen can be delivered by air stones or grates that will sit on the bottom of the tank and which release oxygen in streams of small bubbles from a supply tank. Agitators can be provided that will help bring oxygen into the water as well as assist in driving off CO_2. These need to be planned to ensure that they are properly located in the tank unit for maximum effect. 12-volt motors usually drive agitators used on hauling units. These units need to be considered in sizing the charging system for the vehicle since multiple agitators may require a larger alternator on the delivery vehicle to be purchased. Also, in order to keep the agitators/aerators in operation, the vehicle will have to remain running so that they do not draw down the battery of the vehicle. Vents are frequently provided at the top of hauling tanks so that the gases collecting in the air space at the top of the tank can be exhausted to release the buildup of carbon dioxide during the haul.

In most if not all cases, the hauling vehicle will be equipped with compressed oxygen cylinders to provide the oxygen needed during transport. Flow meters should be built into the delivery system so that flow can be regulated and monitored closely. The oxygen cylinders, which may weight up to 150 pounds (68 kg) each and must be adequately secured in the body of the hauling unit to prevent them from falling over or moving during the trip. State regulations may apply to hauling gas cylinders and you should check with your state authorities on their use, prior to your first haul. Always assume that every piece of equipment in the oxygen delivery system may fail during the trip, which necessitates you have backup or replacement parts on board at all times. Fish die quickly when you run out of oxygen. Be prepared for the worst-case scenario at all times. Non functioning pressure gauge regulators are frequent points of failure that tend to release all of your oxygen about half way into your trip.

Another oxygen device that has much more capacity than the above mentioned gas cylinders are liquid oxygen (LOX) dewar's. These are insulated containers that must be used to contain the liquid oxygen, but the cost of the oxygen is less expensive when compared per cubic unit of oxygen delivered. The LOX dewar's are also heavier, weighing around 780 pounds (355 kg) each, but they contain the equivalent of 4,500 cubic feet of oxygen (127,426 liters or 127 m^3). While compressed oxygen can be stored indefinitely, a LOX dewar will lose about 2 percent daily.

VEHICLES

In order to successfully transport fish, the size of the hauling unit needs to be matched to the vehicle. Remember, you are going to be moving a lot of water, which is heavy. In order to properly size your hauling unit, you will need to calculate the amount of fish that you will normally be moving. You will need to size your hauling tank for that amount of weight. Then, considering the weight of the water, fish, hauling unit, and related equipment and gear that will be moved, a proper size and type of hauling vehicle can be selected. Consultation with sales personnel at reputable truck dealerships can help to calculate the size of truck that will be required to haul the intended loads. Such factors as gross vehicle weight (GVW) rating, engine, and transmission sizing, cooling and charging systems, and operator comfort options such as air conditioning and seating should be considered. If the vehicle exceeds certain limits, different classes of operator licenses may be required as well.

Fish hauling is carried out with everything from pickup trucks to large eighteen wheeled tractor/trailer combinations, depending upon the load and distance, Fig. 11.6. The most commonly used version is a straight body truck with hauling tanks that dump to the side. Half-ton pickups can handle tanks up to around 100 gallon (0.38 m^3) capacity, while larger ¾ and 1 ton versions can be used for trailers that have capacities of up to 750 gallons (2.84 m^3) in two or more units.

The use of trailers that can be towed with a pickup truck is a popular option in fish hauling. Excellent options are available that use a gooseneck or "fifth wheel" trailer. This not only allows a suitably sized trailer to be pulled by the pickup when needed to move a heavy load of fish, but also retains use of the vehicle for general farm use when not on live haul duty.

These types of trailers also place the tanks lower to the ground, which can greatly aid in loading and unloading the fish. Placing the tanks with their loads lower to the ground also helps lower the center of gravity, which increases stability during the hauling operation. Access to the oxygen tanks and related equipment is easier as well and with the use of the gooseneck trailer, the vehicle can negotiate places where straight body trucks would have difficulty going.

OTHER EQUIPMENT

Monitoring equipment is necessary on hauling rigs to ensure measurement of water quality before and during the hauling process, as well as at the delivery site. While chemical test kits provide an inexpensive means of measuring dissolved oxygen, pH, and salinity, the development of electronic meters that are durable and moderately priced has made them a necessity for anyone engaged in hauling as a business. Oxygen meters priced in the $500–750 range are generally very accurate and rugged enough to withstand the rigors of field use. Digital pH meters that provide accurate measurements are now available for under $100. Many of these can fit in a shirt pocket. Salinity can be measured either with a hydrometer (less than $50) or a test kit ($50–60), either of which can provide measurements that will be accurate enough for hauling.

Fittings used in hauling units should be rustproof, with stainless steel, aluminum, or thermoplastic being the most common. Durability is a key feature of anything used in the business and long life should be sought in all features of the hauling unit.

Regulation of bottled oxygen is carried out using a pressure reducing regulator that can drop the pressure of the tank down to one usable in the hauling unit, usually about 10–15 psi (0.7 to 1.0 atmosphere). Plastic tubing, usually 3/8 inch (9.5 mm) diameter, is used to carry the oxygen from the regulator to flow meters placed at each tank unit. These regulate the flow in liters per minute to the air stones or diffusion grates. A single oxygen cylinder holds around 280 cubic feet of gas (7,929 liters) or 23 lb (10.4 kg). Calculation of the drain on the oxygen gas can then be made based on how much oxygen will be flowing over the duration of the trip. Agitators are used to remove excessive CO_2 and to provide oxygen in low density loads (see Fig. 11.7).

Figure 11.7 Agitators Used in Live Haul Tanks.
(Don Webster, University of Maryland, holding agitator).

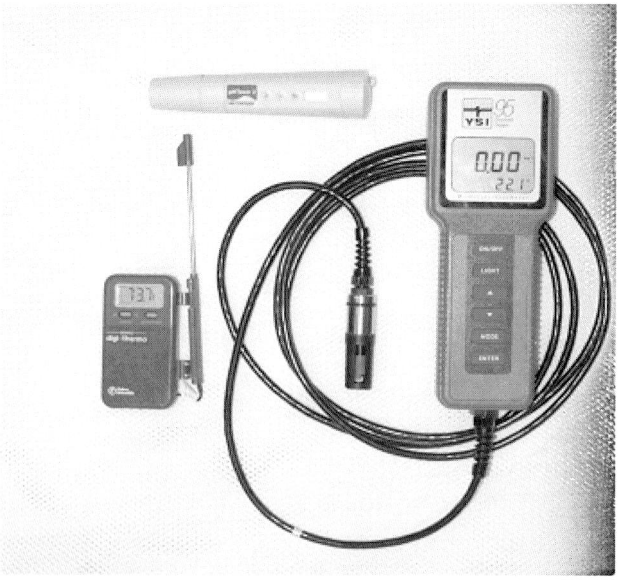

Figure 11.8 Monitoring Equipment for Live Hauling of Fish.

PERMITS

One thing that should not be left out of the hauling operation is permitting. Before moving fish either within or between states, the operator should always make sure that the correct permits have been obtained for the fish being moved. Additionally, the hauler must have the proper paperwork in his possession, in conformance with the requirements of the territory the fish are moving across. This may involve calling the state natural resource agencies in the states that will be traveled through to ensure that the tagging and paperwork are in order. In addition, some states have specifications on the markings that must be on fish hauling trucks. Remember that crossing state lines with fish that may be illegal in another state can not only render you liable to severe penalties in that states but also subject you to prosecution under Federal law. This is serious business. Do not underestimate the vigor that authorities will use to pursue you.

WATER QUALITY

In order to have healthy fingerlings arrive at the culture facility, or to have quality live product arrive at the destination, the fish must be treated properly throughout the hauling process. While catastrophic losses will occur during the loss of life support systems, such as loss of the oxygen system, most hauling losses are more subtle. They occur through disease that is induced through stressing the fish during loading, hauling, and delivery. The effects of these impacts may not manifest themselves for several days after delivery but will almost certainly lead to dissatisfaction on the part of the customer. Many of these problems are related to poor water quality, improper tempering of the animals, or overcrowding them during the hauling process. Several water quality parameters need to be considered when hauling. Since most of these are similar to factors in RAS systems and are covered elsewhere in this book, we will limit discussion in this chapter to factors specifically related to proper fish hauling.

Dissolved oxygen (DO) is important in the life support of your fish. Maintain DO levels at 6 mg/L in the hauling unit. If well water is used to fill the tanks, this must usually be oxygenated before the fish are added to ensure that the water is at saturation. Since saturation occurs at different levels depending upon the temperature of the water and salinity, check the chart in the Appendix to find the level that should be available. DO is placed in the water through the air stones or other delivery systems. Tubing is available that provides small bubbles to dissipate oxygen gas into the water. Generally, the smaller the bubbles, the better they will provide dissolved oxygen to the water. Oxygen meters can be designed to measure tank levels during the hauling process that can be read while the vehicle is moving, or DO can be read at planned stops. Remember the fish can die in 15 minutes if an oxygen delivery component fails. Even if the fish do not die immediately, anoxia (low DO) may stress them to the point where disease will set in within a day or two and kill them later.

Carbon dioxide (CO_2) is a byproduct of respiration of the fish. It must be removed to ensure that the fish arc kept healthy. A buildup of CO_2 can prevent the blood of the fish from effectively carrying oxygen. Agitators are normally used to dissipate CO_2 in the hauling unit, although electric bilge pumps that spray the hauling water up against the top of the tank are also useful in this regard. This is only effective if there is gas exchange or venting from the unit to carry off the CO_2; else the CO_2 dissolves into the water column again.

Ammonia is part of the nitrogen cycle and enters hauling water as excretion products of the fish. It can be toxic to fish in water having a medium to high pH (pH >7.5) and low levels of dissolved oxygen, as it decreases the ability of the fish to take oxygen into the bloodstream and can cause anoxia, or suffocation, of the animal. Ammonia is easily measured with test kits that compare the developed test sample against a standard for a reading. Levels as low as 0.06 mg/L unionized ammonia-nitrogen can damage gills and impair the health of normal fish. (Remember; take fish off feed 24 to 48 hours prior to transport, which reduces ammonia production to nearly zero). In most cases, a pH range of 6.5 to 8.0 is suitable for hauling, although the closer to neutral that one can maintain the water the better. pH is one of the factors that should be known in the source water, the hauling water, and the receiving water and matched as closely as possible to ensure that the fish are kept in the same ranges. Alkalinity will help to maintain a neutral pH and levels above 20 mg/L should be considered beneficial. The ammonia disassociation is pH dependent (see Chapter 2) with un-ionized, or "free" ammonia, increasing as pH increases. Therefore, if ammonia levels are high, waters with lower pH will have less un-ionized ammonia (non-toxic form) to harm the fish.

Alkalinity levels of 50 to 100 mg/L are considered very good for fish, while levels below 20 mg/L are considered soft. Alkalinity and hardness can be increased through the addition of sodium bicarbonate and calcium chloride, respectively. To increase alkalinity by 10 mg/L, add 5 grams of sodium bicarbonate to each 100 gallons (378 L) of the hauling water. To increase hardness by 10 mg/L, add about 6 grams of calcium chloride to each 100 gallons (378 L).

> **"RULE OF THUMB"**
> To increase alkalinity by 10 mg/L:
> Add 5 g sodium bicarb per gallon
>
> To increase hardness by 10 mg/L:
> Add 6 g calcium chloride

Salinity must be considered in hauling operations in several different ways. Fish that are raised in salt or brackish water should always have the hauling water matched to the salinity that they are accustomed to. For other species, the addition of salt to the hauling water is a widely used practice that aids the osmotic balance of the fish. Salt is generally added from 0.5 to 1.0 percent (5 to 10 ppt), depending upon the species. Salinity levels can be easily and inexpensively measured by using a hydrometer calibrated for salt. These are available at most pet stores dealing in fish.

TEMPERATURE AND TEMPERING

Temperature is an important consideration in hauling since there are both minimum and maximum temperatures at which fish can live. Also, warmer waters have lower oxygen solubility, but fish will be more active requiring more oxygen due to their higher metabolism, while at the same time having less oxygen available because of the lower oxygen solubility level. The hauler should be aware that sudden changes in temperature are likely to be harmful to fish. While fish can be taken on longer hauls and at greater densities by reducing the temperature of the hauling water, this must be done gradually to avoid thermal shock. Similarly, when the fish arrive at their destination the temperatures of the hauling and receiving waters may take some time to equalize so that effective transferal can take place.

Tempering is the process in which fish are slowly acclimated to the hauling water or the receiving waters. It must be stressed that this needs to be done slowly and sufficient time needs to be figured into properly matching the water conditions in order to transfer fish that will remain healthy and robust. The lethal temperatures for the species must be known so that they are not exceeded. For example, trout should be hauled in water that is about 52–57°F (11 to 14°C), while tilapia will die in water that cold. They would need to be kept in water that is in the 70–78°F (21 to 26°C) range.

Temperature differences of more than 10°F (5.5°C) can harm fish and smaller ones are more susceptible to these changes. To lower water temperature, ice is frequently used because it is readily available and additional supplies can either be stored in insulated compartments or purchased along the route. It should be noted that the ice should be made from de-chlorinated water since some species are sensitive to this chemical. The addition of ice to the hauling water will lower the temperature and metabolism of the fish as well as raise the point of oxygen saturation of the water. One pound (0.45 kg) of ice will usually lower two gallons of water 10°F (5.5°C) (see Table 11.1 for amount of ice needed to cool water volumes). Temperature changes in fish should not exceed 10°F (5.5°C) in 20 minutes.

Table 11.1 Required Pounds of Ice to Cool Water Volumes

Volume (Gals)	Desired Change In Water Temperature[a] 5°F	10°F	15°F	20°F
50	12	25	38	50
100	25	50	75	100
200	50	100	150	200
300	75	150	225	300
400	100	200	300	400
500	125	250	375	500
600	150	300	450	600
700	175	350	525	700
800	200	400	600	800
900	225	450	675	900
1,000	250	500	750	1,000

[a]NOTE: fish should be tempered at least 20 minutes for EACH 10°F (5.5°C) change in water temperature

When hauling fish for short distances in insulated tanks, the temperature of the shipment water should be matched to that of the destination holding water at the start of the trip. In this way, the fish can be offloaded as soon as the vehicle arrives at the destination preventing costly holding time for the driver and the unit. For longer distances and trips that will take from eight to twelve hours or more, the buildup of temperature can take place beginning several hours before arrival.

Proper planning can help to ensure that deliveries move as quickly as possible while maintaining the fish in top condition.

Consider pH when moving fish since they can be sensitive to extreme changes. When arriving at the destination, a reading should be made of the pH of the receiving water. If there is more than 1 unit difference, the receiving water should be slowly pumped into the hauling tank so that the fish can acclimate properly. Replacing ten percent of the tank water every 10 to 20 minutes until the pH readings are stable will prevent fish from undergoing shock upon transfer.

LOADING

Only healthy fish should be shipped. They should be taken off feed from one to three days prior to shipment, depending upon the water temperature in the culture unit. With higher temperatures, the metabolism of the fish will increase and the fish will utilize their feed faster than in colder waters, when additional time will be necessary for them to empty their gastro intestinal system. Taking the fish off feed will allow many of the waste products to be purged that would otherwise be excreted into the hauling water or regurgitated during handling and shipment thereby affecting the hauling water and causing water quality problems to occur. Fish tend to become excited during transfer, especially when they are netted or crowded, and can develop sores or lesions if not handled quickly and properly. When moving fish with dip nets, care should be taken to use netting that is soft and knotless, especially on species that have scales. These scales protect against disease and, when rubbed off, can allow fungal and other disease organisms to rapidly develop, yielding an unsightly and inferior product for the live market.

Loading should occur in low light conditions and preferable not at the warmest time of the day. During windy conditions, fish can be dried out and, if temperatures are low, the wind chill can have a profound effect upon the fish and leave many dead in a short time. This is especially true of warm water species such as tilapia when moved during winter. Loading rates for hauling fish depend upon several factors:

- Species to be moved
- Size of the fish
- Duration of the haul
- Water temperature

The above factors assume that the fish are in good condition and that the hauling vehicle is adequately designed to provide the necessary life support such as oxygen input, removal of CO_2, minimal buildup of ammonia, with pH, alkalinity, and hardness within proper ranges.

The hauler should be cautioned to start with a known acceptable loading rate and only through experience should this rate be adjusted. Some species can sustain higher loading rates than others. Fewer small fish can be hauled than larger ones and the warmer the water is, the lower the hauling density should be. Also, the greater the hauling time, the lower the density should be.

The loading rates have been worked out for many species such as trout and catfish but there are still gaps for species such as tilapia and yellow perch; see Tables 11.2–4 for specific fish species and use Table 11.5 as a general guide for allowable hauling rates. Adjustment should be made based on the temperature of the water and whether or not efficient oxygen transfer systems are used in the hauling unit. As always, the prudent culturist will begin with a conservative loading and increase the rate only with experience.

Table 11.2 Loading Rates (lbs per gallon) for Catfish in 65°F (18.3°C) Water

Size of Fish (lbs per 1,000 fish)	Hauling Time (hrs) 8	12	16
0.1	0.20	0.20	0.20
1	1.25	1.00	0.70
2	1.75	1.65	1.25
4	2.20	1.75	1.50
8	2.95	2.20	1.80
20	3.45	2.50	2.05
250	5.00	4.10	2.95
500	5.90	4.80	3.45
1,000	6.30	5.55	4.80

NOTES: 1) loading can be increased 25% for each 10°F drop in temperature below 65°F.
2) loading should decrease 25% for each 10°F increase in temperature above 65°F.

Table 11.3 Loading Rates for Warmwater Fish (Largemouth Bass, Bluegills, and Tilapia) in Water of 65°F for Hauls of Less Than 30 Hours Duration

Length (inches)	Number Fish per pound	Approx. Number per gallon	Loading Rate (lbs per gal)
5	10	15	1.50
4	25	25	1.00
3	100	67	0.66
2	400	200	0.50
1	1,000	333	0.33

Table 11.4 Loading Rates for Coolwater Fish (Walleye and Northern Pike) in Water of 55–65°F

Number per pound	Size (in)	Lbs. Of Fish per gal	Transit Time (hrs)
60	3.0	1.30	8.0
500	2.0	0.66	8.0
1,000	1.0	0.55	8.0

Table 11.5 General Guidelines for Loading (lb per Gallon) of Various Types of Fish at 65°F

Type and Length Of Fish (inches)	Duration of transport (hrs) 1	6	12	24
Fingerling food fish				
2	2.0	1.5	1.0	1.0
8	3.0	3.0	2.0	1.5
Adult food fish				
14	4.0	4.0	3.0	2.0
Baitfish				
2	2.0	1.5	1.0	1.0
3	3.0	2.0	1.0	1.0

We can calculate the loading density of a tank using the water-displacement method. For this, we need to know: a) the actual volume of the tank being used; b) the weight of the fish that will be transported; and, c) the volume of the water that will be displaced by the fish (see Table 11.6).

Table 11.6 Volume of Water Displaced by Fish Added to Hauling Vessel

Weight of Fish (Lb)	Water Displaced (gal)	Weight of Fish (lb)	Water Displaced (gal)	Weight of Fish (lb)	Water Displaced (gal)
100	12	1500	180	2800	336
200	24	1600	192	2900	348
300	36	1700	204	3000	360
400	48	1800	216	3100	372
500	60	1900	228	3200	384
600	72	2000	240	3300	396
700	84	2100	252	3400	408
800	96	2200	264	3500	420
900	108	2300	276	3600	432
1000	120	2400	288	3700	444
1100	132	2500	300	3800	456
1200	144	2600	312	3900	468
1300	156	2700	324	4000	480
1400	168				

From these, loading density can be calculated:

$$Loading\ Density\ (lb/gal) = \frac{pounds\ of\ fish}{tank\ capacity(gal) - water\ displaced\ by\ fish(gal)}$$

For example, if we are going to place 900 pounds of fish in a 600 gallon tank, we find from Table 11.6 that the water displaced will be 108 pounds. From our formula we calculate:

$$Loading\ Density = \frac{900lb}{(600 - 108)gal} = 1.83\ lb/gal$$

ADDITIVES

When producing food fish, the aquaculturist always needs to remember that the product is human food and, as such, should be treated in a way that ensures that it is safe and wholesome. Chemicals and additives used in production come under the control of the US Food and Drug Administration (FDA) and are part of a stringent set of laws enacted by Congress. Anything used on food fish comes under the auspices of the FDA. Some substances, such as salt and ice have been deemed to be "Generally Regarded as Safe" (GRAS) and are allowed to be used until such time as they may be objected to. For hauling purposes, ice, salt, and oxygen used to move fish are allowed but anesthetics are not. There is a 21-day waiting period before fish that have been treated with anesthetics can be sold for human consumption. In the normal conduct of business, this makes them impractical for shipments that are destined to market. See Table 11.1 for ice addition considerations and 11.7 for calculations on the use of salt.

Table 11.7 Required Salt Addition to Achieve a Specific Salinity Level in the Hauling Tank

Volume	Salinity in Parts Per Thousand							
(Gals)	1	5	10	15	20	25	30	35
50	0.4	2	4	6	8	10	12	14
100	0.8	4	8	13	17	21	25	29
200	1.6	8	17	25	33	42	50	58
300	2.6	13	25	38	50	63	75	88
400	3.4	17	33	50	67	83	100	117
500	4.2	21	42	63	83	104	125	146
600	5.0	25	50	75	100	125	150	175
700	5.8	29	58	88	117	146	175	204
800	6.6	33	67	100	133	167	200	233
900	7.6	38	75	113	150	188	225	263
1,000	8.4	42	83	125	167	208	250	292

Fingerlings have been transported with anesthetic with great success. Used according to the label, anesthetics such as methane tricainesulfonate (MS-222) are able to reduce the metabolism of fish, thereby lowering their oxygen consumption and allowing them to be transported successfully over longer distances than might otherwise be possible. This is another factor that should be considered when transporting fish, however, since it may also add to the time required to temper the fish when they have arrived at the destination in order to restore them to their normal state.

There are some commercial formulations of additives that may be used for hauling that are approved for use on food fish. Some of these provide buffers for the water so that the pH and alkalinity may be optimized during hauling. Others provide anti-foaming agents for the hauling tanks so that protein foams do not build up during the moving process. Foam can prevent proper ventilation of the tanks, allowing buildup of carbon dioxide due to lack of gas venting. The hauler should always know the intended purpose for the fish being hauled and, if they are being used for food, should use nothing that is not specifically approved for use on human-use products.

SANITIZING

Disease can severely impact the fish farm. Keeping all equipment as clean and sanitized as possible can prevent costly losses of fish. As part of a comprehensive biosecurity plan, the hauling equipment should always be cleaned and sanitized at the conclusion of the transport. There are many excellent ways to carry this out with the use of pressure washers. Sanitizing agents should be used and the equipment well rinsed so that no chemicals remain that might harm future fish.

While regular household bleach may be used as a sanitizing agent, its concentration is only about 5 percent chlorine while HTH is stronger and provides an excellent means of killing potential disease organisms. HTH provides a calcium hypochlorite base with 65 percent of available chlorine. Used at a level of ½ ounce per 25 gallons of water (15 mL/100L), it should be used on all tanks, pumps, lines and equipment for 30 minutes for proper disinfection. Formalin has also been used as a disinfectant on fish farms at a 5 percent solution. Do not forget to include all of the nets, boots, gloves, baskets and other equipment that was used on any haul as well. Do not give disease organisms any opportunity to become established in your business. (See Chapter 13 for more details on biosecurity).

11.8 PURGING AND OFF-FLAVOR

One of the problems often encountered in recirculating systems is development of off-flavors in the fish. One method of eliminating the off-flavor is to hold the fish for 3 to 5 days in a tank with fresh, clean water. The water exchange rate, which should be supplied by fresh water, e.g., well water, should be at least 25% volumetric exchanges per day and preferably 100 to 200% on the first day. If you do not flush out the pollutants from the water, these compounds will not diffuse away from the fish. Withdraw feed at least 24 hours before moving the fish to the purging tank. This will decrease the water exchange requirements in the purging tank. It also decreases the oxygen demand of the fish once they are moved to the purging tank. Off-flavor uptake by fish is rapid and elimination is a slow process.

Fish grown in RAS are very susceptible to off-flavor (muddy taste). There is no more certain way to lose customers than to ship them fish with off-flavor. Before harvesting, take a fish and sample for off-flavor.

Off-flavor is primarily attributed to geosmin and methylisoborneol. These are metabolic compounds produced by actinmycetes (Gerber, 1979), cyanobacteria (Slater and Blok, 1983) or algae (Juttner, 1983). Detection of these compounds is at the fractions of a microgram per kg of flesh (Person, 1979A). A thorough review of off-flavor issues is provided by Brune and Tomasso (1991).

Testing for off-flavor is usually done by taking a fillet and cooking the in a microwave oven without any seasoning and then having someone who has a taste sensitivity to do the taste test. If there is a "hint" of off-flavor, then delay the sale and continue purging. Consider increasing your percentage of fresh water being added.

11.9 POST HARVEST HANDLING

Production facilities selling dead in-the-round or dressed fish must provide refrigeration facilities for short term storage of the product and should in most cases have an ice machine either to provide for all of the cooling or to supplement any mechanical refrigeration. Dead fish retain quality much better if kept as close to freezing as possible. For example, fish held at -2°C (just above their freeing point) have a shelf life several days longer than fish held at + 2°C. Fish shipped on ice retain better quality, in most cases, than if held under standard refrigeration at the same temperature, because the ice provides high humidity, and a washing action, as well as cold temperatures. Large block ice should be avoided. The weight of the ice can crush fish, sharp edges from cracking the blocks can tear the fish, and the large pieces will make poor contact with the fish slowing the cooling rate. If boxes of fish are to be stacked, then fish and ice should not be packed beyond the stacking line. Drainage of melted ice water from the fish should be permitted, and ideally not dripped directly onto fish in the box beneath. Fish should be packed so that dirty melt water does not collect in the belly cavity of the fish. Fillets or other "flesh" surfaces should not be in direct contact with the ice.

The impact of temperature on the shelf-life of fish and other products has been carefully studied in Australia. The results of these studies really highlight the importance of keeping fish as cool as possible. They have developed an equation that relates spoilage rate (SR) to the temperature of the fish (whole body or flesh):

$$SR = b(T - T_0)^2 \qquad (11.3)$$

where: T = temperature of the fish °C,
T_o = reference temperature for the fish spoilage process °C,
b = proportionality constant that depends on the shelf-life of the particular fish.

Surprisingly, T_o is –10°C, because microbes can grow essentially down to that temperature. Although the value of "b" can be calculated, the important term to understand is the $(T–T_o)^2$ term and the impact it has on shelf life. The shelf-life is the reciprocal of the spoilage rate for each species, i.e., the faster the SR, the shorter the shelf-life, or the slower the SR, the longer the shelf-life. It should also be noted that the simple equation above only holds in the "refrigeration" region. At higher temperatures the equation becomes more complicated.

With that introduction, let us look at a table of values for the $(T–T_o)^2$ term assuming that the fish of interest is normally considered to have a shelf-life of 14 days (see Table 11.8).

Table 11.8 Relative Spoilage Rates (SR) as Affected by Storage Temperature

Temperature (°C)	Relative SR $(T – To)^2$	Relative Shelf-life (days)
-2	64	21.8
0	100	14.0
2	144	9.7
4	196	7.1
6	256	5.5
8	324	4.3

As you can see, the impact of small temperature changes is significant. Furthermore, it should be noted that normal "Refrigeration" could legally be defined to go up to 45°F (7°C). Obviously fish kept at that temperature spoil a lot faster! Good "refrigeration" is generally considered to be below about 40°F (4°C), but even this is insufficient for fish (and many other products, as the relative relationship holds). So, for ideal fish storage, temperatures of 33°F or even lower are recommended. Superchilled (unfrozen) fish will have an even longer shelf-life, i.e., fish kept at temperatures between 32°F (0°C) and 28°F (-2°C), often using salt-water ice, which is colder than fresh-water ice. Thus, another benefit of ice is the ability to keep fish colder than most available mechanical refrigeration systems.

11.10 Feed Storage

Feed must be available on a daily basis and therefore must be at least temporarily stored on site. The type and size of the culture operation, the species cultured, the frequency of feed delivery from the vendor, and other factors will determine the type and size of feed storage requirements. Typically, feed is available either in bags or in bulk. Where small quantities are needed, bagged feed is most convenient. Larger operations will find bulk feed more convenient and less costly. However, bulk feed requires handling equipment that may not be necessary for bagged feeds. Feed storage facilities are of two general types: 1) dry, rodent resistant facilities, and 2) refrigerated storage. Most commercial fish feeds can be stored in facilities that are dry and rodent resistant for at least reasonable time periods, i.e., a few weeks, depending on temperature and humidity. Storing feeds too long risks loss of required vitamins and possible spoilage. Some feeds must be refrigerated at all times to prevent spoilage. Feed storage facilities need to be designed to accommodate the feed type and form used. See Chapter 16 for more details on feed degradation.

Under optimal conditions of low temperature and low relative humidity, dry pelleted feeds can be stored 90–100 days. Keep feeds in a room or cooler that is well ventilated and which provides security against rodents and insects. Elevate feed by placing it on pallets and stack feed in bags less than 10 bags high for worker-safety and to prevent crushing the feed at the bottom. High moisture levels and temperatures will rapidly result in poor quality feed. Moisture fosters mold growth and attracts insects. High temperatures break down oils and deteriorate vitamins. If feed becomes rancid or moldy, do not feed it to the fish. The moisture level in pelleted feed after drying will be from 6–8%. Every effort must be made to not let the feed gain moisture once it has arrived at the farm in this dry condition.

It is recommended that samples of the finished products be stored for as long as six months after manufacture, either in refrigerated storage or at ambient temperature. Ambient temperature rather than frozen storage is recommended, so that should concern be raised about a feed, a chemical comparison can be made with feed of the same age that has been stored cool and dry. Complete records should be kept for each batch of feed delivered to farmers. These data should include details such as the date of delivery, batch number, and quantity delivered. Many complaints to the manufacturer about the poor quality of their feed are very often the result of post delivery improper storage conditions on the farm.

Feed Delivery Methods and Delivery Costs

When getting price quotes on feed, also request shipping charges. The shipping charges for the feed can approach the cost of feed. To minimize shipping charges, try to purchase as much feed as possible at once. In addition, purchases of a ton or more often qualify for discounts. Farms using more than 100 tons/yr will generally purchase feed in bulk deliveries of 10 tons or more (a fully loaded truck will haul around 20 ton). Be careful about the way feed is moved from the delivery truck into the bins and from the bins inside the barn to the fish tanks. Each of these processes can increase the fines present, which will severely degrade water quality.

11.11 Handling Fish

Few, if any, aquaculture facilities are designed to optimize handling of any materials used or produced in the facility. Harvesting usually consists of manually dipping the fish out of the tanks. Manual systems work well if there are only a few thousand pounds of fish produced per year. However, if production is measured in tons per year, handling of the fish becomes a major time and labor consideration, and, thereby, a major cost factor. It is also backbreaking work. Adequate planning during facilities design can save much grief later when handling fish. Equipment must be provided where automated handling is to be used. Unfortunately, little research has been done on how best to move fish from one tank to another, when reducing stocking densities. Pumps can be used and they often cause less stress on the fish than do standard dipping methods. How fish handling systems are designed can determine the labor needed for this operation and can determine how much stress the fish will experience.

In small systems, feeding is easily handled by manual methods or by demand feeders, because the weight of feed moved is small. However, in larger systems many tons of feed will have to be fed. Automated methods of handling this feed are necessary and cost effective. The feed distribution facilities needed, will depend on the management system used in a particular operation. Regardless of the system used, some feed bins will be needed as well as some equipment to distribute the feed. Space and funds to purchase or construct and/or install the equipment will be required.

Figure 11.9 Harvesting Artic Char with a Crowder System (Freshwater Institute).

11.12 LABOR

One of the most demanding aspects of aquaculture is the need for labor that is dedicated to the task and has the necessary knowledge and training. When there are fish in the recirculating system, someone has to be present at the facility or on call 24 hours per day, 7 days per week or 24-7. If the operation is small and is operated by one person, that individual must be willing to be on call continuously for long periods of time. The person must be dedicated to the job. Large operations can afford to hire more than one person, so the on-call burden can be spread across different people. Everyone can get some time off, if all employees take their turn being on-call during some weekends and/or holidays. Particularly in small operations, being continuously on-call requires dedication that few people are willing to make, and is one of the most overlooked aspects of aquaculture production.

> **"RULE OF THUMB"**
>
> Being continuously on-call to a fish facility is one of the most overlooked aspects of aquaculture production.

Aquaculture production requires a mixture of knowledge learned in a classical school or college setting, as well as knowledge learned on the job. Currently, there are only a limited number of people who have the needed combination of knowledge and experience to successfully operate a recirculating aquaculture facility. Any sizable recirculating aquaculture operation must make a concerted effort to hire people possessing most of the needed knowledge and to be prepared to train these people on the parts they are missing and to train new people from the beginning of their employment. The manager, owners, and/or investors must be aware that training new people will involve a certain amount of risk and probably will involve loss of at least some crop.

The location of the aquaculture facility will strongly influence the type of people that can be hired. Are the local conditions such that people with the educational backgrounds and experience needed are going to be satisfied to live there? Are the needed people available locally or must they be moved to the site? Obviously, the availability of a suitable labor supply is not the only determining factor in site location, but it is the one most often overlooked.

11.13 ACCESS

Access to the aquaculture site is necessary for employees, feed supply trucks, oxygen supply trucks, fish haulers, and other functions. This implies that there is a road or some method of getting people and vehicle traffic in and out of the site. There must be available access in all weather conditions and there must be space enough to maneuvers trucks and other vehicles. These areas may be gravel covered or hard surfaced. The physical size of these areas will depend on the type of truck expected and the frequency of truck traffic. The turning radius of a tractor trailer is about 65 feet (20 m), which means that the smallest radius on any driveway should also be no less than 65 feet (20 m). Plan for efficient ingress and egress from the farm site for all large trucks. Loading docks may also be necessary. These facilities can be expensive, particularly if not properly planned into the overall facilities design.

11.14 OPERATIONS

Fish production is more likely to be successful and efficient when good fish rearing habits are employed, when techniques are continuously refined, and when realistic long-range planning is done. Many of the techniques discussed below have been adapted from rainbow trout culture methods and can be used for many other fish species.

FISH SAMPLING

Sampling is used to estimate growth by measuring the weight and/or length of a subset of the entire population. Tracking fish growth accurately is necessary for rationing feed, calculating tank densities, and for projecting the time when fish will be ready for sale. By collecting some of the fish from the tank in question, and comparing the information over multiple sampling intervals, growth rates for the entire population can be estimated.

Growth rates are used to:

- predict when the fish will be ready for future production milestones;
- decide whether past performance has been acceptable; and
- determine when some problem that, while not causing fish mortality, may be present and affecting growth.

By analyzing growth rates for different cohorts of fish in conjunction with other production data, e.g., feeding information and water quality conditions, the manager can determine how production efficiencies change from cohort to cohort and how production parameters will affect the fish, and make adjustments to optimize fish production.

PLANNING

Planning the sampling method, organizing the equipment, and having the necessary number of personnel to help can greatly reduce the time it takes to sample fish. While it is possible to sample with one person, two people are preferable – one person to handle the fish while the second records information and operates the balance scale. Teamwork improves accuracy and efficiency, and reduces the tedium of sampling. Efficient and proper technique resulting from teamwork minimizes fish stress and damage that can lead to post-sampling mortality.

Sampling is stressful to fish. Decisions about how often to sample must balance the need for accurate numbers with the desire to minimize fish stress. Fish sampling can be carried out as frequently as every two weeks, but generally, producers sample every four weeks.

Fish larger than 10 grams should be taken off feed for 24 hours prior to sampling. Consumption of oxygen by fish increases during sampling. Oxygen levels should be allowed to stabilize after sampling, and fish should not be fed for several hours post-sampling. Frequent sampling will allow for more efficient feed rationing and prediction of growth, but will also create stressful conditions for the fish and necessitates taking the fish off feed. Activities that disrupt feeding for more than one day, such as grading, sampling, and transporting fish, should be carried out as infrequently as possible. In the case of rainbow trout, fish can reasonably compensate for one day off feed by eating more food the next day, however, they cannot compensate for two consecutive days off feed per week.

A monthly sampling interval is a reasonable compromise between minimizing stress and maximizing accuracy. Be aware of environmental stressors that may affect the fish when they are removed from the tank for sampling. For example, if the building air has large temperature fluctuations during the day, consider sampling at a time when the air temperature is closest to the water temperature. The bucket used for weighing sampled fish should have a diameter at least as large as the fish. The fish should be emptied out of the bucket before they start showing adverse signs of stress, including piping or air gulping and excessive thrashing. The amount of water placed in the bucket and the number of fish you can fit in the bucket before emptying it is subjective, and can be determined with a dissolved oxygen meter and some common sense.

SAMPLING PROCESS

Randomizing the Sample

Random sampling is used to collect a representative group of fish to accurately infer information about population size characteristics.

Except for total weight, almost every statistic used to describe the sample should be able to accurately describe the population, e.g., mean, median, standard deviation, or variance of the weight, length, "height", etc. If a non-random sample is collected, inaccurate production information will be generated. As a result, unreliable predictions and comparisons will be made that will result in costly errors in production management. Crowding, net size, and sampling location are factors that should be considered to ensure that a random sample is obtained.

Net avoidance by larger or smaller fish is a common source of sample bias. Limiting the space in which the fish can move during sampling decreases the chance that fish that are more agile will move away from, or swim out of, the sampling net. Crowding can be stressful and crowded conditions should not be maintained for an extended period of time. During crowding, dissolved oxygen levels should be checked and adjusted. Dissolved oxygen levels should also be monitored if the fish that are removed for sampling are put into a temporary holding tank.

At an absolute minimum, the sampling net should be large enough to contain the largest fish in the population. A small net will bias the sample estimates towards smaller fish as the use of a smaller net increases the chance that a large fish will hit the lip of the net and escape instead of being captured. [Even a net equal to the largest fish is going to have this bias]. Patterns in the distribution of size within a tank can occur because fish develop preferences for tank regions; also, larger or stronger fish can displace smaller or weaker fish. This size assortment should be considered when planning sampling location. If more than one net scoop of fish is needed to obtain an adequate sample size, fish should be taken from more than one location in the tank. If gentle crowding is used, a sample that represents the fish from all levels of the tank should be taken. Consider using a broad cast net in larger tanks. With some practice, you can become pretty efficient at doing this and it is a very good way to grab a sample without pre-alarming the fish. The first cast net thrown will collect the most unbiased sample.

The accuracy required determines the number of fish sampled; conversely, the sample size collected determines the accuracy. For a given accuracy requirement, the factor that influences sample size is the variance in the population. For example, in the case of 10,000 fish ranging in size between 100 g and 1,000 g requires a larger sample for the same accuracy as the same population of fish whose size ranges between 100 g and 150 g. As a rule of thumb for collecting weight data, at least 100 fish should be sampled. However, a minimum sample size of 30 will provide a rough estimate. If the total fish population is small – 200 fish or less – take a census to determine the characteristics of your population of fish without too much additional work. After collecting a sample, all of the fish collected should be measured in order to avoid sample bias. For example, suppose weight and length measurement are collected on 100 fish, but 23 remain in the holding tank. The remaining fish should also be measured to ensure a random sample. It is possible that the first 100 fish measured were the smaller, weaker fish that were the easiest to catch.

Weight measurements are used to: (1) estimate individual fish weight and growth rates over time, and (2) determine how much to feed. Because weight measurements are used for feeding, weighing needs to be done at an earlier age than length. Weight measurements should be done from the time the fish are up and feeding, and producing feces.

A variety of techniques can be used for weight sampling. The choice depends upon the type of information needed and the time available. For weight data, scale sensitivity should be about 1% of fish weight. For example, if fish weight is 100 grams, the balance should display the weight to the nearest gram. If the balance is not sensitive enough to do this, consider bulk weighing multiple fish to improve accuracy. For example, if fish weight is 20 grams and the scale is accurate to the nearest gram, at least five fish at a time should be weighed.

Bulk weighing and counting results in the least amount of information, but is the least stressful for the fish, and takes the least amount of time. This technique may be used when only one person is available for sampling. By collecting bulk weight data, a measurement of total biomass can be obtained without measuring individual fish weights. Average fish weight can be estimated from bulk weight by counting the number of fish included in each sample of fish.

PROCEDURE FOR OBTAINING A BULK WEIGHT

Equipment & Supplies

- Scale – capacity great enough to weigh the mass of fish sampled
- Holding tank or section of the tank divided by netting (check for holes in nets!)
- Transfer Net
- Bucket – capacity large enough to hold the fish and water during weighing

Procedure

1. Add sufficient water from the sample tank to bucket

2. Place the bucket on the scale

3. Tare the scale

4. Collect a random sample of about 100 fish and add the fish to the bucket

5. Record the total fish weight

6. Pour the fish into the holding tank and count them. Once a fish has been counted return it to the original tank

7. Calculate Average Fish Weight $= \dfrac{\textit{Total weight}}{\textit{Number of fish}}$

8. Repeat 3 times to get an average

Individual weighing produces more detailed data than bulk weighing. The data collected with this method can be used to calculate the variance or standard deviation, which are two statistics used for describing size variability in a tank. The additional information collected with this method can be used to decide when fish should be graded.

Example: Statistical Analysis of Fish Sampling

You have a tank of 5,000 fish. You estimate that the typical individual animal will weigh around 100 grams, based upon your last sample and a projected growth ratc. You collect 3 samples of fish according to the above procedures, and the question is whether or not you have sufficient samples to estimate the population weight (average fish weight of the tank). Justify either not taking another sample or that you should take an additional sample (see data in Table 11.9).

As can be seen in the example, Table 11.9, the first sample provided an unrealistic estimate of the average size fish in the tank. The average weight estimated in the tank changed by over 8% between the first and second samples. The third sample resulted in a less than 1% change in the estimate. As a general rule, keep taking samples until the estimate of fish weight changes by less than 2%.

Table 11.9 Statistical Example of Fish Sampling

Sample Number	Sample 1	Sample 2	Sample 3
1	110	92	95
2	90	88	88
3	111	90	98
4	88	65	90
5	99	85	95
6	95	75	101
7	100	95	97
8	125	93	93
9	102	88	96
10	97	77	92
11	135	90	98
12	90	85	101
13	88	77	90
14	97	88	104
15	120	93	99
16	95	70	99
17	130	99	93
18	111	95	97
19	92	89	96
20	99	97	101
Average	103.7	86.6	96.2
St dev	14.2	9.2	4.2
St dev/average	13.7%	10.7%	4.4%
Mean of cumulative sample	103.7	95.1	95.5
Change in Mean	na	-8.3%	0.4%

PROCEDURE FOR OBTAINING INDIVIDUAL WEIGHTS

Equipment & Supplies

- Scale – total capacity enough to weigh several fish
- Holding Tank or Net
- Transfer Net
- Bucket - sufficient to hold all the fish and water

Procedure

1. Add sufficient water from the sampled tank to the bucket
2. Place the bucket on the scale
3. Tare the Scale
4. Place a random sample of fish into the holding tank
5. Take one fish from the holding tank and place it in the bucket. Allow most of the water to drip from the net before putting the fish in the bucket
6. Record the weight and tare the scale
7. Repeat steps 5–6 until all the fish have been weighed
8. The bucket might begin to get full or fish could become stressed before all the fish have been weighed. If so, empty the bucket and put the fish back into the tank. Repeat steps 1 through 7 with fish from the holding tank until all the fish have been weighed.

Length information is used to:

- estimate growth rate in inches (cm), and
- when combined with weight, to estimate the condition factor (CF), which is used to track the body condition.

Length information is not used to calculate feeding rates. As a result, length measurements do not need to be done until fish are about two inches long. The CF is one of the most useful pieces of information obtained during the sampling process. With very few fish, the CF will tell you whether or not this particular cohort of fish is being under or over-fed, particularly when you have some historical data to go by. The CF will reach a fairly consistent value shortly after the fish have gone beyond the early fingerling stage. Use this CF value a lot! (See Chapter 4 for a listing of expected CF values by species).

Both weight and length can be measured very early in the fish growth phase to estimate feeding requirements and growth. Three- to 4-inch fingerlings (8 to 10 cm) and larger fish will be much more resilient after sampling. However, at the sac fry stage and up to about 1 inch (2.5

cm) the fish may need to be euthanized after taking length measurements due to the delicate condition of the young fish. However, early feeding is usually based on a rough estimate, found by multiplying a typical value for sac fry weight times the number of sac fry present. The equation to relate weight (Wt) to length (L) is (see Chapter 4, Eq. 4.11 for more details):

$$Wt = \frac{CF(L)^3}{10^6} \tag{11.4}$$

Length and Weight for Fry

Once the fish reach 2 inches (5 cm), length and weight measurements should be taken once every four weeks. While it is not necessary to sedate fry for weight measurements, they should be sedated for length measurements. Collecting length data is a two-person job—one person handles the fry; the other records data and monitors the fish during anesthesia and recovery. There are many standard definitions used for measuring length. One commonly used definition, the fork length, is measured from the tip of the snout to the base of fork in the tail. (Do not measure to the tip of the tail – fin erosion may shorten the tail and decrease measurement accuracy.)

Equipment & Supplies

- Scale – total capacity enough to weight several fish
- Holding Tank or Net
- Transfer Net

After the equipment is set up, a random sample of about 50 fish is collected and placed into a container of water. Remove a few fish, place them into another smaller container, and slowly add the recommended amount of MS-222 solution (dissolved at a concentration of 1 gram per 100 mL of water). Add enough solution so that the fish are not swimming, but gill movement can still be seen.

Pick up the fish, blot with a paper towel, and place on dry towel or paper towel on top of the tarred scale. Record the weight and fork length (to the nearest mm) and place the fish into a bucket of fresh water for recovery from anesthesia. Length and weight should be matched for individual fish. (This matched data is used to calculate condition factors.) Repeat the process until all fish have been measured. Once all the fish are active in the recovery tank, move them to the rearing tank.

11.15 MISCELLANEOUS OPERATIONS

EUTHANASIA

Living fish pulled from the population should be humanely euthanized when possible. These include not only the fish for sale, but culled fish and fish sampled for laboratory analysis. Three accepted options for euthanasia are anesthetic overdose, decapitation, and pithing. When an approved anesthetic, such as MS-222 is used, a sufficient amount should be added to the water to stop breathing. A ten-minute exposure time following cessation of breathing is recommended. The second euthanasia method, decapitation, can be a messy and is only a partly effective method of euthanasia. Pithing is recommended following decapitation. Pithing, by which the brain or brain and spinal cord are traumatized is the least expensive method of euthanasia, but requires skill to execute properly.

Cranial concussion and cooling are not approved as humane forms of anesthesia, but are often used, especially when preparing the fish for consumption. Do not chemically euthanize fish prior to consumption. (At this time no chemical is approved for this purpose.)

ANESTHESIA

The purpose of anesthesia is to reduce fish stress for transport, sampling, or examination. Fish are generally anesthetized to a state of deep sedation, a condition that exists when the fish do not react to external stimuli except to very strong pressure, and exhibit slight decreased opercular rate.

MS-222 is the only FDA-approved anesthetic for salmonids. FDA-approved dosage limitations are between 15 to 66 ppm for 6 to 48 hours for sedation, and a dose between 50 and 330 ppm for 1 to 40 minutes for deeper anesthesia. The dose is generally fine-tuned on the day of use. A 21-day withdrawal period is required prior to human consumption of fish treated with MS-222. Further details on MS-222 usage can be found in the Appendix.

Two other options for anesthesia are carbon dioxide and sodium bicarbonate (baking soda). These drugs are not approved for treating food fish in the U.S. but are of low regulatory priority. As an anesthetic, the recommended dose for CO_2 is 200 to 400 ppm for 4 minutes. For sodium bicarbonate, the recommended dose is 142–642 ppm for 5 minutes. (FDA has publically indicated that it considers usage of these materials below a specified level a “low regulatory priority.” The doses recommend are below this level. Basically FDA has used this “low regulatory priority" language to permit the use of common aquaculture treatments with a long history of safe use to be used without going through the full regulatory approval process that would be required for each separate species.)

Fish Health Analysis (FHA)

Knowing how your fish behave when they are healthy is one of the keys to maintaining a healthy population of fish. This provides a basis of comparison when observing fish of questionable health. If the fish are sick, they will probably go through a period where they exhibit abnormal behavior in their swimming and feeding habits. By monitoring fish behavior, action can be taken before fish start dying, thereby averting large losses, or expensive chemical treatments. Check the dissolved oxygen, temperature, pH, feed condition, and the other easily monitored and/or controlled parameters before calling for a professional diagnosis. Do not make radical changes to system operation if unsure about the ramifications of the changes.

Sampling fish for a FHA can take various forms. Dead fish, sick fish, or random fish can be collected. These fish should be examined first in-house and, if possible, pictures taken for future reference. Cost increases as the complexity of the testing increases, so it is important to investigate the problem in-house if possible. Consult a fish health specialist for more information (see Chapter 13).

11.16 Record Keeping and Maintenance

Maintain good records. Repeat ***Maintain Good Records***. Each cohort of fish that is grown should have a complete history, e.g., beginning and ending weights, types and quantities of feed fed, manufacturer used, general notes on water quality for the cohort in terms of temperature and oxygen.

Records should be kept of monitored water quality parameters as well as the general observations of the system manager. Recommended frequencies for monitoring water quality parameters are given in Table 11.10. A very simple daily record sheet can be created using a spreadsheet program that would show daily values for: day of month, Feed consumed for day and cumulative feed used, Water temperature, DO, TAN, NO_2-N, Alkalinity, Chlorine, a notes column, and a column for who recorded the data and the time. Spend effort in creating a data base management system so that you can visualize your results. Post your own standard growth curve and compare each cohort to your standard curve. Are they ahead or behind? Why?

Table 11.10 Schedule for Monitoring Water Quality in a Reuse System

Parameter	How measured	Frequency
Temperature	thermometer	daily
Dissolved Oxygen	meter or test kit	daily
Ammonia	test kit	twice per week[a]
Nitrite	test kit	twice per week[a]
pH	meter or test kit	twice per week
Alkalinity	test kit	weekly
Hardness	test kit	weekly
Chloride	test kit	weekly

[a]during startup periods, monitor daily

The various mechanical components in an RAS should be checked frequently, with the frequency period being determined by how critical a component is, e.g., backup generator should be weekly, all pumps weekly, oxygen supply daily. Create a log of the maintenance schedule and have a column for completion of task, condition, action, and person responsible. One of the subtle dangers in RAS is that almost everything works all the time and the tendency is to become lazy in following maintenance protocols. Think of your RAS as a jet airliner that carries several hundred people each flight. You as a passenger on this airplane probably appreciate the fact that a scheduled maintenance procedure is in place. Your fish are passengers in their airplane the RAS. Check the systems regularly.

FISH RECORDS

Every aquaculture facility should have a husbandry plan. This plan lists the conditions under which the species will be raised. For a hatchery operation, details such as temperature for incubation and growout, feed formulation, number of eggs, and other special operational procedures should be listed in this plan. For a growout facility, details such as feed formulation, feeding rate, water quality requirements, expected growth rates, and other operational procedures. Incorporate any changes or improvements you make with each successive production cycle. This plan can be integrated with your predictions for capital expenditures as well. The purpose of the document is to (1) serve as a reference point during the course of the production (2) to serve as a reference document when comparing past production statistics. The details in this plan can help you locate inefficiencies in your operation that can be improved.

It is absolutely critical to keep complete records on fish production, feeding, water quality, fish leaving and entering the building. In addition, it is useful to keep records on system maintenance and visitors. Documentation is a key to gaining a thorough understanding of the system and achieving above average growth rates while minimizing costs. With good records, a producer can detect trends early on, before a problem can significantly affect growth rates or production costs. Such problems might include over feeding or decreased water flows. With this in mind, a series of suggested records are presented are discussed below.

Daily Fish Production Sheet

This data sheet is compiled to track fish production in each tank on a monthly basis. Information to be recorded on this sheet includes: stocking information, grading/sampling activities, fish movements among tanks, feed fed, disease treatments, and fish removed from the tank due to mortalities or sales. These sheets should be kept in a convenient place where the producer can refer to them often. It is best to keep them near each tank but this may not be feasible for some systems.

Monthly Fish Production Summary Sheet

On a monthly basis, after fish have been graded and moved, data from the daily fish production sheets should be transferred to the monthly fish production summary sheet so that it is accessible for easy reference. This summary sheet includes beginning and end-of-month totals for fish numbers and weights in tanks, total movements in and out of tanks, total feed fed, and overall feed conversions.

Water Quality Chart

This data sheet is designed to record water temperature and oxygen levels in the oxygenation system and fish tanks. It is suggested that these records be made twice weekly. Both the air and water temperature should be taken about the same time of day. The dissolved oxygen level should be measured in the oxygenation system as water enters and leaves the tank. Water entering the oxygenation system will generally be 1 ppm below saturation (or about 8 to 10 ppm) and water leaving to the fish tanks should be supersaturated with oxygen to about 15 ppm or so. The producer is also provided space to record the dissolved oxygen level of the water leaving an average tank as well as that leaving a heavily loaded tank. These recordings will give the producer an overview of his system's performance. Space is also provided for comments and other notes such as when the water clouds up following heavy rains or when the weather suddenly changes as such events could trigger reduced feeding, etc. By noting such events, a producer would not be too alarmed if his fish went off feed for example. If the fish went off feed without such a "trigger event", then the observant producer would be on the look out for possible oxygen or disease problems.

Expenses

A data form is compiled to record all expenses during the year for the system. Sections should be provided for expenditures on fingerlings, feed, chemicals, electricity, oxygen, transportation, telephone, advertising, and miscellaneous costs.

Income from Fish Sales

Since this is the purpose of the operation, a form should be set up to help the producer keep track of annual fish sales.

Planning

Although the business plan has been written it is important to have yearly production plans which involve more detail than the business plan. The business plan is often a visionary statement, which may make forecasts for assumed real world conditions. While the initial production plan may have many of those visionary elements, as time progresses a good manager will be able to predict more accurately the timing of the production cycles as well as the profits and losses. Maximizing the difference between the expenditures and gross profit is an important goal in a production environment. Because prices are often dictated by market conditions, it is important for the manager to reduce expenditures, a variable within his or her control. Before starting a production plan, the manager should review the business plan to see what types of profits are expected and when, in addition to capital available for daily operations. As the operation progresses, the plan should be fine-tuned to increase the efficiency of the operation. Perhaps the most important component of your operation but perhaps the most overlooked.

Production planning allows you to find out where your costs are and what you need to do to turn a profit. If you can't figure out how to make a profit in the planning stage, you need to either consult for outside help or let you investors know. As difficult as it seems, it is better to lose $10 than win $100 and lose $1000. On the other hand, your plan may have you losing $100 and then making $1000. Either way, you as a manager need to have a plan of action while always realizing that it is subject to refinement. If you have co-workers or peers, share your plan, they may be able to suggest improvements or discuss dead-ends they have already encountered. As a manager, you need to be optimistic that you can make a profit but also be realistic if there is not. Following are some of the basic planning issues you should consider.

IMPORTANT DATES

Regularly scheduled bill such as those for oxygen delivery, electric bill payments, feed orders, etc. need to be paid promptly to ensure service is not lost. Deadlines for permit applications should be tracked. As a planning tool, think about the fish's needs from birth (egg) to death. If automatically scheduled services are used, it remains the responsibility of the manager to ensure that payments are made in a timely manner

11.17 HOW TO COLLECT, ANALYZE & INTERPRET DATA

DAILY DATA COLLECTION

Use the "Appropriate Forms" to minimize the steps and the complications for getting data into the computer for analysis. The simplest way is to duplicate the format used by the spreadsheet, attach to a clip board, and place in critical locations throughout the facility. These forms should consider:

- Water Quality, Fish Behavior, Feed Consumption
- Accuracy of the data that is being collected

Try to keep data collection to a bare minimum, otherwise it becomes burdensome and is ignored. It is important to ask oneself a few questions before designing your data collection system:

- How does this help me to improve the status of my daily operations?

- Can I use this data in the future to establish trends, which may help me to more effectively predict or improve my production strategy?

- If this data indicates that I have a problem, can I do anything to correct the problem or is it out of my control?

DATA ANALYSIS PURPOSE

Record keeping is important not only for analysis after an event such as keeping track of financial records/keeping track of fish growth rate. Daily monitoring of data is important to forecast for production changes/improvements to head off impending disasters or increase efficiency.

- To integrate the collection of data and later analysis.
- To learn how to see impending disasters.
- To prepare for and recover from emergencies
- Looking at both output and efficiency
- What is the maximum number of fish the facility can produce?
- What is the 10-year production plan?
- What are the current costs? How can they change? Up or Down?
- What type of yearly budget do I have?
- Am I on the way to reaching my production goals?
- What kind of mortality can I realistically expect during the production cycle?

11.18 CARE AND USE OF LABORATORY ANIMALS

For those of you who raise aquatic animals in a university setting or other government sponsored facility, you must be very aware of the Federal requirements on care and handling of animals. The Institute of Laboratory Animal Resources, Commission on Life Sciences, National Research Council, has published the "Guide for the Care and Use of Laboratory Animals" (National Academy Press, Washington DC 1996). This Guide emphasizes performance goals as opposed to engineering approaches. Using performance goals for the cultured fish places greater responsibility on the user and hopefully results in an enhancement to animal well-being. This is serious business. All farms and culture stations should obtain a Guide and review it thoroughly. Any Federally sponsored research must be conducted in accordance with the Guide.

CHAPTER 12

OZONATION AND UV-IRRADIATION[1]

12.0 INTRODUCTION

Bacterial and viral diseases create serious problems in semi-intensive and intensive aquaculture. Use of surface water in flow-through systems represents a risk of contamination by introducing waterborne fish pathogenic microorganisms. Such contamination results in heavy losses in aquaculture worldwide, and has also limited the progress in commercial farming of new aquacultural species. Some commercial operations may be required to disinfect their discharge waters before release into the aquatic environment. Of serious concern are dependable means of controlling pathogens present in the inlet water. Disinfection by ozonation or UV-irradiation are two methods often applied in aquaculture. It is important to distinguish between disinfection of makeup waters (low organic loads) and recirculating aquaculture system (RAS) waters. Both applications are discussed in this chapter. Disinfection by ozonation and UV-irradiation are also used in other aquacultural applications, e.g., reducing or eliminating potential pathogens associated with live prey such as rotifers in marine larval production systems, and surface disinfection of fish eggs. Other methods of disinfection are also discussed at the end of this chapter.

12.1 UV IRRADIATION

Natural and artificial UV light (wavelength of 190–400 nm) may damage microorganisms by directly and indirectly altering nucleic acids. Direct damage is due to absorption of irradiation by DNA with resulting formation of photoproducts. DNA absorbance is high in the UV-C range (190–280 nm), but falls more than three orders of magnitude in the UV-B range (280–320 nm), and is negligible in the UV-A band (320–400

---

[1] Chapter primarily authored by **Helge Liltved,** *Norwegian Institute for Water Research, Branch Office South.* Televeien 3, N-4879 Grimstad, Norway; e-mail: helge.liltved@niva.no

nm) (Miller et al. 1999). The DNA-damaging effect of UV-C is utilized in bactericidal lamps. The low-pressure mercury-vapor lamp emits approximately 85% of its energy output as monochromatic light at a wavelength of 253.7 nm, which is within the optimum wavelength range of 250 to 270 nm for bactericidal effects. Solar UV-B irradiation is responsible for both direct and indirect DNA damage, while UV-A produces only indirect damage. Only direct DNA damage is repairable by photoreactivation.

The UV-C damage often results from dimerization of two pyrimidine molecules. Cyclobutane pyrimidine dimers and pyrimidine-pyrimidone (6–4) photoproducts are the two major classes of photolesions formed by direct DNA absorption of UV-C irradiation (Friedberg et al. 1995). Once the pyrimidine residues are covalently bound together, replication of the nucleic acid is blocked or results in mutant daughter cells unable to multiply (Stover et al. 1986). The moderate energy level of UV irradiation leaves no toxic residuals in the treated water. Although chemical compounds can be altered by the radiation (Gjessing and Kallqvist, 1991; Lund and Hongve, 1994), the UV doses used for disinfection are too low to generate significant amounts of photoproducts (Oliver and Carey, 1976; de Veer et al. 1994). This non-toxicity is of crucial importance when UV is the method of choice for influent disinfection in aquacultural facilities.

UV irradiation units are commonly installed in smolt farms for sea and freshwater disinfection, and are also used for bacteriological control in recirculation systems (Rosenthal, 1981). However, before the UV dose can even reach the target organism, it must be able to transmit through the water. UV applications in highly turbid water as is often the case in RAS will be totally ineffective since the transmission into the water column is very minor, thus killing almost no organisms. Therefore, the lowest expected UV transmittance of the process water should be established and used to predict how much UV intensity must be generated to transmit the desired UV dose through the water between the target organism and the light source.

Ultraviolet filters can be built as nonpressurized open channel units (see Fig. 12.1) or as pressurized tube-and-shell units. The UV bulbs are usually contained within quartz sleeves to allow submergence in the process flow. The quartz sleeves must be kept clean to maintain transmittance. Using UV light does not produce toxic residuals or form byproducts that pose a risk to aquatic organisms.

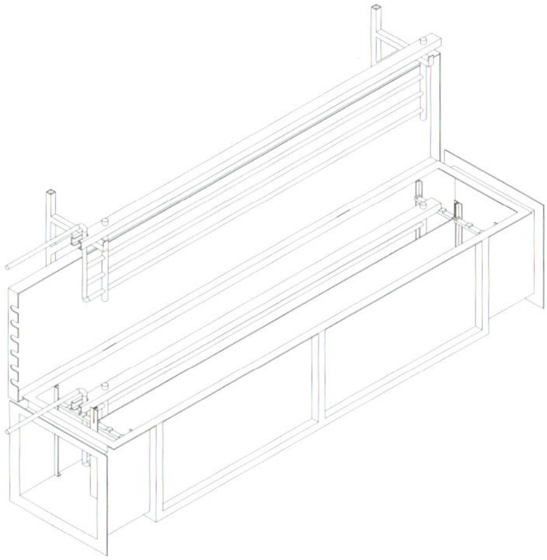

Figure 12.1 A Horizontal UV Channel Filter in Position Mode for Bulb Removal. Drawing courtesy of PRAqua Technologies, Ltd., Nanaimo, British Columbia (from Summerfelt et al. 2001).

UV LIGHT OUTPUT

UV light output strength continually deteriorates from the time you first turn on the machine. Assume in general a 3% per month decrease in output strength or 40% per year. UV output strength is also affected by air temperature. The UV output ratings are 100% at 100ºF (38ºC); at 32ºF (0ºC), the output strength is only 10% of the 100ºF (38ºC) rating. This problem can be addressed by using quartz sleeves that creates a warm air pocket around the UV bulb and this maintains output strength near its maximum potential. The quartz sleeve will reduce output by 5% just due to transmissivity of the sleeve. Bulb replacement in UV units can be of considerable expense, particularly when compared to other forms of sterilization and disinfection. As a rule of thumb, change UV bulbs at least once annually.

> **"RULE OF THUMB"**
>
> Change UV bulbs at least once a year.

APPLICATION EXAMPLE OF UV

Problem

Define a UV unit to provide a minimum lethal dose (MLD) for an organism that requires 100 mWs/cm^2 in an incubator system with a water volume of 1,000 L.

Solution

Select a UV unit from a commercial manufacturer. Such units are rated on maximum flow rate through the unit and their dosage output. Assume you select a unit that is rated at 10 L/min with an output of 40 mWs/cm^2. Now, calculate the allowable flow rate to achieve the specified MLD for the organism we are attacking:

$$Q\frac{L}{min}\cdot 100\,mWs/cm^2 = 10\frac{L}{min}\cdot 40mWs/cm^2$$

$$Q = 4\frac{L}{min}$$

Thus, although the unit is rated for a flow of 10 L/min, you can only operate the unit at a flow rate of 4 L/min. Thus the 1,000 L incubator volume will only be turned over once every 250 minutes or about 6 times per day. This is acceptable. There is clearly judgment involved in determining if you are cycling your water through a UV unit often enough. The turnover has to be sufficiently fast so that organisms do not generate faster than they are being killed off.

12.2 OZONATION

Ozone has seen wide use in aquaculture because it has a rapid reaction rate, produces few harmful reaction by-products in freshwater and oxygen is produced as a reaction end-product. Ozone is an extremely reactive oxidant and a very effective bactericide and viricide. Ozone can also be used to achieve water quality improvements by microflocculating fine particulate matter (making particles that are easier to settle or filter) and oxidizing nonbiodegradable organic molecules (creating smaller and more biodegradable molecules), nitrite, and refractory organic molecules (reducing water color) (Summerfelt and Hochheimer, 1997; Summerfelt et al. 1997).

Application of ozone to aquaculture requires ozone generation, ozone transfer into solution, contact time for ozone to react and disinfect, and possibly ozone destruction to ensure that no ozone residual makes it into the culture tanks (Summerfelt and Hochheimer, 1997).

Ozone Generation

Ozone must be generated on-site. The most efficient method is by the electric corona discharge technique, which involves the passage of oxygen gas, or air, across a gap of narrowly spaced electrodes under high voltage, Fig. 12.1. According to Masschelein (1998), effective ozone generation depends upon the composition of the feed gas, e.g., feed gas impurities, particulates, moisture, and pressure of the feed gas, efficiency of dielectric cooling, characteristics of the electrical current, concentration of the ozone produced, and dielectric design. Protecting the dielectrics within the corona discharge cells requires a feed gas supply that is dry, free from particulate matter and coalescible oil mists, and contains less than 15 ppm hydrocarbons (Dimitriou 1990; Masschelein, 1998). Also, some ozone generators may require some nitrogen gas impurity (>0.5% nitrogen) within the oxygen feed gas in order to achieve maximum ozone production efficiency. To meet this level of nitrogen contamination, some liquid oxygen supplies may require adding a small quantity of nitrogen to the feed gas before it enters the ozone generator.

Purified oxygen is already used to maximize carrying capacity within many intensive aquaculture systems. Corona discharge generators using purified oxygen feed gas require about 10 kWh of electricity to produce 1.0 kg of ozone (Masschelein, 1998). However, ozone production within an air feed gas is 2–3 times less energy efficient than using purified oxygen feed gas (Masschelein, 1998). Also, generating ozone in oxygen feed gas can produce a 10–15% (by weight) concentration of ozone, which nearly doubles the concentration of ozone that can be generated using air as the feed gas. The relatively high concentrations of ozone can be generated to reduce the overall mass of oxygen required to supply ozone. Yet, it is less energetically efficient to produce ozone concentrations of 10–15% (by weight) than to produce ozone concentrations of 4–6% (Carlins and Clark, 1982). Taking all of this into account, ozone production can be optimized according to the demands of the aquaculture system and economic considerations of feed gas cost and energy usage.

Ozone Transfer

Ozone is generated within either air or oxygen feed gas and this ozone/oxygen gas flow must be transferred into water for microbiological inactivation or other oxidative purpose. The ozone gas flow can be co-transferred into the water using any of the typical oxygen transfer devices (Summerfelt and Hochheimer, 1997), which have been discussed in Chapter 8 of this book. Effective transfer of ozone gas into water is important because the cost of producing ozone is not insignificant, especially if the ozone is carried within purified oxygen feed gas that was either purchased or produced on site.

The rate of ozone transfer and the subsequent rate of ozone decomposition depend upon the efficiency of the contacting system used and the rates that ozone reacts with constituents within the water. The rate that ozone reacts depends on the type and concentration of constituents within the water. Rapid reaction with oxidizable inorganics and organics will maintain a low apparent equilibrium concentration of ozone within the liquid film and increase the rate of ozone transfer. The driving force for ozone transfer is maximized when the ozone absorbed is rapidly consumed by reaction with constituents within water. In fact, when ozone reacts very fast, ozone decomposes at the gas surface and no molecular ozone is transferred into the water (Bablon et al. 1991).

Ozone transfer units that have a continuous liquid-phase, i.e., units that disperse gas bubbles within a liquid, -- such as U-tubes, Speece cones, Fig. 12.2, aspirators, bubble diffusers, and enclosed mechanical surface or subsurface mixers -- provide both ozone transfer and some reaction time. Ozone transfer units that have a continuous gas-phase, i.e., units that disperse liquid drops and films within a gas, such as spray columns, packed columns, and multi-stage low head oxygenators, Fig. 12.3, provide efficient transfer but very little time for reaction (Summerfelt and Hochheimer, 1997). Continuous gas-phase transfer units are best suited for use in situations that normally require the transfer of the maximum amount of ozone in the shortest time for economical fixed and variable costs. On the other hand, continuous liquid-phase transfer units are usually selected for situations where reaction is rate limiting and an ozone residual must be maintained for a specific length of time (Bellamy et al. 1991).

Most ozone contactors rely on continuous liquid-phase units that bubble ozone into the liquid (Bellamy et al. 1991). High column bubble diffusers are frequently used for aquacultural applications and can achieve more than 85% ozone transfer to the liquid phase (Helge Liltved, unpublished data). These units are particularly well suited to situations where reaction is rate limiting and an ozone residual must be maintained for a specific length of time, such as during disinfection. Speece cones, Fig. 12.2, and U-tubes are also being used to efficiently and rapidly transfer ozone/oxygen feed gas within RAS, where oxidation of nitrite and organic matter are the primary goals of ozonation (not disinfection).

Figure 12.2 An Ozone/Oxygen Feed Gas Is Injected Into Water Within Three Speece Cones (Plumbed in Parallel for Redundancy and Variable Flow Requirements). System shown is used to disinfect 400–2400 L/min of surface water at the US Fish and Wildlife Service's Lamar National Fish Hatchery (Lamar, PA). Ozone transfer efficiency is typically >99% efficient within these cones when operated at 8–10 psig (0.5 to 0.7 atm) (Summerfelt, unpublished data).

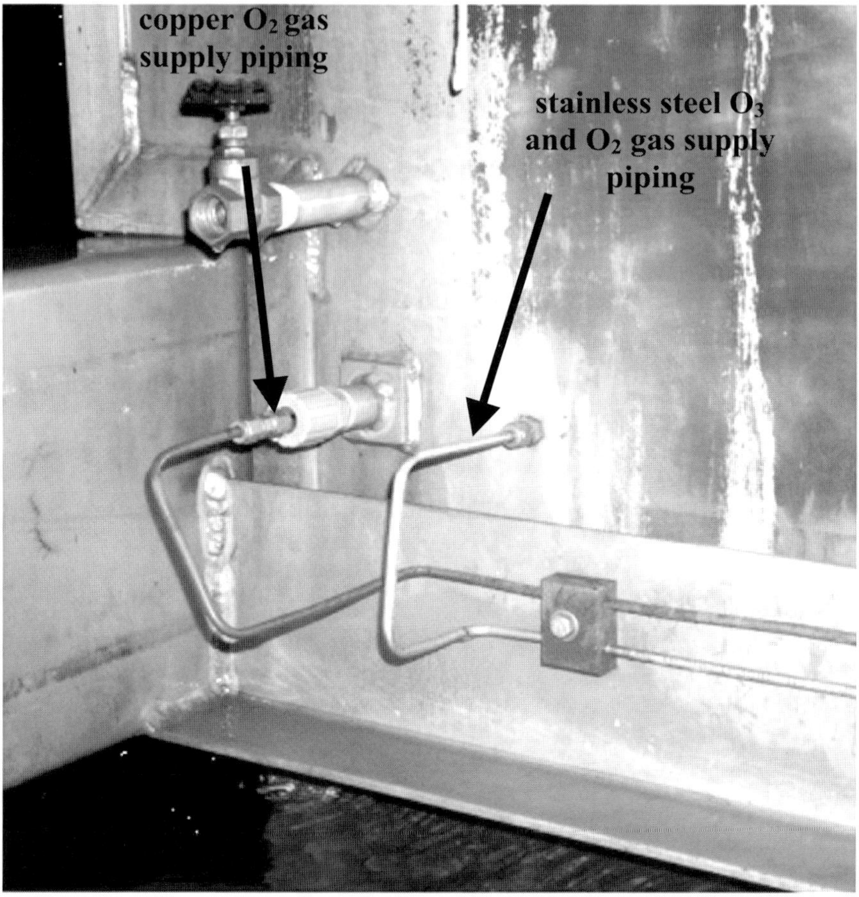

Figure 12.3 An Ozone/Oxygen Feed Gas is Injected Into Water Within an LHO Used to Oxygenate/Ozonate 4800 L/min of Recirculating Water in a System at the Freshwater Institute (Shepherdstown, WV).

Ozone transfer within continuous gas-phase units is not as common as within continuous liquid phase units (Bellamy et al. 1991). When ozone transfer has been reported within continuous gas-phase units, they are mostly packed columns. However, continuous gas-phase units can be designed to efficiently transfer ozone within relatively smaller vessels. Ozone transfer efficiency was 100% in the LHO (Low Head Oxygenator) units evaluated in the recirculating system at The Freshwater Institute. In this system, complete ozone transfer occurred because ozone is 13 times more soluble than oxygen in water according to Henry's law; short circuiting in the gas phase within the LHO was prevented by breaking the chamber into eight separate compartments; gas residence times within the LHO chambers were about 45 min; and, there was nitrite and dissolved and suspended organic material in the water that rapidly reacted with the dissolved ozone (Summerfelt and Hochheimer, 1997).

Ozone transfer into RAS is sometimes accomplished using the same gas transfer unit that is used for oxygen supplementation. This can be done if the transfer unit is fabricated from ozone resistant material (Bullock et al. 1997). In these situations, adding ozone to a recirculating system that is already using purified oxygen only requires installation of an ozone generator and the accompanying ozone distribution, monitoring, and control mechanisms (Summerfelt and Hochheimer, 1997). All of the other necessary equipment (oxygen supply and distribution system, gas transfer units, and control mechanisms) would already be in place, Fig. 12.3.

The off-gas discharged from the transfer unit will contain some ozone if ozone transfer is not 100% efficient. These ozone containing off-gas discharges must be treated to destroy remaining ozone.

Ozone Reaction and Inactivation of Fish Pathogens

Ozone oxidation can kill microorganisms, but requires maintaining a certain dissolved ozone concentration in the water for a given contact time. Disinfecting efficiency depends upon the product of the ozone residual concentration and its contact time. An ozone contact vessel provides the time necessary for the ozone residual to react with and inactivate pathogenic microorganisms. Disinfecting waters may require maintaining a residual ozone concentration of 0.1–2.0 mg/L in a plug-flow type contact vessel for periods of 1–30 min, depending upon the target microorganism (Wedemeyer, 1996). In commercial aquaculture applications, it is extremely difficult to maintain residual concentrations above 1 mg/L; above 2 mg/L is almost impossible with conventionally available equipment. An example system that provides ozone contact time in a two reactor sequence vessel with residual removal is illustrated in Fig. 12.4.

Ozone exposure experiments with bacterial cells have indicated changes in membrane structure which lead to leakage of protein and nucleic acid, and also lipid oxidation, while the intracellular components, protein, and DNA, remain intact (Komanapalli and Lau, 1996). By prolonged ozone exposure, cell viability is reduced with a more significant increase in lipid oxidation and protein and nucleic acid leakage.

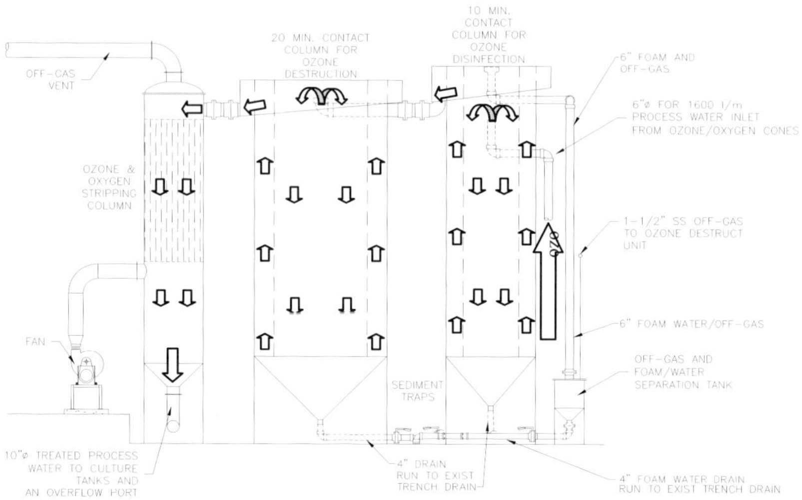

Figure 12.4 Ozone Treatment System for Disinfecting 400–2400 L/min of Surface Water at the US Fish and Wildlife Service's Lamar National Fish Hatchery (Lamar, PA). Drawing courtesy of Oak Point Associates, Biddeford, Maine (from Summerfelt et al. 2001).

In general, bacteria and viruses pathogenic to salmonids are highly sensitive to residual ozone in water (see Tables 12.1 and 12.4). Dose-response estimates of this sensitivity are fairly precise in demand-free water (inorganic buffers, distilled water), indicating 99.9% inactivation or more in the 0.01–0.10 mg/L residual concentration range. In batch experiments with natural water exerting an ozone demand, residual concentration tends to drop rapidly, making reliable dose-response estimates more complicated. As a rule, higher residual concentrations, i.e., 0.1–0.2 mg/L in natural sea-, brackish-, and freshwater, and 0.3–0.4 mg/L in fish farm effluents, seem necessary to obtain the legally required inactivation level.

Table 12.1 Inactivation of Fish Pathogenic Microorganisms in Water by Ozone

Organism	Percent Removal	Time min	Conc. mg/L	C×t, Min ×mg/L	Temp C	pH	Water	References
Bacteria								
A. liquefaciens *A. salmonicida* *P. fluorescens*	99.9	1.2–5.0	0.10	0.12–0.50	20	7	PBS	Colberg and Lingg, 1978
A.salmonicida	100	0.5	0.04	0.02	20	7	PBDW	Wedemeyer and Nelson, 1977
A.salmonicida	99.8	0.33	init. 0.010 res. 0.04		7	7.2	PBS	Liltved and Landfald, 1995
A.salmonicida	99.9	1.0	0.3–0.4		7	7.8	Fish farm effluent	Liltved and Landfald, 1995
A.salmonicida *V. anguillarum* *V. salmonicida* *Y. ruckeri*	99.99	3.0	init. 0.15–0.20 res. 0.05–0.07		9–12	6.3–8.0	Lake, brackish and seawater	Liltved et al. 1995
Enterococcus seriolicida, *Pasteurella piscicida,* *V. anguillarum*	99.9	0.9–1.2	0.089–0.177*	0.084–0.186*	25	7.95	Seawater	Sugita et al. 1992
Yersinia ruckeri	100	0.5	0.01	0.005	20	7	PBDW	Wedemeyer and Nelson, 1977
Yersinia ruckeri	99.9	3.0	0.10	0.30	20	7	Distilled	Colberg and Lingg, 1978
Viruses								
IHNV	100	0.5	0.01	0.005	10	7	PBDW	Wedemeyer et al. 1978
IPNV	100	1.0	0.01	0.010	10	7	PBDW	Wedemeyer et al. 1978
IPNV	99.99	1.0	0.10–0.20	0.10–0.20	9–12	6.3–8.0	Lake, sea, brackish	Liltved et al. 1995
SJNNV	Loss of infectivity	0.5–2.5	0.1–0.5	0.25			Seawater	Arimoto et al. 1996
WSBV	Loss of infectivity	10	0.5*	5			Seawater	Chang et al. 1998

*Concentrations recorded as total residual oxidants (TRO)
SJNNV – striped jack nervous necrosis virus

In dose-response trials, two-stage logarithmic inactivation curves have been obtained (Liltved et al. 1995; Liltved and Landfald, 1995). These were characterized by a rapid inactivation initially, followed by a decreasing inactivation rate with exposure time. Such kinetics could be explained by reduced ozone concentration during the course of exposure, and have also been experienced by other investigators, even in ozone demand-free water (Katzenelson et al. 1974; Colberg and Lingg, 1978; Vaughn et al. 1987). In practical ozonation of fish farm influent and effluent water, it is important that the ozone dose is high enough to account for the initial demand, thereby establishing a sufficient residual concentration for the required contact time.

In demand-free water, dissipation of ozone will still be observed due to the demand exerted by the added microorganisms. This demand will depend on type of organism, the preparation, and washing of the inoculum prior to supplementation, and the density of organisms in the ozonated suspension. In natural waters and in waters found within recirculating systems, additional ozone will be lost in reactions with organics and other compounds. According to ozone demand tests on a high quality trout stream water that is being ozone disinfected at the US Fish and Wildlife Service Lamar National Fish Hatchery (Lamar, Pennsylvania), an ozone concentration of 2–4 mg/L must be transferred to maintain a 0.2 mg/L ozone residual concentration after 10 minutes (Steven Summerfelt, unpublished data). Cryer (1992) reported similar ozone demand results in tests on surface water supplies that were being disinfected at US Fish and Wildlife Service salmonid hatcheries in North America. All of the surface water supplies examined in these studies exhibit relatively high water quality with low concentrations of oxidizable organic material, iron, and manganese (Cryer, 1992; Steven Summerfelt, unpublished data), yet the ozone demand created still reduced the half-life of ozone to less than a few minutes. In comparison, the half-life of ozone dissolved in pure water at 20ºC is 165 minutes (Rice et al. 1981).

The ozone demand of water within RAS, which contains much higher levels of organic material and nitrite, creates a short ozone half-life, e.g., less 15 seconds, and makes maintaining an ozone residual difficult (Bullock et al. 1997). For this reason, it is difficult to add enough ozone to achieve microbial inactivation in recirculating systems. In recirculation systems ozone is most often applied at doses that promote water quality improvement (Brazil, 1996; Bullock et al. 1997; Summerfelt and Hochheimer 1997; Summerfelt et al. 1997). Using ozone in recirculating systems can reduce fish disease simply by improving water quality, which reduces or eliminates environmental sources of stress (Bullock et al. 1997). These studies, as well as experience with ozone application at numerous commercial recirculating systems, indicates that both water quality and fish health can be improved by adding approximately 13–24 g ozone for every 1.0 kilogram of feed fed to a recirculating system.

Ozonation may enhance fine solids removal by changing particle size rather than separating particles from water. As an unstable reactive gas, ozone splits large organics into smaller biodegradable materials that can be more easily removed by heterotrophic bacteria. Conversely, ozone can polymerize metastable organics leading to enmeshment, direct precipitation, bridging, or adsorption (Reckhow et al. 1986). Ozone has been used sometimes with mixed success in a variety of aquaculture systems to remove color and turbidity (Colberg and Lingg, 1978; Williams et al. 1982; Paller and Lewis, 1988; Brazil, 1996; Summerfelt et al. 1997). Effects of ozonation upon particle size change in recirculating systems are still not clearly defined. The function of ozone is complicated; both qualitative and quantitative impacts of ozonation may be specific to a given system (Grasso and Weber, 1988). There are also concerns that even low ozone residuals may cause gill adhesions and mortality in fish exposed to freshly ozonated water (Rosenlund, 1975).

Besides being a method for water improvement in recirculation systems, ozone is valued for its high virucidal activity, Table 12.1. Among fish pathogenic viruses, high sensitivity toward ozone has generally been reported. This also applies to viruses with high UV resistance, i.e., IPNV and WSBV (Liltved et al. 1995; Chang et al. 1998). As viral diseases have become a major threat in worldwide aquaculture, the virucidal properties of ozone will certainly be more valued in the future, both in intake and effluent water disinfection. Ozone-treated water has also proven useful for washing fertilized eggs (Arimoto et al. 1996) and for reducing or eliminating potential pathogens associated with live prey such as rotifers in marine larval production systems (Davis and Arnold, 1997; Theisen et al. 1998).

OZONE DESTRUCTION

Creating an adequate level of residual ozone at the end of the contact chamber to ensure kill-off of bacteria will also necessitate that this same ozone be removed prior to the water reaching the aquatic organisms. Ozone residuals can be lethal to fish at ozone concentrations as low as 0.01 mg/L, but the actual concentration depends upon species and life stage, Table 12.2. Due to the acute toxicity of residual ozone to aquatic animals (Wedemeyer et al. 1979), a de-ozonation unit has to be included. In many cases, residuals are eliminated by water retention within tanks immediately after ozonation, Fig. 12.4, or by applying small doses of a reducing agent, e.g., sodium thiosulphate. Dissolved ozone can also be stripped into air when passed through a forced-ventilation packed aeration column, Fig. 12.4. Air stripping will also remove dissolved oxygen concentrations that are in excess of saturation, which may or may not be desirable. Dissolved ozone can also be destroyed by passing the water through a biofilter or bed activated carbon, reaction with low levels of hydrogen peroxide, or contact with high intensity UV-light (catalyzing the conversion of O_3 to O_2).

Table 12.2 Toxicity of Dissolved Ozone to Fish
(from Summerfelt and Hochheimer, 1997)

Species	Ozone concentration mg/L	Effect
Rainbow Trout	0.0093	96-h LC_{50}[a]
Rainbow Trout	0.01–0.06	lethal
Bluegill	0.01	60% mortality after 4 weeks
Fathead minnow	0.2–0.3	lethal
White perch	0.38	24-h LC_{50}
Bluegill	0.06	24-h LC_{50}
Striped bass (larvae)	0.08	96-h LC_{50}

[a]where LC_{50} is the concentration lethal to 50% of sample fish.

Achieving ozone destruction with UV electromagnetic radiation depends on the wavelength of the UV light source and the quantity of energy transmitted (Rodriguez and Gagnon, 1991; Hunter et al. 1998). Ozone residuals are destroyed at UV light wavelengths ranging from 250 to 260 nm. Ironically, UV wavelengths of 185 nm can be used to generate ozone. The primary ozone treatment components are located on the process pipeline immediately after ozone gas has been transferred into the water. First, the right hand vessel provides 10 minutes (at 1600 L/min flow) of plug flow contacting to achieve disinfection. Next, the middle vessel provides 20 minutes (at 1600 L/min flow) of plug flow contacting to achieve further disinfection and ozone residual destruction. Finally, the left hand vessel (a counter-current, forced-ventilation column) strips any minor ozone residuals and elevated dissolved oxygen levels immediately before the flow is piped to the fish culture systems

Ozonation by-products and their toxicity towards aquatic animals are not well described, especially in seawater. More long-lived reaction products are formed when brackish and seawaters are ozonated. Here, ozone reacts with bromide ions, and to a lesser extend with chloride ions, to form oxidants toxic to fish and shellfish. The most important are hypobromous acid (HOBr) and hypobromite ion (OBr-). Both have strong biocidal effects. By prolonged ozonation hypobromite ion can be further oxidized to bromate (BrO_3-), which is a persistent compound. In addition, small amounts of halogenated organic compounds like bromoform will be formed. Activated carbon filtration has been successfully used for removal of residual ozone and other oxidants in ozonated seawater (Ozawa et al. 1991).

Residual Ozone

Residual ozone in fresh water can be measured by the indigo colorimetric method, which is based on the decolorization of indigo by ozone (American Public Health Association, 1989). The decrease in light absorbance is linear with increasing concentration, and can be measured by a spectrophotometer at a wavelength of 600 nm. Because of the rapid reaction of ozone with bromide in seawater, it is not possible to measure dissolved ozone directly. Methods detecting the total residual oxidants (TRO) formed by ozone have been applied, including the DPD (N, N-diethyl-p-phenylenediamine) method and the iodometric method. As a surrogate parameter, the redox-potential is used for monitoring the oxidizing power of ozone and its oxidants in many practical applications. However, the absence of a linear relationship between residual ozone concentration and redox-potential is a disadvantage of this method.

Material Resistance

Ozone is an extremely corrosive material when used in any water application. Materials that ozonated water will come in contact with must be selected with appropriate resistance properties. Table 12.3 provides a summary of materials and their corrosive resistance properties.

12.3 FACTORS INFLUENCING DISINFECTION EFFICIENCY

A variety of factors have been reported to influence the efficiency of bacterial inactivation by water and wastewater disinfection (Kostenbauder, 1991; Bessems, 1998; Heinzel, 1998). These include type of microbe, growth conditions of the microbe, type, and concentration of disinfectant, ability of the microbe to repair damages induced by the disinfectant, and water quality.

UV REPAIR MECHANISMS

Bacteria may repair and reverse the lethal effects of UV irradiation by two major mechanisms: namely photoreactivation, and dark repair or liquid holding recovery. Both have the potential to increase the post-irradiation culturability by several $\log_{10}$ units as compared with the outcome of immediate plating and dark incubation. The combination of dark repair and photoreactivation generally leads to higher survival than dark repair alone, demonstrating that photoreactivation can revert lesions, which usually are not repaired in the dark (Moss and Davies, 1974).

Photoreactivation

The mechanism of photoreactivation requires visible light as a cofactor. Light in the near UV and visible spectrum (330–480 nm) activates DNA photolyase which reverts certain types of UV-induced damage, i.e., the UV-C and UV-B induced covalent bonds between adjacent thymines, without excising the distorted region (Jagger, 1967; Walker, 1984; Friedberg et al. 1995). The phenomenon was first recognized in the late 1940s when A. Kelner discovered the reversibility of UV damage on the spores of *Streptomyces griseus* by subsequent treatment with visible light (Kelner, 1961).

Conventional Michaelis-Menten enzyme kinetics has been used to describe the process of photoreactivation:

$$E + S \underset{k_2}{\overset{k_1}{\leftrightarrow}} ES \overset{k_3}{\rightarrow} E + P \tag{12.1}$$

In Eq. 12.1, E is the photoreactivating enzyme, S is the substrate (the photorepairable DNA lesion), ES is the enzyme-substrate complex, and P is the repaired DNA lesion. The enzyme binds to the pyrimidine dimers, and monomerises the dimers upon exposure to appropriate light. The reversible ES complex formation is light independent, while the k_3 is zero in the dark.

Table 12.3 Materials Resistance to Ozone Corrosion. From Damez, F., 1982. Materials Resistant to Corrosion and Degradation in Contact with Ozone. In *Ozonation Manual for Water and Wastewater Treatement.* W.J. Masschelein (Ed.), JohnWiley & Sons, Ltd., NY NY)

Materials	Ozonation Type of Exposure			Comments
	Dry air	Moist Air	Water	
Metals				
Chromium-nickel-silver	b*	b	b	
Brass	b	b	b	
Aluminum	a	d	d	
Aluminum alloys	a	d	d	
Pig iron	b	c	c	Slow corrosion
Galvanized steel	b	c	c	Not shock resistant
Stainless steel	d	d	d	Without chlorine present
Sintered stainless steel	d	d	d	
Plastics				
PVC supple	b	b	b	
PVC rigid	a	a	a	
Vinyl ester resin	a	a	a	
Polytetrafluoroethylene	a	a	a	Without pressure
Polyamide (nylon-risen)	–	–	a	
Epoxide (Araldite™)	–	d	d	
Synthetic Rubbers				
Chlorosulfonated polyethylene	a	a	a	With appropriate charge
FPM (viton)	b	c	c	With appropriate charge
Silicone	d	c	c	With appropriate charge
Ethylene propylene	a	a	a	With appropriate charge
Polychloroprene (neoprene)	b	c	c	With appropriate charge
Concrete	a	a	a	
Glass & ceramics	a	a	a	

*NOTE: a = long lasting; b = usable; c = for low ozone concentration; and d = quickly degraded

The rate of photoreactivation is strongly correlated with light intensity. Several hours are normally required for complete reactivation at lamplight intensities, while sunlight intensities have been reported to reduce the time for photoreactivation completion down to fractions of an hour in irradiated *E. coli* and halophilic bacteria (Harm, 1968; Eker et al. 1991). The photorepair mechanism is a widespread, but not universal trait among microorganisms, and there is no clearly defined phylogenetic distinction between the species, which have this ability, and those that have not. Viruses generally do not have the ability to photorepair. However, if the host cell has the required enzymes, photoreactivation of viruses may take place (Asato, 1976).

Dark Repair

Bacteria may use at least three different dark repair, or liquid holding recovery mechanisms: 1) nucleotide-excision repair, 2) SOS-error prone repair, and 3) post replication recombinational repair (Friedberg et al. 1995; Miller et al. 1999). A complex cellular mechanism, named the SOS regulatory system, is involved in all three dark repair processes.

The accurate nucleotide-excision repair process takes advantage of the fact that the information in DNA is present in two copies as a consequence of the complementary double-stranded nature of DNA (Swenson, 1976; Friedberg et al. 1995). An incision is introduced into the damaged strand near the site of the lesion, a DNA fragment which includes the damage is excised, and the missing DNA is then resynthesized by using the opposite strand as a template. In *E. coli*, dark repair is obtained by keeping the bacterium in nutrient-free buffer in the dark for several hours between UV irradiation and plating.

Dark repair processes have been found in some bacteria, including *E. coli* and moderate halophilic bacteria, and it has been proven absent from extreme halophilic Archaea (Fitt et al. 1983; Eker et al. 1991).

WATER QUALITY

Particles

From drinking water and domestic wastewater investigations, it is reported that particles may provide protection against chemical and non-chemical disinfection agents, depending on the particle type and size, and the nature of association between the microorganism and the particle. Increased survival of particle-attached bacteria upon exposure to oxidants has mainly been attributed to: 1) high disinfectant reactivity at the particle surface, and 2) limited mass transfer at the interface. Each of these two factors will lead to a low concentration of oxidant available for inactivation of attached bacterial cells. However, this effect is highly dependent on the chemical demand of the particles. Failure of ozone to reduce the heterotrophic bacterial count more than one $\log_{10}$ unit in a recirculating rainbow trout culture system was attributed to rapid loss of ozone caused by elevated levels of total suspended solids (Bullock et al. 1997).

Several investigators have reported a correlation between suspended solids content and survival of faecal coliforms in UV-irradiated wastewater (Whitby and Palmateer, 1993). Qualls et al. (1983) observed that bacteria harbored by particles were partially protected so that UV disinfection was limited to a 3 to 4 $\log_{10}$ unit reduction in viability. Filtration was required to meet strict bacterial standards.

Scattering and absorption of UV light by particle surfaces may reduce the effect of radiation. Particles of organic origin will absorb more irradiation than mineral particles. Little protection of microbes is provided by clay particles because most of the irradiation is scattered. Shading may limit the exposure of individual bacteria, but has not been a problem in well-designed UV disinfection reactors with lateral dispersion.

In natural fresh and seawater sources, the particle content may rise in periods of high primary production or extreme weather conditions (wind and precipitation). In such periods, the inactivation process may be controlled by two different mechanisms: a rapid inactivation of single cells and cells associated with small particles, followed by a slower decay rate due to the partial protection of microbes associated with larger particles. Effective prefiltration and elevated disinfectant doses may sustain disinfection performance in such periods.

Dissolved Organic Matter

Dissolved organic carbon (DOC) is, by convention, defined as total organic carbon (TOC) after filtration through a 0.45 μm membrane filter. In freshwater supplies, humic substances originating from the terrestrial environment are often the most significant contributor to the DOC, conferring a brownish-yellow color on the water. In wastewater, proteins, carbohydrates, lipids, and organic amines will elevate the concentration of DOC. Oxidizing disinfectants like ozone will lose bactericidal strength through reaction with organic matter. The reaction products will generally have weak or no bactericidal activity. Hoigne (1988) has shown that aqueous ozone reactivity can be ascribed to two mechanisms: direct reactions involving molecular ozone and reactions of active hydroxyl-radical intermediates produced by ozone decomposition.

Humic substances of natural waters are relatively resistant to ozonation. Sufficient contact time will produce small amounts of acetic, oxalic, formic and terephthalic acids, phenolic compounds and carbon dioxide. Generally, ozonated organic matter is more biodegradable than the original compounds. The instantaneous ozone demand of surface waters with DOC content of 2.5–3.5 mg/L has been reported to be in the range of 0.50–0.75 mg/L (Roustan et al. 1998). Surface water ozone demands of 0.4–0.5 mg ozone/mg DOC after 5 min of exposure at pH 7.5 was found by Graham (1999). The ozonolysis of carbon-carbon double bonds in organic molecules is an example of an ozone-demanding reaction.

> **"Rule of Thumb"**
>
> For UV applications, pre-filter water with a 50 micron screen.

Several organic compounds in water, e.g., humic substances, phenols and lignin sulfonates, will absorb UV radiation, thus reducing the doses available for microbial inactivation (Harris et al. 1987). The UV transmission of good quality seawater is generally high. However, the transmission may drop temporarily during bad weather conditions, in periods of planktonic blooms, etc. In general, assume you must use a 50 micron pre-filter to remove most of the particulate matter before applying UV. In shellfish operations, turbidity should not be above 20 nephelometric turbidity units (NTUs).

Inorganic Compounds

Oxidizing disinfectants will react with inorganic compounds in accordance with their oxidation potential. Ozone will be involved in several redox reactions due to its high oxidation potential. Metal and heavy metal ions are oxidized to form stable compounds with low solubility. Ferrous and manganeous ions will react to ferric and manganic ions, respectively, which in turn will react with OH- to form an insoluble precipitate. Bromide will be oxidized to bromate through several intermediary steps, while the reaction with chloride is limited by poor kinetics. The conversion of ammonia to nitrite is a slow, pH-dependent, first order reaction, while the nitrite is rapidly oxidized to nitrate. The latter reaction may have a significant effect on ozone disinfection capacity in wastewater treatment systems with incomplete nitrification. Venosa (1983) reports that as much as 2 mg/L ozone was required to oxidize 1 mg/L of nitrite-N.

pH

Extreme pH values may inactivate microorganisms, or limit their growth. The activity of many disinfectants depends on pH. Small changes in the hydrogen ion concentration may influence the disinfection performance. The pH dependence of the biocidal activity of ozone is not clear. Reduced effect at high pH towards poliovirus and rotavirus as well as the cysts of the parasite *Naegleria gruberi* has been observed (Vaughn et al. 1987; Wickramanayake et al. 1991). However, the opposite relation was evident for *Giardia muris* cysts, which were more sensitive at pH 9 than at pH 5 and 7 (Wickramanayake et al. 1991). Changes in pH have little to no impact upon UV effectiveness.

Temperature

In general, the microbial inactivation rate by chemical disinfectants will increase with increasing temperature. Farooq et al. (1977) found a higher degree of ozone inactivation of *Mycobacterium fortuitum* at elevated temperatures. On the other hand, UV inactivation seems relatively insensitive to temperature. Negligible effects were observed in the range 5–35°C when pure cultures of *E. coli*, *Candida parapsilosis* and bacterial virus f2 were exposed to UV light in batch reactors (Severine et al. 1983).

12.4 Aquaculture Wastewater Characterization

Physical and chemical characterization of the water to be disinfected is of primary importance in order to select the best-suited pretreatment and disinfection method and to determine dose requirements. In the following, various influents and effluents of hatcheries, smolt farms, and processing industry have been characterized in relation to disinfection. These waters have been divided into five types on the basis of origin, concentration range of selected quality parameters and the objectives of disinfection, Table 12.4.

High content of organic matter indicates problems associated with disinfection by oxidants and UV irradiation. More extensive pretreatment than screening is advisable for such waters. In general, total suspended solids (TSS) and dissolved organic carbon (DOC) are important indicators of disinfection performance, and usually there is a close correlation between DOC and the disinfectant demand of the water. In effluents, a more commonly used parameter is chemical oxygen demand (COD), which includes both particulate and dissolved organic matter. COD measurements are not suited for high salinity water due to interference by anions, especially chloride.

12.5 Inactivation Of Fish Pathogens

UV irradiation

Data on inactivation of fish pathogenic bacteria and viruses by UV irradiation are summarized in Tables 12.5 and 12.6. The values show that UV doses of 2–6 mWs cm^{-2} reduce the viable count of the studied species by 99.9% or more in laboratory batch experiments. However, under more realistic continuous flow experiments with particles present, considerably higher doses are required to obtain a high degree of inactivation (Bullock and Stuckey, 1977; Liltved and Cripps, 1999). Therefore, inactivation doses obtained in the laboratory should be used with caution in predicting dose requirements in actual situations.

Studies by Torgersen (1998) indicate that the ISA virus is susceptible to UV light. Loss of infectivity was demonstrated when infected tissue homogenate was subjected to doses of 4–10 mWs cm^{-2}. In contrast, the IPN virus, a non-occluded birnavirus responsible of infectious pancreas necrosis in Atlantic salmon, is UV resistant. A dose of 122 mWs cm^{-2} was required to obtain 99.9% reduction in virus titer in brackish water (Liltved et al. 1995). Doses in the same order of magnitude have been experienced in studies conducted by Japanese investigators (Sako and Sorimachi, 1985; Yoshimizu et al. 1986). In spite of the resistance of IPNV, UV irradiation has become the method of choice for disinfection of supply water in Norwegian smolt farms where the UV units are designed for bacterial inactivation. The IPNV will pass through these installations unaltered since the doses applied are far too low for inactivation. Due to the substantial losses from IPN outbreaks, preventive measures such as upgrading of existing UV units to IPNV inactivating capacity when using water from sources suspected to carry infective levels of IPNV should be considered. Such upgrading will require at least a 5-fold increase in UV dose compared with the present 25 mWs cm^{-2} requirement.

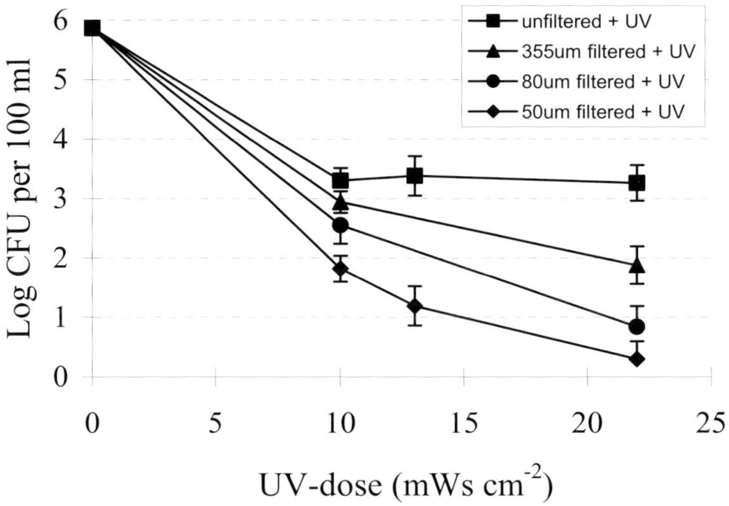

Figure 12.5 The Effect of Pre-filtration on Survival of Aerobic Bacteria in UV Irradiated Water Containing *Artemia* Fragments.

Table 12.4 Water Parameters Affecting Disinfection Performance in Smolt Farms and Effluents From Salmon Filleting Industry & Slaughterhouses

Water type		TSS[a] mg/L	Turbidity NTU	COD mg/L	TOC mg/L	DOC mg/L	UV absorbance	Ozone demand mg/L	References
Influent water	Freshwater (good quality)	0.1–3.0	0.5–1	-	2.5–3.5	2.5–3.5	0.01–0.20	0.5–0.75	SFT 1997, Roustan et al. 1998
	Seawater	0.1–0.5	0.3–0.6	-	-	0.4–2	-	-	Valiela, 1984
Smolt farm effluent (normal tank operation)		0.2–15.3	-	6.0–50	0.3–4.3		-	-	Eikebrokk and Ulgenes, 1993; Bergheim and Åsgård, 1996
		22	3.8	-	9.8	6.6	0.100	3.9	Liltved and Landfald, 1995
Water in recirculating systems for salmonids		5.2–7.4	1.50–1.66	39.3–47.8	-	6.7–7.5	-	-	Summerfelt et al. 1997
Effluent from salmon filleting plant		1600	-	3050	-	-	2.27	-	Liltved, 1997
Effluent from salmon slaughterhouses		40–1375	-	1500–3000	567	-	1–20	-	Fløgstad et al. 1991; Millamena, 1992

[a]Definitions =
TSS total suspended solid
NTU nephleometric turbidity units
COD chemical oxygen demand
TOC total organic carbon
DOC dissolved organic carbon

Table 12.5 Inactivation of Pathogenic Microorganisms in Water by UV Irradiation

Organism	Percent removal	UV dose (mWs/cm^2)	Water	Comments	Reference
Bacteria					
A. salmonicida *A. punctata* *A. hydrophila* *V. anguillarum* *P. fluorescens*	99.99	22			Kimura et al. 1976
A. salmonicida *A. hydrophila* *V. anguillarum* *P. fluorescens* *Y. ruckeri*	99.9–100	21–24	Spring water	UV trans.: 95% 1cm Turbidity: 4 JTU	Bullock and Stuckey, 1977
A. salmonicida *V. anguillarum* *V. ordalii*	99.9	2.9–5.5			Sako and Sorimachi, 1985
A. salmonicida *V. anguillarum* *V. salmonicida* *Y. ruckeri*	99.999	2.7	Brackish	15 ‰ salinity	Liltved et al. 1995
A. salmonicida	99.9	2.4	PBS		Liltved and Landfald, 1995
A. salmonicida	99.9	2.5	Fish farm effluent	TSS 22 mg/L TOC9.8 mg/L	Liltved and Landfald, 1995
A. salmonicida *V. anguillarum* *Y. ruckeri*	99.9	1.2–3.2	PBS		Liltved and Landfald, 1996
A. salmonicida *V. anguillarum* *Y. ruckeri*	99.9	8.5–13.4	PBS	With photo-reactivation	Liltved and Landfald, 1996
HPC	<99.9	22	Seawater		Liltved and Cripps, 1999
Viruses					
IHNV[a]	99.9	2			Sako and Sorimachi, 1985
IHNV	99	1–3			Yoshimizu et al. 1986
IPNV	99.9	150–200			Sako and Sorimachi, 1985
IPNV	99	100–150			Yoshimizu et al. 1986
IPNV	99,9	122	Brackish	15 ‰ salinity	Liltved et al. 1995
ISA-virus	Loss of infectivity	4–10	Tissue homogenat		Torgersen , 1998
WSBV	Loss of infectivity	900	Virus stock solution	20 ppt seawater	Chang et al. 1998

[a]Definitions =
HPC heterotrophic plate count
IHNV infectious hematopoietic necrosis virus
ISA infectious salmon anaemia
WSBV white spot syndrome baculovirus

Table 12.6 UV Energy for a 100% Kill (from Fish Farming News, pg 15, May/June 1999, Paul Hundley author)

Organism	UV Energy (micro-Watt Second per square centimeter, m Ws cm^{-2})
Mold Spores	
Penicillium roqueforti	26,400
Aspergillus glaucus	88,00
Aspergillus niger	330,000
Yeasts	
Brewer's yeast	6,000
Baker's yeast	8,800
Common yeast cake	13,200
Saccharomyces sp.	17,600
Bacteria	
Streptococcus hemolyticus	5,500
Staphylococcus aureus	6,600
Escherichia coli	7,000
Proteus vulgaris	7,500
Bacillus subtillis	11,000
Bacillus subtillis spores	22,000
Sarcina lutea	26,400
Virus	
Bacteriophage (*E. coli*)	6,600
Influenza virus	3,400
Nematode EGGS	40,000

Source: Kelly, Phillips and Hanel.

Even higher doses have been required to reduce the infectivity of the Asian shrimp baculoviruses to zero. The mid-gut glad necrosis virus (BMNV) and the WSBV were inactivated by doses of 410 and 900 mWs cm^{-2}, respectively (Momayama, 1989; Chang et al. 1998). Due to the extreme resistance of these viruses and the excessive water flows required, UV irradiation is not a feasible method for baculovirus inactivation in Asian grow-out shrimp farms. However, high efficiency UV units could be suitable for hatchery and nursery ponds with limited water use. There are other studies providing additional information on the UV levels needed to achieve disinfection (Hunter et al. 1998; Wedemeyer, 1996).

Particle protection mechanism has been indicated for bacteria associated with Artemia fragments, due to lack of a dose dependent inactivation in the dose range of 10–22 mWs cm^{-2} (Fig. 12.5; Liltved and Cripps, 1999). The results obtained suggest a possible transmission of fish pathogenic bacteria to land-based aquacultural installations, even if the influent water is disinfected by UV irradiation. It was further demonstrated that prefiltration improved bacterial removal. Mesh sizes of 50 μm resulted in more than 5 $\log_{10}$ removal efficiency, indicating that influent water to aquacultural systems should be filtered to remove crustaceal fragments and other particles capable of harboring bacteria before UV disinfection.

Absence of toxic residues and reaction products, low costs when treating high quality water, easy operation and maintenance, and minimal space requirements, are main advantages of UV irradiation. Fouling of the quartz sleeves must be dealt with on a regular basis, normally by mechanical or chemical cleaning. The major disadvantage of this method in aquacultural applications is the inefficiency against important fish pathogenic viruses. This might limit the use of UV irradiation in the future as virus control becomes a more important preventive measure in the fish and shrimp farming industry.

PHOTOREACTIVATION

Photoreactivation and dark repair have not previously been studied in fish pathogenic bacteria. In the present study it was demonstrated that *A. salmonicida, V. anguillarum* and *Y. ruckeri* are able to repair UV damage both in the presence and absence of visible light (Liltved and Landfald, 1996; Liltved and Landfald, 2000). Photoreactivation improved survival of these fish pathogens to about the same degree as has been found for common water quality indicator bacteria under similar experimental conditions (Harris et al. 1987).

At a light intensity typical of indoor aquacultural facilities (1500 lx), 10- to 1000-fold recoveries were obtained within the first two hours *in A. salmonicida, V. anguillarum* and *Y. ruckeri* (Liltved and Landfald, 1996). Photoreactivation was essentially completed within 4–6 hours. These trials were conducted in the laboratory with bacterial cells suspended in buffers.

UV irradiation followed by high-intensity light repair was also studied under more realistic conditions. UV exposed *A. salmonicida* and *V. anguillarum* suspended in their natural water qualities, respectively, demonstrated a less pronounced, but more rapid photoreactivation at sunlight intensities than in lamplight. Only 20 minutes and 1 hour respectively was required for complete reactivation (Liltved and Landfald, 2000). These trials show that even in indoor facilities with elevated light fluxes, water retention times, and/or temperatures, recovery processes may elevate the number of viable bacteria after irradiation. More rapid reactivation may occur in UV disinfected supply water and wastewater if exposed to sunlight in outdoor tanks or water recipients.

As compared to drinking water supply systems, the influence of photoreactivation is likely to be more pronounced in aquaculture facilities. In potable water systems, UV disinfection is usually the last step in a succession of treatments. Between this point and the end user there is normally no photoreactivating light available. UV irradiated water in aquacultural facilities will be exposed to artificial light or sunlight (daylight) capable of initiating photoreactivation.

Significant light-induced recoveries should not be neglected, and should be taken into consideration when assessing the efficiency of UV disinfection of aquacultural water. As a rule, the applied UV dose should account for the worst case with respect to reactivation. In Norwegian smolt farms, the veterinary authorities have demanded a minimum UV dose of 25 mWs cm^{-2}. The practical experience with this dose from about 100 UV installations has been good, with only a few outbreaks of bacterial diseases suspected to be transmitted through UV disinfected influent water.

An UV dose of 25 mWs cm^{-2} should be sufficient to suppress detectable repair of the fish pathogenic bacteria studied. However, if the dose is reduced for some reason, i.e., due to reduced lamp intensity caused by aging or fouling of the quarts sleeves, reduced water quality, increased water flow, or partly protected bacteria associated with particles, photoreactivation might elevate the number of survivors to infective levels.

DARK REPAIR

Dark repair, or liquid holding recovery, in fish pathogenic bacteria (*A. salmonicida, V. anguillarum* and *Y. ruckeri*) was found to be less efficient and slower than photoreactivation (Liltved and Landfald, 1996). Even at an elevated temperature of 22°C, 48 hours or more was required for completion. In aquacultural production units, like salmon hatcheries, which are characterized by relatively cold water and short hydraulic retention times, liquid holding recovery alone cannot be expected to have strong impact on post-UV irradiation survivability within the production units. Continuation of the recovery process is probable after disposal of effluent water to the recipient. This could significantly influence estimates of environmental impacts of release of UV treated aquacultural water. Dose-survival curves of UV-irradiated *A. salmonicida* after various post-irradiation treatments for different dark liquid holding periods and illumination levels are shown in Fig. 12.6 (Liltved and Landfald, 1996).

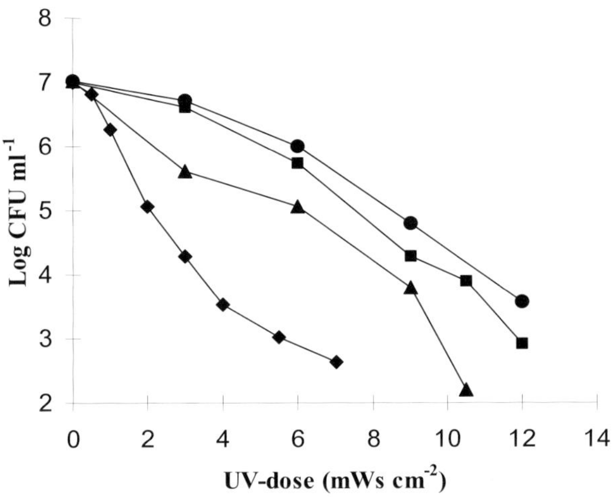

Figure 12.6 Dose-survival Curves of UV-irradiated *A. salmonicida* After Various Post-irradiation Treatments. All incubations were done at 22°C. Symbols: ◆, immediate plating/dark incubation; ▲, dark liquid holding for 48 h; ■, illumination at 1500 lx for 8 h; ●, illumination followed by dark holding in nutrient free buffer.

12.6 OTHER METHODS OF DISINFECTION

There are of course many other means of disinfecting water, equipment and instruments used in an aquaculture setting. Each has its proponents and each has its understood advantages and disadvantages. Some brief recommendations on chlorine dosages and contact times for disinfecting various pieces of equipment are given in Table 12.7.

Table 12.7 Recommendations on Chlorine Dosages & Contact Times (Lawson, 1995)

Equipment Description	Contact Time Minutes	Chlorine Form	
		Bleach (mg/L)	Calcium Hypochlorite (mg/L)
Nets, buckets boots, etc	5	0.70	40
Transport equipment	30	2.64	150
Rearing containers	60	3.51	285

Table 12.8 compares the relative oxidation potential of several known oxidants (Lawson, 1995). Table 12.9 provides an overall summary of common disinfectants.

Table 12.8 Relative Oxidation Potentials of Known Oxidants (Lawson, 1995)

Material	Volts	Oxidation Potential
Fluorine	2.87	Most Reactive
Ozone (O_3)	2.07	↓
Hydrogen Peroxide (H_2O_2)	1.78	
Potassium permanganate ($KmnO_4$)	1.70	
Hypobromus acid (Hobr)	1.59	
Hypochlorous acid (HOCl)	1.49	
Chlorine (Cl_2)	1.36	
Chlorine dioxide (ClO_2)	1.27	
Oxygen (O_2)	1.23	
Bromine (Br_2)	1.09	
Iodine (I_2)	0.54	Least reactive

OSHA PERMISSABLE LIMITS FOR OZONE in AIRSPACES

Time weighted average (TWA):	0.1 ppm
Short-term exposure limit (STEL):	0.3 ppm
Transitional Limit (TWA):	0.1 ppm

More information at the OSHA website:
www.osha.gov/dts/sltc/moethods/inorganic/id214/id214.html

Table 12.9 Comparison of Actual and Ideal Characteristics of Common Disinfectants

Characteristic	Property/Response	Chlorine	Sodium Hypochlorite	Calcium Hypochlorite	Ozone	UV Radiation
Availability	Should be available in large quantities and reasonably priced	Low cost	Moderately low cost	Moderately low cost	Moderately high cost	Moderately high cost
Interaction with extraneous material	Should not be absorbed by organic matter other than bacterial cells	Oxidizes organic matter, also absorbed by organics	Active oxidizer	Active oxidizer	Oxidizes organic mattei	Absorbed by specific organic compounds
Noncorrosive and nonstaining	Should not disfigure metals or stain clothing	Highly corrosive	Corrosive	Corrosive	Highly corrosive	Not Applicable
Nontoxic to higher forms of life	Should be toxic to microorganisms and nontoxic to humans and other animals	Highly toxic to higher life forms	Toxic	Toxic	Toxic	Toxic at high dosages
Penetration	Should have capacity to penetrate surfaces (bacteria)	High	High	High	High	High
Safety	Should be safe to transport, store, handle and use	High risk	Moderate risk	Moderate risk	Moderate risk	Low risk
Solubility	Must be soluble in water or cell tissue	Slight	High	High	Low	Not applicable
Stability	Loss of germicidal action on standing should be low	Stable	Slightly unstable	Relatively stable	Unstable, must be generated as used	Must be generated as used
Toxicity to microorganism	Should be highly toxic at high dilutions	High	High	High	High	High
Toxicity at ambient temperatures	Should be effective in ambient temperature range	High	High	High	High	High

Source: Water Environment Research Foundation

12.7 Conclusions

- Fish pathogenic bacteria are generally susceptible to UV irradiation. A 99.9% or higher inactivation has been obtained by doses of 1.5–3.4 mWs cm^{-2} in high quality natural waters and buffers. Contrasting this susceptibility of bacteria, some viruses (including the IPN-virus and Asian shrimp baculoviruses) are highly resistant to UV irradiation, demanding up to several hundreds mWs cm^{-2} for equivalent bacterial inactivation. This implies that UV units must be designed with high intensity irradiation and low water flow to obtain the required UV dose to inactivate these viruses.

- When bacteria associated with particles are UV-irradiated, protection has been demonstrated (Liltved and Cripps, 1999). Improved overall bacterial removal was demonstrated using particle prefilters with mesh sizes of 50, 80 and 355 µm. More than 5 $\log_{10}$ units total reduction in viable count was obtained with a 50-µm pre-filter prior to a UV dose of 22 mWs cm^{-2}, compared to less than 3 $\log_{10}$ reduction without filtration. Consequently, water for aquacultural purposes should be filtered to remove crustaceal fragments and other particles capable of harbouring bacteria before UV disinfection, thereby reducing the risk of introducing UV-shielded bacteria.

- It has been shown that fish pathogens, including the IPN-virus, are susceptible to ozone. A 99.9% or higher inactivation has been obtained at residual ozone concentrations of 0.15–0.20 mg/L within 60 s in natural lake-, brackish- and seawater. The logarithmic inactivation curves are often characterised by a rapid inactivation initially, followed by a decreasing inactivation rate with exposure time. Such kinetics could be explained by reduced ozone concentration during the course of exposure. In a fish farm effluent, a residual concentration of 0.3–0.4 mg/L was required for *A. salmonicida* inactivation. In practical ozonation of fish farming water, it is important that the ozone dose is high enough to account for the initial demand, thereby establishing a sufficient residual concentration for the required contact time.

- The three fish pathogenic bacteria *A. salmonicida, V. anguillarum* and *Y. ruckeri* are all able to photo-reactivate and exhibit liquid holding recovery when subjected to suitable post-UV-C irradiation conditions. With these two reactivation mechanisms, alone or in combination, the 99.9% inactivation dose has to be raised 3–4 times to obtain the same inactivation as with normal bacterial enumeration procedures, i.e., plating and dark incubation.

- Photo-reactivation is the more efficient reactivation mechanism by causing both a more rapid and a more complete recovery than holding in the dark. Light intensity and temperature influenced the rate of photo-reactivation. 10 to 1000-fold recoveries were obtained within the first two hours at light intensities and temperatures characteristic of indoor salmonid aquacultural facilities. At sunlight intensities, photo-reactivation was completed within 20 minutes in *A. salmonicida*, while 60 minutes was required in *V. anguillarum*. Photo-reactivation should be accounted for when assessing the necessary doses for efficient UV disinfection of aquacultural waters.

- An ozonation system is relatively complex and expensive compared to UV irradiation. The process becomes even more complex if oxygen is produced on-site.

CHAPTER 13

FISH HEALTH MANAGEMENT

Julie Bebak-Williams[1], Alicia Noble[1],
Paul R. Bowser[2] and Gregory A. Wooster[2]

13.0 BIOSECURITY

For commercial success, an aquaculture operation must maintain fish at densities far greater than normally found in nature. The animals must survive and grow rapidly. Regardless of the culture system used, the fish producer must maintain an environment that supports good fish health.

Effective fish health management consists of practices and procedures that emphasize prevention of outbreaks of infectious and non-infectious disease. Implementation of biosecurity practices will reduce operating costs by minimizing the number and severity of infectious disease outbreaks. These practices and procedures should be documented in a facility "Fish Health Management Plan", which is available to the facility staff for periodic review and reference. An effective plan of disease outbreak prevention includes a monitoring protocol that detects fish health problems at an early stage. Running a facility without a prevention plan can be financially catastrophic, as it leads to continual response to disease outbreaks as the fish health management strategy. This method of management results in added and unnecessary costs that include direct losses from mortality, inability to replace stock, facility closure orders, and restriction of movement orders. Market losses include reduced quality of survivors and a restricted

---

[1] Dr. Bebak-Williams and Dr. Noble are veterinary clinicians and research scientists at the Freshwater Institute, Shepherdstown, WV, USA.

[2] Dr. Bowser is a Professor and Greg Wooster is a Research Support Specialist at Cornell University, College of Veterinary Medicine, Ithaca, NY USA. Portions of this Chapter were taken with permission from Bowser, P. R. and J. K. Buttner. 1991. General Fish Health Management. NRAC Bulletin 111. Northeastern Regional Aquaculture Center, USDA/CSREES, University of Massachusetts Dartmouth, North Dartmouth, MA.

market for healthy stock because of damage to a facility's reputation and missed markets. In addition to the direct costs incurred for diagnosis and treatment of an outbreak, additional costs are incurred in the form of diversion of management and labor from other essential tasks and underutilization of the production facility (Paterson et al. 1991).

Biosecurity consists of practices and procedures that:

- reduce the risk that pathogens will be introduced to a facility
- reduce the risk that pathogens will spread throughout the facility
- reduce conditions that can increase susceptibility to infection and disease

Biosecurity cannot completely prevent entry of, or eliminate, all pathogens from a culture facility. Biosecurity accomplishes pathogen reduction rather than pathogen elimination.

Biosecurity is an important part of facility daily operating procedures. Planning should start during the design phase and protocols should be established before the facility comes on line. When biosecurity protocols are implemented in an existing operation, they may be perceived as being inconvenient because they were not considered when operational methods were developed or equipment was selected. Adding biosecurity as an afterthought may introduce an additional layer of complexity to an already inefficient operation. Effective biosecurity need not be an additional layer of work; procedures might be incorporated by modification of procedures already in use. Thinking about biosecurity before production begins allows non-intrusive routines to be developed rather than adding stopgap methods after problems arise. A consistently implemented biosecurity plan will also make it easier to isolate the source of fish health problems (for biosecurity checklist, see Summerfelt et al. 2001).

Biosecure RAS husbandry requires that a system be designed so that it can be cleaned completely, easily and frequently. Any surface can serve as a substrate for microorganisms. All components of a recycle system including biofilters, low head oxygenators, CO_2 strippers, pipes and tanks should be constructed of nonporous materials and arranged to be easily accessible for cleaning and disinfection. Clean-outs should be installed to access in any part of the system for flushing of accumulated biosolids. Because wood cannot be easily and thoroughly disinfected, it should be considered only for fabrication of disposable temporary structures. Equipment and supplies should never be transferred from other locations to the facility.

13.1 PRACTICES TO REDUCE THE RISK OF PATHOGEN INTRODUCTION

WATER SUPPLY

Entry of pathogens through a facility water supply is an important route of introduction, and will increase the risk for infectious disease outbreaks in aquaculture production systems. When possible, a groundwater supply should be used for the facility. Wells and springs do not usually contain resident fish, other aquatic animals, or aquatic invertebrates that could be pathogen carriers. If a pathogen-free water supply is at risk of contamination, or is unavailable, then influent water should be disinfected using ultraviolet radiation or ozonation. Well and spring water may need to be stripped of carbon dioxide and/or nitrogen gas, and oxygen may need to be added prior to using the water for fish culture. For small aquaculture operations of less than 100,000 lb per year (45,450 kg/yr), well water is the best choice because it will be specific-pathogen-free and constant water temperature and flow are more likely than with spring water. The well should be drilled and tested before the property is purchased. If the site does not have an adequate well water supply, another location should be selected. For larger operations, well water is also the best choice if an adequate supply is available. Surface waters harbor fish pathogens, and therefore, should be used only as a last resort, and then, only after effective sterilization (see Chapter 12). If spring water is used, it should be protected from animals that can carry fish pathogens, such as fish, birds, raccoons, salamanders, frogs and snakes.

EGGS AND FISH

Entry of pathogens through the introduction of fish to a culture facility is another important risk factor for disease outbreaks in aquaculture. The risk that pathogens will enter a facility can be reduced by purchasing eggs and fish cultured in a disinfected or specific-pathogen-free (SPF) water supply and certified to be specific-pathogen-free. Certification involves testing a lot or stock of fish for specific fish pathogens relevant to the species and determining, based on statistical probability, whether they are free of those pathogens (Thoesen, 1994). Certification is usually renewed once or twice per year. In general, in the case of egg purchase, the broodstock would be sampled and certified. In the case of fish purchase, a subsample of fish would be examined for certification. Other options are maintenance of a pathogen-free brood stock on site and/or use of quarantine (described at the end of this section) before fish are introduced to the production system.

> **"RULE OF THUMB"**
>
> Every producer that transports fish directly from a pond into a recirculating system will experience catastrophic losses from infectious disease outbreaks.

Even if certified eggs are purchased, they should be disinfected upon arrival at the facility. The U.S. Food and Drug Administration (FDA) regulates the use of egg disinfectants in fish produced as food. Before using an egg disinfectant, verify that it is non-toxic for the intended species. Iodophors (organic iodine complexes with 1% active iodine) are generally used to disinfect salmonid eggs by creating a concentration in the water being treated of 100 mg iodophor/L water (100 ppm) for ten minutes after water hardening (Amend and Conte, 1982; Federal Joint Subcommittee on Aquaculture, 1994). As an example of how to treat 10 L of eggs and liquid, the following sample calculation is provided:

As purchased, 1% active iodine is 10,000 ppm (mg/L). Thus, 0.1 L (or 100 mL of Iodophor) will create a 100 mg/L ppm concentration in the 10 L of egg/liquid solution:

$$0.1L_{Iodophor} \cdot \frac{10{,}000mg_{Iodophor}}{L_{Iodophor}} = \frac{100mg_{Iodophor}}{L_{solution}} \cdot 10L_{solution}$$

One advantage of iodophors is the color change that occurs as it loses its effectiveness. Disinfection occurs while the amber color is present. Yellow or colorless iodophor is no longer effective.

When purchasing eggs and fish, the number of different suppliers should be kept to a minimum to reduce the risk of pathogen introduction. Each supplier should be visited to determine whether the farm practices satisfactory biosecurity. If not, then fish should not be purchased from that source. One indication that the supplier is concerned about biosecurity is if a visitor is provided with pre-visit instructions, e.g., shower, clean clothes and clean footwear before arrival at the facility, no other facility visits the same day, and no contact with other aquatic environments.

Feed

Pathogens may be introduced into a recirculated system along with the fish feed. Commercial dry feeds are processed at high temperatures of about 160–180°F (71–82°C) for steam-pelleted, 180–200°F (82–93°C) for expanded and 220–350°F (104–177°C) for extruded feed, so pathogen introduction from this source is unlikely. However, as each bag (or lot) of feed is used, the lot number, and date manufactured and used, should be recorded in case trace back of feed needs to be carried out. Feed should be stored according to manufacturer's recommendations, usually stacked on a pallet with air circulation on all six sides. To avoid fish health problems related to rancidity or mycotoxins, feed should be used within the time recommended by the manufacturer.

Introduction of pathogens through live food presents a serious risk of contamination. All live food should be cultured in specific-pathogen-free conditions and should never be used from natural aquatic environments, e.g., ponds.

Staff and Visitors

Pathogens can be carried into a facility by staff or visitors (human or animal). An employee and visitor parking area should be established on the periphery of the facility's grounds. Vehicles from an aquaculture facility, e.g., hauling trucks, should be disinfected upon arrival. The number of tours should be minimized and the number of people in a tour should be kept to a manageable size (three to six people; large groups should be broken into smaller groups). Visitors should be recorded in a logbook that includes name, affiliation, date, time and purpose of the visit.

Before they arrive at the facility, all new employees and all visitors should be given detailed instructions on compliance with biosecurity requirements. Personnel working in or visiting the facility should be required to change into clean coveralls and disinfected boots before entering the culture area. Visitors that are not from an aquaculture facility should be required to put coveralls over clothing, put on disinfected boots and wash hands with antibacterial soap before entering the production area. Visitors from an aquaculture facility should be required to remove their clothing, put on clean coveralls and boots and wash hands and arms for one minute. Visitors should be instructed not to touch or lean against anything in the culture room. Barriers should be installed at the entry to areas that are off-limits, e.g., quarantine. After every tour, floors should be cleaned with disinfectant, e.g., 19 mL Roccal per gallon (5 mL/L) of water.

Employees should be discouraged from having aquatic pets at their homes and from working at another fish farm during non-work hours. They should not be allowed to bring pets into the fish culture area. Rodents, birds, other vertebrates and insects should be excluded from the fish culture area.

Foot Baths

Foot baths should be used at the entry to the production area (see Quarantine section for more detailed recommendations). They should be changed at least twice a week, more often if foot traffic is normally heavy. Foot baths should be changed after every tour and floors should be cleaned with disinfectant. See Fig. 13.1 showing Mike Timmons (one of the book authors) using the foot bath at the Freshwater Institute.

Figure 13.1 Entry Area with Foot Bath Prior to Facility Access.

QUARANTINE

Quarantine is the isolation of newly arrived fish. This isolation is imposed to prevent the spread of contagious disease to other fish in the facility. The quarantine facility should be designed for easy cleaning and disinfection. It should be a separate room or facility, not just a tank in the corner of the production facility. Waste discharge should be separate from the overall facilities systems and, if necessary, disinfected with either ozone or ultraviolet radiation (see Chapter 12).

Access to the quarantine facility should be restricted to a minimum number of people. No visitors should be allowed in the quarantine area. The area should be clearly marked with signs and directions as reminders to all company personnel and visitors. Quarantine equipment should be clearly marked and used only in the quarantine facility.

Work in the quarantine area should be saved as the last element of the work day. Personnel should wash hands and arms before going between the quarantine and production areas. Footwear should be disinfected. Clothing should be changed.

Traffic flow of personnel should be designed so that hand washing occurs as soon as personnel enter the area. Three sinks should be installed at the entry to the quarantine room. The first sink should be used for hand washing upon entry. Personnel should wash their hands with antibacterial soap, e.g., triclosan[3], for at least 30 seconds before they enter the quarantine area. After washing, hands should be dried with a clean paper towel and the faucet turned off with the paper towel before it is discarded.

---

[3] Triclosan has been used as an antiseptic since the 1960s. It blocks an enzyme, known as "FabI," that bacteria need to manufacture the fatty acids used in cell membranes. Because animals possess a very different set of enzymes, triclosan does not interfere with this process in humans. This has led to its widespread use in over-the-counter preparations used on the skin or in the mouth. Once inside the cell, triclosan poisons a specific enzyme that many bacteria and funguses need for survival. Triclosan blocks the active site of an enzyme called enoyl-acyl carrier-protein reductase (ENR for short), preventing the bacteria from manufacturing fatty acids it needs for building cell membranes and other vital functions. Humans don't have this enzyme, so triclosan is harmless to them. One molecule of triclosan permanently disables an ENR molecule, which explains why triclosan has powerful antibiotic action even at very low concentrations. Triclocarban's structural similarity suggests a similar mode of action.

The middle sink should contain disinfectant, e.g., 200 ppm chlorine, and be used to disinfect buckets, nets, meters and other equipment.[4] The third sink is used to rinse equipment after it is disinfected. (Regardless of the disinfectant used, it should be treated appropriately before discharge so that aquatic animals and plants are not harmed by the compound.) The doors and sinks should be located at a spot convenient to walk back and forth from tanks to the cleaning area.

Two disinfectant foot baths should be installed at the entry into quarantine. These foot baths should contain disinfectant, e.g., Roccal® at 19 mL per gallon of water, and should be filled at least 2 inches (5 cm) deep. The foot baths must be cleaned and changed as often as necessary to maintain efficacy; at minimum once per week.

Upon arrival, fish should be examined and all (not just a sample) of the shipment placed into quarantine. Fish should arrive in clean, debris-free shipping water and should be at least average in length and weight for their age, have normal skin color and no lesions on the skin or fins. Fish should be feeding and behaving normally within 24 hours after arrival.

An examination for parasites that includes wet mounts of skin scrapings and gill biopsies should be conducted the day of arrival. To determine which, and how many, bags should be sampled, the supplier should be asked if the fish were all collected from the same rearing unit. For each "lot" of fish, sample at least six fish with normal appearance and six fish with abnormal appearance. Throughout the period in quarantine, moribund fish should be examined for parasites and cultured for bacteria and viruses to determine whether pathogens are present that could threaten the remaining population of apparently healthy fish.

The quarantine period is often cited as thirty days. However, quarantine length for an individual facility could be greater or less than 30 days, depending on the species, age, source, and purpose of the fish. It should also account for incubation periods and development times for the pathogens that are known to present a risk, pathogen life cycles and expression of clinical disease in warmwater vs. coldwater conditions. Regardless of the quarantine period chosen, the addition of any fish to ongoing quarantine resets the clock to zero (Harms, 1993).

One objective of quarantine is to increase the probability that, if the fish are infected with pathogens, an outbreak will occur before fish are moved into the production system. Replication time for bacteria, viruses, protozoa and other pathogens is temperature-dependent. Some

---

[4] One liter of 200 mg/L available chlorine is neutralized by 1.5 g of sodium thiosulfate.

recommend maintaining fish at low densities to minimize stress; this practice will not create conditions that allow expression of disease agents that may be present. Consider keeping the water temperature within the optimum range for the targeted pathogen to speed up pathogen life cycles. Expose the fish to the same conditions, e.g., density, feeding, handling, they will encounter in the production systems, so that a problem may be detected before the fish are moved out of quarantine. A subsample of fish can be stressed by exposing them for short periods to low DO concentrations, handling and/or disturbance such as bright lights or motion outside tanks. These conditions will increase the likelihood that an infectious disease outbreak will occur.

Some protozoal pathogens, e.g., Ich, have a life cycle where some stages occur on, and some occur off of (free-living), the fish. In these circumstances, fish can be transferred to a new tank in order to leave behind the free-living stage and reduce the number of parasites that are available to continue the infestation.

If an outbreak should occur while fish are in quarantine, they may or may not be moved into the production system. The decision whether to use these fish in production will depend on the pathogen involved, options for treatment and the ability to detect whether the pathogen is still present after treatment. For example, if a *Trichodina* sp. (ciliated protozoal pathogen) outbreak occurs, it can probably be eliminated by extension of the quarantine period and formaldehyde treatments followed by assessment of the remaining pathogen load through fish sampling and microscopic examination. However, if a viral disease outbreak occurs, there are usually no treatment options, elimination of the viral agent would be impossible and fish would not be moved to the production system.

Prophylactic antibiotics should NOT be used as part of a quarantine protocol. Prophylactic antibiotic use is illegal, will not eliminate pathogens from the fish, and can result in the development of bacteria resistant to the antibiotic.

About one to two weeks before fish are to be moved to the production system, environmental conditions should be gradually changed to mimic conditions in the production system. Alternatively, to allow fish to acclimate, production system water could be introduced before transfer to the production system.

13.2 PRACTICES TO REDUCE PATHOGEN SPREAD

METICULOUS HUSBANDRY

Meticulous husbandry will reduce the risk that pathogens will spread throughout a facility. Foot baths should be installed at the entry to the culture areas (see Quarantine section) and should be cleaned and changed frequently. Personnel should wash their hands with anti-bacterial soap routinely and frequently. Disinfectant and rinse areas should be readily accessible for disinfecting buckets, nets, dissolved oxygen meters, thermometers, and other equipment. Cleaned equipment should be placed in a designated location in an equally clean area. For example, a disinfected net could be placed on a hook or a clean brush could be placed into a disinfected trashcan for storage until the next use. Fish dip nets near tanks should be stored in a disinfectant rinse (see Fig. 13.2).

Decomposing debris provide a substrate for opportunistic pathogens to flourish. Therefore, tanks should be kept free of uneaten feed, feces, algae, and aquatic plants. Sumps, inflow and outflow pipes, aerators, spray bars, and any other equipment inside the tanks should be cleaned frequently. All parts of the system should be inspected, and cleaned if necessary, at least once per month.

Figure 13.2 Fish Dip Nets Near Tanks Should Be Stored in a Disinfectant Rinse. Note: use of life preserver on model Margaret Timmons, since tank depth is 3 meters deep.

CULLING

To reduce the transmission of pathogens from fish to fish, dead and sick fish should be culled. At minimum, fish should be culled once a day. If possible, they should be removed on a continuous basis. Culled live fish should be humanely killed with an overdose of an anesthetic, e.g., tricaine methane sulfonate (MS-222) and not allowed to die from suffocation.

DISINFECTION OF TANKS AND EQUIPMENT

Empty tanks and equipment should be disinfected with 200 ppm chlorine for at least 1 hour. The chlorine should be neutralized with sodium thiosulfate, and rinsed before being used for a new group of fish.

> **"RULE OF THUMB"**
>
> Chlorine, an effective disinfectant, **will kill fish** at even very low concentrations. Chlorine should be neutralized with sodium thiosulfate. One liter of 200 mg/L available chlorine is neutralized by 1.5 g of sodium thiosulfate.

Floors should be cleaned regularly. A clean pesticide sprayer may be used to spray the floors with disinfectant. The disinfectant should be allowed to remain for at least 15 minutes before rinsing off. When spraying disinfectant and rinsing the floor, the spray should be kept as low to the floor as possible to avoid propelling dirt and other debris into the air.

CULTURE ACTIVITIES

Culture activities should be scheduled to minimize the number of different personnel working with a group of fish. If personnel resources are limited, unaffected tanks should be worked on before affected tanks. Younger fish should be cared for before older fish. Equipment that has touched the floor should not contact fish culture water. Any equipment that touches the floor or any contaminated surface should be disinfected before use in the culture system. Fish that have jumped from a tank onto the floor should not be returned to the tank. They should be humanely killed with an overdose of tricaine methanesulfonate (MS-222).

Aerosol Transfer

Managers and designers of aquaculture systems, especially those with tanks aligned in close proximity should be aware that nets, boots, hands, fish, feed or water are not the only source of potential contamination. Aerosols can breach even well designed, isolated quarantine systems. Recent studies have shown that fish pathogens can be spread via the airborne route as aerosol/droplet sprays and can be detected in water downwind from an experimentally generated aerosol/droplet spray (Wooster and Bowser, 1996) and subsequently infect fish (Bishop, Smalls, Wooster and Bowser, 2002). This implies that fish pathogens can be spread via aerosol/droplet spray and could therefore move from tank to tank within an aquaculture facility, if a disease outbreak occurs in one tank. Solutions to help prevent the potential aerosol spread of pathogens include covers for aquaria or tanks, free standing or hanging wall barriers between tanks and changing ventilation systems to limit spray or spread of aerosol plume. It has been reported in wastewater treatment plant literature (Brandi et al. 2000) that using diffusion aeration instead of mechanical agitation reduced dispersal of airborne pathogens from a wastewater treatment plant.

13.3 Reducing Susceptibility to Infection and Disease

Stress associated with handling, low water flow, poor nutrition, poor water quality and other husbandry characteristics will render fish more susceptible to, and aggravate the consequences of, infection with opportunistic and obligate pathogens. There are many strategies that can be used to increase fish vigor.

Optimum Nutrition

Poorly nourished fish are more susceptible to disease. The fish feed schedule and feed characteristics should be such that the fish receive the best nutrition possible.

Control Quality of Fish Stock

Eggs, fry and fingerlings should be purchased from optimum year class brood stock.

Reduction of Fish Stress

A wide variety of pathogens will infect cultured fish. Many disease agents are naturally present in low numbers in soil and water and do not cause problems because the fish are protected by natural defense mechanisms, i.e., undamaged skin, mucus, and cellular components of the immune system. However, when fish already crowded in culture operations are further stressed, their natural disease defense systems may be weakened, which reduces the ability of the fish to protect itself against infectious diseases. Catastrophic mortality that results from an infectious disease outbreak is often the result of, and response to, a stressful experience. Most infectious disease outbreaks can be avoided by proper management.

Culture and harvesting conditions and procedures should be designed to be as stress-free as possible. Poor water quality is a significant stressor that makes fish more vulnerable to disease outbreaks. Poorly nourished fish are also more susceptible. Light (excessive or rapid changes in intensity), noise and movement can stress fish and should be minimized. When netting fish, small rather than large net scoops should be taken. When grading, crowding should be done gently, without keeping too many fish together at one time. Fish must be handled during routine maintenance, stocking, and harvest. When fish are removed and processed, e.g., weighed, transported, they compensate physiologically, but there are limits. To reduce the trauma of handling, make sure all necessary materials, e.g., nets, hauling tanks, scales, and adequate personnel are immediately available. The use of salt (0.1–0.5% by weight in the water, see section 13.6 Treatment), aeration or oxygenation, and anesthetics (use according to FDA regulations) can reduce the stress associated with handling. Handle the fish gently and for as short a time as possible. If possible, do not handle fish that are already stressed or when culture conditions are marginal.

Vaccination

Vaccination should be used to help prevent infectious disease outbreaks. Vaccines are available in immersion or injectable form for many fish diseases.

13.4 Monitoring and Surveillance

By the time the major symptoms of poor fish health are noticeable, mortality has often begun and the problem is more difficult to solve.

Therefore, the key to effective fish health management is a good program of monitoring, surveillance and response. Monitoring is an important part of early identification, isolation and treatment of a problem. If possible, this program should be established with the help of an aquaculture veterinarian.

How monitoring will be accomplished should be considered early on in facility design. Observation should be carried out daily. Fish condition and behavior can be observed through tank windows that are located to allow observation of areas where weak, sick fish normally congregate (at the inflow, in quieter areas of the tank). Culled fish should be periodically assayed for pathogens. Records on growth and feed conversion can be used to detect problems that are costing money, but haven't resulted in mortality.

Routine monitoring of water quality in a production system is imperative. When abnormal behavior is observed, water quality should be checked. If water quality is a problem, e.g., low dissolved oxygen, high unionized ammonia, corrective measures should be initiated quickly to avoid more severe problems later. In an RAS, exchange of extra water can be very helpful ("...the solution to pollution is dilution..."). If abnormal behavior persists for more than two days, or death occurs, the farmer should seek professional assistance. In addition to a description of the history of the problem, water quality data should be provided to the veterinarian and/or diagnostic laboratory.

Even if fish recover quickly after correction of a water quality problem, the producer should think about the possible consequences days to weeks after the event. Behavior and feeding should be carefully monitored. Once or twice a week, a microscopic examination of gill tissue should be carried out to ensure that gills are recovering. Dying fish should be examined and should be cultured for pathogens.

Record Keeping

Record keeping is an important part of a fish health management program. Records should include number of dead and culled fish removed from culture tanks, observations of abnormalities, laboratory results and methods and results of treatment. This information can be used to improve biosecurity protocols and help a facility move toward an effective program of disease prevention and control.

Water Quality

The quality of the culture water has a major effect on fish health. Each species has an optimum range for water quality parameters. Some species are more tolerant than others when water quality deviates from this range. For example, tilapia are more tolerant than rainbow trout of reduced oxygen, high total suspended solids and elevated ammonia concentrations. In a properly managed recirculated system, dissolved oxygen and temperature remain relatively constant throughout the day and growing season, but alkalinity and hardness can drop to dangerously low levels due to consumption of inorganic carbon during the nitrification process (see Chapter 7). In an improperly functioning or overstocked recirculated system, dramatic and rapid changes in dissolved oxygen, ammonia, or nitrite concentrations can result in high mortality of cultured fish or create stresses that can result in poor fish performance for weeks to months after the event.

Behavior

Abnormal behavior is often the first indication of an impending fish health problem. Personnel should become familiar with the normal behavior and appearance of the fish species in culture. All behaviors, including feeding and swimming activity, and response to sudden movement outside the tank must be carefully observed. The fish producer must learn to distinguish nuances in behavior. For example, fish may normally move abruptly when the tank is approached. If they are sick they may not move at all or move abruptly and then become listless.

Healthy fish display "normal" behavior. For example, fish should feed vigorously when food is presented or shortly thereafter. In some culture tanks fish are usually invisible, except when feeding. In this environment, feeding behavior should be observed when automatic feeders are operating.

If feeding rates decline only slightly, fish may appear to be feeding normally. Such declines in feeding rate can be detected by closely monitoring growth rates and by checking for uneaten feed. Growth curves can be used to detect problems at an early stage.

Depending on the species, fish naturally swim in schools or alone. Distribution in tanks will vary by species, but is usually consistent within a species, e.g., some species prefer covered areas while others prefer uncovered areas; some concentrate toward the water inflow while others distribute themselves around the tank.

Behavioral and Physical Abnormalities

Table 13.1 lists abnormalities that may be observed when fish are sick. These signs will aid in the diagnosis of the cause of a problem.

They should be monitored during daily fish observation, while removing dead or dying fish, and during sampling, grading, or carrying out any activity that creates an opportunity to examine fish closely.

Table 13.1 Fish Behavioral and Physical Signs for Stress and Sickness

Fish Behavior	Signs to Observe
Movement	Weak, erratic or lethargic swimming Increased or decreased reaction to external stimuli such as noise or movement Scratching, flashing, or rubbing against tank walls or bottom Twitching, darting, spinning or jumping out of the water Crowding near the influent water supply Swimming upside down Gasping at the water surface
Feeding	Not feeding Reduced feeding (detected by growth curves as well as observation)
Breathing	Decreased rate of opercular movement Increased rate of opercular movement
Physical Condition	Visible lesions or sores Cloudy eyes Protruding eyes Gills swollen, white, pink or pale red, eroded, puffy, bloody, brown Scale loss Swollen abdomen Excess mucous on the skin and/or gills (also, check for excess mucus on tank screens) Spots or fungus on skin Unusual colorations on body surface, including red swollen areas, gray or yellow lesions Flared opercula (gill covers) Frayed fins or tail Bubbles in eyes or in skin

Examination of fish gills is a useful, often underutilized, method for early detection of fish health problems. Gill tissue is a sensitive indicator of poor water quality conditions, i.e., elevated ammonia and nitrite concentrations, is susceptible to attack by parasitic and bacterial pathogens, and provides an entry point for systemic infection by pathogens. Fish producers should become familiar with normal gill appearance and carry out weekly microscopic examination of gill samples. Producers who have adopted this practice have found that gill examination is a worthwhile investment in time in return for the information it provides.

> **"RULE OF THUMB"**
>
> Gill examination is a worthwhile investment in time in return for the information it provides. Know what a "normal" gill should look like!

13.5 DIAGNOSIS

In the event of a fish health problem, an accurate diagnosis is essential. Each fish producer should locate an aquaculture veterinarian who can make a visit to the farm when a fish health problem occurs. A correct diagnosis will determine the treatment regimen, including whether (and which) chemotherapeutics should be used. Attempts to diagnose and treat a problem based on hunches is usually not effective and misdiagnosis will result in wasted money and time, and further degradation in the condition of the fish. An incorrect diagnosis will also prevent the producer from developing an effective strategy to prevent a recurrence of the problem. If a veterinarian is unable to visit during a problem, then live fish may be shipped to a diagnostic laboratory, but the producer should be aware that without an onsite evaluation the recommendations will be unlikely to include a plan for prevention and control of future outbreaks.

Before shipping any fish off site for further diagnosis, the disease diagnostic laboratory should be called. Describe to the laboratory personnel the disease signs that you have observed and determine how they want you to prepare your fish for shipment. Do not expect a diagnosis over the telephone.

By informing the laboratory personnel of your problem and accurately answering their questions, you can facilitate a rapid and accurate diagnosis. The more information that you can provide to the diagnostic laboratory, the better the evaluation of your disease case. Some general guidelines in specimen collection, preparation, and shipment follow.

SPECIMEN QUALITY

The quality of specimen submitted to a fish disease diagnostic laboratory will affect the quality of both the diagnosis and recommendations for corrective action. Dead fish are of little to no value for disease diagnosis because:

- Once death occurs, fish decompose very rapidly. If the disease is caused by a bacterium, other bacteria that grow during the normal decomposition process can quickly overgrow the pathogen and make identification difficult or impossible.

- Parasites require a live host for survival. When a fish dies, the parasites often quickly leave the fish in search of another live host.

- Viruses can be degraded by the compounds that are produced during decomposition. Once the fish dies the viruses may survive for only a limited period of time, sometimes only a few hours.

The veterinarian or diagnostic laboratory can provide guidance regarding the size and number of live fish to be shipped. The best way to transport sick fish to the diagnostic laboratory is for the producer to deliver them directly. Direct transport will provide optimum specimen quality, and will provide the fish health specialist with the opportunity to talk to the producer about the circumstances surrounding the mortalities. If requested, the producer should also bring a water sample from the culture system. This water sample should be collected in a clean container that can be capped tightly. If a chemical contaminant is suspected, the sample must be collected in a glass jar (not plastic) and handled according to instructions provided by the laboratory.

PACKING AND SHIPPING THE SPECIMENS

If fish must be shipped, the following methods (from most to least desirable) can be used:

- Live fish (most desirable)

- Iced fish
- Frozen fish
- Formalin fixed fish (least desirable)

Most diagnostic laboratories prefer specimens that are not preserved in formalin because the formalin will also fix (kill) pathogenic (disease-causing) microorganisms. Reaching a diagnosis may depend on the ability of the diagnostic laboratory to culture a bacterium or virus from the fish. Culture can be done only if the bacteria or viruses are still viable. In addition, a very important aspect of recommending a treatment for bacterial diseases is to determine the antibiotic susceptibility of the bacterium, which requires culture of the microorganism in the laboratory. Formalin-fixed materials can yield important diagnostic information following histological examination, but because of its limitations, most diagnostic laboratories prefer materials from which they can culture pathogenic microorganisms. This issue should be discussed with the diagnostic laboratory.

Fish shipped by any of the above methods should be collected from live fish in the system that are showing disease signs. It is best to collect the fish with a net or trap. Capture of fish by rod and reel will select individual fish that are still feeding actively; the healthiest fish in the population. One of the first signs of many diseases is that fish stop feeding. Accurate diagnosis of the disease may not be possible as these fish may not yet be infected.

Live Fish

1. Obtain a strong, waterproof, insulated shipping container, e.g., a disposable Styrofoam cooler in a sturdy cardboard box, (see Fig. 13.3).

2. The veterinarian and diagnostic laboratory can provide advice about the ratio of fish to water volume. However, in order for sufficient oxygen exchange, the shipping bag should be no more than 1/3 full of clean water from the culture facility. Place the bag in the shipping container and add the fish. Fill the bag with pure oxygen or air. Seal the bag by twisting the open end tightly shut and securing it with several heavy-duty rubber bands or plastic tie-downs. An air tight seal is essential.

3. Water temperature during transport should be maintained within the species tolerance. For coldwater fish, place 3–5 pounds (1.5

to 3 kg) of crushed ice (or ice packs) in a strong plastic bag, seal the bag as described above, and place it in the shipping container next to the bagged fish. Warm water fish should not be packed with ice, and in cold weather hot packs may need to be included in box.

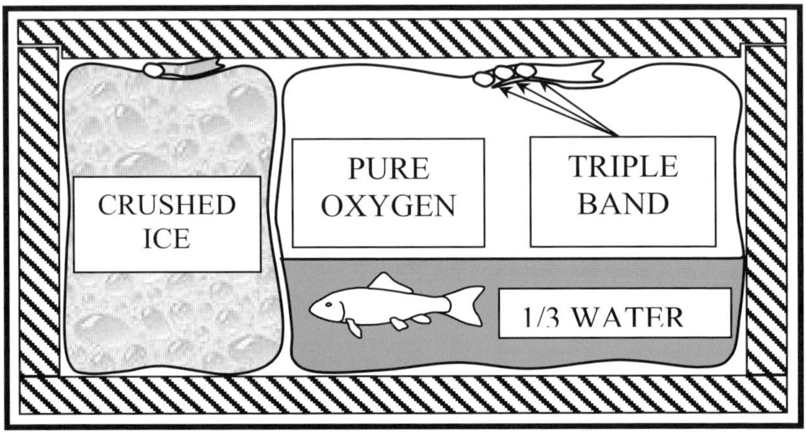

Figure 13.3 Shipping of Live Fish.

4. In a separate, small plastic bag place a note that includes name, address, telephone number, and information describing the fish and the culture system from which they came, e.g., why you suspect a disease; number of mortalities and their appearance; approximate size of diseased fish relative to other fish being cultured; when and how the shipped fish were collected; stock density; any known water quality parameters. Place the bagged note inside the shipping container.

5. Seal the shipping container. Be sure to indicate which end is "up" and that live fish are enclosed.

CAUTIONS: Take extra care to make sure the container won't leak. "Double bagging" can sometimes help. Ship via a carrier that can provide overnight delivery. It is always best to contact the fish diagnostic laboratory prior to any shipment and to coordinate the receipt of the fish with them.

Iced Fish

1. Obtain a strong, waterproof, insulated shipping container, e.g., a disposable Styrofoam cooler in a sturdy cardboard box, (see Fig. 13.4).

Figure 13.4 Fresh Fish Packed for Shipment.

2. Wrap each fish individually with several sheets of newspaper to prevent freeze burns that may obscure signs of disease that could be diagnostically important. Place each fish in a separate plastic bag and seal the bag.

3. Place a larger, strong plastic bag in the shipping container and fill the bag with 2–4 inches (5 to 10 cm) of crushed ice.

4. Place the individually bagged fish on the crushed ice in the larger bag and cover them with an additional 2–4 inches (5 to 10 cm) of crushed ice. Seal the larger bag by twisting the open end shut and securing it with several heavy-duty rubber bands or plastic tie-downs. An air tight seal is essential.

5. In a separate, small bag place a note that includes your name, address, telephone number, and information describing the fish and the culture system from which they came, e.g., why you suspect a disease; number of mortalities and their appearance; approximate size of diseased fish relative to other fish being cultured; when and how the shipped fish were collected; stock

density; any known water quality parameters. Place the bagged note inside the shipping container.

6. Seal the shipping container. Be sure to indicate which end is "up" and that iced (perishable) fish are enclosed.

CAUTIONS: Adequate amounts of crushed ice, usually 10–15 pounds (5 to 7 kg), will be satisfactory to keep the fish chilled during shipment. Ship via a carrier that can provide overnight delivery.

Frozen Fish

1. Obtain a strong, waterproof, insulated shipping container, e.g., a disposable Styrofoam cooler in a sturdy cardboard box, (see Fig. 13.5).

2. Place each fish in an individual plastic bag and seal the bag. Freeze the fish in the individual plastic bags.

3. Place a larger, strong plastic bag in the shipping container and fill the bag with 2–4 inches (5 to 10 cm) of crushed ice.

Figure 13.5 Frozen Fish Packed for Shipment in Styrofoam Box.

4. Place the individually bagged, frozen fish on the crushed ice. Cover the fish with additional crushed ice and tightly seal the bag by twisting the open end shut and securing it closed with strong rubber bands or a plastic tie-down. If possible, use dry ice to insure that the specimens do not thaw during transit. Five

pounds of dry ice will normally keep specimens frozen for 24 to 36 hours, if the shipping container is well insulated.

5. In a separate, small bag place a note that includes your name, address, telephone number, and information describing the fish and the culture system from which they came, e.g., why you suspect a disease; number of mortalities and their appearance; approximate size of diseased fish relative to other fish being cultured; when and how the shipped fish were collected; stock density; any known water quality parameters. Place the bagged note inside the shipping container.

6. Seal the shipping container. Be sure to indicate which end is "up" and that frozen (perishable) fish are enclosed.

CAUTIONS: Check with the commercial carrier for their policy regarding shipment of packages containing dry ice. Ship via a carrier that can provide overnight delivery.

Formalin Fixed Fish

1. Make a 10% formalin solution. Neutral buffered formalin is best. Under practical field conditions, water from the culture facility will usually provide adequate buffering capacity to the solution. (To prepare the desired formalin solution, mix 9 parts water with 1 part formalin). Or, 10% neutral buffered formalin can be purchased and used directly from the bottle.

2. Kill the fish before placing them in the formalin solution. This can be done with an "overdose" of MS-222 (tricaine methane sulfonate at 1 gm per 500 mL H_20). It is important that the fish be rapidly "fixed" by the formalin so that the quality of tissue preservation will yield useful information. Normally, formalin can rapidly fix tissues that are less than 1/2 inch (12 mm) thick. For this reason, the abdomen of larger fish must be opened for its entire length with one continuous cut. Most fish disease diagnostic laboratories will prefer to have the entire fish shipped rather than a limited number of tissues or organs. An "apparently normal organ" may yield valuable diagnostic information when examined microscopically. Questions regarding shipment of whole fish or selected tissues should be directed to the diagnostic laboratory before samples are sent. It is also important that adequate amounts of formalin be used to

preserve the tissues. As a general rule, the **ratio of formalin solution to tissue must be 10:1** by weight or volume; i.e., 1000 mL of the formalin solution: 100 gm fish.

3. The container with the formalin and the tissue must be tightly sealed (see Fig. 13.6). Care should be taken to prevent breakage of the container. Glass containers can break and should be avoided. Use plastic bottles such as empty, clean food containers, e.g., peanut butter, mustard, salad dressing — use food service size for large fish, or containers obtained from scientific supply companies. Seal the container with tape to prevent leakage.

4. The sealed container should be placed in a shipping container filled with Styrofoam pellets or other suitable packing material. Care should be taken to prevent breakage.

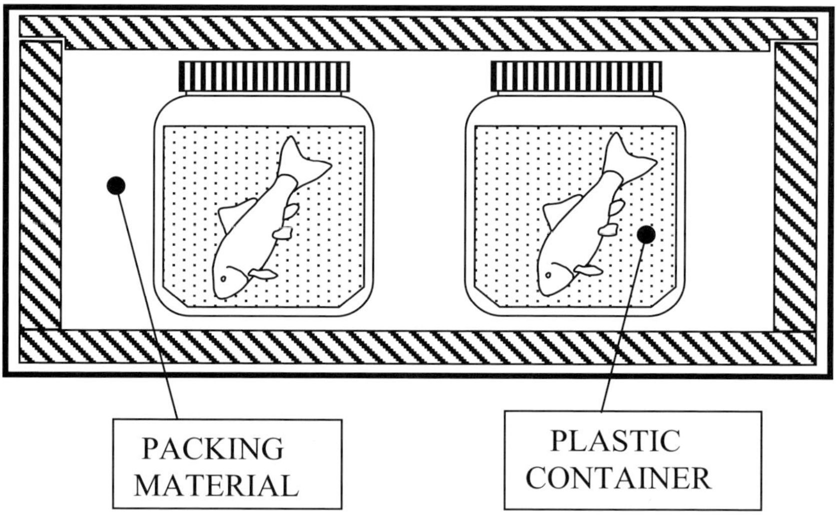

Figure 13.6 Formalin Fixed Fish Packed for Shipment.

5. In a separate, small bag place a note that includes your name, address, telephone number, and information describing the fish and the culture system from which they came, e.g., why you suspect a disease; number of mortalities and their appearance; approximate size of diseased fish relative to other fish being

cultured; when and how the shipped fish were collected; stock density; any known water quality parameters. Place the bagged note inside the shipping container.

6. Seal the shipping container. Be sure to indicate which end is "up" and that preserved fish are enclosed.

CAUTIONS: Formalin is irritating and toxic. Provide and use good skin and eye protection as well as good room ventilation when using this chemical. A good practice when handling formalin or any other potentially irritating or toxic chemical is to use rubber gloves to protect the hands, goggles to protect the eyes, and a breathing mask to reduce inhalation of the chemical.

SUMMARY OF SHIPPING METHODS

The manner in which fish specimens are prepared and shipped will affect the quality of the information that can be obtained when the specimens are examined by the fish health specialist, Table 13.2. Working with a live fish will provide the diagnostician the best opportunity to gain useful information regarding the fish disease. The fish can be examined for live parasites. Microorganisms (bacteria and viruses) can be cultured from specimens that are delivered to the diagnostic laboratory alive and the sensitivity of bacterial pathogens to potential treatment chemicals can be determined. A histopathological examination can be performed on properly prepared organs and tissues. Tissue changes that are indicative of disease can be identified. It is critical that the tissues be preserved properly to insure that they represent the disease process and not decomposition after death.

Table 13.2 Impact of Fish Handling and Preservation Method on Disease Diagnosis

Shipment Method	Parasitology	Bacteriology	Virology	Histopathology
Live	+++	+++	+++	+++
Iced	+	++	+++	+/–
Frozen	–	++ / +	++ / +	–
Formalin Fixed	+ /–	–	–	+++

Legend:

+++	no effect, excellent specimen for examination
++	negligible effect, good specimen for examination
+	moderate effect, specimen may be usable
+/–	substantial effect, specimen may not be useful
–	dramatic effect, specimen not useful

Specimens that are shipped on ice, or frozen, have some important diagnostic limitations. Living parasites may or may not be present on iced fish. When frozen fish are thawed, the shearing action of melting ice crystals will destroy parasitic protozoa, making their identification very difficult, if not impossible. The process of freezing and thawing tissues will also create a great deal of damage, making tissues of limited or no use for histopathological examination. Iced and frozen specimens are normally satisfactory for the culture of bacteria and viruses.

Specimens shipped in formalin are useful for histopathological examination as long as the fish were collected live and were carefully preserved prior to shipment. As mentioned above, the diagnostician must be able to detect tissue changes that are caused by the disease process and not the result of decomposition after death. This can only be done with a carefully processed sample. The formalin fixed sample may not yield the identity of the disease organism and it will not provide information regarding the antibiotic sensitivity of a bacterium.

13.6 TREATMENT

Unless a veterinarian is involved with the investigation, the diagnostic laboratory will contact the farmer with the results of the laboratory tests. The disease will be identified and an appropriate and approved treatment or action will be recommended. In some cases, a change in management techniques will be necessary. In other cases a recommendation to add an antibiotic to the feed (for internal bacterial infections) or a chemical to the water (usually for external parasite infestations) will be made. It is extremely important that the producer follow the recommendations of the veterinarian or diagnostic laboratory and take appropriate precautions before a disease treatment is applied.[5]

One of the most important aspects of chemical treatment is knowing which chemicals are approved for use in food fish production. Each approved chemical can be effectively and legally used to treat a few to several diseases. No one chemical is appropriate for all diseases or situations. For instance, an antibiotic can be very effective in the treatment of a bacterial infection, but is useless if the disease is caused by

[5] Culling the sick fish is an alternative to treating with chemotherapeutics. When deciding whether to cull, treatment costs should be considered. Costs include the compound, labor during treatment and follow-up, risk of exposure to the other fish, additional losses of fish after treatment, possible decreased growth rates and the withdrawal time (waiting period) before fish can be sent to market.

a protozoan parasite. All chemicals have precautions and considerations associated with their use. If a farmer has no experience with a particular chemical, a small group of fish should be treated first, as a test before the entire lot is treated, to avoid potentially heavy losses due to chemical toxicity. Extreme caution should be practiced when applying any chemical treatment. Water quality influences the toxicity of certain chemicals and is adversely affected by some chemicals. The producer should be knowledgeable of the water quality in the culture facility. Of particular interest are dissolved oxygen, alkalinity, and the amount of organic material in the water.

Concentration, length of treatment, and number of consecutive treatments are all important factors to consider when treating fish and calculating the amount of chemical to purchase. Check and double-check calculations. Differentiate between volumetric and weight based calculations, e.g., mL/L or mg/L. Calculations must be done for the active ingredient. Flow rate or water volumes are critical numbers for accurate dosing. Accuracy of measurements should be confirmed to help determine whether or not treatments are being done at effective levels. Know drug safety information for humans (from the Material Safety Data Sheet (MSDS) for the compound) and for fish. For example, antibiotic overuse or skimping on dosages can result in bacterial resistance. Overdosing of antibiotics and sulfas can cause kidney damage in fish. Dilute the chemical before introduction into the tank, as this makes it safer to administer and helps distribute the chemical throughout the water more quickly to prevent overexposure of individual fish as the treatment is introduced into the water.

Fish should be taken off feed for 24 to 48 hours before treatment and, in most cases, should be kept off feed while it is carried out. This practice will reduce waste production and oxygen consumption during treatment. Fish should not be handled for 24 to 48 hours post-treatment.

In general, younger fish are more sensitive to chemical treatments than older fish. When there is any doubt, the treatment should be tested on a subsample of fish before treating the whole population. Treatment should be discontinued and fresh water added if it results in thrashing, jumping, listlessness or any unexpected behavior.

Water quality will be affected by, and will affect, the treatment. For example, D.O. concentrations can decline during formaldehyde treatments. Temperature, pH, dissolved oxygen, alkalinity, organic loading and salinity can decrease efficacy or increase toxicity of a given treatment chemical.

Table 13.3 Treatment Effects on Biofilter Function (Noble and Summerfelt, 1996)

Drug or Chemical	Treatment	Impaired function?
Benzalkonium chloride	2 ppm	yes
Chloramine-T	12 ppm	no
Chloramphenicol	50 ppm	no
Chlorotetracycline	10 ppm	yes
Copper sulfate	1.0 ppm	no
Copper sulfate	5 ppm	no
Copper sulfate	0.75 ppm for 24 hours followed by 0.5 ppm for 30 days	no
Copper sulfate	0.25 ppm for 24 hours twice at 3 day intervals, then a third one at 0.50 ppm, then one with 0.30–0.38 ppm continuously for 30 days	no
Erythromycin	50 ppm	no
Formalin	low conc.	yes
Formalin	1:4000	no
Formalin	25 ppm in 3 indefinite treatments on alternate days	no
Formalin	15 ppm	low
Formalin	149 ppm indefinite	yes
Formalin	50–167 ppm for 1 hour with several treatments	no
Formalin	15–120 ppm indefinite with several treatments	no
Formalin	1:4000 for 1 hour	no
Formalin	35 ppm for 24 hours	yes
Formalin	153 ppm for 40 min, 3 times	no
Formalin and malachite green	25 ppm formalin with 0.10 ppm malachite green on alternate days	no
Furanace	0.1 ppm	low
Hyamine-3500	1.0 ppm	no
Hyamine-3500	2 ppm	yes
Hyamine-3500	1 ppm	yes
Hydrogen peroxide	100 ppm	yes
Malachite green	1.0 ppm	no
Malachite green	0.10 ppm on alternate days	no
Malachite green	0.5 ppm	low
Malachite green	3.0 ppm	no
Methylene blue	5 ppm	yes
Methylene blue	1.0 ppm	yes
Methylene blue	1 ppm	no
Nifupirinol	1 ppm	no
Oxytetracycline	feed	no
Oxytetracycline	50 ppm	no
Potassium permanganate	4.0 ppm	no
Potassium permanganate	1 ppm	yes
Roccal™	0.067 ppm	no
Romet™	50 mg/day per kg fish for 5 days	no
Sodium chloride	3%	yes
Sodium chloride	0.5%	no
Sodium chloride	0.5%	no
Sodium chloride	1.5%	yes
Sulfadiazine	25 ppm	yes
Sulfamerazine	50 ppm	no
Sulfanilamide	25 ppm	yes
Terramycin	feed	no
Trichlorfon	1 ppm	Minimal harm

Some chemical treatments can affect biofilter function by killing the nitrifying bacteria. System design should include plumbing that allows the biofilter to be isolated from the fish culture vessels. If the chemical concentration could be harmful to the biofilter, the tanks should be flushed after treatment with fresh water prior to reestablishing flow to the biofilter. However, it is possible that an untreated biofilter could become a reservoir for the disease organism. Effects of various treatment compounds on biological filtration are summarized in Table 13.3.

If the biofilter cannot be isolated during treatment and the chemical could harm the biofilter function, the chemical could be used at a lower concentration for a more prolonged treatment period. However, with this method a chemical may remain in the water longer than may be tolerated by the fish. An alternative is to treat the system at the therapeutic dose for a shorter period with the assumption that the chemical will remain in the water at an effective dosage as it is recirculated, but will be eliminated before adverse reactions appear in the fish.

Note that not all of the above listed compounds are approved for use in food fish. Check current FDA regulations prior to use of the above compounds.

Units

The units of measure used in the following examples are primarily metric. Concentrations of chemicals are commonly expressed in terms of milligrams (mg) per liter (L) or parts per million (ppm). When making a chemical application in freshwater these two terms are functionally equivalent. One liter of water weighs 1 kilogram = 1,000 grams = 1,000,000 milligrams and 1 mg/1,000,000 mg (or 1 L) = 1 ppm.

Antibacterial compounds are added to the feed as a treatment for systemic (internal) bacterial infections. They are commonly referred to as rates. A generic expression is the weight of antibacterial compound per weight of fish per day for a specified number of days. This may be in terms of mg drug/kg fish weight/day. Historically, some antibacterial treatment rates have been expressed as a combination of English and metric units, e.g., g (gram) drug/lb fish weight/day. A list of conversion factors are provided in the Appendix to assist with calculations.

Methods

Methods for chemical treatment vary depending on the treatment compound, length and type of treatment, and the type of culture system. The supplier of the treatment chemical will provide instructions and precautions regarding its use. As a rule, to avoid adverse reactions between a chemical and metal, do not use metal tanks or metal measuring utensils when treating.

Dip

Dip treatments consist of dipping the fish or eggs into a chemical bath and then quickly removing them. Dips are generally used for treating external problems.

Flush

Flow through or flush treatment ensures that the water quality parameters such as ammonia and dissolved oxygen remain at acceptable levels during the duration of the treatment. This may be necessary for high density systems. The disadvantages of flush treatments are that (1) achieving adequate dilution for proper disposal may be a concern, and (2) the chemical costs can be high.

Baths

Bath treatments are preferred to flush treatments, because they require less of the chemical(s), and are easier to manage. Since no fresh water is added during the treatment (definition of bath treatment), water quality will degrade and supplemental oxygen should be added. If the fish show adverse reactions to the treatment, then water must be exchanged quickly. Drain the treated water out of the system as rapidly as possible, then add fresh water. This procedure will remove virtually all of the treated water and should not overstress the fish.

CHEMICAL STORAGE

Always carefully read and follow the storage instructions for the chemicals you obtain. Never use any treatment chemicals beyond their expiration date. This is for both your safety and the safety of the fish. For example, formalin, when exposed to temperatures below 40°F (4.4°C), will form a white solid precipitate, paraformaldehyde, which is very toxic to fish. At this point, the formalin becomes unusable and must be discarded.

13.7 AQUACULTURE CHEMOTHERAPEUTICS

An aquaculture veterinarian will be able to provide information about regulatory issues surrounding the use of chemotherapeutics (drugs and chemicals used for treatment of disease). The fish producer must be aware of the legal consequences of chemotherapeutic use. A good practice is to maintain only those chemicals that do have specific approval for aquaculture uses at the production facility. To an inspector, presence of non-approved chemicals at an aquaculture facility may imply use for fish treatment. Regulations concerning approved chemicals for use in aquaculture are continuously updated.

The use of these compounds in fish that are intended for human consumption is regulated by the US Food and Drug Administration. Before use, the producer should confirm that chemicals and drugs are approved for treatment of the target species at the intended dose and for the intended time interval.

Rules and regulations are described on the Center for Veterinary Medicine (FDA-CVM) website:

- http://www.fda.gov/cvm/index/aquaculture/aqualibtoc.htm

Veterinary biologics (vaccines, bacterins, diagnostics, etc, which are used to prevent or diagnose animal diseases) are regulated by the United States Department of Agriculture:

- http://www.aphis.usda.gov/vs/aqua/aquaphis.html

Environmental effects, i.e., discharge, of chemical treatments for food fish are regulated by the EPA (Environmental Protection Agency):

- http://es.epa.gov/oeca/ag/sectors/animals/anaqulaw.html#NPDES%20Permits

The FDA regulates chemotherapeutics at three status levels for drug approval: approved, not approved, and low-regulatory priority:

Level 1. Approved drugs have specific usage guidelines such as species, withdrawal time, and dosage and are only approved for specific manufacturers. Tricaine methanesulfonate (MS-222) and formalin are examples of approved drugs.

Level 2. Not approved or prohibited drugs may not be used in aquaculture. An example is malachite green.

Level 3. Low Regulatory Priority (LPR) drugs have specific applications under which they may be applied. There are five conditions that must be met when using these drugs:

- The substances are used for their listed indications.
- The substances are used at prescribed levels.
- The substances are used according to good management practices.
- The product is of an appropriate grade for use in food animals.
- There is not likely to be an adverse effect on the environment.

Note that any local laws may supercede national approval for drug usage. In addition, approval for use does not exempt the user from proper disposal, which must comply with national regulations such as NPDES (National Pollution Discharge and Environmental Safety) and state environmental regulations. Producers should become familiar with these regulations and formulate a compliant response before a fish health problem occurs.

Below are brief descriptions of some aquaculture chemotherapeutics. Approval status for each is listed with the compound along with precautions/considerations associated with their use.

TERRAMYCIN

Terramycin is an antibiotic used to treat systemic (internal) bacterial infections. It is approved by the U. S. Food and Drug Administration (FDA) for the treatment of sensitive bacteria of the genera *Aeromonas, Pseudomonas,* and *Hemophilus* in salmonids and catfish. It is used as a feed additive at a rate of 2.5 grams of drug (active ingredient) per 100 pounds (45.4 kg) of fish weight/day for 10 days. A 21-day withdrawal period is required before the fish may be slaughtered and used for human consumption. It is illegal for a producer to mix his/her own medicated feed.

SULFAMERAZINE

Sulfamerazine is an antibiotic used at one time for the treatment of furunculosis in salmonid fishes. It was used as a feed additive at 10 grams of drug (active ingredient) per 100 pounds (45.4 kg) of fish weight/day for 14 days. A 21-day withdrawal period was required before fish could be slaughtered and used for human consumption. (Note: Old fish health literature implies that sulfamerazine is an approved compound for use on food fish, it is not! Because many individuals were substituting a generic "sulfa drug" for sulfamerazine, the manufacturer allowed its permit for this drug to lapse. Therefore, it is **not legal** to use this drug for food fish at this time.)

ROMET-30

Romet-30 is approved for the treatment of furunculosis in salmonids and enteric septicemia in channel catfish. It is used as a feed additive at a rate of 50 milligrams drug (active ingredient)/kilogram of fish weight/day for 5 days. A 42-day withdrawal period is required for salmonids and a 3-day withdrawal period is required for channel catfish before the fish may be slaughtered and used for human consumption. It is illegal for a producer to mix his/her own medicated feed.

COPPER SULFATE ($CuSO_4$)

Copper sulfate, an effective algicide, is approved for use by the EPA. It can only be used for fish health problems if a producer has signed up to be part of an INAD[6] (investigational new animal drug) through the FDA. Copper sulfate is used to treat a variety of external parasites of fish, and can kill fish if used improperly. The relationship between toxicity of copper sulfate and alkalinity is very important (alkalinity is the total concentration of alkaline substances in the water expressed as equivalent calcium carbonate). In water with an alkalinity less than 50 milligrams per liter (mg/L), copper sulfate can be very toxic to fish and should not be used unless a bioassay has been run in the water first with a limited number of the fish to be treated. The following general guidelines have been established for the use of copper sulfate in Table 13.4:

Since copper sulfate is an algicide, consideration must be given to dissolved oxygen concentrations in the RAS. If the RAS already has low dissolved oxygen, an alternate treatment should be used.

FORMALIN-F, PARACIDE-F, AND PARASITE-S (FORMALIN)

Formalin is approved for use in the treatment of several external parasites. Formalin will remove 1 mg/L dissolved oxygen for every 5

---

[6] An INAD allows a producer to use a drug or chemical while it is undergoing the review process for approval. INAD's are not available for all drugs. An aquaculture veterinarian can help a producer make the necessary contacts to become part of an INAD.

mg/L of formalin used as a treatment. Therefore, if dissolved oxygen in a pond is low, aeration must be provided or a different treatment should be used. Formalin must be stored at temperatures above 40°F (4.4°C) because it will form very toxic paraformaldehyde at low temperatures. Caution should be used when storing this chemical at a hatchery with fluctuating air temperature.

Table 13.4 Permissible Treatment Concentration of Copper Sulfate as Affected by Alkalinity Concentration

Water Alkalinity (mg/L)	Permissible Treatment (mg/L)
0–49	Test for toxicity before use
50–99	0.5–0.75
100–149	0.75–1.00
150–200	1.00–2.00
>200	Ineffective; will precipitate As $CuCO_3$

Potassium Permanganate ($KMnO_4$)

Like copper sulfate, potassium permanganate ($KMnO_4$) is approved for use by the EPA, but not the FDA. The producer must be involved with an INAD through the FDA in order to use potassium permanganate. It has been used effectively against a number of external disease organisms of fish. The normal treatment is 2–8 milligrams per liter (mg/L), depending upon the amount of organic matter in the pond to be treated. Ideally, a "wine red" water color should be maintained for a 12 hour period to ensure an effective treatment. A preliminary test can be performed with a small volume of culture water to determine the appropriate dose for the system.

Sodium Chloride (Salt; NaCl)

Sodium Chloride (Salt; NaCl) is approved for aquaculture use as an "osmoregulatory enhancer". Salt can change the osmoregulatory balance (water balance) of aquatic organisms. It can control external parasitic protozoans by placing them in a condition of severe osmoregulatory shock. Care must be exercised to avoid over treatment, which will place the fish in the same condition of osmoregulatory shock. Sodium chloride is used as a 0.5% to 1.0% concentration in water as an indefinite (long term) treatment or as a 3% concentration in water for 10–30 minutes (stop the treatment earlier if the fish show signs of stress).

13.8 TREATMENT CALCULATIONS

Water treatments are based on water volume. A specified amount of chemical is added to a known quantity of water for a specified time. If too little chemical is added, the treatment will be ineffective; if too much is added or if the fish are left in contact with the chemical for too long, they may become stressed or die.

Feed treatments and fish injections are based on fish weight. A specified amount of drug is added to the feed (at the manufacturer) or injected into the fish. Incorrect dose may result in mortality or ineffective treatment.

Fish farmers should know the water volume of each culture unit, e.g., pond, tank, raceway, before a problem occurs, preferably recording this information when the system is designed or filled with water for the first time. The information should be stored so it is immediately available when needed. To become familiar with the procedure, the producer should practice making water volume calculations, (see Appendix). To obtain an effective, safe treatment, the ability to carry out accurate calculations is critical.

SAMPLE CALCULATIONS

Three hypothetical situations are presented to provide the producer with an opportunity to become familiar with the methodology used to calculate fish disease treatments. For each example, there are several ways to correctly compute the amount of chemical to add or the drug to use. Calculations and steps are shown in detail for one method. These examples are intended to provide illustrate calculations only, not the entire treatment protocol. Chemotherapeutic dose and information about treatment duration or frequency and fish management before, during and after treatment should be obtained from a veterinarian.

Example 1

A tank system of rainbow trout is infected with the parasitic protozoan, *Ichthyophthirius*. The treatment to be used is copper sulfate ($CuSO_4$) and the producer is part of an INAD. The tank system contains 5,000 gallons of water with an alkalinity of 75 milligrams per liter (mg/L or ppm). How much $CuSO_4$ would you use?

Computation steps:

1) Examine Table 13.4 and determine what concentration of $CuSO_4$ should be added to the system to provide an appropriate and safe treatment.

You know that the alkalinity is 75 mg/L. Therefore, an appropriate treatment concentration for $CuSO_4$ is 0.5 mg/L.

2) Determine the quantity of $CuSO_4$ to be added to the tank to achieve the 0.5 mg/L concentration.

2a) *Convert the volume of the tank from gallons (gal) to liters (L).*

$$(5{,}000\,gal)\cdot\frac{3.8L}{gal}=19{,}000L$$

2b) *Determine a correction factor for the proportion of chemical ($CuSO_4$) that is active ingredient. $CuSO_4$ is 100% active.*

$$correction\ factor=\frac{1.00}{1.00\,(active\ ingredient)}=1.00$$

2c) *Compute the amount of chemical ($CuSO_4$) that should be added to the tank.*

$$(volume)\cdot(dosage\ CuSO_4)\cdot(correction\ factor)$$

$$19{,}000L\cdot\frac{0.5mg}{L\,CuSO_4}\cdot 1.00=9{,}500mg\ CuSO_4$$

2d) *Convert milligrams (mg) $CuSO_4$ to grams (g).*

$$9{,}500mg\cdot\frac{1.0g}{1{,}000mg}=9.5g\ CuSO_4\ added\ to\ 5{,}000\ gal\ tank$$

NOTE: *Ichthyophthirius* has a complicated life cycle, which must be considered in its treatment. Multiple treatments are required to effectively control outbreaks of *Ichthyophthirius*.

Example 2

If alkalinity in Example 1 was below 50 mg/L, $CuSO_4$ would not have been the treatment of choice due to potential toxicity to the fish, Table 13.4. In this case, formalin would have been used as an alternative treatment.

Computation steps:

1) *Obtain a recommended treatment concentration for formalin from your diagnostic laboratory.*

   An appropriate dose is 25 mg/L (or 25 ppm formalin)

2) *Determine the quantity of formalin to be added to the tank to achieve the 25 mg/L (25 ppm) concentration.*

   2a) *Convert the volume of tank from gallons (gal) to liters (L).*

$$5{,}000\,gal\,\frac{3.8L}{gal} = 19{,}000L$$

   2b) *Determine a correction factor for the proportion of chemical (formalin) that is active ingredient.* NOTE: *Although formalin is 37% formaldehyde gas dissolved in water, for fish treatment purposes formalin is considered to be 100% active.*

$$correction\ factor = \frac{1.00}{1.00\,(active\ \ ingredient)} = 1.00$$

   2c) *Compute the amount of chemical (formalin) that should be added to the tank.*

   (volume of tank) x (dosage of formalin) x (correction factor)

$$19{,}000L\,\frac{25mg\ \ formalin}{L}\,x1.00 = 475{,}000mg\ \ formalin$$

   2d) *Since formalin is a liquid, it is desirable to convert milligrams (mg) to milliliters (mL).*

$$\frac{475{,}000mg\ formalin}{1{,}000mg\Big/1.0g} = 475g\ formalin$$

$$1.0g\ formalin \cdot \frac{1mL}{1g} = 1.0mL\ formalin$$

$$475g\ formalin = 475mL\ formalin\ (\,added\ to\ tank\,)$$

Example 3

Rainbow trout in a recirculated system are infected with *Aeromonas salmonicida*, the bacteria associated with furunculosis outbreaks. The bacteria is sensitive to oxytetracycline. The fish will be treated by feeding Terramycin-medicated feed at 2.5 gm active ingredient/100 pounds of feed for 10 days. The fish are normally fed at 1% body weight per day and 10,000 pounds of fish need to be fed each day. How much Terramycin (TM) should be mixed with the daily feed ration in order to obtain the desired treatment?

Computation steps:

It is **illegal** for anyone but a feed manufacturer licensed by the FDA to make medicated feed. The producer should contact the feed manufacturer to determine what type of mix is normally available. If, for example, the feed is mixed at 2.5 gm active ingredient per pound of feed, then one lb of feed fed per 100 lbs fish will achieve the desired treatment. With 10,000 lbs fish normally fed at 1% body weight per day, 100 pounds of feed would be fed each day for 10 days. A total of 1,000 lbs of feed will be needed. Terramycin-medicated feed should not be fed at temperatures below 48.2°F (9°C). Fish cannot be marketed for 21 days post-treatment.

POST-TREATMENT REVIEW

When every new fish health problem occurs and when old problems keep recurring, a review should be done to determine the factors contributing to the problems – less than optimum environmental, e.g., water quality, conditions that lead to stress or directly to disease, ineffective biosecurity, and inaccurate measurement and poor management of culture conditions are common causes of recurring fish health problems. Good intuition can help isolate the problem; good record keeping can help to verify its origin so that strategies for early detection and prevention of the same problem in the future can be developed. Financial constraints, the design of the production system or constraints on the production cycle might preclude the development of strategies to prevent recurrence of the problem. However, strategies for early detection should be developed, so that even though the problem occurs again, early response can help mitigate the impact.

13.9 FISH DISEASE DIAGNOSTIC SERVICES

The importance of obtaining an accurate diagnosis for a disease problem cannot be overemphasized. The success of any treatment is absolutely tied to knowing the condition being treated and awareness of complicating factors. Accurate diagnosis of a fish disease requires specialized technical skills and appropriate laboratory facilities. Fish disease diagnostic services are available from a variety of sources in the United States. The fish farmer should establish a working relationship with an aquaculture veterinarian and/or contact a laboratory before a problem occurs to learn of any specific instructions associated with the submission of fish for diagnostic evaluation. For instance, some laboratories can, due to their funding, accept only submissions from restricted geographical locations; some fish health research laboratories accept only diagnostic cases that are referred by other diagnostic laboratories. An annually revised list of agencies and individuals that provide disease diagnostic services is included in the *"Annual Buyer's Guide"* published by Aquaculture Magazine (P.O. Box 2329, Ashville, NC 28802). Further contacts for diagnostic sources can be found by contacting one of the USDA/CSREES Regional Aquaculture Centers (RAC's):

Northeastern Regional Aquaculture Center (NRAC)
http://www.umassd.edu/specialprograms/nrac/

North Central Regional Aquaculture Center (NCRAC)
http://ag.ansc.purdue.edu/aquanic/ncrac/

Southern Regional Aquaculture Center (SRAC)
http://www.msstate.edu/dept/srac/

Western Regional Aquaculture Center (WRAC)
www.fish.washington.edu/wrac/

The Center for Tropical and Subtropical Aquaculture
http://library.kcc.hawaii.edu/CTSA/

CHAPTER 14

BUILDING ENVIRONMENTAL CONTROL

14.0 INTRODUCTION

This chapter provides a short treatment of the basics of heat and mass transfer. The intent is that the reader will be able to calculate the projected heating costs and establish adequate ventilation requirements associated with maintaining some specific thermal environment for their RAS, since most RAS will be housed within a building. Many people instantly assume that since a RAS is inside a building that the heating costs will be large. The assumption is that if water exchanges are high, then of course the water heating costs will also be high. These assumptions are generally overly pessimistic. In most RAS, water exchange will be less than 20% per day of system volume and often less than 10%. Under these conditions, water-heating costs will be a small percentage of the total costs of production.

Many RAS designers make the mistake of ignoring the ventilation requirements associated with a building that houses an RAS. Ventilation requirements for controlling environmental conditions of humidity, temperature, and CO_2 should be approached in the same way that a ventilation balance is done for chickens or cows to maintain an acceptable air quality environment. For an RAS, we have fish instead of these warm blooded animals. In an RAS, the controlling environmental conditions are typically humidity and then CO_2. Just as we need to do a mass balance for controlling water parameters as described in Chapter 4, a mass balance must be performed for the air space in an RAS with the P terms being moisture, heat, and CO_2. Generally speaking, other air quality parameters can be ignored.

14.1 HEAT TRANSFER

There are two forms of heat loss:

- Conductive
- Convective

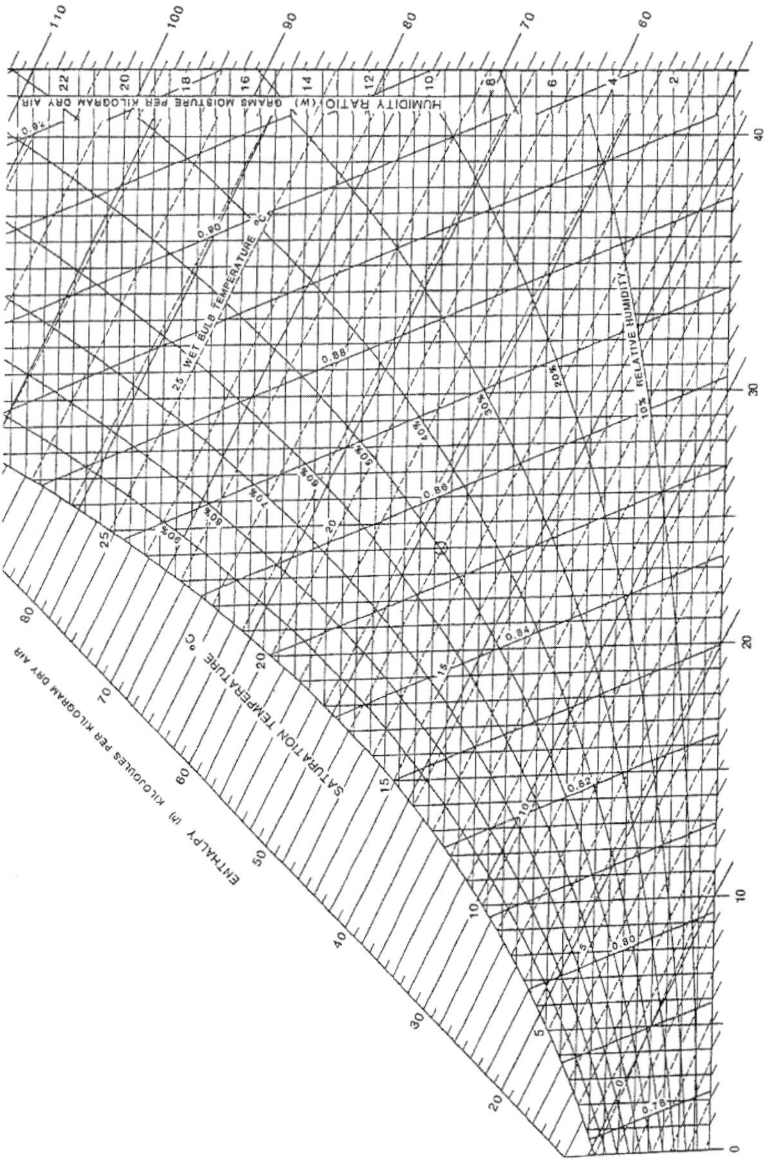

Figure 14.1 Psychrometric Chart.

Conductive heat loss is defined as:

$$Q = \frac{A}{R} \Delta T \qquad (14.1^{a})$$

Convective heat loss is defined as:

$$Q = \dot{m} c_p \Delta T \qquad (14.2)$$

The value for c_p is 0.24 BTU/(lb°F) or 1.0035 kJ/(kg dry air °C). Generally we think of air in terms of volume of air. The volume of air per unit mass is called the specific weight and is approximately 13.5 ft^3/lb (0.84 m^3/kg) dry air. All the important thermodynamic properties of air are shown in the psychrometric chart given in Fig. 14.1. This figure also shows the grains of moisture per lb of dry air as a function of temperature and relative humidity. (Note there are 7,000 grains in a pound.)

Convection heat loss/gain is proportional to the $\dot{m}$ term in Eq. 14.2 and represents the mass of airflow that is being exchanged or ventilated for the building. For most buildings, air exchange rates caused by wind and natural infiltration will be a minimum of 1 to 2 building volumes of air per hour. As a rule of thumb, maintaining inside relative humidity below 80% will require at least an air exchange of 2 air volumes per hour. Therefore, natural infiltration caused by wind may provide the minimum ventilation requirements for the facility. The point to remember is that you cannot eliminate the heat losses caused by ventilation, since these heat losses will occur as a result of either unplanned infiltration or from proactively controlled minimum mechanical ventilation rates. As a further data point, the required ventilation rates for CO_2 control will also be comparable. Where problems sometimes occur is in extremely tight buildings such as concrete block construction with no windows and only two tight passage doors. Infiltration here will be perhaps as low as 0.5 or less air volumes per hour. Under this scenario, CO_2 levels could reach dangerous levels if mechanical ventilation is not used to maintain some minimum threshold (more information is given on minimum CO_2 ventilation rates in Chapter 17 where the CO_2 control program is presented).

---

[a] Variables defined in List of Symbols at the end of the chapter.

Heat loss from water flow is calculated the same as shown in Eq. 14.2, except the c_p value for water is 1.0 BTU/(lb °F) or 4.18 kJ/(kg °C).

In insulated buildings, the heat loss through the floor is less than 10% of total building heat loss. This loss term can be estimated based upon the perimeter of the building and whether or not perimeter or floor insulation is used as follows:

$$\frac{Q}{time} = F\,P\,\Delta T \tag{14.3}$$

The F-term is based upon empirical data and is assigned as:

IP

For Q in BTU, time in hour, P in feet, ΔF in °F

F = 0.81 (4.53 for SI) for floors with no perimeter insulation, BTU/(°F · ft · h)

= 0.55 (3.08 for SI) for floors with 1 inch of rigid insulation, BTU/(°F · ft · h)

SI

For Q in kJ, time in seconds, P in meter, and ΔT in °K

F = 1.38 for floors with no perimeter insulation, W/K·m

= 0.93 for floors with 2.5 cm of rigid insulation, W/K·m

During startup operations, the first winter season will typically cause much higher heat losses than subsequent winter seasons. The first year the soil acts as a large thermal sink and once full, basically eliminates subsequent heat losses or gains. This is why the perimeter insulation is important to prevent heat seepage losses from the thermal envelope.

Yearly Heat Losses or Gains

Determining yearly heating costs or cooling costs requires that an entire year of simulation be used. Heating degree-days (HDD) are used to estimate the amount of energy required for space heating during the cooler seasons. For example, given the value of HDD, the yearly heat loss can be calculated as follows:

$$\frac{Q}{year} = \frac{A}{R} HDD \tag{14.4}$$

To calculate the HDD's you must first find the mean temperature for the day. The average daily temperature can be estimated as the average of the day's high and low temperatures or calculated according to Eq. 14.5. Equation 14.5 requires monthly average temperature data (the monthly maximum, T_{max}, and monthly minimum, T_{min}, for monthly average temperature); some data is provided in Table 17.1. The generic equation to predict any weather variable, U_{day}, for a particular Julian date of the year is based upon the maximum and minimum values of the monthly average values for the particular variable of interest as follows:

$$U_{day} = \left[\left(\frac{U_{max} - U_{min}}{2} \right) (1 + \phi) \right] + U_{min} \tag{14.5}$$

where:

$$\phi = -\sin(\lambda - date); \quad if(date - \lambda) < 0$$
$$\phi = +\sin(date - \lambda); \quad if(date - \lambda) \geq 0$$
$$date = Julian\ day\ of\ calendar\ year$$

The λ value is chosen so that the maximum and minimums occur on the appropriate day of the year. These λ values solar radiation and temperature are:

- $\lambda_{solar} = 83$
- $\lambda_{temperature} = 100$

If the mean daily temperature is at or above 65°F (18°C), then the HDD amount is zero (if 65°F is the base temperature). If the mean temperature is below 65°F(18°C), then the HDD amount equals 65° (18°C) minus the mean temperature. For example, if the mean outside temperature was 55°F (13°C) then the HDD amount equals 10 degree-days (5 degree-days in Celsius). Yearly heat losses can be calculated based upon the yearly HDD's for the location of interest.

In equation form:

$$\begin{aligned} & if\ T_a < T_{base} \\ & HDD = T_{base} - T_a \\ & f\ T_a > T_{base}\ then\ HDD = 0 \end{aligned} \tag{14.6}$$

Cooling degree days (CDD) are used to estimate the amount of air conditioning usage during the summer season. CDD's are calculated similarly to HDD's except that you are calculating the degree-days above some base inside air temperature. Think of cooling degree-days as the flip side to heating degree-days.

For easy reference, the fuel contents of various forms of energy are provided in Table 14.1

Table 14.1 Approximate Fuel Heat Contents

Fuel Type	Heat Content		Units
Natural Gas	1,000	(37,350)	BTU/ft^3 (kJ/m^3)
LP Gas	93,000	(25,800)	BTU/gallon (kJ/Liter)
Fuel Oil	138,000	(38,460)	BTU/gallon (kJ/Liter)
Electricity	3,413	(3,600)	BTU/kWh (kJ/kWh)

14.2 AIR QUALITY CONTROL

Just as we do a mass balance on the fish tanks for pollution loading and the required flow rates to maintain water quality at some minimum target values, we also take the same approach on the thermal air space to control humidity (moisture), temperature, and carbon dioxide (CO_2). For example, a general sensible heat balance on an enclosed space would be depicted in Fig. 14.2

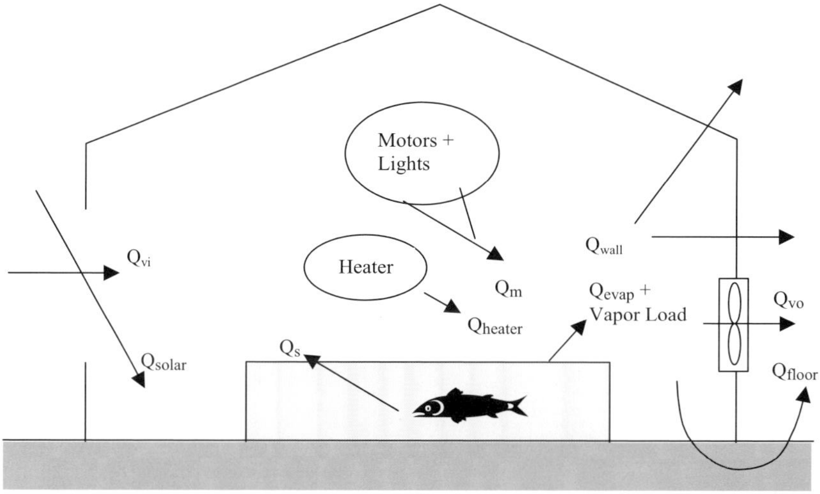

Figure 14.2 General Heat Balance on an Enclosed Ventilated Air Space.

For steady state conditions, the gains of heat must balance the heat losses, or in equation form:

$$Q_s + Q_{solar} + Q_{heater} + Q_m + Q_{vi} = Q_{evap} + Q_{wall} + Q_{floor} + Q_{vo} \tag{14.7}$$

Heat produced by fish, Qs, is approximately 1 BTU/hr per lb (2.2 kJ/kg) of body mass. Evaporation rates from free water surfaces are approximately 0.25 inch/day (6 mm).

Evaporation rate, (E, inches per day), can be predicted as a function of wind speed (S, mph), inside relative humidity (RH, %), and vapor pressures of the water, $e_{s,water}$, and the inside air, $e_{d,air}$ (inches of mercury):

(Eqs. 14.8–14.10 valid for IP units only)

$$E = C_{wind}(e_{s,water} - e_{d,air}) \tag{14.8}$$

$$C_{wind} = 0.44 + 0.118S \tag{14.9}$$

$$e_{d,air} = RH \cdot e_{s,air} \tag{14.10}$$

From Chapter 4, we know the CO_2 production rates are proportional to oxygen consumption rate (roughly 1.4 times the oxygen consumed). The moisture production in a building is very dependent upon the humidity level in the air space (see Eqs. 14.7–14.10). At 100% humidity, there is no evaporation, and the building will rot or rust and deteriorate quickly, as will all your equipment that is inside the structure. The latent heat content of one lb of moisture evaporated is approximately 1,050 BTU/lb (2,440 kJ/kg).

The air quality balance equation can be written as follows for any air parameter of interest "X":

$$P = \dot{m}\,(X_{inside} - X_{outside}) \qquad (14.11)$$

EXAMPLES

Conductive Heat Loss

Wall area is 100 ft^2, R_{value} = 10°F ft^2/BTU, inside temperature is 80°F and outside temperature is 50°F. Calculate hourly heat loss:

$$Q(\frac{BTU}{hr}) = \frac{A}{R}\Delta T = \frac{100\,ft^2}{10\dfrac{hr^{\circ}Fft^2}{BTU}}(80-50)^{\circ}F = 300\frac{BTU}{hr}$$

Metric SI

$$Q(W) = \frac{A}{R}\Delta T = \frac{9.29m^2}{1.761\dfrac{m^2\cdot{}^{\circ}C}{W}}(26.7-10.0)^{\circ}C = 88W$$

Convective Heat Loss-Water

Calculate the heat loss from exchanging 1 gal/min (3.785 L per minute) when the water used to replace the water being discharged (makeup water) is 45°F (7.2°C) (generally this is well water or deep groundwater temperatures) and the inside tank water temperature is 75°F (23.9°C).

$$Q = \dot{m}c_p\Delta T = 1\frac{gal}{\min}\cdot 8.34\frac{lb}{gal}\cdot 1.0\frac{BTU}{lb\cdot{}^{\circ}F}\cdot(75-45)^{\circ}F\cdot 60\frac{\min}{hr} = 15{,}012\frac{BTU}{hr}$$

Metric SI

$$\dot{Q} = \dot{m}\, c_p \Delta T = 3.785\frac{L}{\min}\cdot 4.187\frac{kJ}{kg^{\circ}C}\cdot(23.9-7.2)^{\circ}C\cdot 60\frac{\min}{hr} = 15{,}880\frac{kJ}{hr}$$

Convective Heat Loss-Air

Calculate the heat loss per day due to air ventilation from a building that is 100x100x10 ft (length, width, and ceiling height) when the outside air temperature is 50°F and the inside air temperature is 80°F and the ventilation rate is 3 air volumes per hour.

$$\dot{Q} = \dot{m}\, c_p \Delta T = 3\frac{vol}{hr}\cdot 100\cdot 100\cdot 10\frac{ft^3}{vol}\cdot\frac{lb_{air}}{13.5\,ft^3}\cdot 0.24\frac{BTU}{lb\cdot^{\circ}F}\cdot(80-50)^{\circ}F$$

$$\cdot\, 24\frac{hr}{day}$$

$$= 3{,}840{,}000\frac{BTU}{day}$$

Metric SI

$$\dot{Q} = \dot{m}\, c_p \Delta T = 3\frac{vol}{hr}\cdot 2{,}825m^3\cdot 1.19\frac{kgair}{m^3} +$$

$$1.007\frac{kJ}{kg^{\circ}C}\cdot(\Delta T = 16.67)^{\circ}C\cdot 24\frac{hr}{day} = 4{,}063{,}000\frac{kJ}{day}$$

Cost to Heat Air

Calculate the heating cost associated with the above air heat loss example if the building is heated with LP gas that has a heat content of 92,000 BTU/gal, the burn efficiency is 85%, and the cost of the LPG is $0.80/gal.

$$\$/day = 3{,}840{,}000\frac{BTU}{day}\cdot\frac{gal_{LPG}}{92{,}000BTU}\cdot\frac{\$0.80}{gal_{LPG}}\cdot\frac{1}{0.85} = \frac{\$39.28}{day}$$

Metric SI

$$\frac{\$}{day} = 4{,}051{,}200\frac{kJ}{day}\cdot\frac{L_{LPG}}{25{,}643kJ}\cdot\frac{\$0.21}{L_{LPG}}\cdot\frac{1}{0.85} = \frac{\$39.03}{day}$$

Ventilation Rate for Moisture Control

Calculate the required ventilation rate to maintain an inside relative humidity of 80% at an air temperature is 68°F, if the outside air temperature is 41°F and 70% relative humidity. You have 1,000 ft^2 (93 m^2) of free water surface exposed (assume 0.25 inches, 6 mm, of water is evaporated per day from a free water surface).

$$P_{water}(lb/day) = 1{,}000\,ft^2_{water} \cdot \frac{0.25inch_{water}}{day} \cdot \frac{1ft}{12inch} \cdot \frac{62.4lb_{water}}{ft^3_{water}} = 1{,}300\frac{lb_{water}}{day}$$

$$W_{inside}(68^\circ F \,\&\, 70\%) = 0.0102\, {lb_{water}}/{lb_{air}}$$

$$W_{outside}(41^\circ F \,\&\, 70\%) = 0.0036\, {lb_{water}}/{lb_{air}}$$

$$\dot{m} = \frac{P_{water}(lb/day)}{(W_{inside} - W_{outside})} = \frac{1{,}300\frac{lb_{water}}{day}}{(0.0102 - 0.0036)\, {lb_{water}}/{lb_{air}}}$$

$$= 196{,}970\frac{lb_{air}}{day} \cdot \frac{13.5ft^3}{lb_{air}} \cdot \frac{day}{1{,}440\min} = 1{,}846\ cfm$$

Metric SI

$$P_{water}\left(kg/day\right) = 93m^2_{water} \cdot \frac{6.35mm_{water}}{day} \cdot \frac{m}{1{,}000mm} \cdot \frac{1{,}000kg_{water}}{m^3} = 590\frac{kg_{water}}{day}$$

$$W_{inside}\left(20^\circ C \,\&\, 70\%RH\right) = 0.0102\frac{kg_{water}}{kg_{air}}$$

$$W_{outside}\left(5^\circ C \,\&\, 70\%RH\right) = 0.0036\frac{kg_{water}}{kg_{air}}$$

$$\dot{m} = \frac{P_{water}\left(kg/day\right)}{W_{inside} - W_{outside}} = \frac{590\frac{kg_{water}}{day}}{(0.0102 - 0.0036)\frac{kg_{water}}{kg_{air}}}$$

$$= 89{,}394\frac{kg_{air}}{day} \cdot \frac{0.84m^3}{kg_{air}} \cdot \frac{day}{1440\min} = 52\frac{m^3}{\min}$$

14.3 BUILDING CONSIDERATIONS

MATERIALS

In an aquaculture setting, your building's inside air conditions will often be a warm moist air environment. Under such conditions of high humidity, a drop in inside temperature (referred to as dry bulb temperature or T_{db}) or when warm moist air comes in contact with a cooler surface, the air will drop in dry bulb temperature. If the temperature drop is below the dewpoint temperature (T_{dp}), then condensation will occur. Figure 14.1 provides a psychrometric chart that can be used to determine when condensation would occur for specific inside T_{db} and relative humidity (RH) conditions. For example, if the design inside T_{db} were 80°F (26.7°C) and 75% RH, then the T_{dp} is 71.5°F (21.9°C) at 100% RH. The amount of moisture (sometimes called humidity ratio, W) in the air remains the same between these two conditions (W = 0.0168 lbs moisture per lb dry air or kg/kg dry air).

Using adequate wall and ceiling insulation can prevent condensation. To accomplish this, you would have to work from an expected outside design temperature and then install enough insulation in the wall to prevent the surface from dropping below 71.5°F (21.9°C) in the previous example. Consult just about any heat transfer text for more information in this regard, or see Albright (1989).

For quick reference, the formula to calculate minimum R-value for a wall to prevent condensation is:

$$R_W \geq R_{inside} \frac{t_{db,inside} - t_{db,outside}}{t_{db,inside} - t_{t_{dp}}} \tag{14.12}$$

where R_{inside}

- 0.12 m^2°C/W Vertical wall (high surface emittance)
- 0.11 Horizontal surface heat flow upward
- 0.16 Horizontal surface heat flow downward

For low emittance walls such as foil faced, the R_{inside} will be about twice the above values.

For non-free convection coefficients pertaining to wall surface resistance such as outside walls or inside walls in forced convection conditions, the R_{inside} will be 0.030 to 0.04 m^2°C/W. (Note the conversion from SI to PI is: 5.678 hr ft^2 °F/BTU per m^2°C/W).

We strongly recommend against using any type of fibrous insulating materials, e.g., fiberglass batting, paper based materials, rock wool.

These materials lose 90% of their insulating ability after absorbing only 10% of their initial weight in moisture. We recommend using rigid board insulation, e.g., polystyrene blue board. Ideally, the board material should be covered by a metal surface, but where this is not done, use foil faced board-insulating materials. The Appendix provides a Table of common building materials and R-values (Table A-14 from ASHRAE 1985).

> **"Rule of Thumb"**
> Avoid fibrous insulating materials.
> Use rigid board insulation.

Moisture Control

Consideration should be given to preventing moisture migration into walls and for adequate ventilation in attic spaces. Moisture transfer can be calculated as follows:

$$w = \frac{(p_{wv,inside} - p_{wv,outside})}{R_{H_2 0}} \tag{14.13}$$

The analogy to Eq. 14.13 is steady-state heat transfer by conduction. Vapor pressure difference is analogous to temperature difference, permeance is analogous to thermal conductance, and permeability is analogous to thermal conductivity. The concepts of simple parallel and series resistance circuits also apply to water vapor diffusion as they do to heat diffusion and transfer.

It is generally impossible to prevent moisture migration. Most migration is not through the materials themselves, but from air leakage into the attic space from the warmer area below where animals and people are. During construction and after, pay particular attention to closing all cracks and crevices, particularly those created by installing fixtures into the ceiling. Flush mounting of electrical fixtures is preferred to recess mounting, since no cracks or break in the ceiling materials are created when doing flush mounts.

For attic spaces, 1.0 square unit ventilation area for each 300 square units of ceiling area should be provided. Since louvers will have rain shields and screening, provide 2.25 times the area required or 2.25 square units for each 300 square units of ceiling area.

Table 14.2 provides some permeability values for common building materials.

Table 14.2 Permeability for Various Building Materials

Material	Perm-inches*	Perms**
Air	120	
Gypsum board		50
Interior plywood, 1/4 inch		1.9
Exterior plywood, 1/4 inch		0.7
Pine wood	0.4–5.4	
Concrete	3.2	
Roll roofing		0.05
Aluminum paint		0.3–0.5
Latex paint		5.5
Mineral Wool	116	
Blanket insulation and asphalt paper		0.04
Expanded Polystyrene		
Extruded	1.2	
Bead	2.0–S5.8	
Polyurethane		
Polyethylene 4 mill		0.08
Polyethylene 8 mill		0.04

*To obtain (gram-m)/(24 hr m^2 mm Hg) multiply by 0.017

** To obtain (grams/(24 hr m^2 mm Hg) multiply by 0.66

"RULE OF THUMB"

- Use 1.0 ft^2 of net louver area for each 300 ft^2 of ceiling (1 m^2/300 m^2).
- Increase louver area by a factor of 2.25 to account for screening and louver blockage.

List of Symbols

A	Area, ft^2 (m^2)
c_p	Specific heat of air, BTU/(lb°F) (kJ/kg°C)
date	Julian day of the year as a number
F	Perimeter heat loss term, BTU/°F·ft·h (W/K·m)
$\dot{m}$	Airflow rate, lb/hr (kg/hr)
P	Perimeter length of exposed walls, ft (m)
$p_{wv,inside}$	Vapor pressure inside air, inches mercury water gauge (mm Hg)
$p_{wv,outside}$	Vapor pressure outside air, inches mercury water gauge (mm Hg)
Q	Heat loss or gain, BTU (kJ)
Q_s	Sensible heat production of fish, BTU/h (kJ/h)
Q_{solar}	Solar heat gain, BTU/h (kJ/h)
Q_{heater}	Sensible heat added by space heaters, BTU/h (kJ/h)
Q_m	Sensible heat added by motors and lights, BTU/h (kJ/h)
Q_{vi}	Sensible heat ventilated into air space, BTU/h (kJ/h)
Q_{evap}	Rate of sensible heat converted to latent heat via evaporation, BTU/h (kJ/h)
Q_{wall}	Sensible heat conducted from the space through walls and ceiling, BTU/h (kJ/h)
Q_{floor}	Sensible heat lost through the floor, BTU/h (kJ/s)
Q_{vo}	Sensible heat ventilated out of air space, BTU/h (kJ/h)
R	Thermal resistance, (hr°F ft^2)/BTU (m^2·°K/W) (see the Appendix A-14 for a listing of R-values for common materials, from ASHRAE 1985).
R_{H20}	Resistance to water vapor flow; inverse of permeance or perms, hr ft^2 inch Hg/grains (g-m/(24 h m^2 mm Hg))
RH	Relative humidity, %
T	Temperature, °F (°C)
T_{base}	Temperature base, usually 65°F (18.3°C)
T_a	Average temperature, °F (°C)
T_{db}	Dry bulb temperature, °F (°C)
T_{dp}	Dewpoint temperature, °F (°C)
U_{max} and U_{min}	Maximum and minimum monthly average values for particular weather variable
W	Rate of moisture movement, grains/hr (grams/hr)
X_{inside}	Inside air quality parameter, lbs per lb dry air (kg/kg)

$X_{outside}$	Outside air quality parameter, lbs per lb dry air (kg/kg)
ΔT	Temperature difference inside air film to outside air film, °F (°C)
λ_{solar}	Weather model constant for solar radiation (83)
$\lambda_{temperature}$	Temperature weather model constant for dry bulb temperature (100)
ϕ	Argument used in weather model based upon Julian date

CHAPTER 15

ECONOMIC REALITIES & MANAGEMENT ISSUES

15.0 INTRODUCTION

This chapter is primarily based on the first author's personal involvement in a commercial startup venture that began in 1996. Fingerlakes Aquaculture LLC (FLA) was initially a 200,000 lb (100 ton) per year, fully-integrated tilapia farm that included breeding stock, hatching, and growout operations. The farm was based upon the principles of recirculating aquaculture systems (RAS) technology presented in this text. The farm later expanded to a production capacity of 1.2 million lb/yr (600 ton). During this time, several rounds of financing occurred including capital injection by a venture capital firm. FLA's major market was the Chinese live markets, with a small amount of production targeted for the processed fillet markets. The first part of this chapter reviews some of the experiences associated with this commercial effort and the realities of indoor fish farming, for those who may be considering entering this business. Read this chapter closely.

15.1 CASE HISTORY OF FINGERLAKES AQUACULTURE

Four individuals created Fingerlakes Aquaculture LLC in June of 1996: one of the authors (M.B. Timmons), his brother (marketing executive for a Fortune 500 company), Dr. William Youngs (Professor Emeritus Cornell University Department of Natural Resources), and a local businessman (owner of a lumber yard and apartment units). In addition, Drs. Dave Call (Dean Emeritus Cornell College of Agriculture and Life Science) and Gene German (Professor of Food Marketing, Cornell University) were members of the FLA Board of Managers. At that time, live market prices for tilapia sold at the farm, located in upstate New York, exceeded \$2.00/lb (\$4.40/kg). These market prices did not hold, however. Prices began to fall, and continued to deteriorate, and by late 1999, the market price was just over \$1.00/lb (\$2.20/kg) at the farm.

Recently, the market price has recovered to a farm price of \$1.40/lb (\$3.08/kg) liveweight, but well below the initial farm price levels when the farm was on paper.

In July 1998, FLA received a second major cash infusion from a successful entrepreneur. This individual had extensive and successful startup experience, which included agricultural and seafood types of businesses. Further rounds of financing were required in 1999 and 2000. Each new round of financing brought with it a significant dilution of the founding members' ownership percentage, but the personal guarantees and assignment of personal assets placed by these individuals to enable bank loans remained in place. These cash infusions and loans were simply part of a long struggle to increase production capacity from 200,000 lb/year (91,000 kg/yr) to the targeted 1.2 million lb (545,000 kg) per year. FLA is still planning further expansion to a production volume that will make automated processing cost effective. It remains to be seen if FLA will reach economic success, but if it does, it will be because of:

- large scale operations
- trained and motivated workforce
- improved technology
- adequate water
- greatly reduced electrical costs
- targeted marketing approach

A phased production program, i.e., a ramp up, is the most realistic approach to achieving the goals of the business. Most business plans start with some large scale operation, which is usually necessary to show a positive cash flow. Unfortunately, almost always, these operations, if funded, are unsuccessful. It is extremely difficult to achieve large scale success without having first built this capacity incrementally from smaller operations. This incremental increase in size and complexity of operations enables your management team to learn-by-doing wherein the inevitable mistakes will have a less devastating impact on a smaller operation than they would on a large scale operation. Given a sound technical approach, the ability of the management team to operate the business is the single most important factor in the determination of ultimate success or failure. Simply adding additional like-item systems of a proven and fixed design that is already working well is the best way to increase the size of the operation. If you choose to incorporate untested and unproven designs into your expansion plans, you will undoubtedly face unexpected problems that will compromise your ability to achieve the cost effectiveness that is predicted on paper (as you will be able to do with the programs presented in Chapter 17). On the other hand, once you have a working system and management protocols refined at some reasonable scale, e.g., 200,000 lb (91,000 kg) per year, then this system or farm can be replicated as many times as necessary to reach the desired production levels and obtain the economies of scale necessary to achieve an economically competitive position.

15.2 LESSONS LEARNED

TECHNOLOGY

The FLA team was very confident, perhaps even overconfident, that by using the full-scale systems developed, tested, and proven by Cornell R&D sites in this commercial implementation, there would be no major obstacles. This confidence was inspired by the knowledge that one of FLA principles had spent 15 years in leading a major R&D effort at Cornell University that was completely focused on indoor RAS technology. This was a misplaced confidence, because the Cornell system was based upon a different technology: microbead filters and continuous biomass loading. After approximately one year of working with high rate fluidized sand beds at the Cornell facility and being convinced that sand beds would ultimately prevail as the most economically competitive biofilter system, FLA installed sand beds as the biofilter of choice. The anticipated production level of the FLA's first facility was 400,000 lb (182,000 kg) per year but depended upon the Cornell system of continuous biomass loading, growout, and harvest. In this method, there is a mixture of size cohorts in a single tank with a harvest consisting of approximately 20% of the large fish being removed and then adding an equivalent number of small 75 gram fish back to the tank.

At this point, FLA concluded that because of the lack of accurate and timely data, specifically the feed conversion and growth data associated with this method, FLA abandoned the continuous biomass loading approach in favor of a strict batch all-in-all-out loading. The batch loading method provides very accurate feed conversion and growth data at each tank harvest. FLA management felt that this data was the most critical information needed to define the overall economics of a large scale effort, which was the goal of FLA. The disadvantage of this switch from continuous loading to batch loading was that the batch process yields less production per year from the same sized facility.

FLA also made a change from the Cornell system of relying primarily on settling basins for solids removal to one that used mechanical screen filtration. No one at that time had any experience using mechanical screen filtration with the Cornell double drain design (see Chapter 5) where the center drain concentrates waste by approximately 15 fold. We now know that micro screen filters must be re-sized (larger) if such a type of influent is to be used. Unfortunately, cost considerations frequently over-ride sound engineering judgment, particularly if there is no hard test or prototype results to verify the design. FLA erred by under-sizing their mechanical screen system, and this contributed to high suspended solids levels in the rearing tanks. This, in turn, compromised fish growth performance and overall production levels.

FLA'S LARGEST ERROR—INSUFFICIENT WATER

It seems somewhat obvious that you should build a fish farm only where there is adequate water to support the expected operations. FLA's largest error was in not having an assured source of groundwater to operate the farm. Assuming that groundwater supplies would be adequate, FLA selected their initial site with other criteria in mind. These were important criteria, certainly, but not of the magnitude of the fundamental requirement for adequate water availability. **Never never never** commit to a site until you are convinced you have sufficient water to provide at least 20% water volume exchange per day. Also, you should expect that the initial well recharge capacity will drop approximately 50% from its initial delivery levels (what your well driller demonstrates for one week of constant pumping during the exploratory phase), once the well is used on a continuous basis for a year. So, if you have 100,000 gallons (378 m^3) of standing water in your system, you need a well capacity of at least 20,000 gallons (75 m^3) per day. This means that your initial well capacities should be not less than 40,000 gallons per day (28 gpm, 106 Lpm).

"RULE OF THUMB"

Before construction, identify/verify/establish a water source to replace 20% to 40% of system water volume on a daily basis (warmwater) and 100% of system volume for a coldwater system.

To give you an idea of FLA's sad experience: FLA drilled five separate wells at the first farm site, four of them after the building was constructed when the first well that had initially provided 8 gpm (30 Lpm) of water during tests virtually dried up during construction. The primary author of this chapter believed, at that time, that FLA could successfully run the barn at planned production levels on about 8 to 10 gpm (30 to 38 Lpm) of continuous well flow. This proved to be not possible.

What is typically not accounted for in calculating water needs are:

- Purging requirements
- Cleanup requirements
- Flushing during tank emergencies
- General usage around farm
- Loss of well capacity

FLA pursued a series of efforts to provide more water for the facility, including water recovery from the manure settling ponds and ground filtration and ozone treatment, but none of these attempted solutions were satisfactory. You must have water, in at least the quantities described above. Eventually, FLA abandoned the site at considerable expense, and moved to a new location that had access to municipal water supplies with the capability to deliver several hundred gallons per minute more than the nominal needs of the facility.

MANAGEMENT

Management is the most critical component of any aquaculture venture. From its inception, FLA had an extremely strong management team in place. The General Manager was recruited from his position as a GM of a large (1 million lb/yr or 455,000 kg/yr) integrated salmon farm. Dr. Youngs had started, successfully operated, and eventually sold a large trout farm. The board members were a good mix of business leaders and marketing experts. Even with this experienced management team, FLA barely survived. If FLA had tried a larger scale of effort than their initial farm that consisted of breeding, fingerling, and target growout of 250,000 lb (114,000 kg) per year, we predict they would have failed categorically. FLA produced 168,000 lb (76,000 kg) of sales in their first year of production compared to their target of 250,000 lbs (114,000 kg).

FINANCING

Regardless of all you hear about government guaranteed loans for business startups, in the FLA experience, it came down to equity infusion (self and outside investors) and personal guarantees of loans. The banks will take no risk with their depositors' money. They assume you will fail. They also assume that the value of your assets will be practically negligible. In truth, these are probably reasonable assumptions, particularly regarding the residual value of assets and equipment. The market value for used fish equipment is probably 10 cents on the dollar at best. Yes, you can probably sell common industrial equipment, such as your forklift for a reasonable salvage, but more specialized equipment, such as pumps and other fish equipment are of little to no value. Therefore, the bank will require that any loans to the company be personally guaranteed and may in fact require these loans to be perfected. This means that the guarantors must place assets equal to or exceeding the value of the loans in the bank's control. Then, if necessary, the bank can liquidate these assets if the terms of the loan are not being met.

"RULE OF THUMB"
Assume 10 cents on the dollar for used fish equipment.

The process of finding adequate financing requires considerable time and effort. But be prepared for the realization that generally all mechanisms of financing and leasing will end up at the same point, wherein the loan/lease will require equity guarantees. Also, bear in mind that these activities require tremendous time and effort by someone to pursue potential sources, develop detailed presentation materials, and attending meeting after meeting. Also, assume a successful loan closing will take at least three months before any money is actually available for you to use, and this is after you have found a bank that has said they will loan you the money!

Venture capital (VC) firms are also supposed sources of capital. FLA's experience with these during the initial rounds of seeking capital was that the VC assumed that what you had was probably not unique; that if you could do it, so could countless others, that the overall potential to make money in a food related business was not attractive, especially in comparison to other investment alternatives available to them. So, unless, you have an inside track with a particular venture firm, the likelihood of obtaining equity from them is not high—and even when you do have personal contacts, it is still not a likely outcome.

On the other hand, if you can demonstrate some Phase I success and bring investors to see your operating farm that has demonstrated your capability to breed and produce fingerlings successfully, grow fish with reasonable performance values, and sell fish successfully, then you become much more likely to generate VC funding. The key of course is to achieve success at the first level of an operating farm.

15.3 Investment Choices

In our roles as university researchers, extension professors, and consultants, we are often involved in helping people to decide if entering aquaculture is a wise investment decision. In particular, should an investment be made? Our opinion is that very few groups or individuals are capable of raising fish, using RAS technology or any other method. We would not recommend anyone invest in someone else's aquaculture operation until that group had successfully demonstrated their capability to raise fish. Then, based upon their initial results, assess the likelihood of the company being profitable in an expansion phase. In this assessment, be sure to use realistic values for product selling prices to predict business profits. We have often seen business plans that use retail prices where the farm's only realistic opportunity to sale product will be in the wholesale market. Some business plans will even confuse fillet pricing with in-the-round pricing (whole fish, non-gutted). Be careful and be realistic.

15.4 Species Selection

A general theme for those entering aquaculture is to choose a high value species, such as perch, walleye, or ornamental fish. The primary reason these fish command high market values is that their supply is very limited. As aquaculture develops successfully for a particular species and the market supply dramatically increases, the market price dramatically reduces. Also, do not be misled by these scenarios of apparently lucrative markets. There are reasons that the market is under supplied and driving the high price. The salmon and striped bass industries have experienced market price reductions of roughly 50% in less than a year's time. Similarly, tilapia prices have been falling from the high market prices of a couple years ago and prices may drop further as more production comes on the market.

To best predict price trends, one should compare current market prices for a particular species to some other commodity type fish, such as farmed salmon or catfish. These fish provide alternative choices to the consumer, thus bracketing the range of pricing that a consumer is willing to pay. Other premium meats also provide a substitute for the meat purchaser. For example, the catfish industry produces 600 million lb

(273,000 ton) per year in comparison to the US yearly production of tilapia of 18 million lb (8,200 ton). Pond side prices paid to the farmer for catfish are generally around $0.75/lb ($1.65/kg) on a whole-fish basis. Fillet yields for catfish are 45% and tilapia yields are now at best around 33%. Thus, for tilapia fillets to be competitive with catfish fillets, the equivalent farm price for tilapia would have to be $0.55/lb ($1.21/kg), which may indicate that the current farm prices of $1.00/lb ($2.20/kg) could fall even more. Genetic improvements in tilapia carcass yield will balance the pricing, but until then, large scale tilapia farming will have to be competitive with catfish farm prices as noted above. Similar pricing logic should be applied to other species. Remember, farm tank-side price can drop by half within 6 months. It has happened in the past, and it will happen again. Be forewarned and prepared.

> **"RULE OF THUMB"**
>
> Success in culturing high-value species will result in dramatic drops in market price.

15.5 SCALE EFFECTS AND RISK

The primary consideration that should dictate a particular biomass fish load for an independent production system is associated with catastrophic failure. Systems fail as units, and typically all fish in a particular failed unit will die. Biofilters are one source of system failure. One approach is to failure mitigation is to minimize the number of biofilters on a farm so that each can be designed with substantial redundancy and excess capacity, thus minimizing risk in this regard. The alternative approach is to build several small biomass systems, each with an individual biofilter, so that when a system fails, the severity of the economic loss is small since an individual system represents only a small fraction of the total production capacity. The FLA farm was designed with six independent production systems. Each production system was capable of producing 250,000 lb (114,000 kg) per year of fish; thus system components within a module were designed to handle this biomass load and associated feeding rates. By designing the farm as six independent units, this means that a particular unit could fail without impacting the other five units. It becomes an issue of distributing risk and risk associated with a system failure versus the economics of scale associated with designing larger and larger production modules. Design farms with at least four independent growout systems, so that in the event of a catastrophic failure to a specific system, the farm would only lose 25% of the current crop.

Biofilter characteristics are shown in Tables 15.1 and 15.2 (from Chapter 7). Note the difference in space and volume requirements between the two filter types. For small farm applications, e.g., less than 50,000 lb/yr of production (23,000 kg/yr), the choice of biofilter type probably does not make much difference. It is only in larger farms where the filter selection has a noticeable affect on overall profitability. So, if you are just starting, choose the biofilter that makes the most sense and is simplest for you to operate and service. Trickle filters are often a good starting biofilter.

Table 15.1 Ammonia Assimilation Rates for Different Classes of Biofilters

Filter Type	Rating Basis	Biofilter Ammonia	Oxidation Rates
		15–20°C	25–30°C
Granular	Volume of Unexpanded Media	0.6–0.7 kg TAN/m^3/day	1.0–1.5 kg TAN/m^3/day
Trickling & RBC	Surface area of media	0.2–1.0 g/m^2/day	1.0–2.0 g/m^2/day

Table 15.2 Volumetric Requirements for the Two General Types of Filters for a Unit Nitrification Loading Factor

Filter Type	Surface Area per Unit Volume m^2/m^3	TAN Oxidized per day	Assimilation Rate	Media Volume Required
Granular	1,000 to 10,000	1.0 kg TAN per day	1.0 kg TAN/m^3/day	1.0 m^3
Trickling & RBC	150–300	1.0 kg TAN per day	1.0 g TAN/m^2/day	5.0 m^3

Economically Competitive Scale

You will have to size the scale of your operation to be competitive in your chosen market. If you are going to market your own product, this will allow you to compete at the retail level, but this marketing strategy will require a significant time commitment to serve a series of small customers. For example, a restaurant customer will usually only want 10 to 20 lbs (5 to 9 kg) of fish per week, usually in fillet form. So, get ready to start cutting up fish!

The typical large tilapia farm in the US produces between 200,000 to 1,000,000 lb (91,000 to 455,000 kg) per year of whole fish. As large as this seems to be, these levels of production are not adequate to be competitive with the overseas producers of fresh tilapia fillets. They (the overseas competitors) are able to process fillets for about $0.03/lb ($0.661 kg), and their farms produce between 5 to 15 million lb/year (2,500 to 7,500 ton). You can do some quick pencil calculations on the costs to prepare fillets that will show even if you are really good at it, your hand labor costs will still approach $1.00/lb ($2.20/kg) fillet basis. Expert hand cutters and trimmers might be able to produce a fillet product for a cost as low as $0.50/lb ($1.10/kg). Obviously, this places you at a severe processing cost disadvantage to the large overseas farms, so hand labor processing is not feasible.

Fully automated processing equipment can reduce fillet processing costs to $0.15 to $0.25 per lb ($0.33 to $0.55/kg). Processing costs at this level would allow US producers to be cost competitive with the overseas suppliers. The problem is that the automatic filleting equipment requires volumes on the order of 5 to 10 million lb (2,300 to 4,500 ton) per year to justify the initial equipment expense. Additionally, other economies of scale start to emerge at this level of production as well. You will be able to do cost-benefit tradeoffs associated with in-housc feed manufacturing, oxygen production, electrical generation, and fish by-product utilization. Having a production capability in the range of several 100 thousand lb (45 ton) per year is probably the worst position to be in. Basically, at this level of production, you would have more fish than you could market to local niche local customers, but you would need a fairly large staff and significant fixed costs to operate the farm. All of these factors contribute to high overall production costs. Either you get really big like the successful catfish farmers have, or you stay really small and basically do everything yourself.

15.6 Labor Requirements

Labor costs will become a major factor in your operation. Large farms will provide human coverage 7 days a week, 24 hr per day (24–7). Smaller growout farms rarely employ this type of coverage. The question becomes what is large and what is small. The financial ramifications of catastrophic failure answer this question. If you can economically absorb the loss of all your fish, you might consider limited coverage. Certainly, during the first year of growout operations, we highly recommend as much coverage as possible. It is simply amazing the ways that failure can occur. Catastrophic losses will rarely occur while a human is on-site checking the fish and operation of the system equipment (this on-site human presence is in addition to the automatic sensing and monitoring that is in place, see Chapter 9). As the system matures and built in alarm systems for loss of flow, water level, low oxygen etc. are infrequently activated, then one could consider easing the coverage of the fish farm. Experience shows that about 50% of the problems that arise during farm operations are detected by the automated alarm system. Additionally, the alarm systems will occasionally detect a problem when there is none. Yes, they should be much better, but this is what happens in the real world. There is a delicate balance between monitoring too many things and not enough, and over-reliance on automated warning systems versus having none.

The authors' experience with a variety of tanks ranging in size from 2 to 11 m in diameter is that tanks of various sizes require similar man-hours to manage. In effect, it is the number of tanks and not the size of tanks that is important in determining management hours required. Efficient growers can adequately manage a series of tanks with a time budget of 20 to 30 minutes per day per tank system (11,000 L). Daily activities include daily water chemistry measurements, fish feeding, filter maintenance, and tank cleaning. Weekly maintenance of two to three additional man-hours per tank system for major cleaning activities and preventative maintenance is also necessary. Assuming a 40-hour work week, this indicates that one person could manage seven to nine tank systems (average of 4.3 to 5.3 man-hours per tank system per week). This is also consistent with the Fingerlakes Aquaculture experience, e.g., 5 line staff to manage a 48 tank facility 24–7.

Since many operations on a farm require two people, a facility could be designed assuming two full-time employees/owners to maximize labor efficiency. What is generally overlooked in estimating labor requirements is the additional short-term labor associated with intensive tasks such as harvesting and repair work. It is probably reasonable to estimate that there is about twice as much labor required as you can calculate as being needed. When everything is going well, fish systems require minimal labor, but then there is the rest of the time when things don't go so well! A farm owner might consider using hourly or contract labor for special tasks, e.g., harvesting, hauling, processing, etc. Losordo and Westerman (1994) used 8 hours per day to manage an eight tank facility with a 3 tank nursery (approximately 50% of the labor per tank used in our cost analysis).

15.7 INITIAL COSTS AND ECONOMIC ANALYSIS

The technology being used must support an overall economically competitive operation. The initial cost of your start up equipment will define in part the company's anticipated internal rate of return on the investment. You should assume a five year depreciation schedule for this equipment, but this factor is less important than initial costs in determining your expected economic returns. For larger scale farms (200,000 lb or 91 ton per year and larger), a good ratio of cost to production is that the equipment cost should not exceed $0.50 per lb ($1.10/kg) per year of production and the cost of buildings, land, utilities, etc. should not exceed an additional $0.50 per lb ($1.10/kg) per year. For small farming units, these costs will be at least twice these rates and could conceivably be even fourfold higher. If the up-front investment costs are too high, your farm's internal rate of return on investment will be too low to attract outside investors.

Calculating costs of production, profits and losses require a basic understanding of common economic terms. These are given for ease of reference below. The reader should consult a more complete business text for further help; e.g., Bangs, D.H. 1998. The Business Planning Guide: Creating a Plan for Success in Your Own Business. Upstart Publishing Company, Chicago, IL. The Bangs text is a succinct description of business economics and also covers the basics of writing an effective business plan.

TERMINOLOGY FOR ECONOMIC ANALYSIS

Profit & Loss Statement

This is called a Pro-forma when it projects profits and losses (P&L's) into the future.

Revenue

Self explanatory; usually set up in spreadsheet format where you can easily change price received for products sold. Create as many categories as necessary for products being sold.

Total Revenue = Sum of all revenues

Cost of Goods Sold (CGS)

Sometimes called direct costs, these are proportional to the amount of goods actually sold, e.g., feed, fingerlings, oxygen, processing fees.

Fixed Costs

These costs are incurred whether you sell any product or not (often called your overhead; goal is to make Fixed Costs as small a percentage of total costs as possible). Fixed Costs can be described as:

Fixed Costs = Operating Expenses + Selling, General & Administrative (SG&A)

Operating Expenses = Electricity (pumping) + Heating (air and water) + Hourly labor (front line labor) + Supplies + Insurance (building, key man, liability) + Taxes (property, payroll, sales, NOT Income taxes) + Royalties

SG&A = Management payroll, legal fees, professional services

Total Expenses and Net Income

Total Expenses = Variable Costs + Fixed Costs

Net Income (EBITDA) = Revenue - Total Expenses

Note: at this point Net Income does NOT reflect interest, taxes on profits, depreciation or amortization; hence the acronym EBITDA. EBITDA is very useful since the ITDA can be very dependent upon local banks, equity invested, etc. Financial analysis usually addresses this by calculating Internal Rate of Returns or Net Present Values.

Depreciation

Depreciation is the loss in value over time of a physical asset. From a tax perspective, the IRS regulates depreciation by creating classes of items, e.g., a truck, computer, pump, building. Each item has a maximum allowable depreciation period, e.g., light equipment might be 7 years, a commercial building = 39 years, a residential building = 28.5 years. Having selected the appropriate depreciation period for the class of item, then there are again IRS required methods to calculate depreciation for each year an asset is in service, e.g., straight line, double depreciation, MARCS (modified accelerated cost recovery system).

The Straight Line is the easiest and least confusing method, but accounting firms seem to like MARCS, which requires a look up table approach and accelerates depreciation to the front of the period.

Using straight line, you have to assign a salvage value to the asset or the items value when it would be finally taken out of use at the end of its life cycle. Again, the simple approach is to assign zero, which maximizes allowance depreciation each year, and then you calculate depreciation as:

$$Depreciation\ Cost(\$ / yr) = \frac{Total\ \ Cost - salvage\ \ value}{Years\ of\ life}$$

Units become \$/yr. So, if you purchased a pump for \$1,000, assigned zero salvage value and assigned a useful life of 5 years, your cost per year for depreciation on this item is \$200/yr. In terms of government taxing, the depreciation cost is subtracted from income making taxable income less by the depreciation amount. Now, the tricky part is that if you sell the item being depreciated, this becomes income for your company if sale price is greater than book value. Net income for this sale will be selling price less your current book value of the item, which is initial value less the claimed depreciation to date. So, if you sell the item for its current book value (the value of the asset on the company's tax records), no taxable income is generated.

15.8 ECONOMICS OF RAS

Based upon our experience using RAS, the following table summarizes costs and required inputs for a commercial scale tilapia farm in a northern US climate location.

As mentioned earlier, smaller farm units will incur much higher production costs per unit, and unit input parameters will also be significantly higher. For example, purchasing oxygen in typical welding size cylinders (282 cubic feet, 8 m^3, or 23.5 lbs, 10.7 kg, of oxygen), will cost \$3 or \$4 per 100 cubic feet (2.82 m^3) of product (or 10 fold the costs listed in Table 15.3). Labor is another area where the larger farms achieve great economies of scale, since it takes almost the same labor to service small tanks as it does large tanks and it takes about the same amount of time to for security checking whether there are two tanks or 20.

Table 15.3 Unit Costs and Required Inputs for Large Scale Tilapia Production

Item	1 million lb/yr farm	5 million lb/yr farm
Size of Building, ft^2 (m^2)	40,000 (3,720)	200,000 (18,600)
Total Cost of Facility & Equip	$1,500,000	$7,500,000
Number of Pods	6	30
Volume per Pod unit, gallons (m^3)	76,000 (288)	same
Production per Pod, lb/yr (kg/yr)	250,000 (114,000)	same
Direct Labor Per Building	5 FTE	20 FTE
Growth rate: egg to 700 gram	28 weeks	28 weeks
Feed Conversion Ratio (high quality high protein feed)	1.0 to 1.2	1.0 to 1.2
Electric Use, kWh per Lb (per kg)	1.3 (2.9)	same
Heat Use, Therm[b] per lb (kJ/kg)	0.17 (48,000)	same
Oxygen Use lb/lb (kg/kg) of Feed Fed	1.9 (low efficiency)	1.0 (high efficiency)
Feed Purchase Cost (41% Protein) $/lb ($/kg)	$0.28 ($0.62)	$0.18 ($0.40)
Oxygen Supplied cost[a], $/100 ft^3 ($/m^3)	$0.34 (0.12) + facility fee[a]	same
Electrical Power, $/kWh	$0.04	same
Gas Cost, $/Therm ($/100,000 kJ)	$0.85 ($0.80)	$0.10 ($0.095) (energy plant)[c]
Unit Cost: $/lb whole fish (fillet basis)	$0.78 ($2.52)	$0.55 ($1.77)
Process, Pack, Ice $/lb fillet	$1.50	$0.55
Cost of Fillet FOB, $/lb	$4.02	$2.32

[a]the facility fee will usually add around 20% to the cost of the delivered oxygen
[b]Therm = 100,000 BTU (105,500 kJ)
[c]Large scale farms would most likely be placed at a site with an available, underutilized heat source

15.9 RAS Tilapia, Catfish Pond & Broiler Production

The authors admit that we are enthusiastic about the economic prospects of RAS technology. An often encountered argument against RAS is that they can not compete against large commercial fish ponds such as those used by the US commercial catfish industry. This subsection provides a projection of large scale commercial tilapia produced using RAS (as shown in Table 15.3). A breakdown of anticipated capital costs associated with building a large scale tilapia farm are given in Table 15.4 (based upon actual year 2000 data). We show comparisons of catfish and tilapia to the US broiler industry. Prices, depreciation values, and associated economic factors for the Mississippi catfish farm are taken from Keenum and Waldrop (1988) and were for a 590,000 kg/year farm (summarized in Tables 15.5–7). We make cost predictions of tilapia production, assuming normal technological and management improvements, Table 15.8, and provide a comparison to current tilapia and broiler production costs.

Overall, the tilapia production costs were slightly higher than the catfish production costs, \$1.76 per kg versus \$1.56 per kg. The major point of the comparison shown in Table 15.7 is that when indoor tilapia production is practiced on a scale similar to the large USA outdoor catfish ponds, the costs of production are also very similar. Initial system capital costs for tilapia (\$2.03) are larger than catfish (\$1.44) per kg per yr of production. These investment costs are roughly three to four times the initial capital investments for broiler production of \$0.49 per kg per yr of production capacity.

At this point, there is considerable information available regarding the expected cost of tilapia production for large scale operations and what kind of labor savings could be anticipated to develop over the next five to ten years. For example, many public electric utilities will reduce electrical rate charges by 25% once a load of 500 kW is reached. Large facilities can self-generate electrical power using large electrical generators. This is particularly attractive, since a generator system is required whether the generator is used as a backup system to power outage or as a primary generator. In this case, the cost of electricity to the fish farm is only the operating cost of the unit. Maintenance and depreciation will be slightly more when the generator is used as primary as opposed to backup, but there will not be a lot of difference. Labor requirements will continue to reduce due to automation and tank scaling and we use 50% of the current example in our projected cost scenario. Catastrophic fish insurance would no longer be deemed necessary, since the farmer would become self-insured. System costs are expected to decrease by 25% in comparison to current costs due to improvements and refinements in system designs.

Table 15.4 Capital Cost Characteristics Associated with Tilapia Water Reuse System. Tilapia Production Level of 600,000 kg/yr (1.3 million lb/yr)

Component of Initial Capital Cost	Cost
Growout tanks	
Pumps	
Oxygen and CO_2 control units	
Electronic controller	
Feeders	
Large Scale Biofilter (treat multiple tanks)	
Quarantine hatchery /fingerling area (series of small tanks)	
Subtotal tank costs (approximate)	**$385,000**
Other Equipment (approximation)	
Backup generator (400 kW system)	$60,000
Monitoring system	$20,000
Ice machine (2 ton unit)	$10,000
Feed bin and auger system	$20,000
Harvesting system	$40,000
Water heating system	$30,000
Waste catchment unit	$10,000
Ventilation system	$5,000
Water wells (2)	$10,000
Fish handling equipment	$10,000
Subtotal Other equipment	**$215,000**
Total Equipment Costs (7 year depreciation period)	**$600,000**
Building Costs	
Quarantine area	$15,000
Laboratory and office space	$15,000
Building space	565,000
Septic/restroom	$5,000
Subtotal building Costs (20 year depreciation period)	**$600,000**
Land costs (non depreciated)	**$20,000**
Total funds required (equipment, building, land and contingency)	**$1,220,000**

Table 15.5 Economic Parameters Used for Catfish (Keenum and Waldrop, 1988) and Tilapia Production Analysis

	catfish	tilapia
Feed cost (32% protein, $/kg)	$0.40	$0.40
Feed/gain ratio	2.00	1.40
Harvest weight, kg	0.57	0.68
Fingerlings, $/fingerling	$0.075	$0.060
Cumulative mortality, %	5	5
Electric costs, $/kWh	$0.085	$0.077
Oxygen cost, $/kg)	na	$0.19
Heating fuel costs, $/100,000 BTU	na	$0.62
Depreciation (straight line, zero salvage; 7 yrs equipment, 20 yrs bldg)	–	–
Interest on investment	11%	11%
Interest on operating capital (based upon 50% of total operating costs), %	10%	10%
Repairs and maintenance (% of initial cost)	~5%[a]	5%
Labor ($/man year include 35% fringe benefit rate)		
1- manager	$35,000	$47,000
1- foreman	$26,000	$34,000
3- workers catfish and 2- tilapia	$36,000	$33,000
Total personnel costs	$97,000	$114,000
Total land area, hectares	131.0	2.0
Land value, $/hectare	$1,975	$2,500
Harvesting and hauling, $/kg	$0.09	$0.09

[a]Keenum and Waldrop (1988) perform a very elaborate analysis of repairs and maintenance, but from a simple practical approach, there is minimal difference from using 5% of the initial cost.

Table 15.6 Production System Characteristics Associated with Tilapia Indoor System

Production Characteristic	Value
Size of building	1,780 m^2
Individual growout tank	16 tank facility (7.6 m diameter x 1.4 m deep) 60,000 L
Yearly harvest	590,000 kg
Design Parameters	
Maximum Density	100 kg/cubic meter
Feeding rate (depends on fish size)	2% to 3% body mass per day
Feed conversion rate (feed/gain)	1.40 kg/kg
Supplemental oxygen	0.4 kg oxygen per kg of feed fed
Oxygen absorption efficiency	75%
Power per tank system	9 kW (4,500 Lpm pumping rate per tank)
Fish target harvest size	680 g
Daily water exchange, % of system volume	5%
Temperature difference for water exchange	19.4°C
Building infiltration, minimum air volumes/hr	2

Table 15.7 Comparison of Tilapia, Catfish and Broiler Production Costs ($/kg) for Farms With a Yearly Fish Production of Approximately 590,000 kg

	Costs per kg[a] produced ($/kg)			% of Total Costs		
	tilapia	catfish	broiler	tilapia	catfish	broiler
Ownership costs						
Depreciation	0.28	0.11	contract	15.9	7.1	–
Interest on investment	0.07	0.10	contract	3.9	6.4	–
Catastrophic fish insurance (3%)	0.07	–	contract	3.9	–	–
Liability insurance + land taxes	0.01	0.01	contract	0.7	0.6	–
Subtotal:	0.43	0.22	0.05	24.4%	14.1%	7.7%
Cost of goods services						
Feed	0.55	0.81	0.39	31.2	51.9	60.0
Fingerlings (or chicks)	0.10	0.14	0.09	5.7	9.0	13.9
Oxygen	0.14	–	–	8.0	–	–
Subtotal:	0.79	0.95	0.48	44.9%	60.9%	73.9%
Operating expenses						
Chemicals	–	0.05	0.06	–	3.2	9.2
Repairs & maintenance	0.07	0.04	–	3.9	2.6	–
Heating water	0.02	–	–	1.1	–	–
Heating air	0.03	–	–	1.7	–	–
Electric	0.16	–	–	9.1	–	–
Other utilities	–	0.08	0.02	–	5.1	3.1
Management labor + fringe	0.19	0.17	0.04	10.8	10.9	6.2
Misc.	0.01	–	–	0.7	–	–
Interest on operating capital	0.06	0.05	–	3.4	3.2	–
Subtotal:	0.54	0.39	0.12	30.6%	25.0%	18.5%
Total cost of production	1.76	1.56	0.65	100%	100%	100%

Note: Broiler costs are broken down for comparison based upon contract grower payments and allocation of costs between grower and integrator.

[a]Whole fish basis, non-dressed, tilapia yields approx. 32% and catfish 42%.

Table 15.8 Projected Costs of Large-Scale Tilapia Production in Comparison to Current Production Costs for Tilapia and Commercial Broilers

	Tilapia	Projected Tilapia	Broiler
Ownership Costs, $/kg			
Depreciation	$ 0.28	$ 0.10	
Interest on Investment	$ 0.07	$ 0.05	
Catastrophic Fish Insurance (3%)	$ 0.08	–	
Subtotal:	**$ 0.43**	**$ 0.15**	**$ 0.05**
Costs Goods Services ($/kg)			
Feed	$ 0.55	$ 0.42	$ 0.39
Fingerlings	$ 0.10	$ 0.04	$ 0.09
Oxygen	$ 0.14	$ 0.09	–
Subtotal:	**$0.79**	**$0.55**	**$0.48**
Operating Expense			
Chemicals	–	–	$0.06
Repairs & Maintenance	$ 0.07	$ 0.05	–
Heating Water	$ 0.02	$ 0.01	–
Heating Air	$ 0.03	$ 0.02	–
Electric	$ 0.16	$ 0.10	–
Other Utilities	–	–	$ 0.02
Phone	–	–	–
Management Labor + Fringe	$ 0.19	$ 0.10	$ 0.04
Misc.	$ 0.01	$ 0.01	–
Interest on operating capital	$ 0.06	$ 0.01	–
Subtotal:	**$0.54**	**$0.30**	**$0.12**
Total Cost of Production ($/kg):	**$1.76**	**$1.00**	**$0.65**

Comparison to Broiler Production

Ultimately, fish production from aquaculture will have to compete with other commodity meats such as poultry. It is informative to compare predicted costs of production for fish from indoor and outdoor facilities to those of broilers. Broiler production data is taken from USDA statistics (USDA, ERS 1996A, B, C) and the authors' personal knowledge gained from 20 years working in the industry. USA broiler production is a vertically integrated industry, with the broiler grower being the contract farmer and the foundation of the system. The farmer owns the building, provides husbandry, and pays most of the cost of utilities. For these services, the farmer is paid approximately \$0.09 to \$0.11 per kg of broiler produced. Thus, all costs associated with building ownership, depreciation of capital equipment, labor, and utilities (electric and water and generally about 50% of the fuel heating costs) are borne by the farmer. The productivity per worker has increased from 95,000 kg of broilers per year in 1951 (Watt Publishing, 1951) to 950,000 kg per year in 1991 (Perry, 1991). Similar improvements have been made in equipment, housing, and nutrition and genetics; North (1984) provides an extensive description of all facets of commercial poultry production. It is interesting to note that broiler production in the 1950's was around 5 million kg per year. The productivity per unit of worker and total broiler consumption of the 1950's is very similar to the current productivity standards of the USA tilapia industry (8 million kg per year whole fish basis) and the productivity per person in the fish farming business is approximately 25,000 to 110,000 kg per year.

The primary factor that led the poultry industry to convert from outdoor production to indoor farming was the tremendous savings in labor costs. As mentioned earlier, the labor productivity for indoor broiler production is currently about eight times greater than the labor productivity currently being obtained for indoor tilapia fish farming. Ultimately, indoor fish production has two distinct advantages over poultry production: higher feed conversion efficiency and greater productivity per unit area of building. Broiler production has feed conversion efficiencies of approximately 2.00 (2.09 bird weight, feed to gain ratio on feed energy levels of 3,170 kcal/kg and protein levels of 19.5%), while tilapia conversions are currently in the 1.2 to 1.5 range for feed energy levels of approximately 2,500 to 3,200 kcal/kg and protein levels of 32 to 40%. The yearly meat output per unit floor area from the tilapia system is 255 kg/m^2 compared to 122 kg/m^2 from a broiler house. Thus, net economic productivity per year from a fixed tilapia production facility could be higher than from a broiler facility of similar size, even though the unit costs of tilapia production per weight are slightly higher in comparison to broilers. The advantage for fish production systems is their higher potential rate of return per year from a fixed facility.

CHAPTER 16[1]

FISH NUTRITION AND FEEDS

16.0 INTRODUCTION

This chapter is an overview of fish nutritional requirements and the general physiological considerations that influence the nutrition, feeding, and welfare of fish. It is not meant to be a definitive nutritional text; the intent is to provide basic information to non-nutritionists so they might be able to better understand the inter-relationships between the Recirculating Aquaculture System (RAS) and the cultured animals.

16.1 FEED MANAGEMENT

Feed management involves selecting the proper feed, ordering the proper amounts, scheduling the delivery in a timely manner, storing the food properly, using older feed first, and implementing efficient feeding methods. All of these factors can have an effect on the bottom line. Poor feed management can result in any number of problems. For example, if the proper feed size or formulation is not selected, the fish may not grow efficiently. If a low-phosphorus feed is not selected, effluent problems can result. On the other hand, sound feed management practices can assure optimum growth and lower total costs. The purpose of this section is to discuss the steps needed to properly acquire, store, and distribute feed. Also, for those with limited background in fish nutrition, we review some basic principles of feed nutrition.

---

[1] This chapter was co-authored by Dr. Paul D. Maugle, P. D. M. and Associates, 88 Central Avenue, Norwich, CT., 06360; www.Aquanet.com/pdm/; and Dr. H. George Ketola, Research Physiologist, Tunison Laboratory of Aquatic Science, U.S.G.S., Great Lakes Science Center, Cortland, NY 13045.

Feeding Techniques

The goal of efficient feed management is to calculate the proper ration and present it to the fish. Underfed fish do not reach maximal growth, and may exhibit aggressive behavior during feeding due to limited feed availability, thereby potentially harming themselves or other fish. Underfeeding may also increase the variability of fish sizes within a tank. Overfeeding results in uneaten feed that results in poorer water quality, poorer economic performance, and additional environmental pollution.

Methods

As a percentage of body weight, the quantity of feed required to sustain optimum growth will decrease as the fish grow from fry to fingerlings (see Table 16.1). The fish need to be fed frequently when they are small. By using belt feeders, you can feed continuously or at selected periodic intervals. Generally, the ration for the fish should be the amount they can eat in the 15 to 20-minute period after feed distribution. Crampton et al. (1990) presented an equation to suggest interval times between feedings for salmonids based upon the rate of digestion. Equation 16.1 can be used to develop a scheduled feeding program. As can be seen, the frequency of feeding can be decreased as the fish gain body size. Equation 16.1 should not be applied below 6°C and some maximum interval should be applied, e.g., 24 hours.

$$INTERVAL = 40\,\frac{\sqrt{W}}{T^{1.1}} \qquad (16.1)$$

where

INTERVAL	= time between feedings, hr
W	= body mass, g
T	= temperature, C

Often some feed may exit the tank before the fish are able to consume it. If this occurs, feed more slowly over a longer period of time, or deposit feed closer to the edges of the tank so that the feed takes longer to exit the tank. A rule of thumb for the quantity of ration is to feed a weight of 1% or less of the biomass of the tank at each feeding interval. Feeding multiple times per day increases feed acquisition and feed conversion efficiency. A general guide to feeding rainbow trout is given in Table 16.1. This guide relates the amount fed to the size of the fish and water temperature.

When presenting feed, scattering food is generally better than fixed point feeding, as more of the population has simultaneous access to the feed. This helps reduce wastage, and may also increase uniformity of fish growth. For this reason, when using demand feeders, supplement with hand feeding. Demand feeders are useful because they are inexpensive devices, and make food available to the fish during times when the operator is not at the facility, such as at night. Always do some hand feeding each day for each tank, even when fully automated feeding systems are in place. The visual observation of feeding behavior is extremely important in maintaining a clear understanding of current fish health and whether or not a tank of fish is being over or underfed.

When Not to Feed

Although the fry and fingerlings are normally fed frequently to maximize growth, there are certain conditions that require withholding feed. Take fish off feed completely if any of the following occurs:

- High temperatures
- Fish are sick or stressed
- 24–48 hours before transport
- 24 hours before sampling
- 3–4 days before processing
- Low oxygen levels
- Poor water quality

16.2 SELECTION

Simply stated, optimizing your feed costs would involve minimizing your feed costs per pound of fish growth. A variety of factors will influence this ratio, and the feed with the lowest purchase price may not necessarily result in the best fish growth rates or feed conversion. Some of the factors that will influence how feed should be selected are:

- Growth performance
- Feed quality
- Delivery costs
- Physical characteristics
- Formulation
- Impact on water quality

Table 16.1 Recommended Amounts of Dry Feed for Rainbow Trout per Day, Given as Percent Body Weight, for Different Size Groups Held in Water of Different Temperatures (or Pounds Feed per 100 Pounds of Fish), in Relation to Fish Size and Water Temperature, (Speece et al. 1982).

Water Temperature (°F)	Number of Fish per Pound										
	2,542+	2,542-304	304-88	88-38	38-20	20-12	12-7	7-5	5-3.5	3.5-2.5	under 2.5
	Approximate Size in Inches										
	Under 1	1–2	2–3	3–4	4–5	5–6	6–7	7–8	8–9	9–10	10+
44	3.8	3.1	2.5	2.0	1.5	1.3	1.0	0.9	0.8	0.8	0.6
45	4.0	3.3	2.7	2.1	1.6	1.3	1.1	1.0	0.9	0.8	0.7
46	4.1	3.4	2.8	2.2	1.7	1.4	1.2	1.0	0.9	0.8	0.7
47	4.3	3.6	3.0	2.3	1.7	1.4	1.2	1.0	0.9	0.8	0.7
48	4.5	3.8	3.0	2.4	1.8	1.5	1.3	1.1	1.0	0.9	0.8
49	4.7	3.9	3.2	2.5	1.9	1.5	1.3	1.1	1.0	0.9	0.8
50	5.2	4.3	3.4	2.7	2.0	1.7	1.4	1.2	1.1	1.0	0.9
51	5.4	4.5	3.5	2.8	2.1	1.7	1.5	1.3	1.1	1.0	0.9
52	5.4	4.5	3.6	2.8	2.1	1.7	1.5	1.3	1.1	1.0	0.9
53	5.6	4.7	3.8	2.9	2.2	1.8	1.5	1.3	1.1	1.1	1.0
54	5.8	4.9	3.9	3.0	2.3	1.9	1.6	1.4	1.3	1.1	1.0
55	6.1	5.1	4.2	3.2	2.4	2.0	1.6	1.4	1.3	1.1	1.0
56	6.3	5.3	4.3	3.3	2.5	2.0	1.7	1.5	1.3	1.2	1.0
57	6.7	5.5	4.5	3.5	2.6	2.1	1.8	1.5	1.4	1.2	1.1
58	7.0	5.8	4.8	3.6	2.7	2.2	1.9	1.6	1.4	1.3	1.2
59	7.3	6.0	5.0	3.7	2.8	2.3	1.9	1.7	1.5	1.3	1.2
60	7.5	6.3	5.1	3.9	3.0	2.4	2.0	1.7	1.5	1.4	1.3
61	7.8	6.5	5.3	4.1	3.1	2.5	2.0	1.8	1.6	1.4	1.3
62	8.1	6.7	5.5	4.3	3.2	2.6	2.1	1.8	1.6	1.5	1.4
63	8.4	7.0	5.7	4.5	3.4	2.7	2.1	1.9	1.7	1.5	1.4
64	8.7	7.2	5.9	4.7	3.5	2.8	2.2	1.9	1.7	1.6	1.5
65	9.0	7.5	6.1	4.9	3.6	2.9	2.2	2.0	1.8	1.6	1.5
66	9.3	7.8	6.3	5.1	3.8	3.0	2.3	2.0	1.8	1.6	1.6
67	9.6	9.1	6.6	5.3	3.9	3.1	2.4	2.1	1.9	1.7	1.6
68	9.9	9.4	6.9	5.5	4.0	3.2	2.5	2.1	2.0	1.8	1.7

16.3 GROWTH PERFORMANCE

Being able to specifically evaluate growth is the key to evaluating different feeds. Feed to gain ratio (FG) or its inverse, feed conversion ratio (FCR), and growth rate are two of the best tools for evaluating the growth of your fish. FG is normally expressed as pounds of feed per pound of gain and measures how efficiently the fish are converting the feed into body mass. The goal is to get this number as low as possible.

In aquaculture, FG's of less than one can be obtained, especially when using high energy feeds and concurrently providing good management and water quality. Some are confused when this number drops below one, but this is possible because a large portion of fish weight is water weight (approximately 75–80%). This less than one FG ratio never occurs in warm-blooded animals; the best feed converters are chickens and their FG is around 2.0. One weakness of measuring FG is that the number for pounds of feed fed cannot be accurately adjusted to account for the amount of feed presented to the fish but not eaten by the fish.

The rate of growth, expressed in weight or length gained per day, or Standard Growth Rate (SGR), is a unitless measurement of growth for measuring how fast fish grow. Using a more expensive feed that gives faster growth rates but may reduce overall costs by decreasing the cost of energy and labor required for operating the facility because the production cycle is shortened. Calculating the total costs to raise a pound of fish will show how feed selection can affect other costs. Calculating calorie and protein intake and monitoring fish mortality and disease symptoms are other aspects of benchmarking growth.

To optimize growth, consult the manufacturer's feeding tables, sample data, and use your personal observations. It may be difficult to find feeding tables that give practical instruction on quantities and schedules for feeding. If you are unable to find such a table, you may use the feeding tables provided as examples (see Table 16.1) and the growth equations presented in Chapter 4 to construct an initial feeding regimen, and then collect accurate feed and growth data to be used in refining your feeding tables for reference during future production cycles. By developing customized feeding tables, the feeding regimen will better fit the conditions at your facility. Once you have established rates of growth at your own facility (preferably as inches per month at specific temperatures), you can adjust your feeding charts to show feeding requirements as a function of temperature.

16.4 FEED QUALITY

Feed quality directly influences growth and health of the fish as well as the characteristics of the discharge stream. Although feed quality will vary between manufacturers, it can also vary from lot to lot from the same manufacturer. The exact formulation may change due to price fluctuations or availability of fish ingredients. Save (by photocopy) feed labels so that if you have a problem, you can revert to them as an aid to problem investigation. Although many of the factors regarding feed quality will not be known until after the feed has been tested, there are several quality related factors that can be evaluated before feeding, including date of manufacturer, phosphorous levels, and amount of fines. Compare the date of manufacture to the date of delivery. An example of a lot number stamp is given below. Ideally, the two dates should not differ by more than about a month, and all feed of a specific type should have a similar date of manufacture. Monitor delivery times, as feed sitting in a non-air conditioned warehouse or hot delivery truck can degrade feed quality.

Example: Deciphering lot numbers on a bag of feed
(from Zeigler Brothers, Gardners, PA)

LOT # 051398 066 1546

- The first set of six digits consists of three 2-digit numbers corresponding to the date of manufacture as month-day-year. In the example, the date is May 13, 1998.

- The second set of digits (max of 4 digits) corresponds to the bag identification number for that day. This is bag #66 produced on that day.

- The third set of digits indicates time of production in military time. The bag of feed was produced at 3:46 p.m.

"Fines" are extremely small particles of feed, and may be observed when feeding the fish. These small particles are not utilized efficiently and not only are they wasted food; they contribute to water quality problems. With current feed milling and manufacturing technology, there should be little or no fines in the bag, e.g., less than 1% fines by weight in a bag of feed. Fines are produced by poor manufacturing processes or excessive handling of the bagged feed. Feed of a certain pellet size may exhibit fines, while feed from the same manufacturer of the same formula but having a different pellet size may be free of fines. Fines contribute to a poor quality effluent by increasing the level of suspended solids in the water column. Consult the manufacturer if there is a problem.

16.5 Physical Characteristics

Generally, feed manufacturers have three categories of feeds: starter, fry, and growout. The feed formulations optimize the nutritional characteristics and feed sizes for specific stages of fish development. The general rule for choosing a feed size is to deliver the largest size that the smallest fish will eat. The proper feed size can be selected by comparison of feed size to mouth gape size, gill-raker spacing, and esophagus width. Many manufacturers use fish length as an indicator of proper feed size for specified species. These correlations can be used to determine when to change feed sizes, but generally, observation is also an effective method of assessing when to introduce larger sized feed. In the early stages of development, the fish will be progressing to larger feed sizes rapidly, so purchase only a few bags of the smaller sized feed, and order a variety of sizes ahead of time. Use past experience in making decisions on how much feed of a particular size to order. Consider feeding behavior if selecting between sinking, floating, or slow-sinking pellets. Table 16.2 shows some recommended feed sizes as related to trout size and would be a good starting point for other species of fish.

Table 16.2 Recommended Feed Sizes as Related to Fish Size

Feed Form	US Screen Size (Mesh)	Feed Size mm	Feed Size inch	Fish Size grams	Number per pound
Starter	30–40	0.25		<0.23	2,000^{+}
No. 1 Granule	20–30			0.23–0.6	2,000–800
No. 2 Granule	16–20			0.6–1.8	800–250
No. 3 Granule	10–16			1.8–4.5	250–100
No. 4 Granule	6–10			4.5–15.0	100–30
Pellet		3.2	1/8	15–45	30–10
Pellet		4.7	3/16	45–454	10–1
Pellet		6.3	1/4	>454	<1

Finished feeds are periodically tested to assure that tag specifications are in accordance with national regulations. In the United States, quality control procedures are in place to verify that national standards established for both feed ingredients and finished feeds are met. In this way, feed tags provide some assurance regarding the freshness and quality of feeds. Additionally, the feed manufacturer should periodically monitor feed composition materials and finished products for non-nutrient components such as Mycotoxin, thiobarbaturic acid, histamine. Vitamin E and vitamin C are highly susceptible to oxidation and should be periodically assessed. The fish farmer upon receiving each shipment of feed should (must) take and keep multiple samples (with the feedbag tag included) from every lot of feed. Feed samples must always be kept in a freezer until the cohort of fish that has consumed this food has been sold. If a problem occurs during the growout period, the feed may need to be analyzed by a reputable laboratory to determine if there may be nutritional deficiencies.

The wholesomeness of a finished feed is dictated by how well it is stabilized to retard spoilage using antioxidants, such as ethoxyquin. After a feed is made, the interaction between cationic minerals (such as iron, copper, zinc, magnesium and manganese) catalyze oxidation of vitamins and unsaturated fatty acids, which results in reduced nutritional quality. A well-made feed contains several additives and ingredients that serve to slow the oxidation of essential nutrients in a finished feed. Traditional feed formulation may include one or two of these feed protectants but the production of complete feeds should incorporate a more thorough approach.

Manufacturing wholesome feeds should include the following steps to maintain the quality of the feed:

- Addition of antioxidants, e.g., ethoxyquin
- Ingredient arrival testing
- Stabilized Vitamin E supplementation
- Stabilized Vitamin C supplementation
- Extrusion pasteurization
- Mold inhibitors
- Mycotoxin binders
- Feed quality assurance testing

16.6 PRACTICAL FEED FORMULATION

Practical feed formulations generally contain no more than six to eight macro ingredients. The nutrient data matrix of each ingredient is utilized to meet minimum and maximum species nutritional restrictions. Nutrient restrictions are established based on experience, and generally meet and in some cases, exceed nutritional requirements. Data reports indicating the quantity of nutrients within each ingredient should be periodically updated, as seasonal changes in feedstuffs occur. In addition, a digestibility factor for each nutrient within the ingredient needs to be assigned based on species being fed. The quality of the matrix information greatly impacts the nutritional quality of the finished feed. The minimum and maximum species nutritional restrictions along with the nutrient database of each ingredient, the digestibility of the ingredient, and ingredient costs are then used in linear computer programs to optimize the mixture of ingredients that best meet the nutrients considered. The level of dietary energy is adjusted to provide the optimum protein:energy ratio to meet the nutritional requirements of the size and species being reared. The digestible amino acid profile of the protein is balanced for essential amino acids.

Even though both catfish and tilapia are omnivores, digestibility of feedstuffs is different between the species. Although it is sometimes done, it is not appropriate to feed a diet formulated to rear catfish in ponds to tilapia in recirculating systems. A currently recommended diet makeup for tilapia would follow the guidelines shown in Table 16.3. More information on digestibility of protein, fat, and carbohydrates is provided in Table 16.4 for chinook salmon, rainbow trout, channel catfish, and blue tilapia (NRC, 1993). Table 16.5 provides true amino acid availability and protein digestibility values for certain feed ingredients for Atlantic salmon and channel catfish (NRC, 1993). Table 16.6 summarizes net absorption of phosphorus from various sources by channel catfish, common carp, and rainbow trout (NRC, 1993), and Table 16.7 provides specific information on digestibility coefficients of feedstuffs for tilapia O. niloticus (Lovell, 1989).

Wilson and Poe (1985) have shown that feed manufactured using extrusion as compared to pelleted processing increased the digestibility of energy but had no effect on the digestibility of protein. Popma (1982) described the difference in digestible energy between catfish and tilapia fed the same feedstuffs. Differences in digestible energy of key ingredients used in both catfish and tilapia commercial feeds are summarized in Table 16.8.

Table 16.3 Nutrient Composition of Typical Tilapia Feeds

	Nutrient Composition of Typical Tilapia Feeds			
Fish Weight Nutrient	<2.0 Gram	2 to 10 Gram	10 to 50 Gram	50 to 545 Gram
Energy, digestible – Min	4.0 Kcal/g	3.8 Kcal/g	3.5 Kcal/g	2.9 Kcal/g
Protein – Min	48	45	40	36 or 32
Lipid – Min.	10	10	10	10 to 5
Fiber – Max.	4	4	5	5
Ash – Max.	7	7	9	9
Starch – Min.	12	14	20	24
Ca. – Max.	1	1	1	1.5
Avail. P – Min.	0.6	0.6	0.6	0.6
Lysine – Min.	2	2	1.9	1.7
Methionine – Min.	0.9	0.85	0.85	0.7
Threonine – Min.	1.2	1.2	1.2	1

Ingredients selected for use in tilapia feeds can significantly impact the digestibility of the finished feed. Growth, feed conversion, and subsequently generated pollution are directly related to the degree of digestibility of the finished feed and the amount and type of feces produced. Feedstuff selection for recirculating systems is not only influenced by unit costs for energy, protein, amino acid composition and ingredient digestibility but also the level of phosphorus. The phosphorus/nitrogen (P/N) content of many ingredients such as animal by-product meals, with some exceptions such as herring, feather and blood meal (see Table 16.19) have a high P/N ratio. Generally, plant protein ingredients such as soybean and corn gluten meals have the desirable characteristic of a lower P/N ratio, which makes them appropriate for use in low environmental impact diet formulations.

Table 16.4 Apparent Digestibility of Protein, Fat, & Carbohydrate in Diet Ingredients for Chinook salmon, Rainbow Trout, Channel Catfish, and Blue Tilapia (NRC, 1993)

Ingredient	International Feed No.	Protein (%) Chinook Salmon	Rainbow Trout	Channel Catfish	Blue Tilapia	Lipid (%) Rainbow Trout	Channel Catfish	Blue Tilapia	Carbohydrate (%) Rainbow Trout	Channel Catfish	Blue Tilapia
Alfalfa meal	1-00-023	-	61[a]	-	66[b]	71[c]	51[d]	-	-	12[d]	27[b]
Blood meal	5-00-381	30[e]	69[a]	74	-	-	-	-	-	-	-
Casein	5-01-162	-	95[a]	97[f]	-	-	-	-	-	-	-
Canola meal	5-06-145	79[c]	-	-	-	-	-	-	-	-	-
Corn grain	4-02-935	-	95[c]	60[d]	84[b]	-	76[d]	90[b]	-	66[d]	45[b]
cooked		-	-	66[d]	79[b]	-	96[d]	-	-	78[d]	72[b]
Corn gluten meal	5-28-242	-	87[a]	-	-	-	-	-	-	-	-
Cottenseed meal	5-01-621	-	76[a]	83[g]	-	-	88[d]	-	84[a]	17	-
Fish, anchovy meal	5-0-985	92[e]	-	-	-	-	-	-	-	-	-
Fish, herring meal	5-02-000	91[e]	87[a]	-	-	97[c]	-	-	-	-	-
Fish, manhaden meal	5-02-009	83[e]	-	88[g]	85[b]	-	97[d]	98[b]	-	-	-
Meat and bone meal	5-00-388	85[c]	82[g]	78[b]	-	77[c]					
Poultry by-product meal	5-03-798	74[e]	68[c]	-	-	-	-	-	-	-	-
Poultry, feathers, hydrolyzed	5-03-795	71[c]	58[c]	74[d]	-	68[c]	83[d]	-	-	-	-
Soybean meal, 44%	5-04-604	-	-	-	94[b]	-	-	-	-	-	54[b]
Soybean meal, 48%	5-04-612	77[c]	83[a]	93[g]	-	-	81[d]	-	-	-	-
Starch, corn (uncooked)											
50%		-	-	-	-	-	-	-	24[a]	-	55[f]
25%		-	-	-	-	-	-	-	-	-	61[f]
Starch, corn (cooked)											
50%		-	-	-	-	-	-	-	52[a]	-	66[f]
25%		-	-	-	-	-	-	-	-	-	78[f]
Wheat middlings	4-05-205	86[c]	76[a]	72[d]	-	-	-	-			
Wheat grain	4-05-268	-	-	92[g]	90[b]	-	96[d]	85[b]	-	59[d]	61[b]
extruded, 44%		84[e]	-	-	-	-	-	-	-	-	-

NOTE: Dashes indicate data were not available. [a]Smith (1977) and Smith et al. (1980) determined by metabolism chamber, single ingredient fed.; [b]Popma (1982) determined by indicator method, feces collected by frequent removal from water, ingredient fed in mixed diet; [c]Cho et al. (1982) determined by indicator method, feces collected from a settling column outside the fish tank, ingredient fed in mixed diet; [d]Cruz (1975) determined by indicator method, feces collected by surgical excision, single ingredient fed; [e]Hajen et al. (1992) determined by indicator method, feces collected from water, ingredient fed in mixed diet; [f]Saad (1989) determined by indicator method, feces collected by surgical excision, ingredient fed in mixed diet; [g]Wilson and Poe (1985) determined by indicator method, feces collected by surgical excision, ingredient fed in mixed diet.

Table 16.5 True Amino Acid Availability and Protein Digestibility Values for Certain Feed Ingredients for Atlantic Salmon and Channel Catfish (NRC, 1993)

Feed Ingredient Fish Species	International Feed No.	Protein (%)	ARG (%)	CYS (%)	HIS (%)	ILE (%)	LEU (%)	LYS (%)	MET (%)	PHE (%)	THR (%)	TRP (%)	TYR (%)	VAL (%)
Canola meal	5-06-145													
Atlantic salmon		91.4	96.7	97.1	95.0	87.3	85.0	92.0	99.9	89.2	93.2	-	92.9	83.8
Corn, grain	4-02-935													
Channel catfish		-	-	82.0	90.3	67.9	87.5	96.5	70.5	81.8	69.8	-	77.5	74.4
Corn, gluten meal	5-28-241													
Atlantic salmon		95.0	99.9	90.8	94.5	90.4	88.4	99.9	93.8	91.2	92.0	-	92.0	91.3
Cottonseed, meal	5-01-621													
Channel catfish		-	90.6	-	81.6	71.7	76.4	71.2	75.8	83.5	76.7	-	73.4	76.1
Fish, herring meal	5-02-000													
(flame dried)														
Atlantic salmon		93.8	95.3	86.2	93.8	91.9	94.1	92.3	87.6	92.4	93.2	92.9	95.4	91.4
(steam dried)														
Atlantic salmon		82.6	94.1	94.1	88.2	89.0	89.0	90.1	88.6	88.9	94.9	56.7	90.2	88.3
(low temperature)														
Atlantic salmon		88.8	93.8	95.6	92.6	94.7	94.1	95.8	92.0	93.4	99.8	86.2	96.5	93.5
Fish, menhaden meal	5-02-009													
Atlantic salmon		88.5	86.8	92.0	91.1	88.5	90.1	87.6	83.6	87.4	88.4	89.0	92.1	86.3
Channel catfish		-	91.0	-	84.5	87.1	89.0	86.4	83.1	87.3	87.4	-	88.8	87.1
Meat and bone meal	5-00-388													
Channel catfish		-	87.9	-	82.2	80.8	82.4	86.7	80.4	85.4	76.3	-	83.1	80.8
Peanut meal	5-03-650													
Channel catfish		-	97.7	-	89.4	93.3	95.1	94.1	91.2	96.0	93.4	-	94.5	93.3
Rice bran	4-03-928													
Channel catfish		-	94.2	-	83.4	87.5	90.5	94.7	88.2	89.5	88.2	-	93.7	89.2
Soybean meal	5-04-604													
Atlantic salmon		88.3	86.7	-	86.4	79.2	75.9	83.6	94.0	78.7	84.5	50.3	83.0	77.3
Channel catfish		-	96.8	-	87.9	79.7	83.5	94.1	84.6	84.2	82.2	-	83.3	78.5
Wheat middlings	4-05-205													
Channel catfish		-	95.1	-	94.5	87.8	89.9	96.3	82.8	93.0	89.1	-	89.1	90.1

Abbreviations: ARG = arginine, CYS = cysteine, HIS = histidine, ILE = isoleucine, LEU = leucine, LYS = lysine, MET = methionine & cysti (s) ne, PHE = phenylalanine & tyrosine, THR = threonine, TRP = tryptophan, TYR = tyrosine, and VAL = valine.

Table 16.6 Net Absorption of Phosphorus from Various Sources by Channel Catfish, Common Carp, And Rainbow Trout (NRC, 1993)

Source	International Feed Number	Channel Catfish (%)	Common Carp (%)	Rainbow Trout (%)
Animal Products				
Casein	5–01–162	90[a]	97[b]	90[b]
Egg albumin	–	–	71[b]	–
Anchovy fishmeals	5–01–985	–	–	–
Brown fishmeals	–	–	24[b]	74[b]
Menhaden fishmeals	5–02–009	60[c]	–	–
White fishmeals	–	–	0–18[b]	66[b]
Inorganic phosphates				
Calcium, monobasic	6–01–082	94[c]	94[b]	94[b]
Calcium, dibasic	6–01–80	65[c]	46[b]	71[b]
Calcium, tribasic	6–01–084	–	13[b]	64[b]
Potassium, monobasic	–	–	94[b]	98[b]
Sodium, monobasic	6–04–288	90[c]	94[b]	98[b]
Plant products				
Corn, ground	4–26–023	25[c]	–	–
Phytate	–	1[c]	8–38[b]	–
Rice bran	4–03–928	–	25[b]	19[b]
Soybean meal, dehulled	5–04–612	20[a]	–	–
Wheat germ	5–05–218	–	57[b]	58[b]
Wheat middlings	4–05–205	28[c]	–	–
Yeast, brewers	7–05–527	–	93[b]	–

[a]Data from Wilson, R. P., E. H. Robinson, D. M. Gatlin, III, and W.E. Poe, 1982. Dietary phosphorus requirement of channel catfish. J. Nutr. 112:1197–1202. Values are expressed as percent apparent absorption.

[b]Data from Ogino, C., T. Takeuchi, H. Takeda and T. Watanabe, 1979. Availability of dietary phosphorus in carp and rainbow trout. Bull. Jpn. Soc. Sci. Fish. 45:1527–1532.

[c]Data from Lovell, R. T., 1978. Dietary phosphorus requirement of channel catfish (Ictalurus punctatus). Trans. Am. Fish. Soc. 107:617–621. Values are expressed as percent apparent absorption.

Table 16.7 Digestibility Coefficient of Feedstuffs for Tilapia (*O. Niloticus*) (Lovell, 1989A)

Feedstuff	Percentage Digestibility			
	Protein	Fat	Carbohydrate	Gross energy
Fish meal	**84.8**	**97.8**	–	**87.4**
Meat and bone meal (high grade)	77.7	–	–	68.7
Soybean meal	94.4	–	53.5	72.5
Corn (uncooked, mixed with fish meal)	–	–	65.4	–
Corn (cooked)	78.6	–	72.2	67.8
Wheat	89.6	84.9	60.8	65.3
Wheat bran	70.7	–	–	–
Alfalfa meal	65.7	–	27.7	22.9
Coffee pulp	29.2	–	–	11.4

Table 16.8 Differences in Digestibility of Energy in Feedstuffs by Catfish and Tilapia (*O. Niloticus*) (Lovell, 1989)

	Ingredient Digestible Energy (Mcal/g)	
	Catfish	Tilapia
Alfalfa, 17% Protein	0.67	1.01
Corn Grain		
Raw	1.10	2.46
Processed	2.53	3.02
Menhaden Fish Meal	3.90	4.04
Molasses	3.47	2.94
Soybean Meal, 48% Protein	2.58	3.34
Wheat Flour	2.55	2.89

Table 16.9 Phosphorus to Nitrogen Ratios for a Variety of Feed Ingredients

Ingredient	Protein	Nitrogen	Phosphorus	P/N
Herring Meal	72	11.52	1.00	0.087
Feather Meal	85	13.60	0.70	0.051
Corn Gluten	60	9.60	0.70	0.073
Peanut Meal	47	7.52	0.60	0.080
Soybean Meal	48	7.68	0.65	0.085
Wheat, Soft	11	1.73	0.30	0.174
Yellow Wheat	9	1.42	0.25	0.176
Poultry Meal	58	9.28	2.40	0.259
Menhaden Meal	62	9.92	3.00	0.302
Wheat Middlings	17	2.72	0.91	0.335
Meat/Bone Meal	50	8.00	4.70	0.588
Blood Meal	80	12.80	0.22	0.017

The P/N ratio is an example of derivative criteria, which can be added to a matrix presentation of nutrient specifications. Cho et al. (1994) has recorded the P/N ratio of several ingredients, and this information can be added to the nutrient specification matrix. The nutrient matrix will be discussed later, but it is important to know that actual nutrient levels as well as derivatives such as the P/N ratio can be used to assist in formulating low impact aqua feeds. See Table 16.9 for a summary of P/N ratios for a variety of feedstuffs.

Several states (USA) have established discharge standards to regulate the amounts of phosphorus that may be discharged by fish farms and hatcheries. Properly designed recirculating systems can easily meet these standards by simply discharging 1% or less of the rearing capacity water on a daily basis (see Chapter 10 for more details on managing waste). However, discharge standards for small volume releases are continually under review, and regulations will steadily become more restrictive. The consumer generally wishes to be environmentally sensitive, so marketing efforts for aquaculture produced fish incorporate references to the clean and wholesome image of an environmentally friendly production methodology. This, in turn, can and does result in increased values for aquaculture fish in the open market.

16.7 IMPORTANT ASPECTS OF AQUACULTURE FEEDS

Fish growth and performance can be quickly compromised by poorly digested feed that negatively impacts water quality. The type of feed influences fecal consistency. Dissolved solids from feed affect the color of the culture system water and this indirectly affects fish behavior, sometimes beneficially. The type of pellet and the process used to manufacture the pellet (steam, extrusion, expanded) will have an impact on the water quality as well as on the digestion efficiency of the fish. This is reviewed later in more detail. Designing feeds for low environmental impact is particularly critical when applied to RAS.

Standard nutrient requirements (NRC, 1993) cannot normally be applied to practical feed formulations. These requirements were established for small rapidly growing fish in ideal environments, and are not adequate for high-density commercial rearing situations. Therefore, nutritional requirements must be specifically established for each enterprise, based on the species to be grown and the type of aquaculture system used. Commercial feed formulations are intended to meet or exceed all nutritional requirements with quality products available at reasonable prices, so it is the responsibility of the aquaculturist to know what the species requires and to purchase feed appropriately. While the water column itself contains trace amounts of nutrients independent of the feed that is introduced, these nutritional contributions of vitamins, minerals, key lipids, protein and energy from the environment are negligible and cannot be counted on to contribute to growth.

While many feed ingredients can be used to meet nutritional requirements, not all ingredients are suitable for feeds used in RAS systems. Many are not sufficiently digestible. Some ingredients contribute to the production of excessive solids, increase visceral fat deposits, create high levels of iron, inhibit the absorption of zinc, and can result in undesirable yellowing of fillets.

INGREDIENT QUALITY

Quality feeds can be made only with quality ingredients. Quality assurance begins at the feed mill with ingredient reception, as incoming products are inspected and tested to assure that certain molds or histamines are not present. If ingredients contaminated with these molds are used in the feed, the animals consuming the feed could develop problems. Mycotoxin is a particularly dangerous contaminant that can be found in fish food if affected ingredient lots are used to make the feed.

One of the major ingredients in fish food is fishmeal. This ingredient is tested to determine the levels of histamine in the fishmeal, and the lot will be rejected if levels are beyond acceptable limits. Histamine is formed by bacterial decarboxylation of the amino acid histidine. Histamine is frequently found in brown fish meals (from sardines and mackerel). Feeding diets rich in histamine to trout causes discoloration of skin and erosion of the stomach (Watanabe et al. 1987). Coldwater species, such as trout, are more sensitive to Aflatoxin than are warmwater species.

The other basic ingredients in fish food are grains, with corn and wheat being predominant. Corn is rejected if aflatoxin levels exceed 20 ppb or if fumonisin[2] levels are found to exceed 1 ppm. Wheat is tested to determine the levels of deoxynivalenol (DON), which may also be referred to as vomitoxin, and is rejected if levels exceed 5 ppm. DON originating with wheat reduces growth and growth rates, increases morbidity, and promotes some tumors in salmon livers (author's personal experience).

The standards for mycotoxin tolerance were established for terrestrial animals, not for fish. Fish are more sensitive to mycotoxin than the federal standards would indicate. Therefore, do not rely on the federal standards to gauge the safety of fish food. The feed manufacturers are aware of this, so they generally incorporate one or more specific mycotoxin binders into the feed, such as bentonite, zeolite, yeast, ball clay, and other type additives. However, no specific data on the efficacy of these additives in fish food is available. Feed manufacturers routinely add 0.05% (450 g per ton) of feed with a

---

[2] Fumonisins are environmental toxins produced by the molds *Fusarium moniliforme* (*F. verticillioides*), *F. proliferatum*, and other *Fusarium* species that grow on agricultural commodities in the field or during storage. These mycotoxins have been found as contaminants, mainly in corn, worldwide. More than ten types of fumonisins have been isolated and characterized. Of these, fumonisin B_1 (FB_1), fumonisin B_2 (FB_2), and fumonisin B_3 (FB_3) are the major fumonisins produced in nature. The most prevalent of these mycotoxins in contaminated corn is FB_1, which is believed to be the most toxic.
Referenced from:
1) Thiel, P.G., Marasas, W.F.O., Sydenham, E.W., Shephard, G.S. and Gelderblom, W.C.A., 1992. The implications of naturally occurring levels of fumonisins in corn for human and animal health, Mycopathologia 117:3–9.
2) Musser, S.M. and Plattner, R.D., 1997. Fumonisin composition in cultures of *Fusarium moniliforme*, *Fusarium proliferatum*, and *Fusarium nygami*. J. Agric. Food Chem. 45:1169–1173.
3) Ashley, L. M., 1970. A symposium on diseases of fishes and shellfishes. Am. Fish. Soc. Spec. Pub. 5:366–379.

mycotoxin absorbent to prevent these conditions, such as MTB-100, which is a cell wall derivative from Saccharomyces.

In addition to the lack of specific standards for fish and the unquantified results of the mycotoxin mitigating additives, the quality of the final nutritional content of feed will vary based on the natural differences of the ingredients from seasonal variations and location of origin. Therefore, the aquaculturist must continually assess the performance of the nutritional matrix by carefully monitoring fish health and growth rates. The quality of the nutrient matrix in feed formulation programs is indispensable. Seasonal nutrient variability within key ingredients can impact extrusion characteristics, and create variability in final feed characteristics.

Fishmeal is used to meet essential amino acid requirements as well as maintain overall palatability of the finished feed. Fishmeals are sold by protein levels, ash levels, and the temperature of processing. Principal indicators of fishmeal quality are that it be Salmonella negative, contain less than 500 ppm histamine and be stabilized during manufacture with at least 250 ppm Santoquin. Lower levels of histamine are preferable, particularly in starter feeds.

Fish species vary in their sensitivity to Aflatoxin. Coldwater species such as trout are more sensitive than warmwater species. Under experimental conditions, long term rearing of trout, over a 6 month period on a diet containing 0.01 ppb pure Aflatoxin produced hepatocarcinoma (Ashley, 1970). Before a mill accepts grains, several screening tests for mycotoxin as well as moisture are performed as standard operating procedure. A prudent approach to feed formulation in RAS should include the use of approximately 450 g (1.0 lb) per ton of a mycotoxin absorbent such as MTB-100, a cell wall derivative from Saccharomyces (available from Alltech Inc).

Formulation

The organic materials that provide energy to the fish are either proteins, carbohydrates, or lipids. Vitamins and minerals are also necessary dietary ingredients.

Proteins

Proteins are composed of amino acid chains of varying proportions; most proteins contain approximately 22 amino acids. Ten of these amino acids are deemed to be "essential" and cannot be synthesized by fish, therefore, they must be added to the diet. Two other amino acids, cystine and tyrosine, are conditionally required because they can be synthesized from two other essential amino acids if supplied in sufficient quantities. Cystine can be synthesized (by way of cysteine) from the essential amino acid methionine. Tyrosine can be synthesized from the essential amino acid phenylalanine. These transformations are irreversible and represent only a part of the functions for methionine and phenylalanine; therefore these conditionally required amino acids (cystine and tyrosine) can spare part of the quantitative requirements for methionine and phenylalanine.

Fishmeal, which generally contains all the essential amino acids, is a common protein source. Supplements can be added to feed containing alternate proteins as necessary to meet the minimum requirements. Proteins comprise a significant portion of the salmonid diet, generally ranging between 35 and 55%, depending on the age of the fish, with smaller fish having a higher metabolic rate, thereby requiring the higher protein levels.

Fishmeal is used to meet essential amino acid requirements as well as maintain overall palatability of the finished feed. Commercially available fishmeals are graded and sold by protein levels, ash levels, and the temperature of processing. Fishmeal must be Salmonella negative, contain less than 500 ppm histamine and stabilized during manufacture with at least 250 ppm ethoxyquin (Santoquin) to be of acceptable quality. Lower levels of histamine are preferable, particularly in starter feeds.

Lipids

Lipids are the most efficient source of energy for fish, but do not entirely replace the other two sources. Nutritionally important lipids include triglycerides and phospholipids. Triglycerides include fats and oils. Lipids are categorized into three groups: fats, oils, and waxes. Fats and oils are the primary energy source in feed and are important for a balanced ration. Fats and oils are nearly the same chemically and nutritionally, with the main difference being the melting point. Three phospholipids are lecithin, cephalin, and sphingomyelin. The available energy content of lecithin was found to be 6.5 kcal/g for simple-stomached animals. This represents the maximum theoretical energy value of lecithin, since approximately 25% of the molecule is phosphoric acid and choline. Little work has been done on the energy value of cephalins and sphingomyelins. These compounds, however, are not found in significant amounts in foods or feeds (Scott, 1986). Trout feed typically contains between 8% to 15% fat content. Some farmers advise feeding lower fat feed before transport, as fish fed a low fat ration fare better.

Carbohydrates

Carbohydrates are not normally included as a large part of the salmonid diet due to their low nutritional content and poor digestibility of energy. Although they are an inexpensive source of energy, they do not supply any essential nutrients that cannot be obtained elsewhere in the diet. Additionally, excess carbohydrates can cause liver problems leading to death in salmonids. Trout diets should contain no more than 12–20% of maximum digestible carbohydrates. However, carbohydrates are important components of the feed for other reasons. In the form of starches, carbohydrates play an important role in the binding of the feed. Cooking, extrusion, and expansion are methods used to improve digestibility and binding capabilities of the salmonid pellets.

16.8 Physiological Relationships

Fish are poikilothermic (cold-blooded) animals. Their body temperature and metabolic rate are largely determined by water temperature. Species responses differ somewhat, but the concepts are similar. The goldfish (Carassius auratus) provides a typical example, Fig. 16.1, of the relationship between metabolic rate and temperature (Fry and Hart, 1948). The top bottom graph in Fig. 16.1 represents the difference between active and standard (resting) metabolism. Metabolic rate increases as temperature increases, up to a point, beyond which there is a plateau. At even higher temperatures, their metabolic rate decreases, and eventually the fish die.

Another factor that influences metabolic rate is body size. As size increases, metabolic rate per unit body weight decreases (Hoar, 1966). Because the metabolic rate is a primary consideration in determining the amount of feed required, there is a predictable relationship, Fig. 16.2, between body size (in length), temperature, and amount of feed required (Wheaton, 1977). Chapter 4 (Loading and Growth) presents equations to predict growth and feeding requirements as affected by body size, water temperature, and fish characteristics. At low temperatures the amount of feed that a fish needs is relatively small; but as the temperature increases for a given size of fish, the feed requirements increase, up to a point. For each species of fish, there is a size and temperature relationship that determines how much to feed that fish. Ideally, one should understand this relationship or use the relationship for a similar species as a model and make adjustments from experience. For further information, see Buterbaugh and Willoughby (1967), Klontz et al. (1985), Lovell (1989), Stickney (1979), Piper et al. (1982), and Westers (1987).

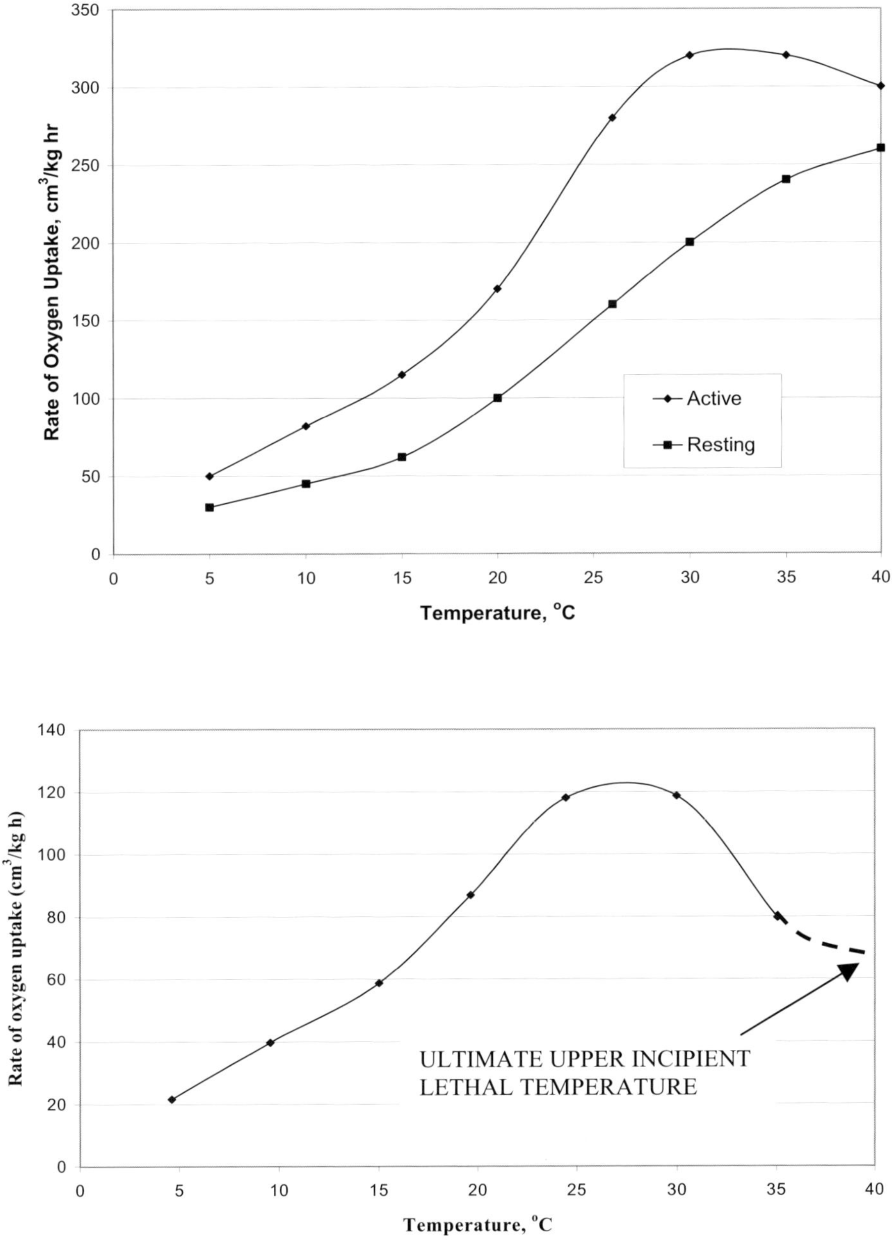

Figure 16.1 Metabolic Rate as Measured by Oxygen Consumption in the Goldfish (Fry And Hart, 1948; Taken from Wheaton, 1977, pg. 106).

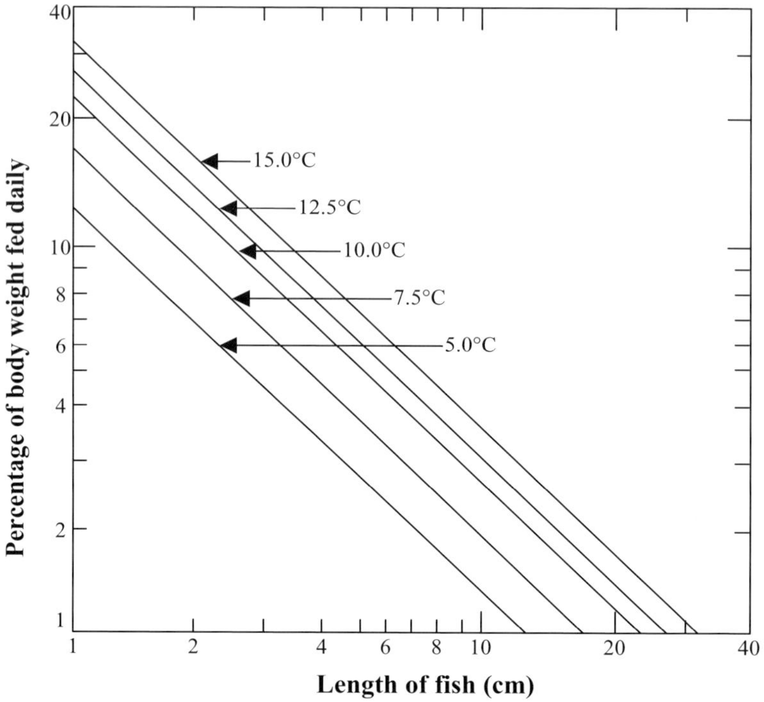

Figure 16.2 Feed required by Trout as Related to Temperature (C) and Body Size (from Wheaton (1977) Fig. 13.62 pg. 577).

16.9 WATER CHEMISTRY AND DIETARY NEEDS

Water chemistry influences physiology and nutrition of fish. For example, rainbow trout (*Oncorhynchus mykiss*) living in seawater experience osmotic pressure that causes water to be continually lost from the body fluids which poses a danger of dehydration. To counteract osmosis and avoid dehydration, the trout drinks saltwater and eliminates excess salts by active transport across the gills and by urinary excretion. The opposite is true for trout in fresh water. Water moves osmotically into the tissues of the fish, through the skin, gills, etc. Therefore, freshwater fish do not drink but rather excrete copious amounts of dilute urine. This difference in physiological behavior significantly affects nutritional requirements. The drinking of seawater contributes noticeably to nutrient intake as can be seen by examination of data by Shehadeh and Gordon (1969), Table 16.10. In fresh water, there was no intake of minerals by drinking. When rainbow trout were reared in a 30% mixture of sea water to fresh water (30/70), there was a marked increase in drinking (42 mL/kg of body weight per day); and at full strength (100%) sea water there was an even greater rate of drinking (129 mL/kg of body weight per day).

Table 16.10 Drinking Seawater Contributes to Nutrition of Rainbow Trout

Nutrient Intakes per kg of Body Weight/Day Water (fresh, sea or mixture**)						
		Fresh water (100%)	SW/FW (30/70)	SW/FW (50/50)	Sea water (100%)	Nutrient Intake from Feed ***
Intake by Drinking*						
Fluid*	(mL)	0	42	95	129	1
Mineral intake (computed)****						
Calcium	(mg)	0	5	19	52	100
Magnesium	(mg)	0	19	64	174	15
Sodium	(mg)	0	147	499	1355	100
Chloride	(mg)	0	266	902	2451	100
Adsoprtion of minerals*						
Calcium	(mg)	-	3	4	17	-
Magnesium	(mg)	-	1	9	74	-
Sodium	(mg)	-	92	437	1288	-
Chloride	(mg)	-	195	845	2400	-

*Data from Shenadeh and Gordon (1969).

** FW = fresh water; SW = sea water.

*** Estimated nutrient intake by way of feed containing 1% calcium, sodium, and chloride, and 0.15% magnesium. Feed intake = 1% of bw/day.

**** Mineral intakes computed from fluid data and concentrations of minerals in mixtures of 30%, 50% and 100% sea water. Concentrations for 100% sea water were: calcium, 400 mg/Liter; magnesium, 1,350 mg/Liter; sodium, 10,500 mg/Liter; and chloride, 19,000 mg/Liter (Spotte, 1970).

Table 16.10 compares estimates of mineral intakes by feed to intakes by drinking seawater. Estimates of mineral intakes by feed assume a hatchery diet fed to trout at a normal feeding rate of 1% of body weight/day. These estimates show that intake of calcium by drinking represented about one-half the amount consumed in feed, Table 16.10. The amounts of magnesium and sodium consumed by drinking seawater were about 10 to 13 times higher than by feed. Intake of chloride was much higher than obtained from feed. Absorption of these minerals from drinking was considerable, Table 16.10. Calcium absorption was moderate, 3–17 mg. The amounts of magnesium, sodium, and chloride absorbed from water were much higher than obtained from feed. This shows that when trout are held in seawater, drinking contributes substantially to the nutrient intake.

The gills also contribute to the absorption of calcium and other divalent metal ions even in fresh water. A study of brook trout (*Salvelinus fontinalis*) by McCay et al. (1936) involved three replicates of 50 fish. Twenty-five trout from each group were analyzed for calcium at the start and the rest of the trout were fed a diet of liver for 12 weeks. Liver contains very little calcium. Chemical analyses of carcasses at the beginning and at the end of the study, Table 16.11, showed that feed provided only about 1.1 milligrams of calcium per fish, whereas the total calcium deposited in the carcass was much more. Computations showed that the feed could have only provided about 19% of the calcium accumulated. Since these fish do not drink in fresh water, the rest of the calcium was likely obtained by direct absorption through the gills and skin. The gills constantly process water, and there is a significant uptake of divalent metal ions by the gills in the fresh water (Lovelace and Podoliak, 1952; Reid et al. 1959; Boroughs et al. 1957; Smith, 1930).

Table 16.11 Uptake by Brook Trout of Calcium (Ca) from Food and Water (43 mg/L Ca) (McCay et al. 1936)

Measurement		Start	End	Gain
Body weight	(g/fish)	3.70	4.43	0.73
Carcass Ca	(mg/fish)	14.2	20.0	5.8
Feed fed	(g/fish)		18.7	
Feed Ca fed	(mg/fish)		1.1	

Percentage of Ca deposited (an indication of uptake) from: $\frac{1.1\,(100)}{5.8} = 19\%$

From water (calculated as remainder): 100–19 = 81%

16.10 FUNCTIONAL ANATOMY OF DIGESTION

The gastrointestinal tract influences the fish's nutritional needs. The intestinal tract of a 2 1/2-week-old embryo of a brown trout (*Salmo trutta*), developing at 12°C, is basically a straight tube, very simple, with the yolk sac attached to provide nutrition as shown in Fig. 16.3

(Burnstock, 1959). Brown trout embryos hatch in about 5 ½ weeks, at which time the stomach is only partially developed. At this stage of development, nutrition is provided by the yolk sac. At seven weeks, the intestinal tract is still not fully capable of processing food. At nine weeks, the fish swim up and begin to feed for the first time when their pyloric caeca are still rudimentary. There is still yolk material remaining that provides part of their nutrition. Because of under-development of the digestive tract, exogenous food provided for first-feeding should be highly digestible and contain high levels of energy and nutrients. Different species develop at different rates. Some species begin to feed in less than one week after hatching.

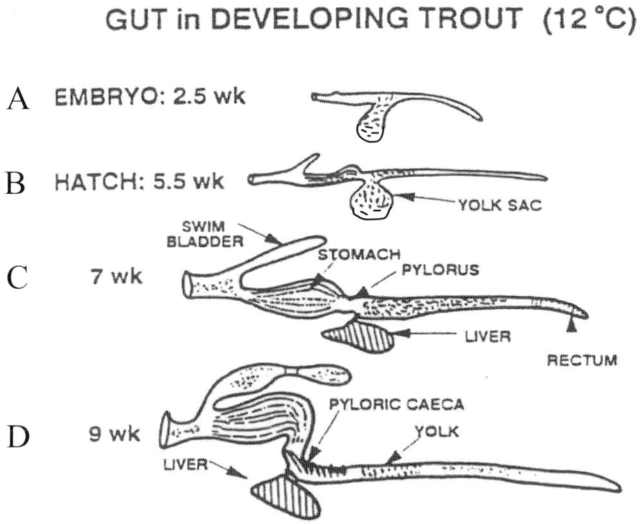

Figure 16.3 Gut Development in Trout (12°C) from Early State of Its Development at A) 2.5 Weeks, B) 5.5 Weeks, C) 7 Weeks, and D) 9 Weeks. (from Burnstock, 1959).

It is important for fish to start feeding properly. If first-feeding fish do not receive adequate nutrition, the rest of that fish's life may be negatively impacted. There are a number of species of fish that are very difficult to feed initially. Recent studies with Atlantic salmon (*Salmo salar*) show a negative effect of oxidized dietary lipid on first feeding of these fry (Ketola et al. 1989). These salmon are most susceptible to oxidized lipids during the first few week of feeding, but with age they develop greater capabilities of handling oxidized lipids. There may be other fishes similar in this regard. Carnivorous fish, especially their fry, which do not ordinarily eat, oxidized lipids, appear to be most sensitive to oxidized dietary lipids in their diets. Therefore, the quality of fish oil and fat-containing ingredients is especially important for the diets of young fish, especially the first-feeding fish.

Variations in Intestinal Tracts of Different Species

Figure 16.4 (Halver, 1959) shows the comparative diagram of gut length as a percent of body length for several different species of fishes. The gastrointestinal tract of rainbow trout is a relatively simple and short system, roughly 60 to 80% of its total body length. In catfish (*Ictalurus punactus*), the digestive tract is about 1.5 times the length of the fish. In carp (*Cyprinus carpio*) the length of the digestive tract is about 3 times the body length; in milkfish (*Chanos*), which is an herbivore, the digestive tract is about 5 times the body length.

There are other important differences in digestive capabilities between these fishes. The trout and the catfish both have acid producing stomachs; the carp, which is a member of the minnow family, does not have an acid producing stomach. Minnows in general do not have true stomachs and cannot produce hydrochloric acid. Acid affects their nutritional responses and needs. Later, we will see how the lack of acid in carp affects their ability to digest phosphorus, Table 16.12. The milkfish has a gizzard, apparently to grind its food. These fish have very long gastrointestinal tracts. Milkfish and carp tend to graze continuously, whereas carnivorous fish tend to eat fewer meals less frequently and to retain food in their digestive tracts for longer periods.

Table 16.12 Comparative Availability of Phosphorus for Carp, Catfish and Trout (NRC/NAS, 1983)

Source of P	Solubility (g/100 cc)	Availability Carp	Catfish	Trout
$NaH_2PO_4 \cdot H_2O$	60	94	90	98
KH_2PO_4	33	94	–	98
$Ca(H_2PO_4)_2 \cdot H_20$ (mono)	1.8	94	94	94
$CaHPO_4 \cdot 2H_20$ (di)	0.03	46	65	71
$Ca_3(PO_4)_2$ (tri)	0.002	13	–	64
Fish meals	–	18–24	40	66–74
Casein	–	97	30	90
Yeast	–	93	–	91
Rice bran	–	25	–	19
Wheat germ	–	37	–	37
Phytate	–	3–18	Nil	0–19

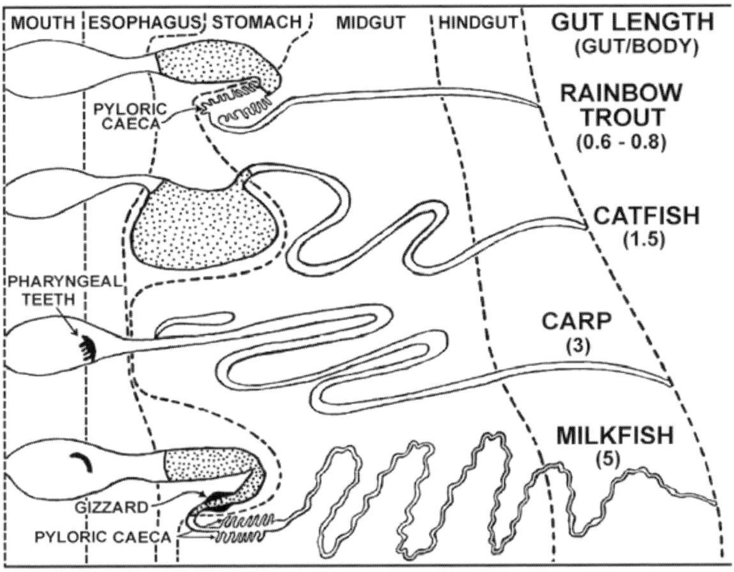

Figure 16.4 Variation in Digestive Tracts. Stippling indicates an acid secreting stomach. Rainbow trout (carnivore). Catfish (omnivore emphasizing animal sources of food, pouched stomach). Carp (omnivore emphasizing plant sources of food, pouched stomach). Milkfsh (microphagous planktivore, tubular stomach with muscular gizzard).

16.11 MINERALS

Phosphorus is an essential nutrient for fish. Figs. 16.5 and 16.6 show the results of an experiment in which Atlantic salmon (*Salmo salar*) were fed a phosphorus-deficient diet supplemented with different forms of phosphorus— either phytate phosphorus, meat and bone meal phosphorus, or a highly-soluble sodium dihydrogen monohydrate phosphate ($NaH_2PO_4 \cdot H_2O$) (Ketola, H.G., unpublished). Plant phosphorus in general is about two-thirds phytate phosphorus. Figures 16.5 and 16.6 show that phytate phosphorus supplements did not benefit salmon, whereas supplements of phosphorus in meat and bone meal and sodium phosphate significantly improve both growth and bone ash measurements. The same is true of trout and probably other fishes as well (NRC, 1983).

Table 16.12 summarizes reports reviewed by the National Research Council (1983) that compare the availability of various sources of phosphorus to common carp, channel catfish, and rainbow trout. For trout, highly soluble sodium phosphate ($NaH_2PO_4 \cdot H_20$) and potassium phosphate (KH_2P0_4) are both highly available (98%), and monocalcium phosphate ($Ca(H_2P0_4)_2 \cdot 2H_2O$) is also readily available (94%). However, dicalcium phosphate ($CaHPO_4 \cdot 2H_20$) and tricalcium phosphate ($Ca_3(P0_4)_2$) are less available (71 and 64%), following the trend of decreasing solubility. In carp, this decrease in availability is much more dramatic due to the lack of hydrochloric acid production necessary to solubilize di- and tri-calcium phosphate. Trout and catfish, both of which produce acid, were better able to solubilize and digest these phosphates. These differences in nutritional capabilities relate to basic differences in physiological characteristics of these fishes.

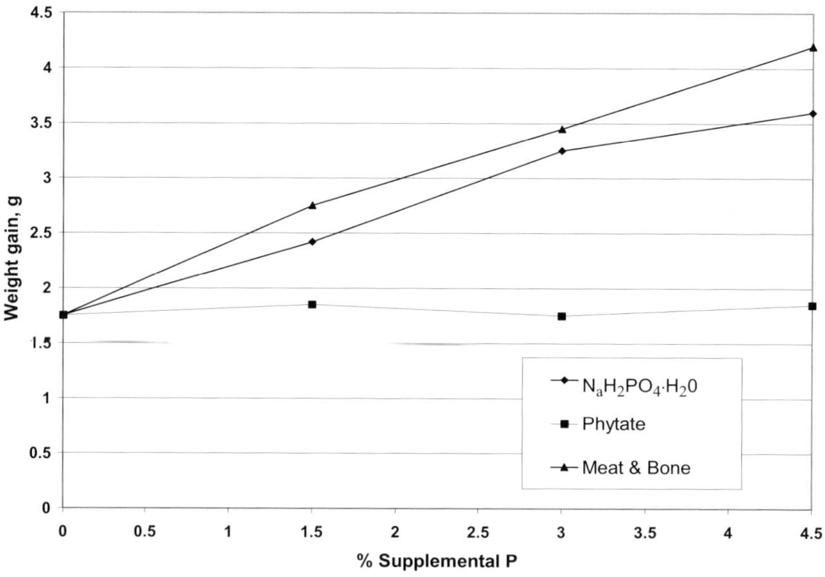

Figure 16.5 Availability of Phosphorus to Atlantic Salmon and Effect on Growth.

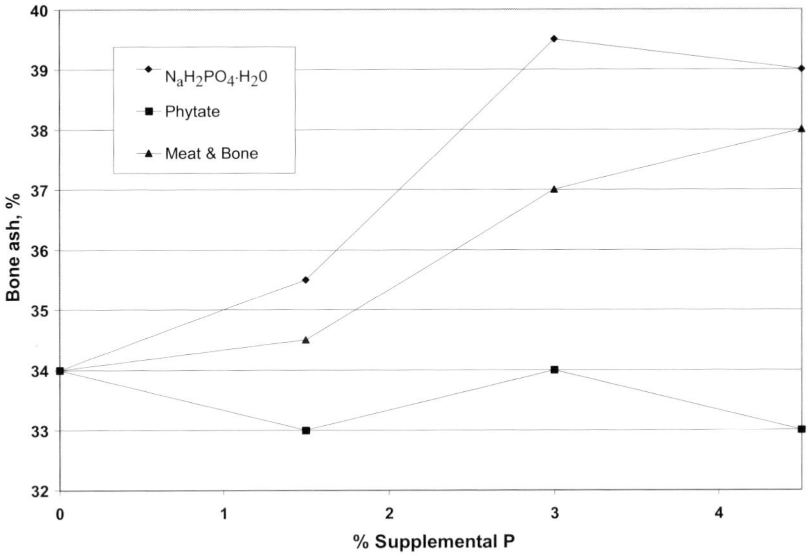

Figure 16.6 Availability of Phosphorus (P) to Atlantic Salmon and Effect on Bone Ash.

Table 16.13 General Summary of the Mineral Requirements of Various Fishes

Mineral	Levels*	Possible Signs of Deficiency
Ca %	0.1 - .3 (s)	—
P %	0.5–0.8	reduced calcification of bones, reduced growth
Mg	400–800	renal calculi, sluggishness, mortality
Na	ND	osmoregulation functions
K %	0.5 to 1.0**	reduced growth
Cl	—	osmoregulation functions
Fe	30–170	anemia
	150–400 (s)	
Zn	15 to >60	cataracts, reduced growth
	100–150 (s)	shortened body
Mn	2.4–13	abnormal tail
	10–100 (s)	shortened body
Cu	3–5	reduced cytochrome C oxidase, slow growth
I	1–5 (s)	goiter
Se	0.15 to 0.4	reduced glutathione peroxidase, muscular degeneration, anemia

*Tentative requirements (NRC/NAS 1983, 1993) or estimated safe (s) levels in mg/kg diet unless otherwise noted.
**Austic et al. (1989)

Table 16.13 shows a general summary of the mineral requirements of various fishes. The minerals requirements that must be met by feed are relatively difficult to determine because the water often contributes a significant portion of the nutritional minerals needed, with few exceptions, i.e., phosphorus. Some of the values presented in Table 16.13 are estimated safe levels, while others are estimates of the minimum requirements. Calcium, phosphorus, and several other minerals have been studied quite extensively. For zinc, there is a wide range in the estimates of the minimum requirements and safe levels because there are components in natural ingredients that markedly affect zinc absorption. For instance, if diets contain fishmeals with a very high level of ash (about 20%), there will be a marked decrease in the absorption of the zinc in that diet (Ketola, 1979). Even though such a diet may contain 60 mg zinc/kg, reduced absorption of zinc may result in cataracts, a clouding of the eye lens protein leading to blindness. Zinc deficiency in rats is manifested by a marked increase in inter-animal aggressive activity (Halas et al. 1975). In fish, zinc deficiency results in poor fin condition that may suggest aggressive fin nipping, but that has yet to be demonstrated. Deficiency of iodine causes goiter, first identified as thyroid carcinoma in wild brook trout (*Salvelinus fontinalis*) in waters low in iodine (Marine and Lenhart, 1911). Addition of iodine to the water reversed the condition. For further information on mineral deficiencies in fish, see Halver (1989), Lovell (1989), and NRC/NAS (1981 and 1983).

Table 16.14 Fish Vitamin Requirements and Recommended Levels of Concentration in Diet

Vitamin	Level* (IU or mg/kg)	Signs (partial list)
A, IU/kg	2,500–5,000	retinal degeneration exophthalmia, corneal edema
D, IU/kg	500–2,400	poor growth fatty liver, tetany, droopy tail
E, IU/kg	25–100	anemia muscular degeneration
K, mg	1	prolonged clotting
C (Ascorbic acid), mg	50–100	scoliosis, lordosis hemorrhages, dark coloration of skin
Choline, mg	400–1,500	slow growth fatty liver

*Tentative minimum requirements as noted are expressed in IUs (International Units of activity) or mg/kg finished feed (NRC 1983, 1993). This table includes no margin of safety to compensate for losses due to pelleting, oxidation during storage or losses by leaching in water. Practical formulations must include appropriate overages to compensate for such losses. Stabilized forms of vitamins should be used especially for vitamins C and E and others (vitamins A, K and D).

16.12 VITAMINS

High-density rearing conditions make it necessary for the commercial feed manufacturer to provide a complete and in some cases over formulated vitamin package in the content of their feed. The commercial feed manufacturer must provide a guarantee that the feed will retain the stated nutritional values for at least 3 months after production. Tables 16.14 and 16.15 summarize the requirements of fishes for vitamins and identify typical signs of deficiency. Vitamin C deficiency causes scoliosis and lordosis, which are deformities in the backbone that are not reversible. Early signs of vitamin C deficiency include hemorrhages in fins, skin, and internal organs. Halver (1989) specifically itemized the vitamin requirements for trout. Vitamin requirements for tilapia have been determined for only vitamin C (Stickney et al. 1984; Soliman et al. 1986), vitamin E (Satoh et al. 1987), and riboflavin and pantothenic acid (N.R.C., 1993). For further details on various forms of vitamin deficiency, see reports by NRC/NAS (1981, 1983), Halver (1989), Lovell (1989), Poston (1986), Piper (1982), Stickney and Lovell (1977), Hilton and Slinger (1981), and Ketola (1976).

Table 16.15 Requirements of Fishes for Vitamins and Some Typical Signs of Deficiency

Vitamin	Level* (mg/kg)	Signs (Partial List)
Thiamin (B_1)	1–15	anorexia, convulsions
Riboflavin (B_2)	3–25	anorexia, dark coloration, incardination, cataracts, high mortality
Pyridoxine (B_6)	3–20	epileptic-like seizures, spiral swimming, rapid breathing
Pantothenate	10–50 100–200 (s)	clubbed gills sluggish
Niacin	10–28 150–200 (s)	anorexia, sunburn, ascites
Biotin	0.1–1.5	anorexia, anemia, mortality
Folic acid	1–2 5–15 (s)	anemia
Cyanocobalamin (B_{12})	0.015–0.05	anemia (?)

*Tentative minimum requirements are expressed in mg/kg finished feed (NRC 1983, 1993). This table includes no margin of safety to compensate for losses due to pelleting, oxidation during storage or losses by leaching in water. Practical formulations must include appropriate overages to compensate for such losses.

Stability of Vitamins in Feeds

(The following discussions, figures and tables until the Section on Vitamin Requirements and Stability – were extracted from a report by BASF, 1992)

Vitamin C (ascorbic acid) deficiency is one of the most common problems encountered in feeding fish, because vitamin C is normally very unstable during manufacture, Table 16:16A, and storage, Table 16:16B, of the diet. Losses of crystalline ascorbic acid in the diet may be great, especially during the pelleting process. For this reason, assuring an adequate delivery of vitamin C represents one of the major concerns in feeding fish. Table 16.16 shows the average stability of vitamins in feed stored at room temperature. Of particular importance is the rapid reduction of the natural forms of vitamin C (crystalline ascorbic acid) and vitamin E (alcohol form) when incorporated in feed. Therefore, most feed manufacturers use more stable forms, such as ascorbyl phosphate and vitamin E acetate.

Vitamin C is extremely difficult to maintain in premixes or feeds since it is susceptible to destruction by so many environmental factors, especially oxidation. Phosphorilation of ascorbic acid (Ascorbyl phosphate) produces a highly stable product.

Vitamin A Stability

New technology has further improved Vitamin A stability by a cross linking process, such as the reaction between the gelatin and the sugar, that makes the beadlet insoluble in water, giving it a more resistant coating that can sustain higher pressure, friction, temperature and humidity (see Fig. 16.7).

Stability studies conducted with new cross linked vitamin A indicate higher stability than with soluble beadlet. Chen (1990) measured the stability of the 3 cross linked vitamin A beadlets on the market in trace mineral premixes and feeds. After 3 months storage at high temperature and humidity, the vitamin A retention varied from 30 to 80%, depending on the antioxidant present in the beadlet. In a 30% dairy concentrate pelleted at 200°F, retention at pelleting varied from 78 to 96%. After three months storage at high temperature and humidity, retention varied between 57 and 62%. The improvements in vitamin A stability through extrusion, in the last decade, increased by 35% mainly due to the use of cross linking processing.

Table 16.16A Equivalent Average Retention of Vitamins During Pelleting as a Percent Retained (%) (BASF, 1992)

	Pelleting Temperature, °F/Conditioning Time, Min.				
Various equivalent combinations of temperature and time	140/2	160/2	180/2	200/2	220/2
	150/1	170/1	190/1	210/1	230/1
	160/0.5	180/0.5	200/0.5	220/0.5	
Vitamin	170/0.3	190/0.3	210/0.3	230/0.3	
A 650 beadlet	95	93	90	85	79
D_3 325 beadlet	97	95	93	91	89
D_3 400 M.S.	95	92	88	82	77
E acetate 50%	99	98	97	96	95
E alcohol	75	65	54	43	23
MSBC*	80	72	65	56	44
MPB*	82	74	68	60	50
Thiamin hydrochloride	93	89	82	74	63
Thiamin mononitrate	95	93	89	84	77
Riboflavin	95	93	89	84	78
Pyridoxine	94	92	87	82	75
B_{12}	99	97	96	95	94
Calcium pantothenate	95	93	89	84	78
Folic acid	95	93	89	84	77
Biotin	95	93	89	84	77
Niacin	96	94	90	86	80
C, Ascorbic acid	75	65	55	45	35
C, Ascorbyl phosphate	97	95	93	91	89
Choline	99	98	97	96	95

*MSBC and MPB: menadione sodium bisulfite complex and menadione dimethyl pyrimidinol are stable synthetic forms of vitamin K.

Table 16.16B Average Stability of Vitamins in Feed Stored at Room Temperature (BASF Corp., 1992)

	Vitamin retention (%) Month				Loss/Month
Vitamin	0.5	1	3	6	%
A (beadlet)	92	83	69	43	9.5
D_3 (beadlet)	93	88	78	55	7.5
E acetate	98	96	92	88	2.0
E alcohol	78	59	20	0	40.0
MSBC	85	75	52	32	17.0
MPB	86	76	54	37	15.0
Thiamin hydrochloride	93	86	65	45	11.0
Thiamin mononitrate	98	97	83	65	5.0
Riboflavin	97	93	88	82	3.0
Pyridoxine	95	91	84	76	4.0
B_{12}	98	97	95	92	1.4
Calcium pantothenate	98	94	90	86	2.4
Folic acid	98	97	83	65	5.0
Biotin	95	90	82	74	4.4
Niacin	93	88	80	72	4.6
C, Ascorbic acid	80	64	31	7	30.0
C, Ascorbyl phosphate	93	88	78	55	>7.5
Choline	99	99	98	97	1.0

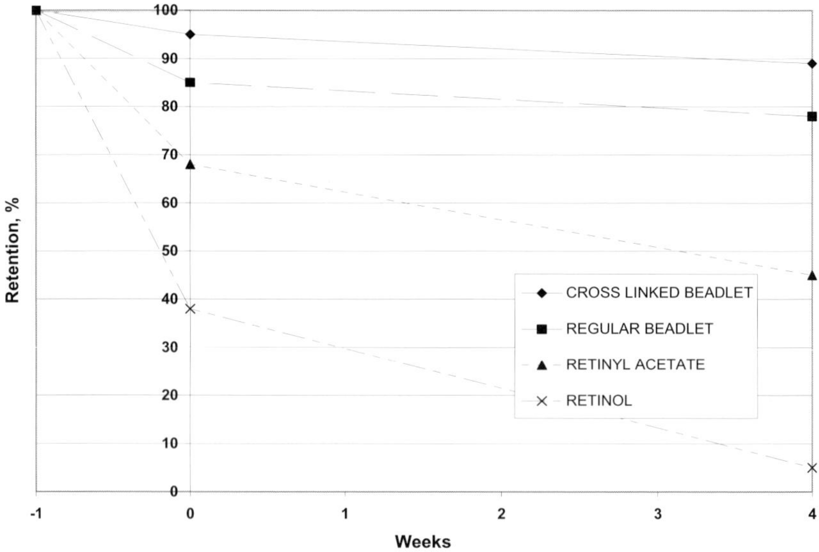

Figure 16.7 Stability of Vitamin A on Extrusion (37°C, 65% RH).

Vitamin E and Minerals

Vitamin E, as dl-alpha-tocopherol, is an antioxidant by itself and, therefore, if applied directly to feeds, is consumed rapidly. The free phenolic hydroxy group in this molecule is responsible for the antioxidant activity. When the hydroxy group is protected by formation of an ester, as in tocopheryl acetate, the compound obtained is resistant to oxygen, since it has no double bonds or free hydroxy groups. Vitamin E acetate is stable in feeds with neutral or slightly acidic pH. However, even slightly alkaline conditions may affect the stability, such as when limestone carrier is used or in the presence of large quantities of magnesium oxide. Under these conditions, some of the protective acetate groups split off and free tocopherol is formed, which can be rapidly oxidized. Dove and Ewan (1986) determined the stability of alpha-tocopherol in feeds without and with trace minerals. At the end of 3 months storage at 25–30ºC, alpha-tocopherol retention was 50% and 30%, respectively. The further addition of 245 ppm copper as copper sulfate, produced 0% retention after 15 days. Tocopherol, the most concentrated form of vitamin E activity, is such an unstable vitamin form that it should not be considered for any animal nutrition application.

Schneider (1988) determined the stability of tocopheryl acetate and tocopherol in vitamin-trace mineral premixes stored at ambient and stressful conditions. At the end of 1 month storage at ambient conditions, the retention was 95% and 44%, respectively, and at high temperature and humidity, the retention was 90% and 13%, respectively.

Vitamin K

Menadione, pure vitamin K_3, is a crystalline yellow powder that is unstable and irritating to skin and mucous membranes. It is not utilized in pure form, but is formulated with sodium bisulfate and derivatives thereof. Menadione sodium bisulfate complex (MSBC) and Menadione dimethyl pyrimidinol bisulfate (MPB) are more stable than MSB.

B-Vitamins Stability

B-vitamins are also unstable to a certain extent. Vitamin B_1 and B_6 are more stable under acidic conditions, while pantothenic and folic acids are most stable in a slightly alkaline environment. pH of the medium is far less important than the aggressiveness of moisture and trace elements. Thiamin hydrochloride is destroyed rapidly in a choline/trace mineral premix (high moisture, pH 4–5) while it is fairly stable in a basemix (low moisture, pH 7–8). Vitamin solubility in water is inversely correlated to stability. Thiamin mononitrate with a solubility of 10g/100mL, is significantly more stable in premixes than thiamin hydrochloride with a solubility of 100g/100mL (Adams, 1982), Fig. 16.8.

Vitamin B_6 is more rapidly destroyed in a choline chloride/trace mineral premix (high moisture) than in a basemix (low moisture). Calcium–d–panthothenate is quite stable. Losses occur only after prolonged storage at acidic pH. Riboflavin in stable in all premixes and also under climatic stress.

Vitamin B_{12} and choline are very stable compounds, but B_{12} is slightly sensitive to strong acid, alkali, reduction, light, ascorbic acid and ferrous sulfate.

Folic acid is stable to heat and air, but unstable in acid and alkaline solutions. It is light sensitive, slightly sensitive to moisture and sensitive to oxidation and reducing agents.

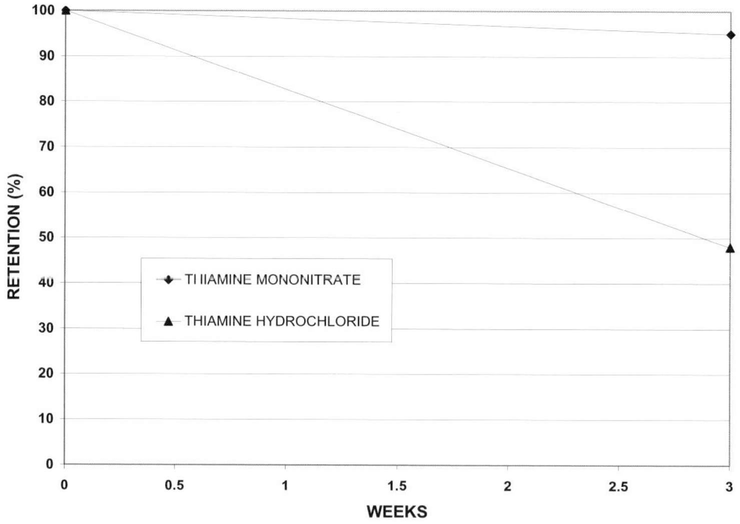

Figure 16.8 Storage Stability of Thiamine in a Vitamin-T.M. Premix (110°F or 43°C, 85% RH; BASF Corp, 1992).

Zhuge and Klopfenstein (1985) determined the stability of riboflavin and niacin in a broiler premix without and with trace minerals. At the end of 7 months storage, riboflavin retained 50% and 46%, respectively. Niacin retained 96% and 91%, respectively. Schaaf (1990) reported retentions of 100%, 100% and 93% for pyridoxine, riboflavin and folic acid, respectively, in vitamin premixes stored at ambient temperature for 3 months. Christian (1983) in a basemix study, determined riboflavin and calcium pantothenate stability after 3 months storage. Riboflavin retained 72% at low temperature and humidity, and 35% at high temperature and humidity. Calcium pantothenate retained 52% and 16%, respectively. Adams (1982) reported the stability of pyridoxine and thiamin in premixes without and with trace minerals. After storage for 3 months under stressful conditions, pyridoxine retained 100% and 45%, respectively. After 21 days under stressful conditions, thiamin hydrochloride retained 48% and thiamin monoitrate, 95%. BASF (1986) compared the stability of crystalline ascorbic acid and ethyl cellulose coated ascorbic acid through pelleting. Crystalline retained 85% and ethyl cellulose, 82%. A follow up study determined the stability of ascorbyl-phosphate. This compound not only is very stable, but also maintains the bioavailability. Ascorbyl phosphate retained 95% through extrusion, Fig. 16.9.

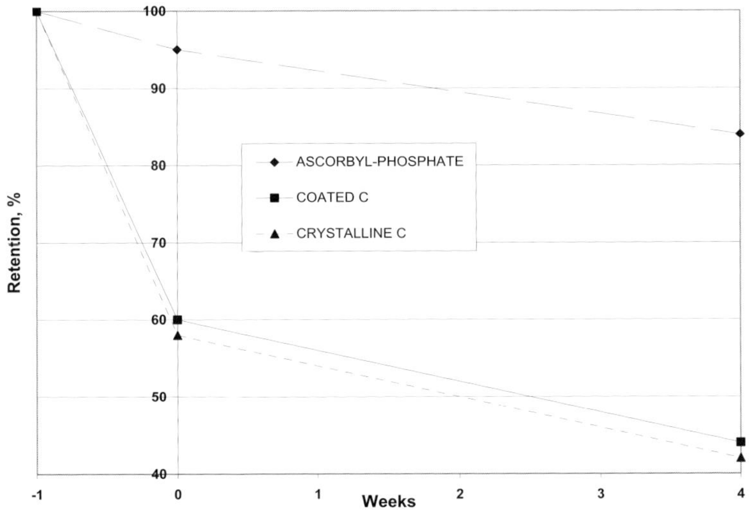

Figure 16.9 Stability of Ascorbic Acid on Extrusion at a Major Catfish Feed Manufacturer (37°C, 65%RH; Source BASF Corp., 1992).

Extrusion manufacture increases digestibility of fish feeds over pelleted feeds. Additionally volatile components are flashed off the exudate. At the same time, vitamin stability is weakened during ingredient mixing, grinding, and exposure to high temperatures resulting from the extrusion process. The relative stability varies with the vitamin (see Table 16.17; Coelho, 1991).

Table 16.17 Stability of Vitamins to Feed Extrusion and Expected Vitamin Loss Per Month for the Various Classes of Vitamins (Coelho, 1991)

Very High	High	Moderate	Low	Very Low
Choline	Riboflavin	Thiamin	Thiamin	Menadione
Chloride		Mononitrate	HCl	
	Niacin	Folic Acid		Ascorbic Acid
Ascorbic				
Polyphosphate	Pantothenic	Pyridoxine		
Sulfate	Acid			Vit. E alcohol
Monophosphate	Vit. E acetate	Vit. D_3		
	Biotin	Vit. A		
	B_{12}			
1%/month	*6%/month*	*11% /month*	*17%/month*	*50%/month*

VITAMIN REQUIREMENTS AND STABILITY

To compensate for instability, feed manufacturers generally use relatively high levels of vitamin C supplementation (250–500 mg/kg), even though the minimum requirement is quite low (50–100 mg/kg). However, new protected or stabilized forms of vitamin C have been developed and are now commercially available. Recent advances in the stabilization of vitamin C made by Hoffman LaRoche, Pfizer and BASF, have resulted in chemically stabilized forms, ascorbic 2-polyphosphate, ascorbic 2-sulfate and ascorbic 2-monophosphate. These chemically stabilized forms of vitamin C are highly stable during extrusion manufacture. Earlier efforts to stabilize vitamin C using lipid encapsulation did not result in a product able to withstand the temperatures and moisture levels of extrusion manufacture. Commercial manufacture of aqua feeds should not rely on unstabilized forms of vitamin C (De Antonis et al. 1993).

While new forms of vitamin C have greatly improved the cost effectiveness of vitamin C fortification, over formulation of other vitamins for extruded aquafeeds is still commonly practiced today. Manufacturers generally add vitamins in excess of actual requirements to allow for losses during manufacture, shipping, and storage prior to feeding, referred to as overages. Overages for coldwater and marine species are higher than those for warmwater fishes such as tilapia. Further losses may occur in the water by leaching. For practical recommendations for tilapia and warmwater fishes (see Table 16.18). These overages can be viewed as being a recommended supplemental vitamin addition to the natural vitamins that would be obtained by the fish from ingesting the rest of their diet.

16.13 FEEDSTUFF SELECTION

Feedstuffs for water restricted systems are selected based on unit costs for dietary energy, protein, and amino acid composition and ingredient digestibility and also by the level of phosphorus. Many potential feed ingredients can be eliminated from consideration based on their total phosphorus content. Of special concern are those ingredients that are high in ash and indigestible fiber. High ash ingredients such as some fishmeals, meat and bone meal, poultry meal, and soybean meal contain high amounts of calcium and phosphorus. Fish and animal bone and phytin are not completely digested and when discharged in feces contribute to phosphorus pollution of the water column. High levels of calcium and phytin-sequestered phosphorus have been implicated as a cause of cataracts in young salmonids and reducing the bioavailability of dietary zinc. Phosphorus in the form of phosphate is a contributor to eutrophication, and therefore should be controlled to the extent possible. The amount of phosphate entering the effluent can be reduced by purchasing feeds containing low levels of phosphorus.

Table 16.18 Recommended Supplemental Overage as a Percent for Specific Vitamins That are Used in Extruded Type Feeds Fed to Warmwater Fishes

Vitamin	Percent Overage
A - Acetate	150
D_3 - Cholecalciferol	130
E - Acetate	150
Thiamin mononitrate	250
B_{12}	130
Biotin	150
Folic Acid	200
Riboflavin	150
Niacin & Choline	150
Stabilized Vitamin C	110

Unpublished data, Paul Maugle (author)

The phosphorus/nitrogen (P/N) ratio of many ingredients such as animal by-product meals has been established. Many animal by-products meals have high P/N ratio. Generally, plant protein ingredients such as soybean and corn gluten meals have a lower P/N ratios, which is a desirable characteristic for inclusion in low environmental impact diet formulations. However, this beneficial characteristic is offset by other undesirable characteristics: indigestible phytin in the case of soybean meal and high levels of the carotenoid lutein associated with corn gluten meal preclude high inclusion levels in trout feeds.

An example of a derivative criteria, the P/N ratio of several ingredients provided by Cho et al. (1994) can be added to the nutrient specification matrix. The nutrient matrix will be discussed at a later point but it is important to know that actual nutrient levels as well as derivatives such as the P/N ratio can be used to assist in formulating low impact aqua feeds (see Table 16.9).

Ketola and Richmond (1994) have shown that rainbow trout requirements for dietary non-phytin phosphate are greater for bone mineralization than for weight gain. A summary of their findings, Table 16.19, suggest that the minimum level of non-phytin phosphorus in rainbow trout diets ranges between 0.51 and 0.6% depending on size of fish. As a safety margin, in commercial diets, actual levels are somewhat higher.

Table 16.19 Dietary Requirement for Non-phytin Phosphate in Trout (Ketola and Richmond, 1994)

	Small Trout	Large Trout
Maximum Growth	0.41%	0.34–0.54%
Maximum Bone Ash	0.51%	>0.54
Whole Body P content	0.51–0.61%	--

Most discharge standards in the USA, with a few exceptions such as Idaho and New York, do not currently require low phosphorus discharge levels. However, future discharge standards may require a reduction in the amount of phosphates that are discharged. Additionally, aquaculture produced fish can and will continue to command increased values in the open market, due in part to the clean and wholesome image that has been created by recent marketing efforts, some of which emphasizes the benefits of RAS in particular to the environment.

ENERGY

The requirement for energy is an important aspect of fish nutrition. Fig. 16.10 shows the overall process of utilization of energy by fish. The first fraction of energy loss occurs during the process of digestion, followed by losses through urinary and gill excretions and sloughing off of body tissues. During the process of digestion and absorption, there is an expenditure of energy called heat increment. The energy that is retained is called net energy. Part of the net energy is used for metabolism and voluntary activity, leaving that which is actually recoverable in tissue growth and reproduction.

Overall, the major nutrients in fish feed (protein, fat, and carbohydrates) vary in their digestible energy contents, Table 16.20. The digestible energy content of protein is about 4.5 kilocalories per gram, when digestibility is about 90%. The digestible energy content of fat is about 8 to 9 kcal/gram. Values for carbohydrates are variable. For instance, starch, which is a complex carbohydrate, has a fairly high gross energy content (4.18 kcal/g), but its digestibility decreases as the level of starch increases in the diet, thus its digestible energy also decreases. Simple sugars do not require digestion, are readily absorbed and therefore have high digestible energy values regardless of level in the feed.

FISH OILS

For diets that include oil as a dietary component, fish oil is still the major oil component, although other oil sources can be substituted depending upon the fish species, e.g., tallow, white grease, poultry fat, and soy oil. However, the successful use of fish oil requires special standards and criteria to be implemented. In the commercial arena, a key to quality is to purchase cold processed fish oil. Upon arrival of fish oil at the mill, additional antioxidants should be added. The addition of 250–500 ppm BHA & BHT is necessary to assure the stability of fish oil through the manufacturing process and subsequent shipment to the end user. These antioxidants can be added by the fish oil supplier and / or by the feed manufacturer at ingredient reception but long term stability of oils in the feed require its utilization. Fish oils that are acceptable for use in fish feeds should contain less than 3% free fatty acids, less than 1% moisture, less than 1% nitrogen and less than 20% Totox defined as (2 X (peroxide value) + (Anisidine Value). Fish oil should not be stored in heated tanks as elevated storage temperature can contribute to increased oxidation.

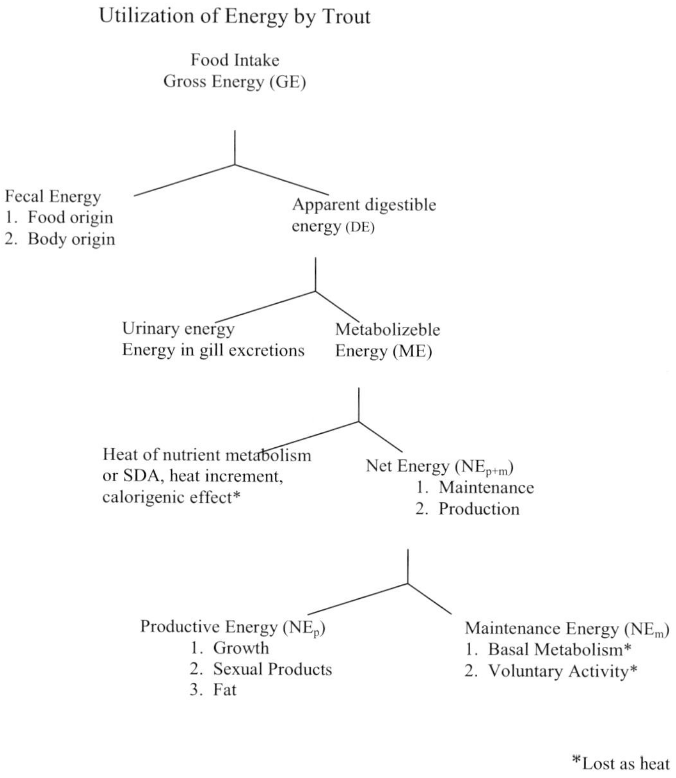

Figure 16.10 Overall Process of Utilization of Energy by Fish.

ESSENTIAL FATTY ACIDS

Fish require specific fatty acids in their diets. Fish, as a rule, require n-3 fatty acids, i.e., eicosapentaenoic (20:5 n3), docosahexaenoic (22:6 n3), or linolenic acid (18:3 n3) (NRC, 1983). Some tilapia may require only n-6 fatty acids such as linoleic (18:2 n6) and arachidonic acids (20:4 n6) (Takeuchi et al. 1983; and Kanazawa et al. 1980). Common carp and chum salmon (*Oncorhynchus keta*) may require n-3 and n-6 fatty acids (Watanabe et al. 1975; and Takeuchi and Watanabe, 1982). Generally, fatty acids should compose about 1% of the diet. Deficiency signs include reduced growth and sometimes fatty liver degeneration, anemia, fin erosion, increased susceptibility to bacterial infection such as flexobacteria, and fainting syndrome or stress shock when agitated.

Table 16.20 Major Nutrients in Fish Feed (Protein, Fat, and Carbohydrates) vary in their Digestible Energy Contents

			Energy (kcal/g)	
Nutrient Class		Digestibility	Gross	Digestible
Protein		90[1]	–	4.5[2]
Fat: (PUFA)		90[3]	9.3	8.4
Carbohydrate[3]				
Starch	20%	69	4.18	2.9
	40%	53	4.18	2.2
	60%	26	4.18	1.1
Glucose	20%	99	3.72	3.7
	40%	99	3.72	3.7
	60%	99	3.72	3.7
Sucrose	20%	99	3.94	3.9
	40%	99	3.94	3.9
	60%	99	3.94	3.9

[1] Cho & Slinger (1979)
[2] Available energy, Smith (1971)
[3] Singh & Nose (1967)

PROTEIN REQUIREMENTS

Figure 16.11 shows the results of feeding coho salmon (*Oncorhynchus kisutch*) a series of diets containing no carbohydrate and varying ratios of protein and fat (Ketola, H.G., unpublished). These data demonstrate a protein/energy relationship in the diet. As the protein increases up to 54%, the growth rate of coho salmon significantly increased. Beyond that level, however, there was no further benefit from increasing the level of protein. Also, fat and protein contents of the carcass shows marked changes as the ratios of dietary protein/fat changed. The optimum fat and protein content in the carcass is probably that associated with the minimum protein needed for maximum growth (54%). Above that level, the deposition of fat is further decreased, and the fish becomes leaner.

The protein requirements of various species of fishes are shown in Table 16.21. The so-called "protein requirement" is a summation of amino acid requirements that must be provided to the fish. There are about 10–12 amino acids that are essential to different species of fish. Other non-essential amino acids serve as non-specific nitrogen sources in the synthesis of other amino acids and proteins. Each species has a rather specific ratio of essential amino acids. Deficiency of any one of these amino acids causes a reduction in the rate of growth. Also, excesses of some amino acids caused reduced growth. Therefore, the balance of amino acids is important. There are two amino acids that are unique: methionine and phenylalanine. While these amino acids are essential, there are two other amino acids, which can substitute for part of their needs. Cysteine and cystine can replace part of the need for methionine, and tryosine can replace part of the need for phenylalanine. The information on the requirements for amino acids is not extensive.

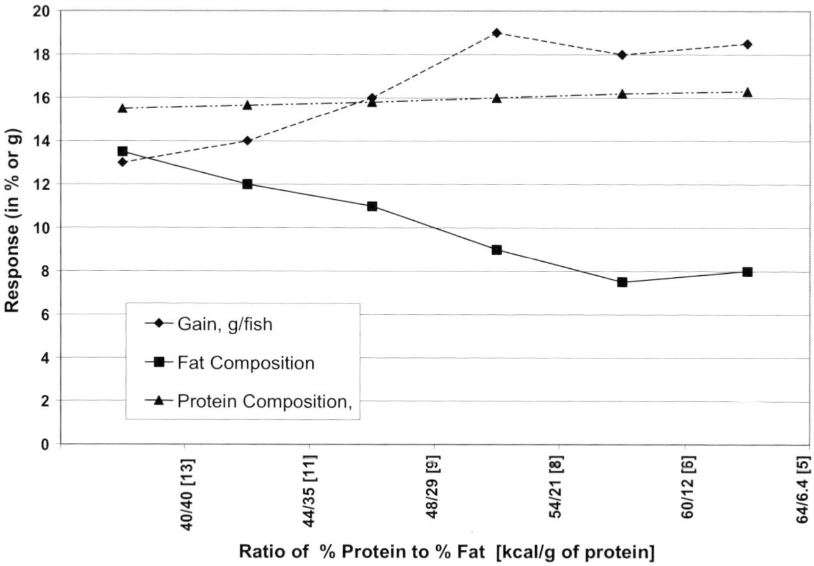

Figure 16.11 Effect on Weight Gain and Composition of Coho Salmon *(Oncorhynchus kisutch)* Fed Diets Containing Varying Ratios of Protein and Fat (Ketola, H.G., unpublished).

Table 16.22 summarizes quantitative amino acid requirements for several species of fish. A few specific signs of deficiencies have been observed. For instance, deficiency of lysine causes caudal fin erosion (Ketola, 1979). Deficiency of methionine causes cataracts (Poston et al. 1967). Deficiency of tryptophan causes cataracts and scoliosis (Poston and Rumsey, 1983). For further information on protein and amino acid requirements and deficiencies, see reports by Cowey and Sargent (1972), Halver (1989), Lovell (1989), Stickney and Lovell (1977), MAFES (1982) and NRC/NAS (1981, 1983).

When a researcher simply requires a diet for adequate laboratory research, probably the best thing to do is purchase an appropriate premixed feed from a commercial feed manufacturer. There are many such diets available for catfish, trout, and salmon. For names of commercial fish feed manufacturers, consult advertisements in Salmonid (U.S. Trout Farmers Association, Harpers Ferry, WV), Northern Aquaculture (Victoria, B.C., Canada) or Aquaculture News (Jonesville, LA) or other aquaculture magazines or newspapers. When feed of a specifically defined composition is required, there are several federal specification diets such as the ASD and Abernathy federal formulations (see Tables 16.23 and 24). These diets may be also be purchased from several fish feed manufacturers and may satisfy many research needs. Compositions for purified diets may be found in several sources (Halver, 1989; NRC, 1981 and 1983).

Table 16.21 Protein Requirements of Various Species of Fish

Species (Scientific name)	Requirement (% of diet)
Rainbow trout (*Oncorhynchus mykiss*)	40–46
Chinook salmon (*O. tshawytscha*)	40
Plaice (*Plauronectas platessa*)	50
Guilthead bream (*Chrysophys aurata*)	50
Common carp (*Cyprinus carpio*)	31–38
Channel catfish (*Ictalurus punctatus*)	31–38
Japanese eel (*Anguilla japonica*)	32–36
Grass carp (*Ctenopharyngodon idella*)	44.5
Puffer fish (*Fugu rubripes*)	50
Estuary grouper (*Epinephelus salmonides*)	40–50
Milkfish (fry) (*Chanos chanos*)	40
Red sea bream (*Chrysophrys major*)	55
Smallmouth bass (*Micropterus dolomiaui*)	45
Largemouth bass (*Micropterus salmonides*)	40
Tilapia	
Tilapia aurea (fry)	56
Tilapia aurea	34
Tilapia mossambica	40
Tilapia zillii	35

NRC/NAS (1981; 1983)

There are some species and life stages (early, swimup) that cannot be successfully maintained by feeding a typical trout, salmon, or catfish diet. For these cases, you may choose among many excellent aquarium fish foods that are available in a variety of forms such as flakes, pellet, crumbles, and natural and live food organisms. Some species may require a special formula feed that can be developed with the aid of a nutritionist, feed manufacturer or aquarium fish specialists. Others may require a closer imitation of their natural foods.

Table 16.22 Quantitative Amino Acid Requirements for Several Species of Fishes

	Species				
Vitamin	Salmon	Eel	Carp	Catfish	Trout
Arginine	6.0	4.5	4.2	4.3	5.9
Histidine	1.8	2.1	2.1	1.5	1.6
Isoleucine	2.3	4.0	2.3	2.6	2.1
Leucine	4.0	5.3	3.4	3.5	3.7
Lysine	5.0	5.3	5.7	5.0	6.1
Methionine & Cysteine	3.8	5.0	3.1	2.3	3.0
Phenylalanine & Tyrosine	5.3	5.8	6.5	5.0	3.1
Threonine	2.3	4.0	3.9	2.0	3.4
Tryptophan	0.5	1.1	0.8	0.5	0.6
Valine	3.2	4.0	3.6	3.0	2.2

Recommendations based upon: NRC (1981; 1983). Trout: Ketola (1983), Ogino (1980), Poston and Rumsey (1983), Rumsey et al. (1983), Hughes et al. (1983).

Table 16.23 Federal Specified Diet for Atlantic salmon ASD2–30 (for Granules and Pellet Sizes)

1.	Fish food shall carry the following guaranteed analysis:	
	Crude protein,	≥ 55.0%
	Fish meal protein	≥ 33.0%
	Crude fat	≥ 17.0%
	Moisture (at sack-off)	≤ 10.0%
2.	Herring meal	≥ 50%
3.	Dried shrimp meal: minimum protein 38%	5.0%
4.	Soy flour: defatted, minimum protein 48.5%, maximum fat 1%	20.3%
5.	Dried blood flour, minimum protein 80%	10%
6.	Trace mineral premix No. 2 (USFWS)	0.5%
7.	Vitamin premix No. 30 (USFWS)	0.6%
8.	Choline chloride, 50%	0.22%
9.	Ascorbic acid	0.075%
10.	Herring oil: stabilized with 0.04 BHA-BHT (1:1) or 0.01% ethoxyquin	12.0%
11.	Lignin sulphonate pellet binder, e.g., Ameribond, Orzan, or equivalent	2.0%
12.	Fish oil stabilized with 0.04% BHA-BHT (1:1) or 0.01% ethoxyquin OR soybean lecithin may be used at no more than 2% with fish oil	12.0%

Table 16.24 Federal Specifications: Abernathy Diet S8–2 (84) (for Starter Crumbles)

1.	Fish food shall carry the following guarantee:	
	Crude protein, not less than 48%	≥ 48.0
	Fish meal protein, not less than 40.5%	≥ 40.5
	Crude fat, not less than 17% or more than 19%	≥ 17.0
	Moisture, not more than 10% at sack-off	≤ 10.0
2.	Herring meal, minimum protein 70%, 8–12% fat, maximum salt 3%, maximum ash 15%. Stabilized	50%
3.	Dried whey product or dried whey, minimum protein 12%	10%
4.	Spray dried blood flour or flash dried blood meal	10%
5.	Condensed fish solubles, minimum protein 30% OR poultry byproduct meal, protein 60% to 68%, fat maximum 12%, ash maximum 16%	1.5%
6.	Wheat standard middlings, wheat mill run, or wheat shorts	Remainder
7.	Vitamin premix No. 2 (USFWS)	1.5%
8.	Choline chloride, 60% product	0.58%
9.	Ascorbic acid	0.1%
10.	Trace mineral mixture No. 1 (USFWS)	0.5%
11.	Lignin sulphonate pellet binder	2%
12.	Fish oil stabilized with 0.04% BHA-BHT (1:1) or 0.01% ethoxyquin OR soybean lecithin may be used at no more than 2% with fish oil	12%

16.14 FEEDSTUFF DIGESTIBILITY

Feed should be selected with the thought in mind that the feed will contain indigestible nutrients that become either biological waste products, or which contribute to nutrient pollution because they are not completely digested. Ammonia, nitrate, phosphate, and organic matter are some of the primary dissolved pollutants. With that in mind, feeds should meet the nutritional requirements of the species for optimal growth while at the same time minimizing nutrient excess. Feeds formulated with excessive protein or made with poorly digestible ingredients should be avoided as much as possible. Though fish can utilize protein for energy, this approach results in increased ammonia production. Net protein utilization (NPU) is the term used to describe the amount of body protein gain divided by protein fed. NPU can be improved by reducing the amount of protein used for energy metabolism. This is accomplished by optimizing the protein to energy ratio of the feed. Improvement of NPU can also be accomplished through feeding an ideal protein or a protein source of high biological activity, preferably both. Feeding an ideal protein presents an amino acid pattern that best meets the species requirements.

Differences in digestible energy of key ingredients used in commercial fish feeds are summarized in Table 16.25 for trout and Table 16.26 for catfish and tilapia (note each fish species is different). The ingredients selected for use in trout feeds significantly impact the digestibility (Kaushik and Medale, 1994) of the finished feed.

The digestibility of carbohydrates varies among species and the degree of digestibility of different feed components can be quite wide (see Table 16.27 for trout). Raw or partially cooked carbohydrates are more easily digested by catfish and tilapia than by salmonids. However, undigested carbohydrates increase the quantity of feces excreted. During the feed manufacturing process, the cooking of starches during extrusion improves the durability and water stability of the pellet, and can reduce the amount of solid waste produced.

A number of management practices can also affect the digestibility of a feed. As meal and feed pellet size increases, digestive and absorptive efficiencies decrease (Solomon and Brafield, 1972 and Windell et al. 1978). More frequent feedings of smaller meals tend to increase the digestibility of a given diet formulation.

Table 16.25 Apparent Protein Digestibility of Ingredients in Diets for Trout

Ingredient	Digestible Protein (%DM)	Digestible Energy (kJ/g DM)
Fish Meal	66.1	19.4
Corn Gluten Meal	58.9	17.5
Soy Bean Meal	46.5	13.5
Triticale	14.4	13.6
Rapeseed	39.6	14.9

*4.186 kJ = 1 kcal

Table 16.26 Apparent Protein Digestibility of Ingredients in Diets for Catfish and Tilapia

Ingredient Digestible Energy (Mcal/g)		
Ingredient	Catfish	Tilapia
Alfalfa, 17% Protein	0.67	1.01
Corn Grain		
Raw	1.10	2.46
Processed	2.53	3.02
Menhaden Fish Meal	3.90	4.04
Molasses	3.47	2.94
Soybean Meal, 48% Protein	2.58	3.34
Wheat Flour	2.55	2.89

Table 16.27 Apparent Digestibility of Carbohydrate in Ingredients in Diets for Trout

Ingredient	ADC of Starch (%)
Rice	39
Wheat	54
Potato	<5
Tapioca	<15
Corn	33
Extruded Wheat Starch	96

Many of the differences in ingredient or nutrient digestibility between species are a function of whether a particular species possesses an endogenous source of certain digestive enzymes (Maugle, 1982). At present, there are a number of commercially available exogenous enzymes that can increase the availability of nutrients of select ingredients. Trout are not insulin deficient and are able to utilize mono and diglycerides. In preliminary studies with extruded trout feed made by Southern States, a 0.2 (as 1.2 reduced to 1.0) Feed Conversion Unit improvement was found in fish reared with hemicellulase and xylase supplements, (ChemGen) when compared to trout reared with the same diet but without the enzyme supplement.

METHODS OF EVALUATION DIGESTIBILITY

There is no more accurate way to determine how a particular feed is suited for your management style and particular strain of fish than to perform some simple digestibility studies. Depending upon your staff qualifications, you can do this internally or arrange to have a trained fish nutritionist help you with the study (such assistance can often can be arranged through your feed supplier). The basic methodology of this study is that a non-digestible marker, such as chromic oxide (Cr_2O_3) is included in the diet at a low concentration (known value, e.g., 1% by weight). The amount of the marker going through the GI tract will remain fixed as mass, but other nutrients will be extracted by the fish as the feed is digested, thus increasing the percentage concentration of the marker in the feed. Working with your feed manufacturer, you can then use this data to customize a feed to your operation. See the following detailed example.

APPARENT DIGESTIBILITY IN FISH

Apparent digestibility of a nutrient is the net digestibility of the nutrient. Thus, these two terms, apparent and net, are used interchangeable. We will define apparent digestibility $D_{\%}$, mathematically as:

$$D_{\%} = 100\% \cdot \frac{Amount\ of\ nutrient\ fed - Amount\ of\ nutrient\ in\ feces}{Amount\ of\ nutrient\ fed} \qquad (16.2)$$

$D_{\%}$ can be determined by use of indicators without measuring feed intake or excreta output. Indicators may be natural indigestible constituents in feed or they may be added to it. Suitable indicators should:

- be totally indigestible and unabsorbable
- be easy to assay
- not be synthesized by the animal
- pass through the digestive tract at the same rate as feed
- be biologically inert so that its presence in the feed should not influence digestion or physiology.

A number of digestibility indicators have been proposed such as lignin or acid-insoluble ash. Added indicators include chromic (III) oxide (Cr_2O_3), ferric oxide or oxides of rare earths. Chromic oxide is commonly used.

If the concentrations of the indicator substance [Cr] and a nutrient are determined for a feed and for representative samples of unleached excreta, then the $D_{\%}$ of the nutrient can be calculated from the following equation:

$$D_{\%} = 100 - 100 \cdot \left(\frac{Cr \ in \ feed}{Cr \ in \ excreta} \right) \cdot \left(\frac{Nutrient \ in \ excreta}{Nutrient \ in \ feed} \right) \tag{16.3}$$

One way to obtain unleached excreta from aquatic animals such as fish is by manual stripping, Austreng (1978). For example, fish should be fed the test diets containing the indicator (chromic oxide) for sufficient time to ensure normal consumption and complete replacement of all other non-indicator feed residues in the digestive tract. Fish are netted one at a time, and firmly but gently held in a moistened towel or glove while gentle pressure is applied to the abdominal region of starting just behind the pelvic fins and progressing toward the anus to expel excreta. Uncontaminated (without milt or urine) excreta are collected. If excreta are not readily expelled, the fish are rejected. The procedure should not be done on abnormal trout that show obvious signs of reduced growth, general weakness or lethargy. Samples of excreta should be collected from enough fish and pooled until the amount is sufficient for required analyses. Samples of feed and excreta are analyzed for concentration of indicator [Cr] and [nutrient] of interest. For further information about determination of apparent digestibility, the reader is referred to Austreng (1978), Bondy (1987), and Halver (1989).

Example

As an example of calculating $D_{\%}$, suppose a diet containing 5.4% nitrogen (N) and 0.5% indicator chromium (as Cr_2O_3) was fed to trout for 10 days after which excreta were sampled, dried, and analyzed for the percentage concentrations of N and indicator. Results of the analyses and calculations for apparent digestibility of nitrogen and dry matter are shown in Table 16.28 below:

Table 16.28 Example Data for Collecting the Percentage Apparent Digestibility of Nitrogen and Dry Matter

Trout	Cr [indicator] (%)		N [nitrogen] (%)	
Feed	Feed	Excreta	Feed	Excreta
Test feed	0.5	1.8	5.4	2.0

Calculations:

$$D_{\%,DM} = 100 - 100 \cdot \left(\frac{Cr\ in\ feed}{Cr\ in\ excreta} \right) = 100 - 100 \cdot \left(\frac{0.5}{1.8} \right) = 72.2\%$$

$$D_{\%,N} = 100 - 100 \cdot \left(\frac{Cr\ in\ feed}{Cr\ in\ excreta} \right) \cdot \left(\frac{Nutrient\ in\ excreta}{Nutrient\ \ in\ feed} \right)$$

$$= 100 - 100 \cdot \left(\frac{0.5}{1.8} \right) \cdot \left(\frac{2.0}{5.4} \right) = 89.7\%$$

In this example, the apparent digestibility of dry matter ($D_{\%,DM}$) is 72.2% and that for dietary nitrogen (N) is 89.75 ($D_{\%,N}$).

Note, an example calculation demonstrating how to determine if there is a significant difference in mean results is given in the Appendix.

16.15 PELLETED, EXPANDED, AND EXTRUDED FEEDS

Fish feeds today are manufactured in three types: extruded, expanded, or steam pellets. Extruded and expanded pellets have improved durability and digestibility, in comparison to steamed. Extruded pellets can be produced to be heavier or lighter than water, providing excellent management options to the culturist. Extruded feed typically produces feed with the best feed conversion numbers as well as providing the most uniform product in size/shape/quality. Expanded pellets are about as good as extruded pellets and at times are very difficult to tell apart from extruded. Steam pelleting is preferable when adding medication, but is generally not as desirable because of its limitations concerning buoyancy and feed conversion. There is considerable variability between the feeds of different manufacturers in important feed characteristics, such as grind size, digestibility, and the amount of fines. The grind size should be no larger than 1.2 mm (3/64-inch) for general fish diets and no larger than 0.8 mm (2/64-inch) for larval diets (one way to compare quality and price among manufacturers is the grind size). A key mistake that many make is to think the cost of the pellet is the most important cost parameter. This is not so; rather, the important cost parameter is the cost to produce a kg of fish. The most expensive feed to buy may produce fish growth at the lowest total cost when conversion and water quality, which affects feed conversion, are considered. The culturist should create a strong working relationship with feed manufacturer, sales representative, and nutritionist.

> **"RULE OF THUMB"**
> The lowest cost feed usually does not produce the lowest feed cost per unit gain.

Each pelleting process can be defined as to its advantages and disadvantages. Today, most aquaculture feeds are purchased as extruded pellets, primarily it seems for the ability to create a floating pellet. This is not necessarily the best choice; it depends upon your farm. Review your options closely before deciding. One of the main disadvantages of the expanded feed pellet is that they slowly sink. RAS operators often prefer the floating pellet because the high turbidity in RAS's usually means you cannot "see" what is happening very far down into the water column. Cornell successfully used a slow sink expanded pellet for many years in clear but dark water in a trout RAS.

You should probably pay more attention to the overall quality of the feed and particularly the percent of fines than whether it is a sinking or floating pellet. Perhaps using a combination of floating and sinking would be a good alternative. Remember, fish adapt slowly going from one type of feed to another in terms of sinking or floating.

STEAM PELLETING

Steam pelleting is the oldest of the three methods. After the ground mash is premixed, steam is added to condition and partly gelatinize the starches and to assist in pellet binding. The mash is typically exposed to steam for short periods of time (less than 35 seconds), and to processing temperatures of 100–180°F (38–82°C). The conditioned mash is forced through constricting tapered holes in a die and the resulting pellets are cut by a series of knives to a desired length. After the pellets are cut, they are blown dry with a final moisture content of 9–10%. Steam pellets are cylindrical in shape typically having lengths 1.5 to 2 times longer than their diameter. They are hard and have a glassy exterior. Steam pellets are fairly dense and will sink; their advantages are:

- Less energy is required during manufacture.
- Less heat means less destruction of heat sensitive nutrients, medications, and vitamins.

- Initial cost is less.

Now the disadvantages of steam pelleting:

- Greater incidence of fines.
- Pellets sink.
- Smallest sized pellet available is roughly 2.4 mm (only good for larger fish).
- Total fat content cannot exceed 20%.
- Marginal pellet durability in the water column.

EXTRUDED

Pre-mixed mash is introduced into an extruder barrel and a significant amount of water is added. This high moisture mash is then placed under intense pressure, heat and friction. This causes starch gelatinization two to three times that found in steam pelleting. Some manufacturers preconditioning the mash to improve palatability, digestibility, and durability. Processing temperatures during extrusion can reach up to 300°F (149°C). This super heated mixture is then forced through a die. This causes a rapid reduction in pressure, which in turn causes the pellets to expand. The expansion reduces pellet density that creates a pellet that is less dense than water, i.e., a floating pellet. Pellets leaving the die have moisture contents 10–15% higher than steam pellets, which then requires additional energy to dry the product to 10% moisture. Advantages of extruded feed are:

- Expansion can be controlled to allow the product to sink, slow sink, or float (allows fish management options to observe feeding behavior).

- Fat levels higher than 20% are possible (allows higher energy feed pellets).

- Higher temperature process improves availability of nutrients (arguable) resulting in better feed conversion ratios.

- Better digestibility means less waste for the system to handle.

- Carbohydrates are used as binders so "nutritionally blank" ingredients are not required to bind the product. Better feed conversions and reduction in total fecal load is the result.

- Structural integrity allows for smaller, more consistent sizes.
- Pellets are durable, uniform in size, and have few fines.

Everything is not a plus using extruded feeds, disadvantages are:

- Nutrient, medication, and vitamin degradation is higher because of the additional heat used in the manufacturing process. This requires that these ingredients be supplemented at higher levels.
- Costs are higher to the purchaser due to higher manufacturing equipment costs, the need for nutrient supplementation, and a slower production time (average cost of 0.2 to 5% higher than steam pelleted feed).
- Truck hauling costs are higher since pellet density is lower (typical bulk trucks might be able to hold 40,000 pounds (18,180 kg) of extruded feed compared to 42,000 pounds (19,090 kg) of steam pelleted feed).

Rangen Feeds (Buhl, Idaho) and Melick Feeds (Catawissa, PA) are two companies that manufacture an extruded feed.

Expanded

The expanded feed manufacturing process is similar to that of extruded feed except the cooked mash must be sent through a pellet mill to form the pellets, which typically results in sinking pellets (rarely and with great skill, it is possible to produce a floating feed). Expansion does not require as much moisture as extrusion, which permits the pellets to be dried without heat, which reduces operating costs. Expanders produce denser pellets than extruders but not as dense as a steam pellet. There are only a few expanders currently in use in the United States to produce fish feed (Zeigler Brothers, Gardners, PA is an example).

16.16 Floating Feeds

Whether a feed pellet floats, partially sinks, or sinks is determined by the level of starch in the formula as well as the degree of striation of the starches and the bulk density of the exudate. For example, a 20.5% starch diet will yield a slowly sinking feed in which approximately 50% of feed initially floats when first immersed in water but sinks within minutes of being fed. In general, maintaining the diet between 20 to 22% starch will promote a floating pellet and support the proper formation of a feces casing[3] for fish that produce a casing around their feces, e.g., tilapia. Diets that are too high in protein will result in diarrhea.

The floating or sinking characteristics as well as pellet durability can be adjusted by the selection of feedstuffs, the amount of water and oil used internally on the mix, and the amount of cook during extrusion. The amount of water and oil added internally to the feed can either enhance or inhibit the striation of starches. In general, a higher starch diet will reduce feces specific weight; thus making the feces more apt to float. This knowledge can be applied proactively to work in your favor. For example, if dissolved air flotation is a primary means of removing feces and suspended solids, then a higher starch diet would be an advantage. On the other hand, in an RAS that uses a central drain system for feces and sedimentation removal, a higher protein and lower starch diet would be advantageous, as the feces would tend to sink.

16.17 SUMMARY

Summaries of the three feed manufacturing processes and their key characteristics are given in Table 16.29. Use it as a guide and not a bible. Fish performance as affected by the feed will depend upon a lot of things besides the type of pellet.

[3] The casing around a fecal strand is a glycoprotein that is a mixture of starch and protein. That is why the starch protein ratios have to be maintained within some range in order to produce a casing that will retain

Table 16.29 Comparisons of Feed Manufacturing Techniques

	Compressed Pellets	Extruded	Expanded
Initial feed cost	lowest	highest	intermediate
Starch gelatinization, %	<40	>80	60–80
Max. temp., °F (°C)	180 (82)	300 (149)	300 (149)
Max. fat level, %	20	40	30
Digestibility	good	best	better
Sinking available	yes	yes	yes
Floating available	no	yes	possibly
Slow sink available	no	yes	possibly
Fines upon receipt, %*	1 to 6	<1	<1
Vitamin and nutrient degradation	lowest	highest	intermediate
Feed conversions	worst	best	intermediate
Uniformity of feed	variable	excellent	good
Availability	most mills	some mills	few mills

*Fines can result from poor handling practicing of finished feed or minimal grinding

CHAPTER 17

SOFTWARE PROGRAMS

17.0 OVERVIEW

There have been a lot of equations presented in this book. By now, you should have a firm grasp on the principles involved in and the logic behind the equations. Someone (probably an engineer!) once said that aquaculture is mostly doing calculations. Well, to make your life easier, we have a generally comprehensive set of spreadsheets and DOS programs that can take a lot of the drudgery out of the calculation steps.

The spreadsheet programs (referred to as workbooks) are as follows:

- Oxygen Design: Calculates all parameters for water flowing from a biofilter, through a spray tower, and back into the tanks. These values include oxygen demands, inflow and outflow concentrations of oxygen and nitrogen, and other water quality parameters.

- Fish Tank Design: Calculates values needed for the design of the tank and the biofilter, including flow rates, and the areas and volumes required for different production rates. There is also a worksheet in the Fish Tank Design workbook that calculates feeding rates and associated growth rates based upon water temperature. This worksheet can be used to forecast growth rates that can be expected, and serve as a basis of comparison of expected versus actual growth rates in the fish tank program.

- Pipe Flow Design: Calculates values relative to the design of the pipes, including drainage pipes and flow volumes. Calculations include velocities, areas, and flow rates for different pipe sizes and the corresponding head pressure requirements.

- Cost Analysis: Calculates, based on specifications given by the user, the fixed and variable costs of facility operation and costs per pound of fish produced for the specified aquaculture facility.

The DOS based programs are:

- CO_2 Control Options: An extensive program that analyzes several control options and associated costs for removing CO_2, plus the additional cost of restoring heat to the building that is lost during ventilation of the removed CO_2.

- LHO Design & Management: **L**ow **H**ead **O**xygenators can be designed with this quick-basic software. Calculates operating efficiencies, total gas pressures, and quantity and cost of oxygen transferred. This program can be used to optimize design of LHO's or to predict consequences of changing operating parameters.

- AIRPUMP: This program can be used to design airlift and aeration systems, including the proper pipe dimensions and airflow rates for achieving the desired oxygen transfer and water flow rates.

Disclaimer and Warning

We do NOT accept any liability for application of these programs. We urge you to do your own independent calculations before applying the results to your particular situation. There is **no substitute for some hand calculations** to verify certain critical design or management values.

This software is provided "as-is". There are **no warranties** expressed or implied by the publisher, distributor, or authors of this book. In **no event** shall these parties be liable for any special, consequential, incidental, or other damages related to the use of the software, documentation, or diskette.

The copyright laws that pertain to computer software protect this product. It is illegal to make copies of the software for any purposes other than backups. It is illegal to give copies of the software to any other person by any means (diskette or electronically). Any modification, adaptation, or creation of derivative works based upon the software or source code is also prohibited.

All product names in this documentation are trademarks or registered trademarks of their respective holders and should be treated as such.

17.1 SPREADSHEET PROGRAMS[a]

Four workbooks, written in Microsoft Excel, have been developed to facilitate calculation of the many values and parameters used in aquaculture. This guide is designed to explain the format of these workbooks, as well as define and clarify the values contained in the cells of the worksheets. The four workbooks and their functions are briefly described below.

Each of these workbooks is made up of several work*sheets*. Each worksheet is simply a separate page in the workbook. The workbooks are divided into different worksheets to separate the different parts of the design in an organized manner. This also makes the presentation easier to read and follow. You can switch between worksheets in a particular workbook simply by clicking on its tab at the bottom of the screen.

Here is an example of a worksheet inside a workbook:

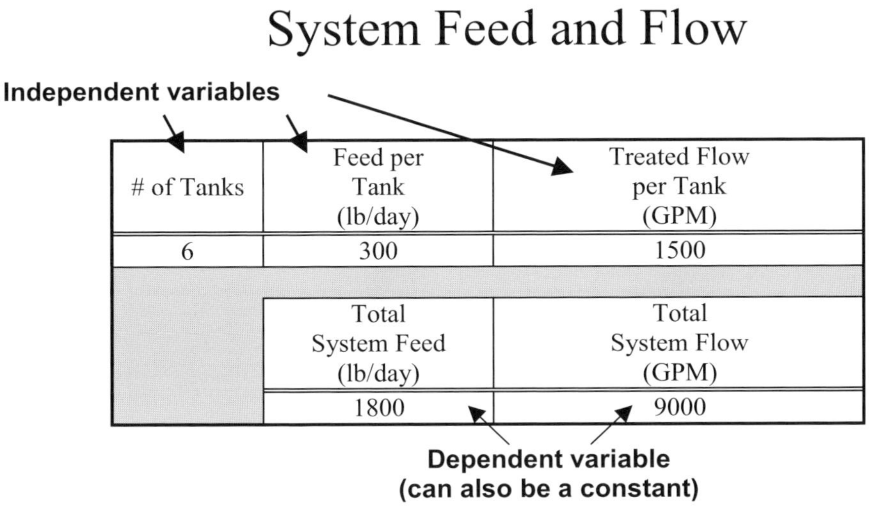

Notice that information inside the cells is in red and blue colors (you have to open a spreadsheet to see this; the book text is black and white).

[a] The CD software provides the majority of the Excel and DOS programs in both IP and SI units. The example in the text are mostly done using IP system.

Blue cells contain independent variables. These are user-changeable cells and contain the values that you can enter and/or edit. They are independent and do not contain any formulas that depend on any other cells in the workbook.

Red cells contain either dependent variables or constants. Dependent variables are values that are calculated from formulas that depend on other cells in the workbook. Constants are values that never change (or for simplicity are assumed never to change), such as the barometric air pressure at sea level. You could not directly change the contents of the red cells even if you tried, because they are locked. The only way these will change is if you edit a value in a blue cell that affects a red cell, or unprotect the sheet using the password (aben435, note case sensitive). If you do this, re-protect the sheet before proceeding.

OXYGEN DESIGN (WORKBOOK)

The following is a brief description of the function of each worksheet in the workbook, followed by a listing and definitions (if necessary) of the contents of each cell:

System Feed and Flow (Worksheet)

Calculates total feed and water flow for a system consisting of a given number of tanks. The user enters how many tanks there are of each size, the quantity of feed per tank, and the flow for each tank.

System Feed and Flow		
# of Tanks	Feed per Tank (kg/day)	Treated Flow per Tank (LPM)
5	75	5,500
Oxygen Cost (delivered) $/kg	Total System Feed (kg/day)	Total System Flow (LPM)
0.18	375	27,500

Independent Variables (Blue)

- Number of tanks

- Feed per tank (lb/day)
- Treated Flow per tank (gpm): Flow rate through each tank.

Constants (Red)

No constants in this sheet.

Dependent Variables (Red)

- Total System feed (lb./day): Depends on # of tanks
- Total System Flow (gpm): Depends on # of tanks

Oxygen Demand (Worksheet)

Using the values calculated in Feed and Flow sheet and a user-provided oxygen demand factor, this sheet will calculate the oxygen demand for the system and the required change in oxygen concentration in mg/L.

Oxygen Demand		
Total System Feed (kg/day)	Oxygen Demand Factor (kg Oxygen/kg Feed)	Total Oxygen Demand (kg/day)
375	0.60	225
Total Oxygen Demand (mg/min)	Total System Flow (Liter/min)	Required Delta DO (mg/L)
156,000	27,500	5.7

Independent Variables (Blue)

There are no independent variables on this sheet.

Constants (Red)

- Oxygen Demand factor (lb. Oxygen/lb. feed): This value is the quantity of oxygen required per pound of feed being fed. This oxygen demand includes the oxygen required by the fish (approximately 0.25 lb per lb), the nitrifying bacteria (0.12 lb per lb) and any additional demand by heterotrophic bacteria in the water column or system (BOD). Note: This worksheet is based on the quantity of oxygen required in the water column, and does not address or compensate for inefficiency of the oxygen transfer system.

Dependent Variables (Red)

- Total System Feed (lb./day): This is taken from Feed and Flow sheet

- Total Oxygen Demand (lb./day): Total pounds of oxygen that will be used by the fish in one day; this depends on the amount of feed

- Total Oxygen Demand (mg/min): Another way of expressing the oxygen demand; this one gives it to you in milligrams of oxygen consumed per minute.

- Total system flow (Liters/min): Another way of expressing the flow rate through the system. Computed using the conversion from gallons to liters.

- Required Delta DO: Required change in dissolved oxygen concentration between the flow into the tower and the flow out of it.

Water Quality Parameters (Worksheet)

Calculates all necessary water quality parameters using previously calculated values and user provided temperature, barometric pressure, tower influent dissolved oxygen (DO), and tower influent dissolved nitrogen (DN).

Water Quality Parameters				
Temperature (C)	Theta (dimensionless)	Water Vapor Pressure (mmHg)	Barometric Pressure (mmHg)	Atmospheric Oxygen (%)
25	0.92	23.77	760	20.95%
Atmospheric Nitrogen (%)	Bunsen Coefficient for Oxygen (L/L□atm)	Bunsen Coefficient for Nitrogen (L/L-atm)	O2 Molecular Wgt/Vol Ratio (mg/mL)	N2 Molecular Wgt/Vol Ratio (mg/mL)
78.08%	0.0284	0.0144	1.42903	1.25043
DO Solubility in 100% Oxygen (mg/L)	DN Solubility in 100% Nitrogen (mg/L)	DO Solubility in atmosphere (mg/L)	DN Solubility in atmosphere (mg/L)	
40.64	18.03	8.25	13.64	
DO Tension (mmHg/(mg/L))	DN Tension (mmHg/(mg/L))	Tower Influent DO (mg/L)	Tower Influent DN (mg/L)	
18.70	42.16	4.0	15.0	
Alpha (dimensionless)	Phi for Oxygen (dimensionless)	Phi for Nitrogen (dimensionless)		
1	1	0.94		

Independent Variables (Blue)

- Temperature (°C): Temperature of the water

- Barometric Pressure (mmHg): Atmospheric pressure, measured in millimeters of mercury, approximately 760 mm at sea level.

- Tower Influent DO (mg/L): Concentration of oxygen entering the tower

- Tower Influent DN (mg/L): Concentration of nitrogen entering the tower

Constants (Red)

- Atmospheric Oxygen (%): Percent of oxygen found in the atmosphere (~21%)

- Atmospheric Nitrogen (%): Percent of nitrogen found in the atmosphere (~78%)

- O_2 Molecular Wgt/Vol Ratio (mg/mL): Ratio of the molecular weight of oxygen to its volume (~1.43 mg/mL)

- N_2 Molecular Wgt/Vol Ratio (mg/mL): Ratio of the molecular weight of nitrogen to its volume (~1.25 mg/mL)

- Alpha (dimensionless): Ratio of the mass transfer coefficient G at 20°C for field water to G at 20°C for clean water; generally a value near 0.9

- Phi for Oxygen (dimensionless): Diffusivity of oxygen relative to oxygen (1.00)

- Phi for Nitrogen (dimensionless): Diffusivity of nitrogen relative to oxygen (0.94)

Dependent Variables (Red)

- Theta (dimensionless): (273.16)/(273.16 + Temperature)

- Water Vapor Pressure (mmHg): Calculated vapor pressure in millimeters of mercury

- <u>Bunsen Coefficient for Oxygen (L/L*atm)</u>: Liters oxygen/liter water at 760 mmHg absolute pressure

- <u>Bunsen Coefficient for Nitrogen (L/L*atm)</u>: Liters nitrogen/liter water at 760 mmHg absolute pressure

- <u>DO Solubility in 100% Oxygen (mg/L)</u>: This is at a partial pressure of 760 mmHg

- <u>DN Solubility in 100% Nitrogen (mg/L)</u>: Same as above

- <u>DO Solubility in atmosphere, DN Solubility in atmosphere (mg/L)</u>: Saturation concentrations of these gases in the atmosphere

- <u>DO Tension, DN Tension (mmHg/(mg/L))</u>: Tension per mg/L of the dissolved gas in mm of mercury.

Tower Parameters (Worksheet)

Calculates all tower parameters, including the cost of oxygen per day, using previously calculated values and user-provided tower flow rate, flow rate through nozzle, and Tower/Nozzle G20.

Tower Parameters

Delta DO (mg/L)	Delta DN (mg/L)	Tower Flow (LPM)	Nozzle Flow (LPM)	Tower/Nozzle G20 (dimensionless)
5.7	-2.5	20000	1000	0.60
G_T for Oxygen (dimensionless)	G_T for Nitrogen (dimensionless)	Saturation DO (mg/L)	Saturation DN (mg/L)	Oxygen Partial Pressure Ratio (decimal)
0.68	0.64	15.6	9.6	0.38
Nitrogen Partial Pressure Ratio (decimal)	Column Pressure (mmHg)	Tower Oxygen Mole Fraction (decimal)	Tower Nitrogen Mole Fraction (decimal)	Oxygen Molar Flow (mole/min)
0.54	721	0.42	0.58	4.8
Oxygen Volumetric Flow (liter/min)	Oxygen Mass Flow (kg/day)	Oxygen Cost ($/day)	Volumetric G/L (%)	Absorption Efficiency (%)
116.7	222.6	$40.07	0.58%	73%

Independent Variables (Blue)

- <u>Tower Flow (gpm)</u>: Flow rate through the tower

- Nozzle Flow (gpm): Flow rate through each nozzle

- Tower/Nozzle G_20): Mass transfer coefficient at 20°C. This is a dimensionless value used to calculate the mass transfer coefficients for oxygen and nitrogen at the temperature you entered.

Constants (Red)

- Delta DO (mg/L): Required change in dissolved oxygen concentration, taken from the Oxygen Demand sheet.

Dependent Variables (Red)

- Delta DN (mg/L): Change in dissolved nitrogen concentration, calculated using Delta DO and tensions for oxygen and nitrogen.

- G_T for Oxygen, G_T for Nitrogen (dimensionless): Mass transfer coefficients for these gases at the temperature you entered.

- Saturation DO, Saturation DN (mg/L): Dissolved gas saturation concentration.

- Oxygen Partial Pressure Ratio, Nitrogen Partial Pressure Ratio (decimal): Partial pressure of each gas as a fraction of the total gas pressure.

- Column Pressure (mmHg): Absolute pressure in column.

- Tower Oxygen Mole Fraction, Tower Nitrogen Mole Fraction (decimal): Moles of each gas in the tower as a fraction of the total moles.

- Oxygen Molar Flow (moles/min): Moles of oxygen flowing through tower into tanks per minute.

- Oxygen Volumetric Flow (liter/min): Liters of oxygen flowing through the tower into tanks per minute.

- Oxygen Mass Flow (kg/day): Kilograms of oxygen flowing through the tower into tanks per day.

- Oxygen Cost ($/day): Cost per day of the above flow of oxygen, based upon the delivered oxygen cost to the farm in cents for each kilogram of oxygen to flow through the tower into the tanks.

- Volumetric G/L (%): Volumetric ratio of oxygen gas to liquid.

- Absorption Efficiency (%): Percentage of oxygen absorbed by the water.

Flow Diagram (Worksheet)

This shows a diagram of flow from the biofilter, through the tower, and into the tanks with values calculated in the spreadsheets.

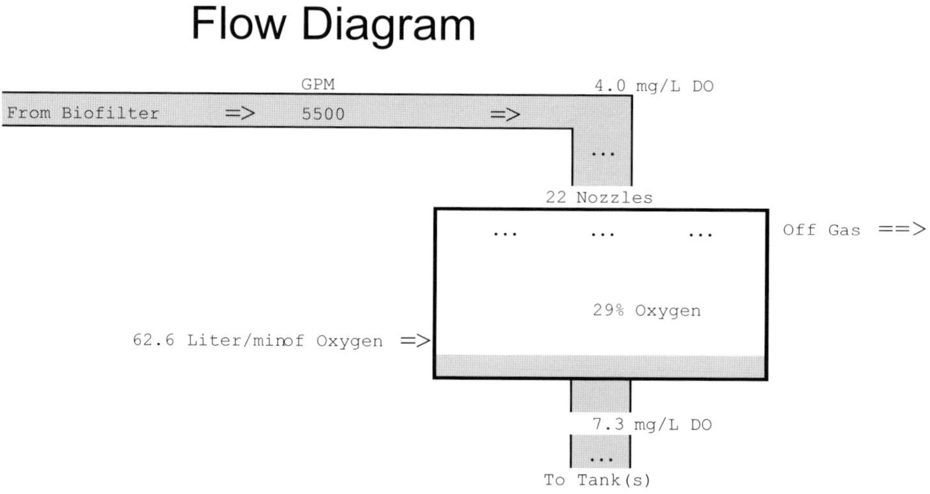

Unit Conversions (Worksheet)

For simplicity in writing formulas, this sheet has been added with the necessary unit conversions for the calculations done in the workbook. There are no dependent or independent variables here.

Unit Conversions		
Liters per Gallon	Minutes per Day	Milligrams per Pound
3.785	1,440	453,590
mmHg per Atmosphere		
760		

Constants (Red)

- Liters per gallon = 3.7853
- Minutes per Day = 1440
- Milligrams per pound = 45390
- mmHg per Atmosphere = 760

FISH TANK DESIGN (WORKBOOK)

The following is a brief description of the function of each worksheet in the workbook, followed by a listing and definitions (if necessary) of the contents of each cell:

Summary Data (worksheet)

The data contained in the cells of this sheet pertains to various parts of the design, or is restated from another part of the workbook (because of its importance). Therefore, it has been all grouped together under the title of "summary data". The Summary Data Sheet will look like the following:

Summary Data

% oxygen by aeration	BioNitr Rate gms/m2/day	Surface area for media m2/m3	Depth of unexpanded sand, ft
100%	0.15	4,100	6.0
Ammonia C in tank mg/L	% NH3 remov acr filter	Pump power per biofilter, kW	Protein Conc Feed, %
2.00	20.00	8	32%
Delta Oxygen across Pumps (mg/L)	Tank Ox Level (mg/L)	Sand Bed Upward Velocity cm/sec	lbs of fish/gal
5.5	5	3.50	0.85
Fish size, lbs	Feed to gain ratio, f/g	% feed/day per fish	
1.5	1.30	1.50	
growth/day %	Expected Column Ht ft (1.33D*1.4+2.0)	Nitrification gms/day/ft2 of x-sect	Flow rate biofilter gpm/x-sect ft2
1.15	13.2	104.6	51.5
Ox lbs/day/100gpm	FARM total yearly production, lbs/yr		
6.6	496,664		

Independent Variables (Blue)

- % oxygen by aeration: How much of the oxygen added to the water will come from aeration (as opposed to liquid oxygen).

- BioNitr rate, gms/m^2/day: Rate at which nitrifying bacteria oxidize/consume ammonia; expressed on an areal basis or per unit area of media.

- Surface area for media, m^2/m^3: How much surface area per unit volume of media is provided (manufacturer usually provides).

- Depth of unexpanded sand, ft: In the bioreactor (we are assuming it is a sand bed). The spreadsheet can be used for any form of media, just provide the proper surface area.

- Ammonia C in Tank, mg/L: Concentration of ammonia in the fish tank, which is assumed to be the entering concentration into the biofilter.

- <u>%NH_3 remov acr filter</u>: Percent of NH_3 content in the tank that is removed as the water moves across the filter (basically an efficiency value for the filter).
- <u>Pump power per biofilter, kW</u>: How much power the pumps for each biofilter require.
- <u>Protein conc feed, %</u>: Percentage of protein contained in the fish feed.
- <u>Delta oxygen across Pumps, mg/L</u>: The change in the water's oxygen concentration across the pumping station or oxygen addition device.
- <u>Tank Ox level, mg/L</u>: Concentration of oxygen in the tank, used to calculate oxygen lbs. per day being added to the water (1 cm/s = 14.7 gpm/ft^2).
- <u>Sand Bed Upward Velocity, cm/sec</u>: Velocity of water as it moves upward in the sand bed (essentially the hydraulic loading; this value required even if you are not designing as a sand filter, since used to calculate flow rate through the biofilter).
- <u>lbs of fish/gal</u>: Design biomass loading level, lb/gal.
- <u>Fish size, lbs.</u>: Average weight of the fish for the current design calculation (think of it as the harvest size of the fish, since this will create the maximum loading).
- <u>Feed to gain ratio, f/g</u>: Ratio of feed used to fish biomass gain.
- <u>% feed per day</u>: How much feed is fed to the fish as a percentage of their weight (obtained from the worksheet called "Growth data").

Constants (Red)

No constants in this sheet.

Dependent Variables (Red)

- <u>Growth/day, %</u>: Amount that fish grow in one day, as a percentage of their weight the previous day.

- Expected Column Ht., ft (1.33D*1.4+2.0): Calculated height of the sand bed column; this can be ignored if you are not using an expanded bed, e.g., trickling tower.

- Nitrification gms/day/ft^2 of x-sect: Nitrification rate per foot squared of cross-sectional area of the medium.

- Flow Rate Biofilter gpm/x-sect ft^2: Flow rate through the biofilter per square foot of cross-sectional area.

- Ox lbs/day/100 gpm: Oxygen added per day per 100 gpm of water flow.

- Farm Total yearly production, lb/yr: Pounds of fish produced by farm in one year (is calculated from the data in the Tank Design Parameters Sheet).

Tank Design Parameters (Worksheet)

This sheet is set up in a table format, and generally looks like this:

Tank Design Parameters

Tank Diam ft	Depth ft	# of tanks of each size	Gallons	lbs/tank	lbs Feed per day
8	2.0	1.0	752	639	10
14	3.0	2.0	3,454	2,936	44
31	4.5	6.0	25,405	21,594	324
31	6.0	1.0	33,873	28,792	432

Tank Diam ft	aeration flow req, gpm	sand bed x-sect ft2, req'd	sand bed Nitr flow rate gpm	Flow Rate for NH3 control	O2 lb/day Based on O2 x filter	O2 needed x filter lb/day	Gain/mon	lbs/yr	gm NH3 per day
8	54	1.2	63	59	2.3	1.2	221	2,655	128
14	247	5.6	290	270	10.4	5.6	1,016	12,196	589
31	1,814	41.4	2,133	1,985	76.9	41.4	7,475	89,698	4,329
31	2,419	55.2	2,844	2,647	102.5	55.2	9,966	119,598	5,772

As always, you can change the values in blue but not the ones in red. In this table, you specify the tank size by giving the diameter and depth, and the sheet will tell you how much production is possible, among the other things described below. Cells are described below.

Independent Variables (Blue)

- Tank Diam, ft: Diameter of a fish tank.
- Depth, ft: Depth of a tank.
- # of tanks of each size: For each tank diameter and depth that you entered, indicate how many of these there are.

Constants (Red)

- No constants in the table.

Dependent Variables (Red)

- Gallons: Number of gallons of water that are in a particular tank (line by line).
- Lbs/tank: Pounds of fish that fit in each tank.
- Lbs feed per day: Amount of feed per day added to each tank.
- Aeration flow req, gpm: Flow rate required to provide the needed oxygen for fish and biofilter.
- Sand Bed x-sect, ft^2, req: Required cross-sectional area of the sand bed, for each size of tank.
- Sand Bed Nitr Flow Rate, gpm: Actual flow rate through the sand bed, for nitrification based upon required cross sectional area and depth of bed and specified sand bed fluidized velocity.
- Flow rate for NH_3 control: Flow rate *required* through the sand bed or biofilter, for proper nitrification to maintain NH_3 concentration in fish tank at or below design target value. Compare this to the value above to make sure you are getting proper flow.
- O_2 lb/day based on O_2 x filter: Actual lb/day of oxygen being added to the water.

- O$_2$ needed x filter, lb/day: Required lb/day of oxygen for proper aeration. Again, compare to value above; if there is not sufficient oxygen being delivered to the biofilter, it will go anaerobic and not perform (plus do other bad things, like produce hydrogen sulfide which will kill the fish).

- Gain/mon: Total increase in fish biomass expressed in pounds of fish per month. Based of amount of feed and feed to gain ratio for the particular tank size per individual tank.

- Lbs/yr: Total gain per year per individual tank of specific size.

- Gm NH_3 per day: Total grams of ammonia produced per day per individual tank size.

Note: there are three flow rates calculated in the above worksheet:

1. Aeration flow requirements,

2. Sand bed Nitr flow rate, and

3. Flow Rate for NH_3 control.

Ideally, these three values will be the same. If they are substantially different by more than 25%, then reconsider your design, e.g., change the oxygenation system or media used on the biofilter.

Growth Data (Worksheet)

This is another sheet with tables: it has two short ones and are demonstrated and described following the sample tables.

Growth Data			
Base Temp		TU's/inch	
65.0		15	Choose a rate and place in the block of cells below.
Temp	Growth in/month	Growth inches/day	
70	0.3	0.0111	
75	0.7	0.0222	
80	1.0	0.0333	
85	1.3	0.0444	

Condition Factor			Growth Rate Chosen (in/day)		
760			0.0444		
Size, inches	wt, lbs	wt, grams	wt after 1 day of growth	grams growth per week per fish	%/day
0.75	0.000	0.1	0.000	0.2	18.8
1	0.001	0.3	0.001	0.3	13.9
2	0.006	2.8	0.006	1.3	6.8
4	0.049	22.1	0.050	5.2	3.4
8	0.389	176.7	0.396	20.7	1.7
12	1.313	596.2	1.328	46.5	1.1

Values in first table:

Independent Variables (Blue)

- Base Temp: Base temperature used in MTU calculation. For tilapia, this value is 65°F and trout it is 32 °F.

- TU's/inch: Thermal units required to grow one inch (see Chapter 4).

- Temp: Enter a range of temperatures in degrees F, higher than the base temperature.

Constants (Red)

No constants in this table.

Dependent Variables (Red)

- Growth, in/month: Inches of growth per month for each temperature you entered.

- Growth, inches/day: Growth per day for each temperature you entered.

Values in second table:

Independent Variables (Blue)

- Growth Rate Chosen, (in/day): Choose a growth rate from the last column in the first table and enter it here; this value will then define the growth rates demonstrated in the succeeding table.

- Condition Factor: Enter the condition factor associated with the fish of interest, e.g., tilapia is 760 or trout is 400 (see Chapter 4).

- Size, inches: Enter a range of sizes important to you, e.g., to develop a feeding chart, to see how this affects their weight gain.

Constants (Red)

No constants in this table.

Dependent Variables (Red)

- Wt, lbs: Typical weight for each size of fish you entered for the particular condition factor.

- Wt, grams: Fish weight in grams.

- Wt. after one day of growth: Weight of a fish after it has grown for one day at the rate you have chosen.

- Grams growth per week per fish: This calculation is useful to see if your values are reasonable. Also, for some temperatures and lengths, you eventually will calculate unreasonably high values so you should assign some maximum value above which the growth rate is assigned the default value.

- %/day: Weight gain as a percentage of the previous day's weight. Useful in creating daily feeding charts.

PIPE FLOW DESIGN (WORKBOOK)

This workbook is made up almost entirely of tables. The following is a brief description of the function of each worksheet in the workbook, followed by a listing and definitions (if necessary) of the contents of each cell:

Pipe Flows (worksheet)

Used to calculate dynamic losses, as are the other sheets in this workbook. E.g., one 90 degree elbow requires one dynamic head, $V^2/2g$.

Pipe Flows		
	Pipe Diam (inches)	
	4.0	
Desired Flow gpm	Pipe Velocity ft/s	Head, inches v^2/2g
1300	33.19	205.3
1000	25.53	121.5
500	12.77	30.4
100	2.55	1.2
50	1.28	0.3
25	0.64	0.1
10	0.26	0.0

Independent Variables (Blue)

- Pipe Diam (inches): Diameter of pipe to be used for the flow desired.

- Desired Flow, gpm: Enter a range of flow rates through the pipe.

Constants (Red)

No constants in this table.

Dependent Variables (Red)

- Pipe Velocity, ft/s: Velocity of water in the pipe for each desired flow rate.

- Head, inches $v^2/2g$: Head produced as a result of the velocity for each flow rate. This is useful to quickly show you the magnitude of pressure loss associated with each elbow.

Perforated Plate Calculations (Worksheet)

Calculates values needed in the design of the perforated plates as used in LHO designs or other trickling filters or similar devices.

Perforated Plate Calculations					
area,ft2	Diameter of Flooded Round Plate, Inch	Total Flow Required gpm	Hole Diameter inches	gpm/ft2	
15.32	53	500	0.50	32.64	
Head Available inches	Dynamic Heads Req	Velocity Created ft/s	Flow Rate,gpm per hole	Number holes needed	Holes per square foot
5.00	2.00	3.66	2.2	223	15
7.50	2.00	4.49	2.7	182	12
10.00	2.00	5.18	3.2	158	10

Independent Variables (Blue)

- Diameter of Flooded Round Plate, Inch: Diameter required for the plate (assumes it is a circular device; if not override the cell that has "area" and enter area of plate directly; you will need to unprotect the worksheet to do this).

- Total Flow Required, gpm: Total flow rate required through the holes in the plate.

- Hole Diameter, Inches: Desired diameter of each hole in the plate.

Constants (Red)

No constants in this table.

Dependent Variables (Red)

- Area, ft^2: Area of perforated plate.

- Gpm/ft^2: Flow rate per square foot of the perforated plate.

Values in table:

Independent Variables (Blue)

- Head Available, Inches: Enter a range of amounts of head available for the flow through the plate.

- Dynamic Heads Required: Number of dynamic heads that are needed.

Constants (Red)

No constants in this table.

Dependent Variables (Red)

- Velocity Created, ft/s: Velocity that results from the available head and the number of dynamic heads required.

- Flow Rate, gpm per hole: Flow rate through each hole that results from velocity created and the size of the hole.

- Number holes needed: Holes needed to satisfy the required flow rate.

- Holes per square foot: Number of holes per square foot of the plate.

Drain Flows (worksheet)

Calculates values for flow through pipes where the flow is driven by some specified amount of water column head being available; you must define the dynamic heads required for the flow from the beginning through the point of discharge (note: the discharge is one dynamic head).

Drain Flows				
		Dynamic Heads Required		
		3.00		
Head Available inches	Velocity Created ft/s	Pipe Diam Inches	Area ft^2	Flow Rate gpm
2.0	1.89	2.00	0.022	19
6.0	3.28	2.00	0.022	32
12.0	4.63	2.00	0.022	45
24.0	6.55	2.00	0.022	64
2.0	1.89	3.00	0.049	42
6.0	3.28	3.00	0.049	72
12.0	4.63	3.00	0.049	102
24.0	6.55	3.00	0.049	144
2.0	1.89	4.00	0.087	74
6.0	3.28	4.00	0.087	128
12.0	4.63	4.00	0.087	181
24.0	6.55	4.00	0.087	257
2.0	1.89	6.00	0.196	167
6.0	3.28	6.00	0.196	289
12.0	4.63	6.00	0.196	408
24.0	6.55	6.00	0.196	577
2.0	1.89	8.00	0.349	296
6.0	3.28	8.00	0.349	513
12.0	4.63	8.00	0.349	726
24.0	6.55	8.00	0.349	1,026

Independent Variables (Blue)

- Dynamic Heads Required: Number of dynamic heads needed from beginning reference point through point of discharge.

- Head Available, Inches: Enter a range of amounts of head available for the flow through pipe.

- Pipe Diam (inches): Enter a range of diameters for which flow rates will be calculated.

Note: This spreadsheet is useful for exploring pipe design options and the effects of pipe diameter related to required/desired flow rates and the amount of water elevation that will be required to produce the desired flow. A typical use of this spreadsheet would be to calculate the water elevation (head) required to produce a desired flow entering the tank or leaving the tank. Generally, one tries to minimize the amount of elevation required to produce the desired flow, even when more elevation head is available to you.

Constants (Red)

No constants in this table.

Dependent Variables (Red)

- Velocity Created, ft/s: Velocity that results from the available head and the number of dynamic heads required to produce the flow.

- Area, ft^2: Cross sectional area of pipe for the diameter entered.

- Flow Rate, gpm: Flow rate through the pipe that results from velocity created and the size of the pipe.

Drainage Pipe Calculations (Worksheet)

Calculates values for flow through the drain pipes using an assumed velocity entered by the user. For conventional floor drains, the 2 ft/s value corresponds to a 3/1,000 grade.

DRAINAGE PIPE FLOW	
Assumed velocity in pipe, ft/s*	
2.00	
Pipe diam, inches	Resulting Flow Rate, gpm
3	44
4	78
5	122
6	176
8	313
10	490
12	705
16	1,253
24	2,820
30	4,406

*velocity used for sewer pipe cleaning to prevent siltation assumes a 0.3% fall rate (3/1000)

Independent Variables (Blue)

- Assumed Velocity in pipe, ft/s: Enter a design velocity.

- Pipe Diam (inches): Enter a range of diameters for which flow rates will be calculated.

Constants (Red)

No constants in this table.

Dependent Variables (Red)

- Resulting Flow Rate, gpm: Flow rate through the pipe that results from the assumed velocity and the size of the pipe.

COST ANALYSIS (WORKBOOK)

This workbook is set up a little differently. The first sheet is entirely for input, containing only independent variables. The second sheet contains the resulting dependent variables, and is therefore completely locked. You cannot enter values in this sheet. Please examine the following sample worksheets and brief descriptions of the input and output cells in each worksheet of the workbook. This workbook is very useful for calculating the costs associated with planned or on-paper farms. The approach taken is that a farm is made up of pods that are independent of each other. A farm may have one or more pods but each pod is to be considered to be identical. Costs and other input information is typically entered for a single pod, since construction estimates would be done in this way for the farm, e.g., design a pod and then multiply the number of pods to achieve the desired total farm production.

Input Information

	Feed to Gain Ratio	Days of Operation per Year		Total Labor Cost $/yr-building	Daily Water Exchange Rate
	1.25	365		$125,000	10.0%
Heat Exchanger Efficiency	O2 Cost $/lb	Feed Cost $/lb	lb. O2 req/lb feed (includes efficiency)	Electrical Cost $/kWh	Gallons in Pod Unit, Tank, Settling, Biofilter
0.0%	$0.04	$0.25	0.80	$0.05	100,000
Number Pods in Building	Building Backup and Monitoring Costs				Cost $/complete Pod system w/access
6	$100,000				$80,000
Equip. depr. period, yrs	Building depr. period, yrs.	kW/pod system (power requ'ts)	Mortality over growout period	Fingerling Cost $/fingerling	Selling Size lb
7.0	39.0	20.0	5.0%	$0.10	1.50
Water temp deg F	Temp makeup water, deg F		Initial Cost of Building	Average Ceiling height, ft	Total floor area ft^2
80.0	50.0		$500,000	20.0	40,000
Overall R value for Building hr*ft^2*F/Btu	Minimum Air Volume Changes, ft^3/hr	Interest Cost decimal	Cost of Fuel $ per 100,000 BTU	Min Feed per pod lb/day	Max Feed per pod lb/day
20.0	2.0	0.10	$1.00	0	100

Input Information

The user uses this sheet to enter all the pertinent input data. The values are used to calculate the costs listed on the Cost Summary sheet. Note the use of the pod approach to defining input variables.

Independent Variables (Blue)

- Feed to Gain Ratio: Feed to gain ratio.

- Days of Operation per Year: Number of days in the year that the facility is in operation.

- Total Labor Cost, $/yr-building: Cost of labor per farm per year.

- Daily Water Exchange Rate: Percentage of water that is replaced per day.

- Heat Exchanger Efficiency: Percentage of heat that is absorbed by the water.

- O_2 Cost $/lb: Cost per pound of oxygen added to the water; raw cost. There are 12 cubic feet per lb of oxygen at atmospheric pressure. Suppliers will generally charge per 100 cubic feet supplied and there will be a facility charge for rent of the oxygen unit; include all these in the cost per lb.

- Feed Cost $/lb: Cost per pound of feed.

- Lb. O_2 supp/lb feed (includes efficiency): Amount of oxygen that needs to be supplied per pound of feed fed to the fish. This value should reflect the entire demand divided by the efficiency of delivery. Biological demand includes fish, biofilter and BOD loads.

- Electrical Cost, $/kWh: Cost per kWh of energy used.

- Gallons in Pod Unit, Tank, Settling, & Biofilter: Gallons of water in one pod system.

- Number Pods in Building: How many pod units are there are in one building.

- Building Backup and Monitoring Costs: Initial cost of these systems.

- Cost $/complete Pod system w/ access: Initial investment cost.

- Equip. depr. period, yrs: Useful life of the equipment, for accounting purposes.

- Building depr. period, yrs: Useful life of a building, for accounting purposes.

- KW/pod system (power requ'ts): kW of power used by each pod system.

- Mortality over growout period: Percentage of fish that die over the whole growth period, from fingerling to market size.

- Fingerling Cost, $/fingerling: Cost of each fingerling.

- Selling Size, lb.: Weight at which fish are sold.

- Water Temp, deg F: Temperature of the fish tank water in degrees Fahrenheit.

- Temp makeup water, deg F: Temperature of the makeup water, in degrees Fahrenheit.

- Initial cost of Building: Price paid for the building at purchase time (does not include land cost).

- Average Ceiling height, ft.: Height (average) for the ceiling in the barn; used to calculate ventilation exchange rates.

- Total Floor Area, ft^2: Floor area of one building.

- Overall R value for Building, $hr \cdot ft^2 \cdot F/Btu$: Heat transfer constant for building walls and ceiling.

- Minimum Air Volume Changes, ft^3/hr: This is the minimum ventilation rate assumed for the barn, typically this will be at least 2 air volume exchanges per hour.

- Interest Cost, decimal: Interest rate, in decimal form.

- Cost of Fuel $/100,000 BTU: Heating costs are based upon the supplied cost of the energy source; listed here per Therm (100,000 BTU's).

- Min Feed per pod lb/day: Minimum feeding rate that will be considered in the analysis per pod.

- Max Feed per pod, lb/day: Maximum feeding rate considered per pod, e.g., when pod reaches maximum feeding capacity. This should also be viewed as the average daily feeding rate that will be achieved over the complete production cycles, i.e., the average feeding rate per day over the entire year will define the quantity of fish produced over the year.

Constants (Red)

No constants in this table.

Dependent Variables (Red)

No dependent variables in this table.

Summary of Costs

Feed/Tank-day	Total Production lb/week	Electric Cost/lb	Feed Cost/lb	Water Heating Cost/lb	Air Heating Cost/lb	Oxygen Cost/lb	Labor Cost/lb	Fingerling Cost/lb	Repairs Cost/lb	Depreciation Cost/lb	Total Cost $/lb
100	10,950	$0.13	$0.18	$0.06	$0.03	$0.08	$0.18	$0.01	$0.05	$0.10	$0.82
130	14,235	$0.10	$0.18	$0.05	$0.02	$0.08	$0.14	$0.01	$0.04	$0.08	$0.70
160	17,520	$0.08	$0.18	$0.04	$0.02	$0.08	$0.11	$0.01	$0.03	$0.06	$0.62
190	20,805	$0.07	$0.18	$0.03	$0.01	$0.08	$0.09	$0.01	$0.02	$0.05	$0.57
220	24,090	$0.06	$0.18	$0.03	$0.01	$0.08	$0.08	$0.01	$0.02	$0.05	$0.53
250	27,375	$0.05	$0.18	$0.02	$0.01	$0.08	$0.07	$0.01	$0.02	$0.04	$0.50
280	30,660	$0.05	$0.18	$0.02	$0.01	$0.08	$0.06	$0.01	$0.02	$0.04	$0.47
310	33,945	$0.04	$0.18	$0.02	$0.01	$0.08	$0.06	$0.01	$0.02	$0.03	$0.46
340	37,230	$0.04	$0.18	$0.02	$0.01	$0.08	$0.05	$0.01	$0.01	$0.03	$0.44
370	40,515	$0.04	$0.18	$0.02	$0.01	$0.08	$0.05	$0.01	$0.01	$0.03	$0.43
400	43,800	$0.03	$0.18	$0.02	$0.01	$0.08	$0.04	$0.01	$0.01	$0.03	$0.42

Assumptions:

- Air temperature equal to target water temperature + 4°F; this is reasonable if the air is heated to heat the water. If the water is heated directly, then the air temperature can be maintained a few degrees cooler than the tank water temperature.

- Average outside air temperature for the year is assumed to be the same as the makeup water temperature.

- Heating calculations assume a 100% burn efficiency.

- Repairs and maintenance is 5% of original building and equipment costs.

- Costs do not include taxes, interest, accounting or legal costs

Independent Variables (Blue)

- No independent variables.

Constants (Red)

No constants in this table.

Dependent Variables

- Feed/Tank-day: Creates a series of feeding rates between the minimum and maximum values. This should be viewed as what "can" happen, particularly in a startup venture where maximum feeding levels may not be achieved on a sustained basis.

- Total Production lb/week: Pounds per week of fish produced; primarily this is output information, so that you can easily see what your weekly sales should be.

The following costs are all applicable to each pound of fish produced:

1. Electric: Total electrical cost/total production.
2. Feed: Total feed cost/total production.
3. Water Heating: Total water heating cost/total production.
4. Air Heating: Total air heating cost/total production.
5. Oxygen: Total oxygen cost/total production.
6. Labor: Total labor cost divided by total production.
7. Fingerling: Total fingerling cost divided by total production.

8. Repairs: Repairs is based upon 5% of the initial capital costs divided by total production.
9. Depreciation: Straight line depreciation based upon number of depreciation years specified for equipment and the building and their initial capital costs divided by total production.
10. Cost/lb: Total cost of producing one pound of fish; this total cost is made up of the receding cost columns in the worksheet; one could selectively determine a particular subset of costs, e.g., direct costs plus labor cost and neglect other component costs, e.g., depreciation and repairs.

Note how some of the above costs will not change as feeding rate is increased, e.g., feed, oxygen, fingerling, while the fixed overhead, e.g., electric, heating, labor and repairs, will reduce dramatically on a per unit basis as feeding rates are increased. This is the ***key to minimizing total cost*** per unit of production. The cost analysis does not include taxes, interest, accounting or legal expenses because of the individual site and farm specific nature of these expenses. These costs can be significant and must be considered for your own particular farming situation.

17.2 DOS BASED PROGRAMS

CO_2 CONTROL MODE

USER'S GUIDE (BOOK VERSION 1.0)[1]

Introduction

Carbon dioxide control in intensive recirculating aquaculture systems (RAS), Fig. 17.1, is an important consideration in the design of commercial-scale systems and in systems where high carbon dioxide levels have contributed to adverse fish health. Both situations present the problem of maximizing carbon dioxide removal from the water column and venting it to the atmosphere at minimal cost. The use of pure oxygen absorption equipment in intensive aquaculture systems has exacerbated potential carbon dioxide problems. This type of equipment may not provide adequate gas exchange for stripping of carbon dioxide, therefore carbon dioxide may accumulate to toxic levels. Additionally, pure oxygen absorption units support higher fish densities, which also

[1] Dr. Barnaby J. Watten, U.S. Geological Survey, Leetown Science Center, Kearneysville, WV 25430 contributed to the original modeling effort of this CO_2 software.

increases the total carbon dioxide production within the system and increases the potential for carbon dioxide problems.

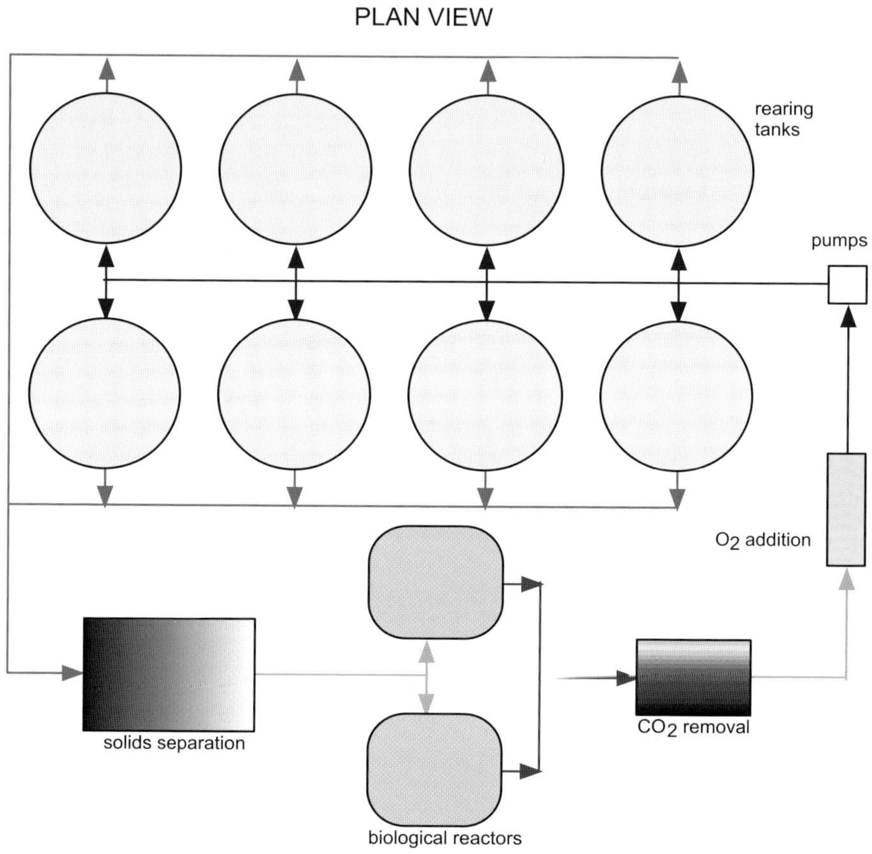

Figure 17.1 "Generic" Recirculating Aquaculture System (RAS) Layout.

The current CO_2 CONTROL Model (Book Version 1.0) provides an interactive design tool that gives side by side comparisons of seven carbon dioxide control strategies, Figs. 17.2–8:

1. packed tower carbon dioxide stripping unit
2. addition of sodium hydroxide
3. water exchange
4. in-tank surface aeration
5. side-stream surface aeration

6. in-tank diffused aeration
7. side-stream diffused aeration control strategies

The model is based on mathematical models of each carbon dioxide control strategy and gives the user economic and feasibility comparisons between all options. Facility heat and moisture balances are included in the program to account for typical production system concerns of ventilation for heat, moisture, or carbon dioxide control. It is important to note that all models employed work well under typical aquacultural conditions. However, in cases of high alkalinities (>500 ppm), the in-tank aeration options may not perform to full design capabilities. In these cases, the design should allow for an appropriate safety margin, e.g., an additional 10% of aeration capacity should be designed into the system.

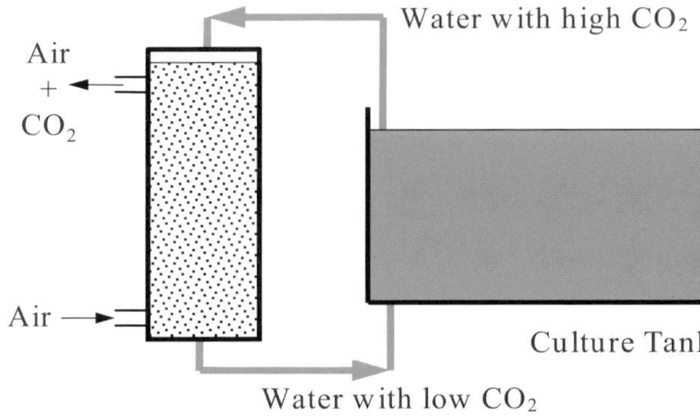

Figure 17.2 Diagram of Packed Tower Stripping Carbon Dioxide Control Option.

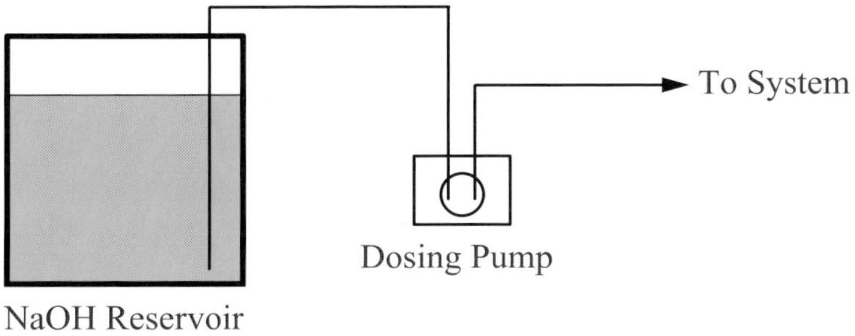

Figure 17.3 Diagram of Sodium Hydroxide Addition Carbon Dioxide Control Option.

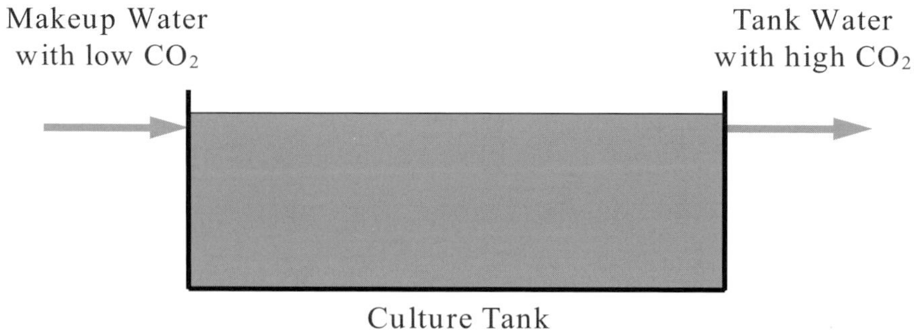

Figure 17.4 Diagram of Water Exchange Carbon Dioxide Control Option.

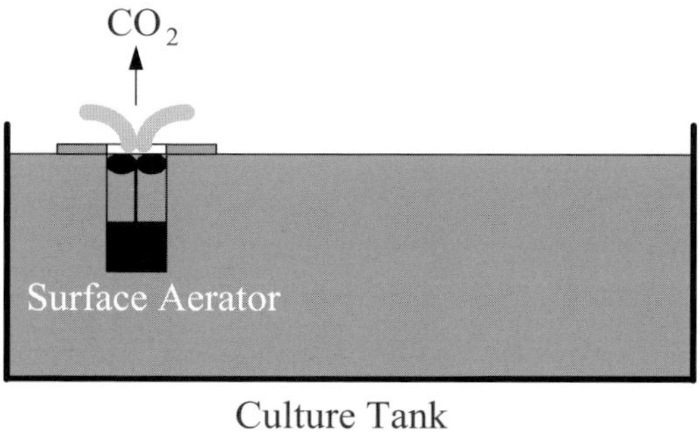

Figure 17.5 Diagram of In-Tank Surface Aeration Carbon Dioxide Control Option.

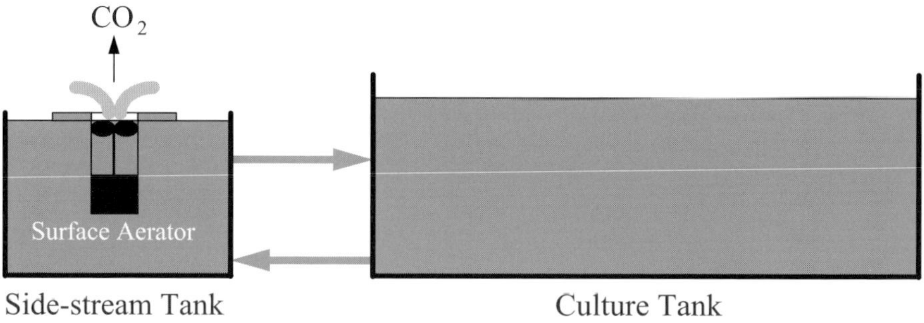

Figure 17.6 Diagram Of Side-Stream Surface Aeration Carbon Dioxide Control Option.

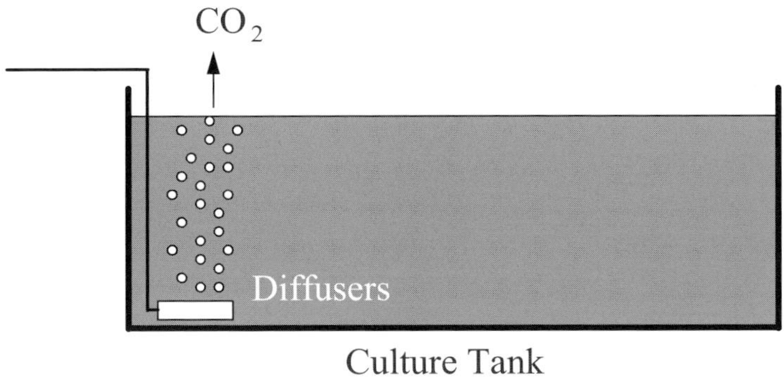

Figure 17.7 Diagram Of In-Tank Diffused Aeration Carbon Dioxide Control Option.

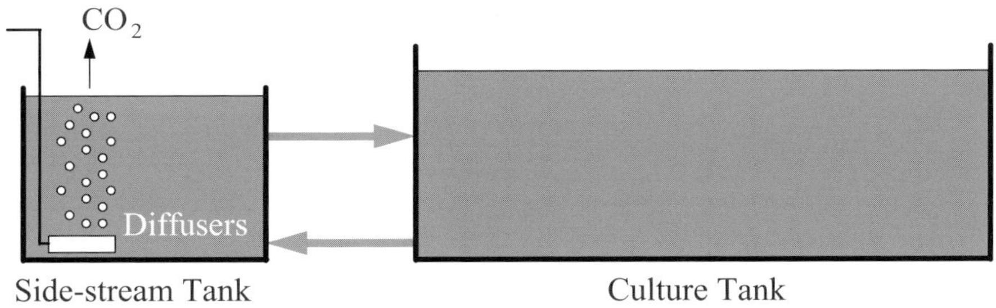

Figure 17.8 Diagram Of Side-Stream Diffused Aeration Carbon Dioxide Control Option.

Installation

CARBON DIOXIDE CONTROL Book Version 1.0 requires the following as a minimum (PC's purchased in the last couple years should have all the following as minimum features):

- ☑ Intel microprocessor-based PC or compatible (80486 or higher recommended)
- ☑ Color monitor
- ☑ CD ROM drive
- ☑ 640K RAM
- ☑ DOS version 3.3 or higher

CARBON DIOXIDE CONTROL is DOS-based and runs from the DOS command line, including the MS Windows DOS Shell within MS Windows 3.x or higher, MS Windows 95 or higher, and MS Windows NT or higher.

The program CD-ROM contains the file required to run CARBON DIOXIDE CONTROL Version 2.1:

co2contr.exe

To install the program on a hard disk simply copy the program file from the program CD-ROM to the hard disk (example provided for the DOS command line):

1. Insert the program CD-ROM into the computer's CD-ROM drive:

2. Create a directory on the hard disk (**c:**) for the files to reside in (or you can create a folder on your desktop):
   c:\>**md co2contr**
   (press **ENTER**)

3. Copy the file from the program CD-ROM to the hard disk directory:
   c:\>**copy a:co2contr.* c:\co2contr**
   (press **ENTER**)

The program may also be run from the program CD-ROM instead of the hard disk.

Getting Started

To run the program:

1. Change the current directory to the program directory:
   c:\>**cd c:\co2contr**
   (press **ENTER**)

2. Once in that directory, execute the program file:
   c:\co2contr>**co2contr.exe**
   (press **ENTER**)

The program will begin with an introduction screen, displaying the following:

****

CARBON DIOXIDE CONTROL

Book Version 1.0 -Copyright April 1998

****

Program by

B.J. Vinci, M.B. Timmons, S.T. Summerfelt, B.J. Watten

The Conservation Fund

Freshwater Institute

Shepherdstown, WV

****

The objective of this program is to evaluate options and costs for controlling CO_2 in water reuse systems. Treatment options evaluated include the packed stripping tower, the addition of a caustic (NaOH), water exchange, surface aeration and diffused aeration. The program allows a user to view current default settings and change these settings to configure the simulation to approximate their specific situation. Simulation output provides for side-by-side comparison of the options. Please refer to the User's Guide for more detailed information.

Press the ENTER key to proceed. Have fun!!!!

MENU The options menu screen displays the following choices:

OPTIONS

1. View/Change Building Inputs

2. View/Change Fish Culture Inputs

3. View/Change Control Option Inputs

4. View/Change Economic Inputs

5. Start the Simulation

6. Exit the Program

To select any of the six options enter the appropriate number and press the **ENTER** key.

Options 1, 2, 3, and 4 allow viewing and changing of current values for system variables. Refer to the INPUTS section of this manual for more detail on each of these option choices.

Option 5 will start a simulation using values for the system variables as they were displayed in options 1, 2, 3, and 4. Choosing option 5 will display a simulation progress bar as the program is working:

Calculating

Once all simulation calculations are complete an output summary is displayed. The output summary shows the results of the simulation and is discussed in the OUTPUTS section of this manual.

Inputs
Interactive program variables are divided into four categories:

1. General Building Inputs

2. Fish Culture Inputs

3. Control Option Inputs

4. Economic Inputs

Changing Variable Values
Each of these categories has interactive variables that can be changed by the user. To view the variables, go to the option menu, and choose either 1, 2, 3, or 4 to select the desired category. This will display the interactive variables that may be subsequently changed to approximate a user's own situation. After choosing a category, a new screen will be displayed showing the variables and the current value for each. To change a variable value:

1. Press the **y** character key and the **ENTER** key.

2. Enter the number of the variable of which you would like to change and press the **ENTER** key.

3. A window will pop up asking for the new value of the variable – enter the new value and press the **ENTER** key.

4. The screen will be refreshed showing the new value that has been changed as the current value (check to be sure that the displayed value is in fact the desired value).

Once finished with the viewing or changing of variables for a selected category, press the **n** character key and **ENTER** key. The program then returns to the options menu.

General Building Inputs
The variables that may be changed under this option are defined as follows:

1. Start Date (MM/DD) – The day of the year the simulation will start. Entered in month/day format.

2. Number of Simulation Days – The number of days for which the simulation will be carried out. Entered as number of days.
3. Yearly Maximum Monthly Temperature – The maximum monthly average outside temperature for a year. Entered in degrees Fahrenheit. See Table 17.1 for sample data.
4. Yearly Minimum Monthly Temperature – The minimum monthly average outside temperature for a year. Entered in degrees Fahrenheit. See Table 17.1 for sample data.
5. Elevation Above Mean Sea Level – The facility's elevation above mean sea level. Entered in feet.
6. Target Building Temperature – The target ambient air temperature inside the building. Entered in degrees Fahrenheit.
7. Target Relative Humidity – The target relative humidity level inside the building. Entered as a percent.
8. Maximum Available Ventilation Capacity – The maximum ventilation that is possible in the building. Entered as total building air volumes exchanged per hour.
9. R-Value for Ceiling – The average unit area thermal resistance for the building's overhead covering or ceiling. Entered in degrees Fahrenheit-square feet-hour per BTU. See Table 17.2 for sample data.
10. R-Value for Wall – The average unit area thermal resistance for the building's walls. Entered in degrees Fahrenheit-square feet-hour per BTU. See Table 17.2 for sample data.
11. R-Value for Curtain – The average unit area thermal resistance for the building's curtains or windows. Entered in degrees Fahrenheit-square feet-hour per BTU. See Table 17.2 for sample data.
12. Heat Load – The total heat load generated in the building from lights, pumps, and other mechanical equipment. Entered in kilowatts.
13. Building Width – The width of the building. Entered in feet.
14. Building Length – The length of the building. Entered in feet.
15. Building Ceiling Height – The height of the building ceiling (assumes a flat ceiling). Entered in feet.
16. Infiltration Air Changes – The approximate complete building air changes that occur through infiltration. Entered in building volumes per hour. See Table 17.3 for sample data.
17. Ambient CO_2 Level – The allowable concentration of carbon dioxide in the building ambient air. Entered in parts per million. See Table 17.4 for sample data.

Table 17.1 Maximum and Minimum Monthly Average Outside Temperatures For Selected USA Locations In Degrees °F (Kreider, J. F., & Kreith, F. ,1975, Solar Heating And Cooling: Engineering, Practical Design, And Economics, Washington, DC: Hemisphere Publishing)

Location	Max	Min		Location	Max	Min
Mobile, AL	82.6	53.0		Great Falls, MT	69.4	22.1
Phoenix, AZ	89.8	49.7		Omaha, NE	78.5	22.3
Little Rock, AR	81.9	40.6		Reno, NV	67.7	30.4
Los Angeles, CA	69.1	54.4		Concord, NH	69.6	21.2
Denver, CO	72.9	28.5		Atlantic City, NJ	75.1	34.7
Hartford, CT	73.4	26.0		Albuquerque, NM	78.5	35.0
Wilmington, DE	76.0	33.4		Buffalo, NY	69.8	24.1
Washington, DC	78.2	36.9		Raleigh, NC	77.9	41.6
Miami, FL	82.3	66.9		Bismark, ND	71.7	9.9
Atlanta, GA	78.9	44.7		Columbus, OH	74.8	29.9
Honolulu, HI	79.4	72.4		Oklahoma City, OK	82.8	37.0
Boise, ID	75.2	29.1		Portland, OR	67.2	38.4
Chicago, IL	75.6	26.0		Pittsburgh, PA	72.1	28.9
Indianapolis, IN	75.2	29.1		Providence, RI	72.1	29.2
Des Moines, IA	76.3	19.9		Columbia, SC	81.6	46.4
Wichita, KS	80.9	32.0		Sioux Falls, SD	74.3	15.2
Louisville, KY	77.6	35.0		Memphis, TN	81.3	41.5
New Orleans, LA	81.9	54.6		Dallas, TX	85.0	45.9
Portland, ME	68.1	21.8		Salt Lake City, UT	76.9	27.2
Baltimore, MD	76.8	34.8		Burlington, VT	69.0	16.2
Boston, MA	73.7	29.9		Richmond, VA	78.1	38.7
Detroit, MI	74.4	26.9		Seattle, WA	64.9	38.3
Minneapolis, MN	72.3	12.4		Charleston, WV	74.9	36.6
Jackson, MS	82.3	47.9		Milwaukee, WI	68.7	20.6
St. Louis, MO	78.1	31.9		Cheyenne, WY	70.0	25.4

Table 17.2 Unit Area Thermal Resistance's (R-Values) of Typical Building Construction and Insulation Materials (ASHRAE, 1985, Handbook of Fundamentals. Atlanta, GA: American Society of Heating, Refrigerating, and Air-Conditioning Engineers)

Building Material	**R-value °F-ft^2·hr/BTU (W/m^{2}K)**
Plywood (Douglas Fir)	
0.25 inch thick	0.3 (0.05)
0.50 inch thick	0.6 (0.11)
Particleboard (Medium Density)	
0.25 inch thick	0.3 (0.05)
0.50 inch thick	0.5 (0.09)
Spruce-Pine-Fir – 1.75 inch thick	2.4 (0.42)
Cinder Block	
4 inch thick	1.1(0.19)
8 inch thick	1.7 (0.30)
12 inch thick	1.9 (0.33)
Asphalt shingles	0.4 (0.07)
Blanket and Batt Mineral Fiber Insulation	
4 inch thick	11.0 (1.94)
6.5 inch thick	19.0 (3.35)
7 inch thick	22.0 (3.87)
9 inch thick	30.0 (5.28)
12 inch thick	38.0 (6.69)
Expanded Polystyrene Insulation Board	
1 inch thick	3.4 (0.60)
2 inch thick	6.7 (1.18)
Convection Coefficients for Wall Surfaces[1]	
Winter time 15 mph (24 kph) wind	0.17 (0.029)
Summer time 7.5 mph (12 kph) wind	0.25 (0.044)
Still air (less than 100 ft/min or 0.5 m/s)	1.00 (0.176)

[1]Convection coefficients are the inverse of these R values; to convert (°F-ft^2-hr/BTU) to W/m^{2}K divide by 5.678

Table 17.3 Approximate Typical Infiltration Air Changes for Different Building Types

Building Type	Infiltration Air Changes[a] (volumes/hour)
Cinder Block	0.7–1.5
Aluminum Sided	1.0–2.0
Wood Frame	1.0–3.0

[a]Number of windows and doors greatly affect infiltration

Table 17.4 Allowable Concentrations of Carbon Dioxide at Standard Temperature and Pressure to the Nearest 25 ppm (ASHRAE, 1991, Handbook - Applications. Atlanta, GA: American Society of Heating, Refrigerating, and Air-Conditioning Engineers)

Toxicity Standard	Allowable Concentration (ppm)
Immediately dangerous to life and health[a]	50,000
Acceptable ceiling concentration, not to be exceeded during an 8 hour shift[b]	30,000
Time-weighted average, not to be exceeded in any 8 hour shift of a 40 hour week[b]	5,000
Atmospheric concentration	350

[a]National Institute for Occupational Safety and Health standard
[b]Occupational Safety and Health Administration standard

Fish Culture Inputs

The variables that may be changed under this option are defined as follows:

1. Number of Culture Tanks – The total number of culture tanks in the building. Entered as number of tanks.
2. Water Volume per Tank – The water volume per culture tank. Entered in liters.
3. Water Surface Area per Tank – The water surface area exposed to ambient air for each tank. Entered in square meters.
4. Total Feed Fed per Day – The total quantity of fish feed fed into all tanks in the building. Entered in kilograms per day.

5. Daily Water Exchange Rate – The amount of culture water exchanged per day as a percent of the total water volume. Entered as a percent.
6. Tank Water Dissolved CO_2 Concentration – The target steady state concentration of dissolved carbon dioxide in the culture tank water. Entered in milligrams per liter.
7. Tank Water Dissolved O_2 Concentration – The target steady state concentration of dissolved oxygen in the culture tank water. Entered in milligrams per liter.
8. Tank Water pH – The target steady state pH of the culture tank water. Entered as a dimensionless number.
9. Tank Water Temperature – The target steady state temperature of the culture tank water. Entered in degrees Fahrenheit.
10. Tank Water Alpha Correction Factor – Coefficient to correct mass transfer coefficients for the presence of dissolved solids in the culture water. Entered as a dimensionless number.
11. Tank Water Beta Correction Factor – Coefficient to correct dissolved gas solubility for the presence of salts, particulates and surface active substances in the culture water. Entered as a dimensionless number.
12. Makeup Water Dissolved O_2 Concentration – The concentration of dissolved oxygen in the makeup water. Entered in milligrams per liter.
13. Makeup Water pH – The pH of the makeup water. Entered as a dimensionless number.

Control Option Inputs

The variables that may be changed under this option are defined as follows:

1. Packed Tower Height – Height (or depth) of the packed media in the packed tower utilized for carbon dioxide stripping. Entered in meters.
2. Hydraulic Loading to a Packed Tower – The water loading to the packed tower expressed per unit packed tower cross sectional area. Entered in kilograms per square meter-second.
3. PT Volumetric Gas-Liquid Ratio – The ratio of the volume of air blown through a packed tower to the volume of water pumped through a packed tower. Entered as a dimensionless ratio.
4. Packing Specific Surface Area – The total surface area per unit volume of packing material (from manufacturer's

specifications). Entered in square meters per cubic meters. See Table 17.5 for sample data.

5. Packing Surface Tension – The critical surface tension for the packing material (from manufacturer's specifications). Entered in newtons per meter. See Table 17.5 for sample data.
6. Packing Nominal Diameter – The nominal diameter (or size) of the packing material (from manufacturer's specifications). Entered in meters. See Table 17.5 for sample data.
7. Packing Factor – The packing factor of the packing material (from manufacturer's specifications). Entered as per meter. See Table 17.5 for sample data.
8. Surface Aerator SAE – Standard aeration efficiency (SAE) of the surface aeration equipment. Entered in kilograms oxygen per kilowatt-hour. See Table 17.6 for sample data.
9. Diffused Aerator SAE – Standard aeration efficiency (SAE) of the diffused aeration equipment. Entered in kilograms oxygen per kilowatt-hour. See Table 17.6 for sample data.
10. Volume of Side-stream Aeration Tank – The volume of the side-stream basin used in the surface and diffused side-stream aeration options. Entered in liters.
11. Flow to the Side-stream Aeration Tank – The water flow to the side-stream basin used in the surface and diffused side-stream aeration options. Entered in liters per minute.
12. Diffuser Depth in the Culture Tank – The depth of the air diffusers under water in the culture tank. Entered in centimeters.
13. Diffuser Depth in the Side-stream Tank – The depth of the air diffusers under water in the side-stream basin. Entered in centimeters.
14. Airflow into the Diffusers – The flow rate of air through the diffusers in both the side-stream and in-tank diffused aeration options. Entered in liters per minute.
15. Normality of Sodium Hydroxide Solution – The normality of the sodium hydroxide solution used. Entered as normality (equivalents per liter).

Table 17.5 Packing Parameters for Common Packing Materials Used in Stripping Towers (Perry, R. H., & Green, D. W., 1992: Perry's Chemical Engineers' Handbook. New York: McGraw-Hill.)

Packing Type	Nominal Size (m)	Critical Surface Tension (N/m)	Specific Surface Area (m^2/m^3)	Packing Factor (m^{-1})
Intalox saddles, plastic	0.025	0.033	206	105
Intalox saddles, plastic	0.050	0.033	108	69
Intalox saddles, plastic	0.075	0.033	88	50
Intalox saddles, ceramic	0.025	0.061	255	320
Intalox saddles, ceramic	0.050	0.061	118	130
Intalox saddles, ceramic	0.075	0.061	92	70
Pall Rings, plastic	0.025	0.033	205	170
Pall Rings, plastic	0.050	0.033	100	82
Pall Rings, plastic	0.090	0.033	85	52
Pall Rings, metal	0.025	0.075	205	157
Pall Rings, metal	0.050	0.075	115	66
Pall Rings, metal	0.090	0.075	92	53
Rashig Rings, ceramic	0.025	0.061	190	510
Rashig Rings, ceramic	0.050	0.061	92	215
Rashig Rings, ceramic	0.075	0.061	62	120
Rashig Rings, metal	0.025	0.075	185	450
Rashig Rings, metal	0.050	0.075	95	187
Rashig Rings, metal	0.075	0.075	66	105

Table 17.6 Standard Aeration Efficiency (Sae) for Surface and Diffused Aeration Equipment (Lawson, T. B. 1995, Fundamentals of Aquacultural Engineering, NY: Chapman & Hall)

Aeration Device	SAE (kg O_2/kWh)
Surface Aerators	
Low speed	1.2–2.4
Low speed with draft tube	1.2–2.4
High speed	1.2–2.4
Vertical pump	0.7–1.8
Paddle wheel	1.1–3.0
Diffused Aerators	
Fine bubbles	1.2–2.0
Medium bubbles	1.0–1.6
Coarse bubbles	0.6–1.2

Economic Inputs

Variables that may be changed under this option with their definition follow:

1. Fuel Cost – The cost per gallon of liquefied petroleum gas. Entered in dollars per gallon of liquefied petroleum gas (LPG; a gallon of LPG is approximately 92,000 BTU's of heating energy; a Therm is 100,000 BTU and 1.055 kJ/BTU).
2. Electric Cost – The cost per kilowatt-hour. Entered in dollars per kilowatt-hour ($/kWh).
3. Packed Tower Capital Cost – The total cost to construct a packed tower system including costs for the tower, packing material, pumps, and blowers. Entered in dollars.
4. Sodium Hydroxide Capital Cost – The total cost to construct a sodium hydroxide system including costs for the holding vessels, dosing pumps, and safety equipment. Entered in dollars.
5. Water Exchange Capital Cost – The total cost to incorporate the water exchange option including costs for additional piping and pumps. Entered in dollars.
6. In-tank Surface Aeration Capital Cost – The total cost to incorporate the in-tank surface aeration option including costs for the surface aeration equipment. Entered in dollars.
7. Side-stream Surface Aeration Capital Cost – The total cost to incorporate the side-stream surface aeration option including costs for the side-stream tank, pumps, piping, and the surface aeration equipment. Entered in dollars.
8. In-tank Diffused Aeration Capital Cost – The total cost to incorporate the in-tank diffused aeration option including costs for the diffused aeration equipment and blowers/compressors. Entered in dollars.
9. Side-stream Diffused Aeration Capital Cost – The total cost to incorporate the side-stream diffused aeration option including costs for the side-stream tank, pumps, piping, diffused aeration equipment, and blowers/compressors. Entered in dollars.
10. Equipment Depreciation Period – The depreciation period for the capital equipment. Entered in years.
11. Heater Efficiency – The efficiency of the heating system used. Entered as a decimal.
12. Fan Efficiency – The efficiency of the fans used for building ventilation (from manufacturer's specifications). Entered in cubic feet of air per minute per watt.

13. Pump Efficiency – The efficiency of the pumps used for water pumping (from manufacturer's specifications). Entered in gallon of water per minute per kilowatt.
14. Blower Efficiency – The efficiency of the blowers used in the packed tower and diffused aeration options (from manufacturer's specifications). Entered in gallon of air per minute per kilowatt.
15. Sodium Hydroxide Solution Cost – The cost of the sodium hydroxide solution per liter. Entered in dollars per liter.

Outputs

After a program simulation run is complete, an output screen is displayed showing the results of the simulation using the current set of variable values. A sample output screen is shown below:

			Output Summary							
Option	**Start**	**Days**	**Ventilation Rates (lb/hr)**				**Cumulative Costs ($)**			
#	**Date**		**Temp**	**Humid**	**Infil**	**CO_2**	**Option**	**Heat**	**Vent**	**Total**
1	1/1	1	5467	643	7378	1977	16	4	0	21
2	1/1	1	2852	546	7378	0	29	8	0	38
3	1/1	1	2852	546	7378	0	4382	8	0	4390
4	1/1	1	2975	546	7378	1967	1	8	0	9
5	1/1	1	4271	546	7378	1977	9	6	0	15
6	1/1	1	1554	550	7378	1967	-8	10	0	2
7	1/1	1	3650	550	7378	1977	5	7	0	12

Option #1: Packed Tower
Option #2: NaOH Addition
Option #3: Water Exchange
Option #4: In-tank Surface Aeration
Option #5: Side-stream Surface Aeration
Option #6: In-tank Diffused Aeration
Option #7: Side-stream Diffused Aeration
Choose an Option (1–7) for details or 8 to return to the OPTIONS menu

The first part of the above screen displays an output summary table with the following columns:

- Option #
- Start Date
- Days

- Ventilation Rates (Temp, Humid, Infil, CO_2)
- Cumulative Costs
  (Option, Heat, Vent, Total)

Viewing Design Details

Following the summary table is a legend explaining each option number. The user may select an option to get a detailed output displaying the design specifics associated with each option. To select an option to get the detailed design output:

1. Enter the number of the option you want to view and press the **ENTER** key.
2. A window will pop up displaying the detailed output for the chosen option. Refer to the DETAILED OUTPUTS section for more information about each option's detailed output.
3. When you are finished viewing the detailed output press the **ENTER** key.

The screen will be refreshed showing the original output summary.

Once you are finished with viewing all detailed outputs and the output summary press the **8** number key and **ENTER** key. The program then returns to the options menu.

Summary Outputs

- Option #
  Carbon dioxide control option; defined by the following:
  - #1. Packed Tower
  - #2. Sodium Hydroxide Addition
  - #3. Water Exchange
  - #4. In-tank Surface Aeration
  - #5. Side-stream Surface Aeration
  - #6. In-tank Diffused Aeration
  - #7. Side-stream Diffused Aeration

- Start Date
  The date (month/day) the simulation was started.

- Days
  The number of days for which the simulation was run.

- Ventilation Rates

  Temp – The required building ventilation for temperature control on the last day the simulation was run in pounds of air per hour.

  Humid – The required building ventilation for moisture control on the last day the simulation was run in pounds of air per hour.

  Infil – The building ventilation from infiltration air changes on the last day the simulation was run in pounds of air per hour.

  CO_2 – The required building ventilation for carbon dioxide control on the last day the simulation was run in pounds of air per hour.

- Cumulative Costs

  Option – The cost of employing a carbon dioxide control option over the simulation period. Includes costs for capital, equipment depreciation, water pumping, air blowing, and motor operation.

  Heat – The cost of heating the building to the target ambient air temperature over the simulation period.

  Vent – The cost of ventilating the building according to the maximum required ventilation over the simulation period.

  Total – The total cost of any one carbon dioxide control option. The total includes the option cost, the heating cost, and the ventilating cost for the entire simulation period.

Detailed Outputs

The following is an example of a detailed output for a control option:

Control Option #1: Packed Tower

Height: 1.0 m	Area: 2.1 m^2
Flow: 2477 liter/min	GL Ratio: 5.0
Return Flow CO_2: 8.07 mg/L	Return Flow pH: 7.39
Tower Effluent CO_2: 8.01 mg/L	Tower Off-gas: 0.29%CO_2
Tank CO_2: 20.0 mg/L	Tank pH: 7.00
Tank Alkalinity: 100.9 mg/L	Flooding Ratio: 4.7%
Hackney&Colt K: 1.0/m	Onda Kla: 78.5/h

Press ENTER to return to the OUTPUT SUMMARY

This window identifies which option is being viewed and the various design parameters related to the control option. Definitions and units of the output parameters are presented for each option:

Control Option #1: Packed Tower

- Height – The height (or depth) of the packed media in the packed tower in meters.
- Area – The required cross sectional area of the packed tower in square meters.
- Flow – The required water flow through the packed tower in liters per minute.
- GL Ratio – Volumetric ratio of air blown through the tower to water pumped through the tower.
- Return Flow CO_2 – The concentration of dissolved carbon dioxide in the water returning to the culture tank in milligrams per liter (accounts for liquid phase carbonate chemistry equilibrium reactions).
- Return Flow pH – The pH of the water returning to the culture tank (accounts for liquid phase carbonate chemistry equilibrium reactions).
- Tower Effluent CO_2 – The concentration of dissolved carbon dioxide in the water leaving the packed tower in milligrams per liter.

- Tower Off-gas – The mole fraction of carbon dioxide in the gas exiting the packed tower as a percent.
- Tank CO_2 – The concentration of dissolved carbon dioxide in the culture tank water in milligrams per liter.
- Tank pH – The pH of the culture tank water.
- Tank Alkalinity – The alkalinity of the culture tank water in milligrams per liter calcium carbonate equivalents.
- Flooding Ratio – Parameter that expresses how close the packed tower is to a flooding condition. Based on a model put forth by Eckert. 100% is the flooding condition and a safe design number is 60% or less.
- Hackney & Colt K – Estimate of the mass transfer coefficient at 20 degrees Celsius for carbon dioxide based on the Hackney and Colt transfer model in per meter.
- Onda Kla – The overall mass transfer coefficient at 20 degrees Celsius for carbon dioxide based on the Onda equations in per hour.

Control Option #2: NaOH Addition

- NaOH Addition – The required dosage of sodium hydroxide that must be added to the system in liters per day.
- NaOH Normality – The normality of the sodium hydroxide solution used.
- Tank CO_2 – The concentration of dissolved carbon dioxide in the culture tank water in milligrams per liter.
- Tank pH – The pH of the culture tank water.
- Tank Alkalinity – The alkalinity of the culture tank water in milligrams per liter calcium carbonate equivalents.

Control Option #3: Water Exchange

- Makeup Flow – The required rate of water exchange in liters of makeup water per day.
- Makeup CO_2 – The concentration of dissolved carbon dioxide in the makeup water in milligrams per liter.
- Makeup Alkalinity – The alkalinity of the makeup water in milligrams per liter calcium carbonate equivalents.
- Makeup pH – The pH of the makeup water.
- Tank CO_2 – The concentration of dissolved carbon dioxide in the culture tank water in milligrams per liter.
- Tank pH – The pH of the culture tank water.

- Tank Alkalinity – The alkalinity of the culture tank water in milligrams per liter calcium carbonate equivalents.

Control Option #4: In-tank Surface Aeration

- Motor Power – The required motor power of the surface aeration equipment in kilowatts.
- Aerator SAE – The standard aeration efficiency of the surface aeration equipment in kilograms oxygen per kilowatt-hour.
- SOTR – The standard oxygen transfer rate of the surface aeration equipment in kilograms oxygen per hour.
- AOTR – The actual oxygen transfer rate of the surface aeration equipment under the given conditions in kilograms oxygen per hour.
- CO_2 Kla – The overall mass transfer coefficient at 20 degrees Celsius for carbon dioxide in per hour.
- O_2 Kla – The overall mass transfer coefficient at 20 degrees Celsius for oxygen in per hour.
- Tank CO_2 – The concentration of dissolved carbon dioxide in the culture tank water in milligrams per liter.
- Tank pH – The pH of the culture tank water.

Control Option #5: Side-stream Surface Aeration

- Motor Power – The required motor power of the surface aeration equipment in kilowatts.
- Aerator SAE – The standard aeration efficiency of the surface aeration equipment in kilograms oxygen per kilowatt-hour.
- SOTR – The standard oxygen transfer rate of the surface aeration equipment in kilograms oxygen per hour.
- AOTR – The actual oxygen transfer rate of the surface aeration equipment under the given conditions in kilograms oxygen per hour.
- CO_2 Kla – The overall mass transfer coefficient at 20 degrees Celsius for carbon dioxide in per hour.
- O_2 Kla – The overall mass transfer coefficient at 20 degrees Celsius for oxygen in per hour.
- SST Flow – The water flow through the side-stream tank in liters per minute.
- SST Volume – The volume of the side-stream tank in liters.
- Return Flow CO_2 – The concentration of dissolved carbon dioxide in the water returning to the culture tank in milligrams

per liter (accounts for liquid phase carbonate chemistry equilibrium reactions).

- SST CO_2 – The concentration of dissolved carbon dioxide in the side-stream tank water in milligrams per liter.
- Tank CO_2 – The concentration of dissolved carbon dioxide in the culture tank water in milligrams per liter.
- Tank pH – The pH of the culture tank water.
- Tank Alkalinity – The alkalinity of the culture tank water in milligrams per liter calcium carbonate equivalents.

Control Option #6: In-tank Diffused Aeration

- Motor Power – The required motor power of the diffused aeration equipment in kilowatts.
- Aerator SAE – The standard aeration efficiency of the diffused aeration equipment in kilograms oxygen per kilowatt-hour.
- SOTR – The standard oxygen transfer rate of the diffused aeration equipment in kilograms oxygen per hour
- AOTR – The actual oxygen transfer rate of the diffused aeration equipment under the given conditions in kilograms oxygen per hour
- Air Flow – The air flow through the diffused aeration equipment in liters per minute.
- Diffuser Depth – The depth of the diffusers under water in centimeters.
- CO_2 Kla – The overall mass transfer coefficient at 20 degrees Celsius for carbon dioxide in per hour.
- O_2 Kla – The overall mass transfer coefficient at 20 degrees Celsius for oxygen in per hour.
- Tank CO_2 – The concentration of dissolved carbon dioxide in the culture tank water in milligrams per liter.
- Tank pH – The pH of the culture tank water.

Control Option #7: Side-stream Diffused Aeration

- Motor Power – The required motor power of the diffused aeration equipment in kilowatts.
- Aerator SAE – The standard aeration efficiency of the diffused aeration equipment in kilograms oxygen per kilowatt-hour.
- SOTR – The standard oxygen transfer rate of the diffused aeration equipment in kilograms oxygen per hour.

- AOTR – The actual oxygen transfer rate of the diffused aeration equipment under the given conditions in kilograms oxygen per hour.
- Air Flow – The air flow through the diffused aeration equipment in liters per minute.
- Diffuser Depth – The depth of the diffusers under water in centimeters.
- CO_2 Kla – The overall mass transfer coefficient at 20 degrees Celsius for carbon dioxide in per hour.
- O_2 Kla – The overall mass transfer coefficient at 20 degrees Celsius for oxygen in per hour.
- SST Flow – The water flow through the side-stream tank in liters per minute.
- SST Volume – The volume of the side-stream tank in liters.
- Return Flow CO_2 – The concentration of dissolved carbon dioxide in the water returning to the culture tank in milligrams per liter (accounts for liquid phase carbonate chemistry equilibrium reactions).
- SST CO_2 – The concentration of dissolved carbon dioxide in the side-stream tank water in milligrams per liter.
- Tank CO_2 – The concentration of dissolved carbon dioxide in the culture tank water in milligrams per liter.
- Tank pH – The pH of the culture tank water.
- Tank Alkalinity – The alkalinity of the culture tank water in milligrams per liter calcium carbonate equivalents.

LOW HEAD OXYGENATOR: TIMMONS LHO MODEL[2, 3]

Low Head Oxygenators (LHO)[3] consists of a distribution plate positioned over multiple (5 to 10) rectangular chambers, Fig. 17.9. Water flows over the dam boards at the end of a raceway or is pumped upwards from an indoor fish tank, through the distribution plate, and then falls through the rectangular chambers. These chambers provide the gas-liquid interface needed for mixing and gas transfer. The streams of falling water impact a collection pool at the bottom of each chamber

[2] The CD Rom provides this program using both SI and IP units.

[3] The LHO program Input Screen Values assumes the PC computer is using American settings. European users need to go to their computer's "Control Panel" and choose "Regional Settings" and then "Numbers" and select the "American style", e.g, comma becomes a period. Chapter 8 discusses the mathematical basis for the LHO model in depth.

where the effluent water flows away from each chamber equally in parallel. All of the oxygen is introduced into the outer or first rectangular chamber, passes through the series of individual chambers, and finally is vented to the atmosphere at a much lower concentration. Each of the rectangular chambers is gas tight and the orifices between the chambers are properly sized and located to reduce back-mixing between chambers.

The operating performance of an LHO is primarily affected by the hydraulic head on the distribution plate (Y_1), the orifice hole size (Y_2), the depth of the receiving pool (Y_3), and the drop or fall distance from the orifice plate to the receiving pool of water (Y_4). Timmons et al. (2001) developed a mathematical model to predict the performance of LHO devices. The Timmons LHO model can be used to optimize the design of an LHO for better operating efficiency or of how best to operate an existing LHO unit to achieve the desired outputs. The model can predict the effects changes to oxygen and water flow rates, and then predict the exit conditions and performance parameters. By iterating the inputs, the user can eventually determine if a particular LHO unit can deliver the required oxygen transfer for a particular facility and the associated costs. The model allows for atmospheric air, oxygen, or a combination of the two to be used as the influent gas. Some facilities might not require pure oxygen, and can satisfactorily use a blower to deliver air to the LHO. In periods of high oxygen demand, oxygen can be metered into the LHO in appropriate quantities with a solenoid valve. The use of atmospheric air with a high G/L ratio will also allow certain applications of LHOs to be used for carbon dioxide removal. The way to minimize the use of costly pure oxygen, is to try to optimize the geometry of an LHO for the greatest capture. A designer of a LHO will also need to make a compromise between pumping costs and LHO efficiency. As the overall height of an LHO increases to accommodate a greater drop height and pool depth, so will the associated pumping cost due to the pump working against a higher pressure head. Sample input and output screens are reviewed below.

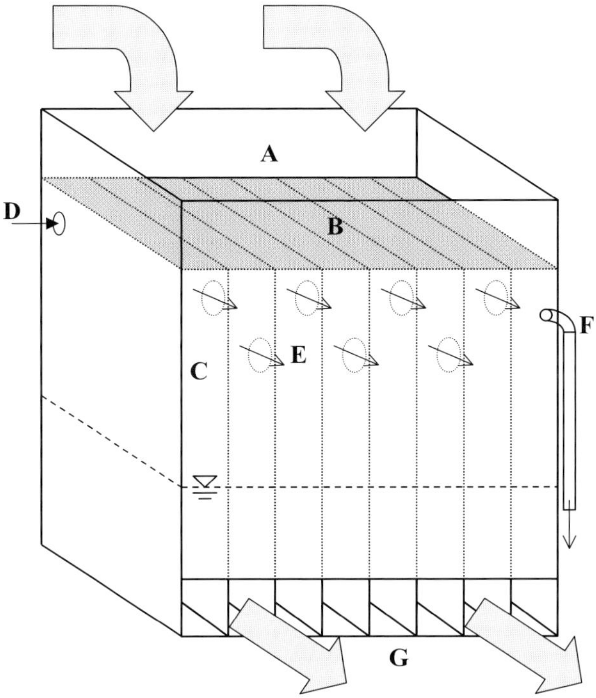

Figure 17.9 Typical LHO Configuration and Components Showing Water Flowing Into a Collection Trough or Plate (A), Through a Perforated Distribution Plate (B), and is Oxygenated in the Chambers (C), as Gas Flows From Inlet Gas Port (D), Through Holes Between Chamber to Chamber (E), to the off Gas Port (F), Where Excess Gas is Bubbled off Under Water. Water exits at the bottom of the unit (G).

The LHO model can also be used to predict CO_2 stripping for a single chamber without packing material. Remember that CO_2 stripping will require G/L ratios of 500 to 1,000%, which is G/L value of 5 to 10.

Model Results

The model neglects the effects of argon (approximately 7 mm Hg). The output screen provides total gas pressure (TGP) of the effluent water in mm Hg; this value does not include pressure due to Argon gas. The percent TGP in the effluent water is calculated as the sum of the excess gas tensions (oxygen, nitrogen and CO_2) plus the atmospheric pressure divided by the atmospheric pressure. Note that the total atmospheric pressure includes Argon. The change in argon gas pressure across the

LHO device will be approximately zero and thus does not contribute to any change in total gas pressure. The program output gives Final TGP for oxygen, nitrogen, and carbon dioxide and this value also includes water vapor pressure, e.g., at 20°C, water vapor pressure is 17.54 mm Hg, but, as mentioned, does not include Argon gas pressure.

Inputs

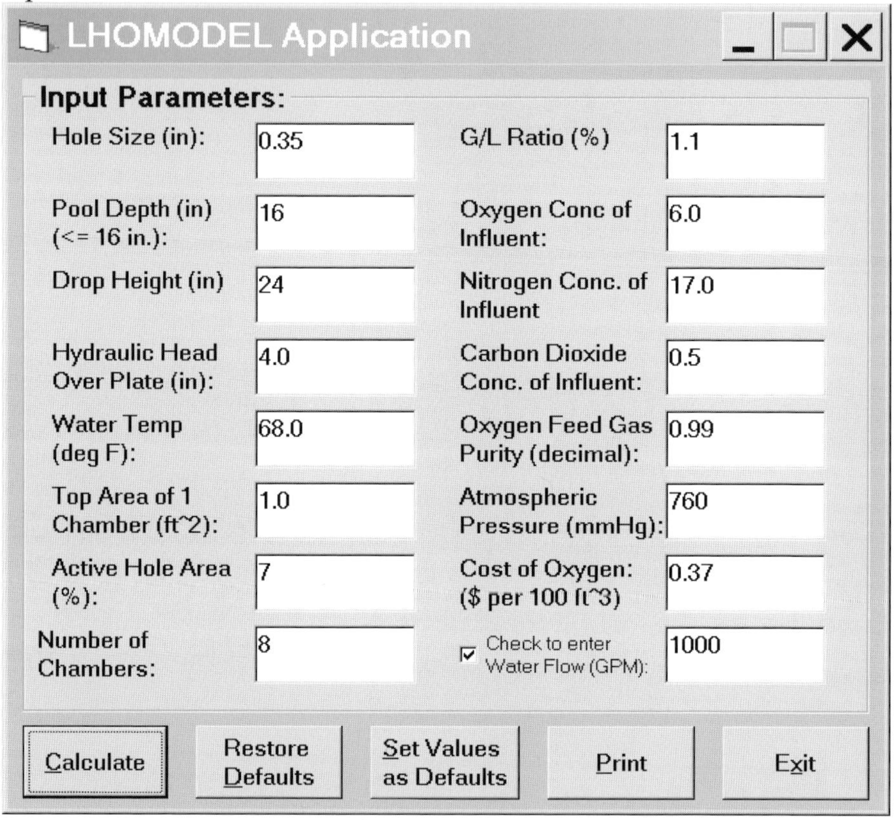

Figure 17.10 Sample Input Screen for the LHO Interactive Model.

Input Variables

Note: If you are using this program in Europe, your computer is configured to read "decimals or periods" as "," and periods "." as commas ",". Reconfigure your computer to read in American notation; else you will see a runtime error on the screen.

- Hole Size (in). The diameter of an individual orifice in inches (note this value can NOT be greater than 0.75 inches (19 mm) (screen may take the a larger value, but will make results INVALID).

- Pool depth (in) (<= 16 in). The depth of the receiving pool in inches; program assumes there is not effect from pool depth for depths greater than 16 inches.

- Drop Height (in). Distance from bottom of orifice plate to the surface of the receiving pool, (in inches).

- Hydraulic Head Over Plate (in). The amount of water head that is on the orifice plate in inches.

- Water Temp (F). Water temperature in degrees F.

- Top Area of 1 Chamber (ft^2). This is the area of flooded plate associate with a single chamber. The program calculates total area of flooded plate as product of the number of chambers and the area of one single chamber.

- Active Hole Area (%). This is a common way of defining hole area as a percentage opening of the entire flooded plate; typical values are 5 to 10%. Sheet material that is pre-punched is also specified based upon percent hole area.

- Number of Chambers. Usually LHO's will have from 5 to 8 chambers.

- G/L Ratio (%). This is the gas to liquid ratio on a volume to volume basis, which is a gaseous volume and a liquid volume. Percent G/L's are in the 0.5 to 2% range typically.

- Oxygen Conc of Influent. Concentration of oxygen of the incoming water in mg/L.

- Nitrogen Conc. of Influent. Concentration of nitrogen in the incoming water, mg/L.

- Carbon Dioxide Conc. of Influent. Concentration of carbon dioxide in the incoming water, mg/L.

- Oxygen Feed gas Purity (decimal). The purity of the feed gas in terms of oxygen, entered as a decimal equivalent.

- Atmospheric Pressure (mm Hg). The atmospheric pressure in mm of mercury (atmospheric pressure at sea level is 760 mm Hg).

- Cost of Oxygen ($ per 100 ft^3). Cost of delivered oxygen in terms of 100 cubic feet of gas (there are approximately 12 cubic feet of oxygen per lb at atmospheric pressure).

- Check to enter Water Flow (GPM). Sometimes the flow rate that will occur or will be forced to occur is known, e.g., a 1,000 gpm pump is used to supply water to the LHO flooded plate. Checking the "box" and entering the known flow rate will base oxygen transfer and cost data based upon the entered value, instead of calculating a flow from a specified hydraulic head on the flooded plate. You still need to enter the geometric data and hydraulic head since these variable affect the oxygen transfer rates.

Boxes at Bottom of Input Screen

- Calculate. Push this button once you have entered all the input data; will take you immediately to the output screen.

- Restore Defaults. The LHO program reads certain data from a file called "default.txt". IF you want to return to the last set of stored data, push this button.

- Set Values as Defaults. If you have changed the set of default conditions, often you may want to save this new data into the default.txt.

- Print. Prints the current screen information.

- Exit. Shuts down the program.

Outputs

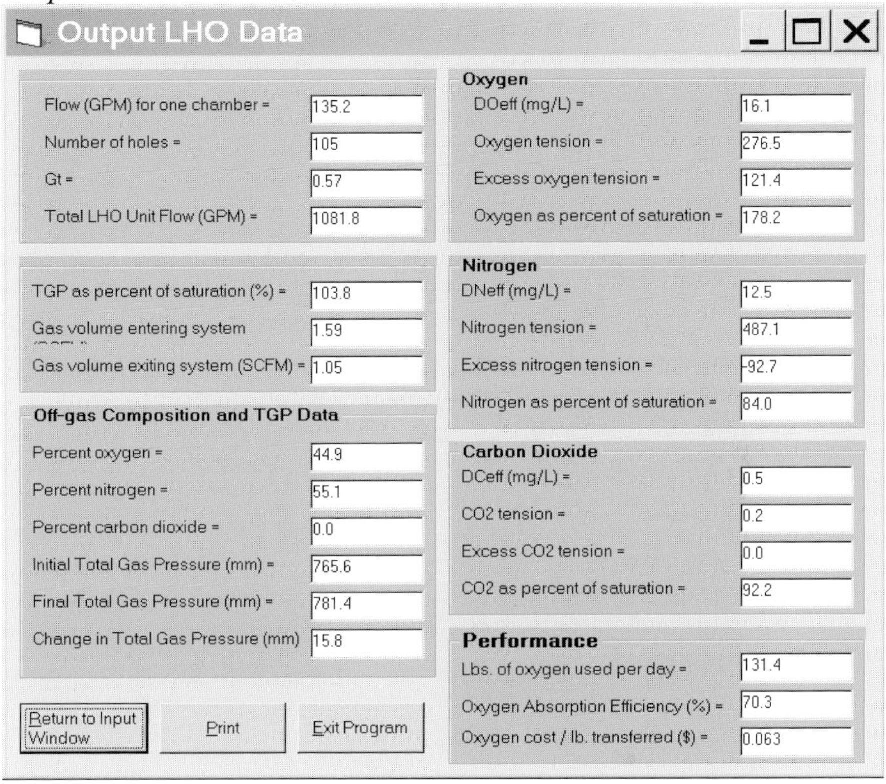

Figure 17.11 Sample Output Screen From the LHO Program.

Output Values

- Flow (GPM) for one chamber. This is the calculated flow based upon the input data; note that if flow rate is specified, it will be the total flow defined divided by the number of chambers.

- Number of holes. Calculates the number of holes per chamber (useful if you need to know how many to drill).

- Gt. This is the overall gas transfer coefficient in its temperature dependent form (see Eqs. 8.14 and 8.15).

- Total LHO Unit flow (GPM). Total flow for all the chambers passing through the LHO unit.

- TGP as percent of saturation (%). This is the total gas pressure of the departing effluent from the LHO as a percentage of atmospheric pressure.

- Gas Volume entering system (SCFM). The amount of gas entering the LHO expressed in standard cubic feet per minute (SCFM).

- Gas volume exiting system (SCFM). Amount of gas that exits the LHO expressed in standard cubic feet per minute (SCFM).

- Off-gas Composition and TGP Data

  Percent oxygen. The percent oxygen in the off gas,%.

  Percent nitrogen. The percent nitrogen in the off gas,%.

  Percent carbon dioxide. The percent carbon dioxide in the off gas, %.

  Initial Total Gas Pressure (mm). The TGP of the incoming water in mm Hg.

  Final Total Gas Pressure (mm). The TGP of the effluent in mm Hg.

  Change in Total Gas Pressure (mm). Change in TGP of the water in mm Hg.

- Oxygen

  DO eff (mg/L). Dissolved oxygen level in the effluent water.

  Oxygen tension = Pressure due to oxygen in the effluent, mm Hg.

  Excess oxygen tension. Pressure above saturation due to oxygen, mm Hg.

  Oxygen as percent of saturation. Percent representation of pressure due to oxygen relative to atmospheric saturation, %.

- Nitrogen

  DN eff (mg/L). Dissolved nitrogen level in the effluent water.

  Nitrogen tension. Pressure due to nitrogen in the effluent, mm Hg.

  Excess nitrogen tension. Pressure above saturation due to nitrogen, mm Hg.

  Nitrogen as percent of saturation. Pressure due to nitrogen relative to atmospheric saturation, %.

- Carbon dioxide

  DC eff (mg/L). Dissolved Carbon dioxide level in the effluent water.

  CO_2 tension. Pressure due to Carbon dioxide in the effluent, mm Hg.

  Excess CO_2 tension. Pressure above saturation due to CO_2, mm Hg.

  CO_2 as percent of saturation. Pressure due to CO_2 relative to atmospheric saturation, %.

- Performance

  Pounds of oxygen used per day. The quantity of liquid oxygen that is used by the LHO device per day, i.e., the amount of oxygen gas that a flow meter on the oxygen tank would record and you would pay the supplier for.

  Oxygen Absorption Efficiency (%). Transfer efficiency for oxygen into the water; ratio of the amount of oxygen that is actually transferred into or absorbed into the water and the quantity of oxygen that is used.

  Oxygen cost / lb. Transferred ($). Cost of oxygen transferred into the water, reflects the efficiency of the LHO device, $/lb.

Buttons at Bottom of Screen:

- Return to Input Window. Takes you back to the input screen.
- Print. Prints the output data page.
- Exit Program. Shuts down the current execution.

AIRPUMP

AIRPUMP is a DOS program that assists in the design of airlift pumping and aeration systems. AIRPUMP will help you choose the proper pipe dimensions and airflow to meet the desired levels for oxygen transfer and water flow rate for your particular application. A review of the theoretical aspects of airlift pumping and associated references is given in Chapter 3.4. AIRPUMP can be used on most PC type computers, but not Macintosh.

There are four variables that influence the volumetric water flow rate of an airlift pump (see Fig. 17.12):

- tube diameter
- tube length
- lift
- volumetric airflow rate

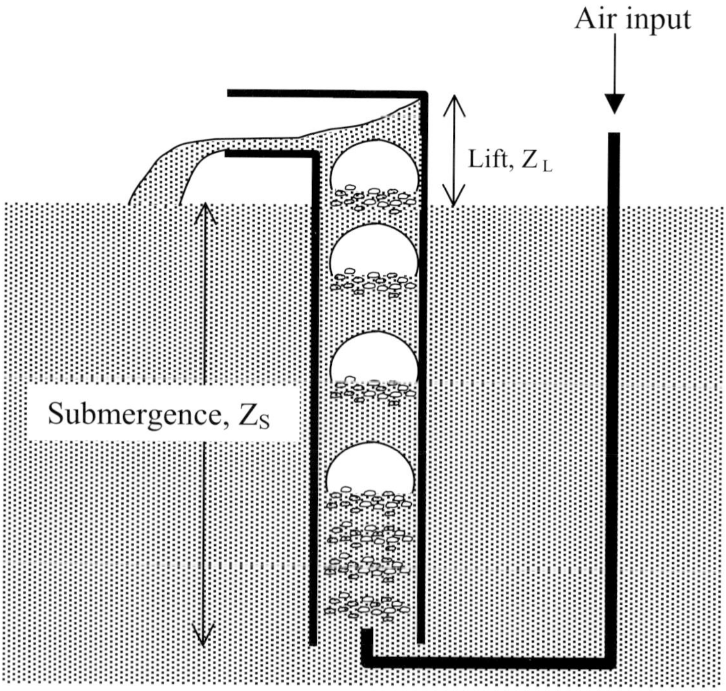

Figure 17.12. Typical Airlift Pump Application.

An additional variable that may influence the flow pattern is whether or not a diffuser is used in the introduction of air. Fine bubble diffusers or airstones generally produce bubbles from 1 to 3 mm in diameter. If such a diffuser is used, bubble flow may exist in tubes with diameters greater than about 10 mm and at low airflow rates and high submergence ratios. The transition from bubble to bubbly-slug flow is indicated in the AIRPUMP output information table (B = bubble flow, S = slug flow when pipe diameter D is <25 mm or S = bubbly-slug flow when D >25 mm).

Five more variables are required to determine the oxygen, nitrogen and total gas saturation:

- water temperature
- salinity of water
- initial oxygen concentration
- initial nitrogen concentration
- atmospheric pressure

Default values are assigned to all of these variables when the program is started. These values may be changed interactively during program operation.

For given pump geometry and water conditions, the volumetric water flow, oxygen transfer rate and output gas saturations are determined for a range of airflow rates. The gas transfer calculations are done using a plug flow model to take into account the changing pressure, gas saturation concentration, and gas mole fraction along the riser tube. This information is then presented in the AIRPUMP output table. A user choice is given as to whether English or metric units are used. The input information and output table can also be written to an output file for future use.

Instructions and Example

Begin by copying the AIRPUMP.exe file onto your desktop. Then double click on the program; user prompts are shown in BOLD.

After you are completely familiar with the definitions of the terms used in the interactive input screen, proceed by entering the appropriate values for the airlift pump of interest. See Chapter 3.4 for a more complete discussion on flow types and regimes.

Enter AIRPUMP <RETURN> to start the program.
The first screen you see will look like this:

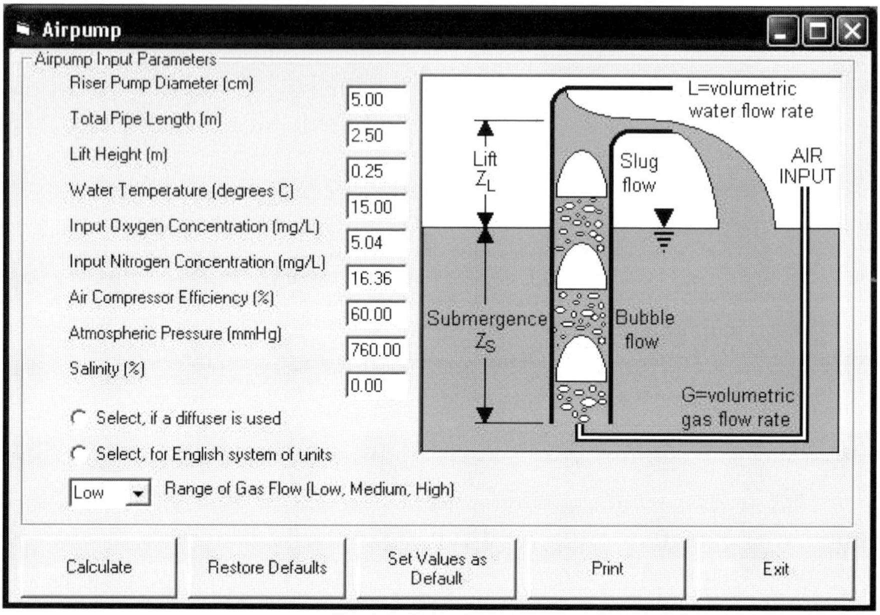

After entering your input data, place the cursor on the Calculate tab and Enter. The output screen will appear as shown below.

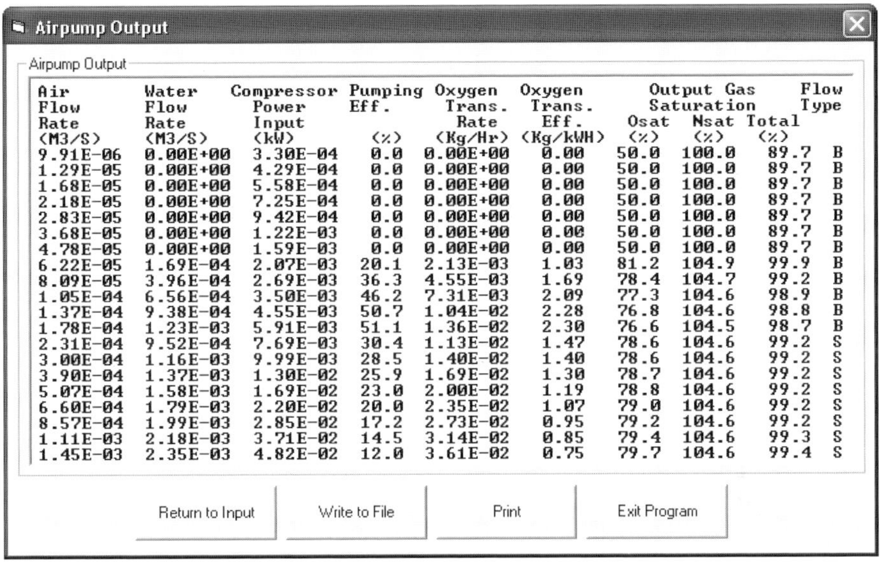

Air Flow Rate (M3/S)	Water Flow Rate (M3/S)	Compressor Power Input (kW)	Pumping Eff. (%)	Oxygen Trans. Rate (Kg/Hr)	Oxygen Trans. Eff. (Kg/kWH)	Output Gas Saturation Osat (%)	Nsat (%)	Total (%)	Flow Type
9.91E-06	0.00E+00	3.30E-04	0.0	0.00E+00	0.00	50.0	100.0	89.7	B
1.29E-05	0.00E+00	4.29E-04	0.0	0.00E+00	0.00	50.0	100.0	89.7	B
1.68E-05	0.00E+00	5.58E-04	0.0	0.00E+00	0.00	50.0	100.0	89.7	B
2.18E-05	0.00E+00	7.25E-04	0.0	0.00E+00	0.00	50.0	100.0	89.7	B
2.83E-05	0.00E+00	9.42E-04	0.0	0.00E+00	0.00	50.0	100.0	89.7	B
3.68E-05	0.00E+00	1.22E-03	0.0	0.00E+00	0.00	50.0	100.0	89.7	B
4.78E-05	0.00E+00	1.59E-03	0.0	0.00E+00	0.00	50.0	100.0	89.7	B
6.22E-05	1.69E-04	2.07E-03	20.1	2.13E-03	1.03	81.2	104.9	99.9	B
8.09E-05	3.96E-04	2.69E-03	36.3	4.55E-03	1.69	78.4	104.7	99.2	B
1.05E-04	6.56E-04	3.50E-03	46.2	7.31E-03	2.09	77.3	104.6	98.9	B
1.37E-04	9.38E-04	4.55E-03	50.7	1.04E-02	2.28	76.8	104.6	98.8	B
1.78E-04	1.23E-03	5.91E-03	51.1	1.36E-02	2.30	76.6	104.5	98.7	B
2.31E-04	9.52E-04	7.69E-03	30.4	1.13E-02	1.47	78.6	104.6	99.2	S
3.00E-04	1.16E-03	9.99E-03	28.5	1.40E-02	1.40	78.6	104.6	99.2	S
3.90E-04	1.37E-03	1.30E-02	25.9	1.69E-02	1.30	78.7	104.6	99.2	S
5.07E-04	1.58E-03	1.69E-02	23.0	2.00E-02	1.19	78.8	104.6	99.2	S
6.60E-04	1.79E-03	2.20E-02	20.0	2.35E-02	1.07	79.0	104.6	99.2	S
8.57E-04	1.99E-03	2.85E-02	17.2	2.73E-02	0.95	79.2	104.6	99.2	S
1.11E-03	2.18E-03	3.71E-02	14.5	3.14E-02	0.85	79.4	104.6	99.3	S
1.45E-03	2.35E-03	4.82E-02	12.0	3.61E-02	0.75	79.7	104.6	99.4	S

If you change the value of a variable, the variable list will be revised and presented again. When you have completed your revision of the variable list and started the program, you will be asked if you would like the information written to a file. If you would like a permanent record of input and output information from the run, simply list the name of an output file. Include which drive you would like the file to be created on, e.g., A: Filename.Dat. If you do not specify that the information be written to an output file, it will appear only on the screen. If you request that an output file be created, the output information will appear on the screen and be written to the named output file.

If you would like to do another run, place the cursor over the Return to Input tab, and you will be back at the input screen. Change what variables you wish, and the place cursor over the Calculate tab as before.

CHAPTER 18

AQUAPONICS: VEGETABLE HYDROPONICS IN RECIRCULATING SYSTEMS

James E. Rakocy[1]

18.0 INTRODUCTION

Aquaponic systems are recirculating aquaculture systems (RAS), which produce both fish and plants. The simultaneous production of fish and plants is possible because the system requirements for growing fish are very similar to those required for growing plants. RAS's are designed to raise large quantities of fish in relatively small volumes of water by treating the water to remove toxic waste products and then reusing the treated water. During the continual process of treatment and reuse, non-toxic nutrients and organic matter accumulate in the water. These metabolic by-products are potentially valuable and can be used to grow plants.

Plants grow rapidly in response to the high levels of dissolved nutrients that are either excreted directly by fish or generated from the microbial breakdown of fish wastes. In RAS's that have daily water exchanges of less than 5%, the accumulation of dissolved nutrients approaches the concentrations found in hydroponic nutrient solutions. Nitrogen, in particular, can occur at very high levels in recirculating systems. Fish excrete waste nitrogen through their gills in the form of ammonia. Bacteria convert ammonia to nitrite and then to nitrate. Ammonia and nitrite are toxic to fish, but nitrate is relatively harmless and is the preferred form of nitrogen by higher plants, such as fruiting vegetables. It is the symbiotic relationship between fish and plants that makes the consideration of an

[1] Dr. James E. Rakocy is a Research Professor of Aquaculture at the University of the Virgin Islands and the Director of the Agricultural Experiment Station, RR 2, Box 10,000, Kingshill, VI 00850, USA, jrakocy@uvi.edu.

aquaponic system a reasonable system design criteria.

Aquaponic systems offer several advantages. In RAS's, the disposal of accumulated waste is always a major concern. Recirculating systems are promoted as a means of reducing the volume of waste discharge to the environment. Certainly the volume is reduced but the pollution load (organic matter, dissolved nutrients) per unit of discharge is correspondingly higher. This more concentrated discharge may pose a threat to the environment in some situations, or an additional expense if the wastewater is discharged to a municipal sewer system for further treatment. Effluent is discharged from the system to eliminate organic sediment and prevent nutrient buildup. In aquaponic systems, the plants recover a substantial percentage of these nutrients, thereby reducing the need to discharge water to the environment and therefore extending water use, i.e., by removing dissolved nutrients through plant uptake, the water exchange rate can be reduced. Minimizing water exchange reduces operating costs of aquaponic systems in arid climates and heated greenhouses where water or heated water represents a significant expense.

Profitability is always a major concern when considering a recirculating system. Recirculating systems are expensive to construct and operate, and profitability often depends on serving niche markets for live fish such as tilapia, whole fresh fish on ice, or other high value products. A secondary plant crop, which receives most of its required nutrients at no additional cost, improves system profit potential. The daily feeding of the fish provides a steady supply of nutrients to plants, which reduces or eliminates the need to discharge and replace depleted nutrient solutions or adjust nutrient solutions as is required in hydroponics. The carbon dioxide vented from fish culture water can increase plant yields in enclosed environments. The plants purify the culture water and can, in a properly sized and designed facility, eliminate the need for separate and expensive biofilters. Biofiltration represents a major capital expense and a minor operational expense. In well-designed aquaponic systems employing sand, gravel, or raft hydroponics, the hydroponic component can provide sufficient biofiltration for the fish, and therefore the cost of purchasing and operating a separate biofilter is avoided. These costs are charged to the hydroponic subsystem, which, in the case of lettuce, generates approximately two thirds of the system's income. The profitability of recirculating systems can thus be improved substantially with aquaponics, if there is a good market for the vegetable crop.

The expense of water quality monitoring is reduced in aquaponic systems as waste nutrients are generated daily at uniform levels and there is generally excess wastewater treatment capacity. An aquaponics system also generates savings in several areas of construction and operation by sharing operational and infrastructural costs for pumps, blowers, reservoirs, heaters, and alarm systems. Initial capital investment is reduced in that an aquaponics system can be erected with a modest increase in acreage over that required for a hydroponic facility. Aquaponic systems do require high capital investment, moderate energy inputs, and skilled management. The premium prices available in niche markets may be required for an aquaponic facility to be profitable.

There are, of course, disadvantages to aquaponic systems. The most obvious of these is the large ratio of plant growing area in comparison to the fish rearing surface area. A large ratio of plant surface to fish surface is needed to achieve a balanced system where nutrient levels stay relatively constant. Ratios vary from 2:1 to 10:1 or greater depending on the degree of solids removal, with larger ratios needed as solids removal efficiency decreases. In essence, aquaponic systems emphasize plant culture, which is an advantage if viewed by a horticulturist. Most of the labor expended in the facility is devoted to seeding, transplanting, maintaining, harvesting, and packing plants. Additionally, a new set of skills is required for the plant component, so a commercial operation would do better with both an aquaculturist and horticulturist on staff. Another disadvantage is that the horticulturist must rely on biological control methods rather than pesticides to protect the plants from pests and diseases. However, this restriction can be viewed as an advantage in that the plant products can be niche marketed as “pesticide free”.

18.1 System Design

The design of aquaponic systems is generally based on the design of a recirculating system, with the addition of a hydroponic component and the possible elimination of a separate biofilter and devices for fine and dissolved solids removal such as foam fractionators. Suspended solids and dissolved organic matter generally do not reach levels that require foam fractionation in aquaponic systems with the correct design ratio of plant to fish area. The essential elements of an aquaponic system consist of a fish rearing tank, a suspended solids removal component, a biofilter, a hydroponic component and a sump (Rakocy and Hargreaves, 1993), Fig. 18.1. Effluent from the fish rearing tank is first treated to reduce suspended solids and BOD. Next, the effluent is treated to oxidize ammonia and nitrite by fixed-film nitrification. Then, the water flows through the vegetable production unit where plant uptake recovers dissolved nutrients. Finally, the water collects in a reservoir or sump, from which it is returned to the rearing tank. The system can be configured so that a portion of the flow is diverted to a particular treatment unit (Naegel, 1977; Wren, 1984; Quillere et al. 1993). For example, a small side-stream flow may go to a hydroponic component after solids removal while most of the water passes through a biofilter and returns directly to the rearing tank.

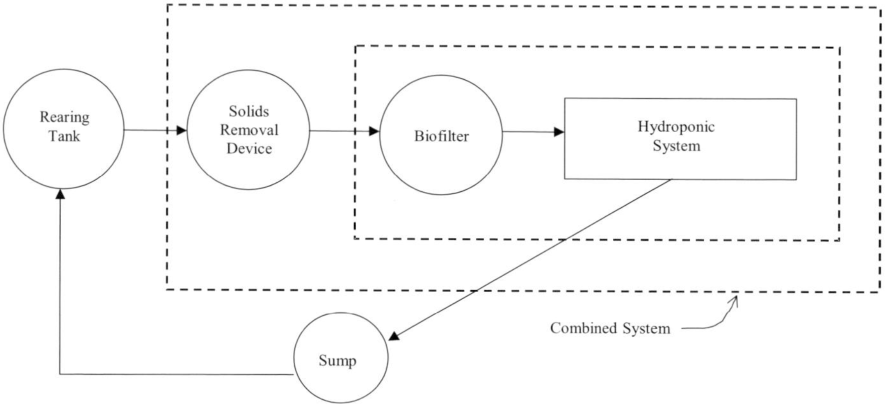

Figure 18.1. Optimum Arrangement of Aquaponic System Components. Components enclosed in rectangular boxes can be combined in a multifunctional unit.

There are several alternatives available to the system designer depending on the size of individual units and the project goal; whether it is for commercial production or a hobby. The biofiltration and hydroponic components can be combined by using a plant support media, such as gravel, (Lewis et al. 1978; Sutton and Lewis, 1982; Rakocy, 1984; Watten and Busch, 1984) or sand (McMurtry et al. 1990), which also functions as biofilter media. Raft hydroponics, which consists of floating sheets of polystyrene for plant support, can also provide sufficient biofiltration if the plant production area is sized properly (Rakocy, 1995). Combining biofiltration with hydroponics is a desirable goal as eliminating the expense of a separate biofilter is one of the main advantages of aquaponics. An alternative design combines solids removal, biofiltration, and hydroponics in one unit. The hydroponic support media (pea gravel or coarse sand) captures solids and provides surface area for fixed-film nitrification, although with this design it is important not to permit the unit to become overloaded with suspended solids. An overload of suspended solids is always a threat due to variations in fish feeding activities and efficiency of the solid removal component. For these reasons, gravel or sand beds should be avoided for large commercial-scale operations.

Aquaponics Research at the University of the Virgin Islands (UVI)

Aquaponics research at the University of the Virgin Islands (UVI) has focused on the culture of tilapia in outdoor tanks equipped with raft hydroponics. As the UVI system developed, there were many design evolutions. Most of the experimental work was conducted in six replicated systems that consisted of a rearing tank (12.8 m^3), clarifier (1.9 m^3), two hydroponic tanks (13.8 m^2), and a sump (1.4 m^3), Fig. 18.2. The hydroponic tanks (28 cm deep) were initially filled with gravel supported by wire mesh above a false bottom (7.6 cm). The gravel bed, which served as a biofilter, was alternately flooded with culture water and drained. Coarse gravel proved to be difficult to work with, so it was removed in favor of a raft system, consisting of floating sheets of polystyrene (2.44 m x 2.54 m x 3.8 cm). A rotating biological contactor (RBC) was then used for nitrification. Effluent from the clarifier was split into two flows, one going to the hydroponic tanks and the other to the RBC. These flows merged in the sump, from which the treated water was pumped back to the fish rearing tank.

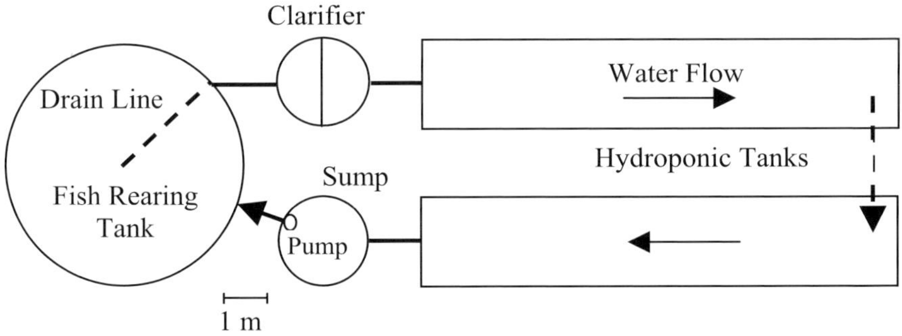

Figure 18.2. Design of UVI Experimental Aquaponic System.

The fish rearing tank was situated under an opaque canopy and the clarifier and sump were covered with plywood. Shading inhibits algae growth, lowers daytime water temperature, and creates more natural light conditions for the fish. The rearing tank in this particular design proved to be too large relative to the plant growing surface area of the hydroponic tanks, or, conversely, the hydroponic tanks were too small relative to the size of the rearing tank. When the rearing tank was stocked with tilapia at commercial densities (107 fish/m^3), the daily feed ration to the system was so high that nutrients rapidly accumulated to levels that exceeded the recommended upper limits for hydroponic nutrient solutions [2,000 mg/L as total dissolved solids (TDS)] (Rakocy et al. 1993). The optimum ratio between the fish feeding rate and plant growing area was determined, using Bibb lettuce as the baseline plant, to be 57 g of feed/day/m^2 of plant growing area, (Rakocy, 1989A). At this ratio, the nutrient accumulation rate decreased, and the hydroponic tanks were able to provide sufficient nitrification. This success enabled the RBCs to be removed when the fish stocking rates were reduced to levels that allowed feed to be administered near the optimum rate for good plant growth. The optimum ratio will vary depending on plant species and the production method, i.e., staggered vs. batch culture.

Batch Culture of Sweet Basil in UVI's Aquaponic System.

The experimental system was scaled up two times. In the first scale-up, the length of each hydroponic tank was increased from 6.1 m to 29.6 m. The optimum design ratio of 57g feed/day/m^2 of plant growing area allowed the rearing tank to be stocked with tilapia at commercial levels (for a diffused aeration system) without excessive nutrient accumulation. In the second scale-up, the number of hydroponic tanks (29.6 m in length) was increased to six and the number of fish rearing tanks was increased to four, Fig. 18.3. This production unit design represents a realistic commercial scale, although there are many possible size options and tank configurations. For example, the number of hydroponic tanks could be reduced from six to two, as in the experimental unit, by increasing the width or length as long as the total area remained the same. Likewise, the rearing tanks (7.8 m^3) could be increased in size and/or number, provided there is a corresponding increase in the surface area of the hydroponic tanks. There is more flexibility in sizing and configuring outdoor systems, which are restricted to tropical or subtropical climates for year-round production of fish, than is available for indoor systems. In temperate indoor systems, the components must be configured tightly to conserve space, and the unit cannot exceed the width of a greenhouse bay, which ranges from 6.7 to 9.45 m. UVI's commercial-scale unit (second scale-up) could be configured to occupy as little as 0.05 ha of land.

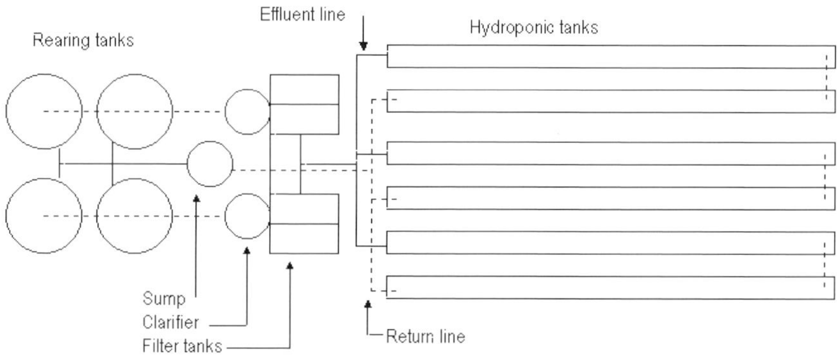

Figure 18.3. UVI Commercial-Scale Aquaponic System (not to scale). Between the two sets of filter tanks is the degassing tank. Water is pumped from the sump to the fish rearing tanks, from which effluent flows by gravity through the clarifiers, filter tanks, degassing tank and hydroponic tanks before retuning to the sump.

18.2 FISH PRODUCTION

An aquaponic system must be operated continuously at near maximum capacity for both the fish rearing component and the hydroponic plant component to be profitable. It is important to consider the different methods of fish management that could be used when selecting a design for the aquaponic system. The fish biomass must be maintained close to the system's carrying capacity to make most efficient use of the available space, maximize production, and arrive at a constant rate of feed application. Constant feed rates are necessary to accurately size the hydroponic component. The basic methods of fish management are sequential rearing, stock splitting and multiple rearing units. It is important to determine the best method for varying circumstances prior to designing a commercial system, as each method has different requirements. Changing rearing methodology in a production mode is costly and interruptive to steady fish production

Sequential rearing involves the culture of several age groups (multiple cohorts) of fish in the same rearing tank (Watten, 1992). When one age group reaches marketable size, it is selectively harvested; using nets and a grading system, and an equal number of fingerlings are immediately restocked in the same tank. There are three disadvantages with this system. The periodic harvests stress the remaining fish, increase mortality and may trigger disease outbreaks. Stunted fish avoid capture and accumulate in the system, wasting space and feed. Over time it is difficult to maintain accurate stock records, which leads to a high degree of management uncertainty and unpredictable harvests. Nevertheless, sequential rearing has been used successfully on a commercial scale with tilapia, a very hardy species, despite its drawbacks. Sequential rearing may be risky for species less tolerant to the repeated stress of partial harvest.

Stock splitting involves stocking very high densities of fingerlings, then periodically reducing the population as the carrying capacity of the rearing tank is reached (Van Gorder, 1991). A typical stock splitting system would divide the initial population three times so that the fish would go from one tank to two tanks and then from two tanks to four tanks and finally from four tanks to eight tanks, from which marketable fish are harvested. Alternatively, single cohorts can be moved to successively larger tanks, which reduces stress on the fish and generally takes less time than trying to physically divide a single cohort into two equal populations, based on weight or number, to be placed in tanks of the original size.

If the splitting cohort technique is chosen, a total of 15 rearing tanks are required to be able to harvest eight tanks at the end of each growth interval, a period of 5 or 6 weeks for tilapia. This method of fish management avoids the carryover problem of stunted fish and improves the accuracy of stock inventory assessment. However, the moves can be very stressful on the fish unless a swimway is installed, which connects all the rearing tanks to a narrow channel and the fish can be herded into it through a hatch in the wall of the rearing tanks and maneuvered into another rearing tank by movable screens. With swimways, the division of the populations in half involves some guesswork because the fish cannot be weighed or counted. An alternative method involves crowding the fish with screens and pumping them into a hauling tank or directly into another tank using a pescalator.

With the single cohort approach using multiple rearing units, the entire population is moved to larger rearing tanks when the carrying capacity of their current rearing tank is reached. The fish are either herded through a hatchway between adjoining tanks or into swimways connecting distant tanks. Multiple rearing units usually come in modules of two to four tanks and are connected to a common filtration system. After the largest tank is harvested, all of the remaining groups of fish are moved to the next largest tank, and the smallest tank is restocked with fingerlings.

A Portion of UVI's Aquaponic System Showing (counterclockwise from upper right): a Fish Rearing Tank, Clarifier, Two Filter Tanks, Degassing Tank with Internal Standpipe Well, and Return Sump.

A variation of this concept is the division of a long raceway into compartments with movable screens. As the fish grow, their compartment is increased in size and moved closer to one end of the raceway where they will eventually be harvested. However, raceways present problems with respect to removing solids and ensuring uniform water quality. Another variation is simply to use several of the largest size tanks. Each rearing tank contains a different age group of fish, but they are not moved during the production cycle. This system does not utilize space efficiently in the early stages of growth, but the fish are never disturbed and the labor involved in moving the fish is eliminated.

UVI's current commercial-scale, aquaponic system uses multiple rearing tanks to simplify stock management, as the fish are not moved during their 24-week growout cycle. The system consists of four fish rearing tanks (7.8 m^3 each, water volume), two cylindro-conical clarifiers (3.8 m^3 each) with a 45° slope, four filter tanks (0.7 m^3 each), one degassing tank (0.7 m^3), six hydroponic tanks (11.3 m^3 each, 29.6 m x 1.2 m x 0.4 m), one sump (0.6 m^3) and one base addition tank (0.2 m^3). The total hydroponic surface area is 214 m^2, and the total system water volume is 110 m^3. Tilapia production is staggered in the four rearing tanks so that one rearing tank is harvested every 6 weeks.

The system is used to culture Nile tilapia (*Oreochromis niloticus*) and red tilapia. Fry are sex-reversed with 17α-methyltestosterone according to the INAD (Investigations in New Animal Drugs) protocol to obtain a consistently high percentage (~99%) of male fingerlings. Nile tilapia fingerlings are stocked at a rate of 77 fish/m^3 to obtain a harvest size of 900–1,000 g. These large fish are processed for the fillet market. Red tilapia fingerlings are stocked at 154 fish/m^3 to achieve an average weight near 500 g for the whole fish West Indian market. Every six weeks approximately 600 kg of fish are harvested. Annual production of tilapia from the system is approximately 5 mt or 160 kg/m^3 of rearing space. Production results for year 2001 are shown in Table 18.1.

Tilapia grow well at high densities if good water quality is maintained. In the commercial-scale system, dissolved oxygen (DO) levels in the rearing tanks are maintained at 5–6 mg/L by high DO in the incoming water and by diffused aeration with air delivered through 22 air stones around the perimeter of the tank. A 1-hp (0.7 kW) blower provides air to the rearing and degassing tanks. Vigorous aeration vents carbon dioxide gas into the atmosphere and prevents its buildup. A high water exchange rate quickly removes suspended solids and toxic waste metabolites (ammonia and nitrite) from the rearing tank. A ½-hp in-line pump produces a flow of 380 Lpm and an average retention time of 1.4 hours/rearing tank. However, flows to the individual rearing tanks are adjusted so that the tank with the highest biomass receives the highest flow rate, which exceeds 130 Lpm for a retention time of less than 1 hour. The other rearing tanks receive proportionately lower flow rates relative to their biomass. Values of ammonia-nitrogen and nitrite-nitrogen in the rearing tanks are approximately 1–2 mg/L and <1 mg/L, respectively.

Table 18.1 Stocking Rate, Initial Size, Final Size, Growth Rate, Final Biomass, Feed Conversion Ratio, and Survival of Male Nile Tilapia (*Oreochromis niloticus*) and Red Tilapia in UVI's Commercial-Scale Aquaponic System in 2001

Variable	Nile Tilapia	Red Tilapia
Stocking Rate (number/m^3)	77	154
Initial Size (g)	89	29
Final Size (g)	989	515
Growth Rate (g/day)	5.3	2.9
Final Biomass (kg/m^3)	76	64.5
Feed Conversion Ratio	1.4	1.7
Survival (%)	99.9	81.5

Through careful attention to management of the water quality parameters of DO, ammonia-nitrogen and nitrite-nitrogen, it has been possible to grow tilapia at high densities. Other water quality variables of importance to the system are water temperature, pH, and alkalinity. Water temperature ranges from a low of 23.0°C in the winter to a high of 29.0°C in the summer. The average water temperature has been 27.0°C, which is lower than the optimum temperature (30°C) for tilapia and higher than the optimum temperature (20–22°C) for many vegetables. The system water temperature is lower than that in nearby ponds because none of the system's surface area is exposed to direct sunlight. The pH is generally maintained in the range of 7.0 to 7.5 by adding equal amounts calcium hydroxide and potassium hydroxide. Total alkalinity averages approximately 100 mg/L as calcium carbonate ($CaCO_3$).

Suspended Solids

Fish generate large amounts of fecal waste, most of which should be removed from the waste stream before it enters the hydroponic tanks. Other sources of particulate waste are uneaten feed and organisms, e.g., bacteria, fungi and algae that grow in the system. If this organic matter is allowed to accumulate in the system, it will depress DO levels as it decays and produce large amounts of carbon dioxide and ammonia. If deep deposits of settleable solids form on the tank floors, they will decompose anaerobically (without oxygen) and produce methane and hydrogen sulfide, which are very toxic at concentrations less than 1 mg/L.

Suspended solids have special significance in aquaponic systems. If a large amount of suspended solids enters the hydroponic component, it may accumulate on plant roots and produce a deleterious effect by creating anaerobic zones and blocking the flow of water and nutrients to the plant.

However, some accumulation of solids may be beneficial, as this stimulates mineralization, the process of releasing inorganic nutrients into the water as organic solids decompose. Mineralization increases the levels of several nutrients essential to plant growth. Without sufficient solids available for mineralization, more nutrient supplementation would be required, thereby increasing the operating expense and management complexity of the system. Another benefit of solids is brought about by the action of the microorganisms that decompose it. There is evidence that microbes associated with decomposing solids are antagonistic to plant root pathogens and help maintain healthy root growth. Therefore, it appears that a delicate balance must be reached between too much accumulation of suspended solids and too little.

Chapter 6 describes the common devices used for removing solids from recirculating systems. These include settling basins, tube (plate) separators, the combination particle trap and sludge separator, centrifugal separators, microscreen filters and bead filters. Sedimentation devices, e.g., settling basins and tube separators, primarily remove settleable solids (>100 microns) while filtration devices, e.g., microscreen filters and bead filters, remove settleable and non-settleable solids. Solids removal devices vary in regards to their efficiency, solids retention time, effluent characteristics (both solid waste and treated water) and water consumption rate. While many devices may be appropriate for aquaponic systems, there is no research on the relationship between techniques for solids removal and the performance of hydroponic vegetables. Given the importance of solids to vegetable production, this is an area that needs to be explored.

Sand and gravel hydroponic substrates are sometimes used to remove solid waste from the water flow stream. The solids remain in the system to provide nutrients to the vegetables through mineralization. As solids accumulate in the media, there is an increase in the cation exchange capacity (CEC), i.e., the ability of the media to adsorb and retain cations, positively charged nutrients, which are available for plant growth. Since cation concentrations are often high in aquaponic systems, CEC is generally not an important factor to plant growth. Sand and gravel media become easily clogged with solids, so bed tillage or periodic media replacement may be required. The use of sand is becoming less common, but one popular aquaponic system uses small beds (2.4 m by 1.2 m) containing pea gravel ranging from 3 to 6 mm in diameter. The hydroponic beds are flooded several times daily with aquaculture wastewater and then allowed to drain completely, the effluent being returned to the rearing tank. During draining phase, air is brought into the gravel. The high oxygen content of air speeds the decomposition of organic matter in the gravel. The beds are also inoculated with red worms that assimilate organic matter and improve bed aeration.

The organic waste produced in aquaponic systems does not break down completely. The fraction of organic waste that resists microbial decomposition is referred to as being refractory. Particulate refractory compounds will slowly accumulate in substrates such as pea gravel or on the bottom or raft system troughs. Dissolved refractory compounds give the culture water a brown or tea color, which contains tannic acid, humic acid and other humic substances. These compounds have mild antibiotic characteristics and are beneficial to the system's fish and plants. Humic compounds form metalo-organic complexes with Fe, Zn, and Mn and thereby increase the availability of these micronutrients to plants.

The most appropriate solids removal device for a particular system may depend primarily on the organic loading rate (daily feed input and feces production) and secondarily on the plant growing area. For example, if large amounts of fish (high organic loading) are raised relative to a small plant growing area, then a highly efficient solids removal device such as a microscreen drum filter is desirable. Microscreen drum filters capture fine organic particles, e.g., 60 micron and larger, which are retained by the screen for only a few minutes prior to backwashing and removal from the system. In this system, the dissolved nutrients, excreted directly by the fish or produced by mineralization of very fine particles and dissolved organic matter, are sufficient for the size of the plant growing area. At the other extreme, if small amounts of fish (low organic loading) are raised relative to a large plant growing area, then solids removal may be unnecessary, as more mineralization is needed to produce sufficient nutrients for the large plant growing area. However, the solids cannot be allowed to accumulate on the bottom of a tank and form anaerobic zones. A reciprocating pea gravel filter, subject to flood and drain cycles, may be the most appropriate device in this situation because solids are evenly distributed in the gravel and exposed to high oxygen levels (21% in air as compared to 0.0005–0.0007% in fish culture water) on the drain cycle, thereby enhancing microbial activity and increasing the mineralization rate.

UVI's experimental-scale aquaponic systems have relied on cylindro-conical clarifiers to remove settleable solids (Rakocy, 1984). The clarifiers have a volume of 1.9 m^3 and are constructed with fiberglass-reinforced rigid plastic sheeting (1 mm thick). The cylindrical portion of the clarifiers is situated above ground and has a central baffle that is perpendicular to the incoming water flow, Fig. 18.4. A second, smaller baffle is located in front of the effluent port to retain floating solids. The lower conical portion has a 60° slope and is buried under ground. The apex of the cone is caulked into a conical depression in a concrete pedestal in which a PVC drain line is embedded. The drain line rises vertically out of the ground to the midpoint of the cylinder and is capped with a ball valve. Rearing tank effluent enters the clarifier just below the water surface. The incoming water is deflected upward by a 45° pipe elbow to dissipate the current. As water flows under the baffle, turbulence diminishes and solids settle on the sides of the cone. The solids accumulate there and form a thick mat that eventually rises to the surface of the clarifier. To prevent this, approximately 30 male tilapia fingerlings are required to graze on the clarifier walls and consolidate solids at the base of the cone. The fingerlings also prevent solids buildup in the rearing tank drain lines by swimming through them and control the growth of filamentous bacteria, which constrict the drain lines and reduce flow. Solids are removed from the clarifier two to three times daily. Hydrostatic pressure forces the solids through the drain line when the ball valve is opened.

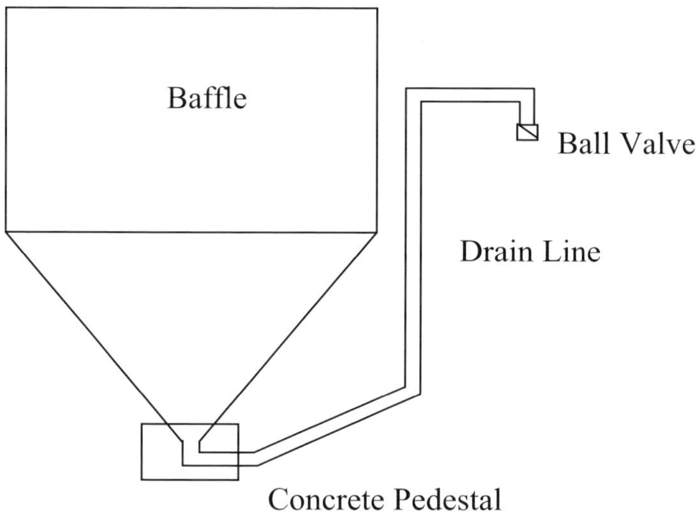

Figure 18.4. Cylindro-Conical Clarifier in UVI's Experimental Aquaponic System.

In UVI's current commercial-scale system, the clarifier volumes have been increased to 3.8 m^3 to achieve a 20-minute retention time, which resulted in much more effective removal of settleable solids. The clarifiers were constructed with standard fiberglass material. The slope of the conical portion was reduced to 45°. Although a 60° slope is more efficient for solids removal, installation of a heavy fiberglass tank with such a large cone is not feasible. Each clarifier is stocked with 40 male tilapia fingerlings to enhance clarifier performance, e.g., for grazing on surfaces and resuspending solids from upper reaches of the clarifier.

Although the cylindro-conical clarifier can remove 21% of the dry weight of feed added to the system during a production cycle (Rakocy et al. 1991), large quantities of solids are not removed. Twarowska et al. (1996) determined that 35.3% of feed input to a tilapia culture system can be captured as settleable and non-settleable solids (based on volatile solids analysis) by using a particle trap and particle separator (17.6% removal) in combination with a microscreen drum filter (17.7% removal). These results indicate that the clarifier removes approximately 59% of the total removable solid waste. Although tilapia fingerlings are needed for effective clarifier performance, their grazing and swimming activities are also counterproductive in that they resuspend many fine solids, which leave the clarifier in the waste stream. As the fingerlings become larger (>200 g), clarifier performance diminishes. Therefore, periodic replacement of clarifier fish with small fingerlings (50 g) is necessary.

With clarification as the sole method of solids removal, large quantities of solids were discharged into the hydroponic tanks where they settled out and formed sludge deposits more than 5-cm deep at the influent end. This was a very undesirable condition, adversely affecting the plant component, as the sludge would float to the surface, engulf the plant roots and either kill the plants or greatly reduce their growth. A series of experiments resulted in arriving at a design solution that incorporated another water treatment stage consisting of additional tanks filled with orchard netting for the removal of fine solids. Two rectangular filter tanks (0.7 m^3) were installed after each clarifier. Effluent from the clarifier flows through these tanks in series. The netting is washed once or twice a week with a high-pressure water sprayer, and all the water in the filter tanks is discharged.

Effluent from the rearing tanks is highly enriched with dissolved organic matter, which stimulates the growth of filamentous bacteria in the drain lines, clarifiers, and filter tanks. The bacteria appear as translucent, gelatinous, light-tan filaments. Tilapia consume the bacteria and control its growth in the drain lines and clarifiers, but bacteria accumulate in the filter tanks. Without the filter tanks, the bacteria would grow on the plant roots. It does not appear to be pathogenic, but it does interfere with the uptake of dissolved oxygen, water, and nutrients, thereby affecting the plant's growth. The feeding rate to the system and the flow rate from the rearing tank determine the extent to which bacteria manifests itself, but it can be contained by providing a sufficient area of orchard netting, either by adjusting filter tank size or using multiple filter tanks.

Organic matter accumulates on the orchard netting between cleanings, forming a thick sludge. Anaerobic conditions develop in the sludge, which leads to formation of reduced gasses such as hydrogen sulfide, methane and nitrogen. Therefore, a degassing tank is used to receive the effluent from the filter tanks. A number of air diffusers vent the gasses into the atmosphere before the culture water reaches the hydroponic plants. The degassing tank has an internal standpipe well which splits the water flow into three sets of hydroponic tanks.

Okra, which Is Not Generally Cultured Hydroponically,
Thrives in UVI's Aquaponic System.

In UVI's commercial-scale system, the combination clarifier and filter tanks have maintained very low levels of suspended solids. In one trial with leaf lettuce, suspended solids values averaged 4.2 mg/L in the filter tank effluent (Rakocy et al. 1997). The hydroponic tanks also contribute to suspended solids removal. There is no phytoplankton in the water, which is clear but darkly colored with humic substances.

Effluent Treatment and Disposal

Solids that are discharged from aquaponic systems must be disposed in an environmentally acceptable manner (see Chapter 10 Waste Management and Disposal). There are several methods for effluent treatment and disposal. Effluent can be stored in aerated ponds and applied as relatively dilute sludge to land after organic matter has stabilized. This method is advantageous in dry areas where sludge can be used to irrigate and fertilize field crops. The solid fraction of the sludge can be separated from the water and used with other waste products from the system (vegetable matter) to form compost. As an example, solids from the UVI commercial-scale aquaponic system are discharged through drain lines into two lined 16-m^3 ponds, which are continuously aerated with diffused air. As one pond is being filled over a 2 to 4-week period, water from the other pond is used to irrigate and fertilize field crops. Urban area facilities might have to discharge solid waste into sewer lines for disposal at the municipal sewage treatment plant.

BIOFILTRATION

A major concern in aquaponic systems is the removal of ammonia, a metabolic waste product that is excreted through the gills of fish. Ammonia will accumulate and reach toxic levels unless it is removed by the process of nitrification (referred to more generally as biofiltration), in which ammonia is oxidized first to nitrite, which is toxic, and then nitrite is oxidized to nitrate, which is relatively non-toxic. Aquaponic systems have used biofilters with sand, gravel, shell or various plastic media as substrate (Rakocy and Hargreaves, 1993). Biofilters perform optimally at a temperature range of 25–30°C, a pH range of 6.0–9.0, saturated DO, low BOD 's < 20 mg/L, and total alkalinity of 100 mg/L or greater. Nitrification is an acid-producing process that destroys alkalinity. Therefore, base must be added almost daily to maintain relatively stable pH values. Some means of removing dead biofilm from the biofilter media is necessary to prevent media clogging, short circuiting of water flow, decreasing DO values and an associated decline of biofilter performance. See Chapter 7 for a discussion of biofiltration principles and a description of various biofilter designs and operating procedures.

Ideally, aquaponic systems should be designed so that the hydroponic subsystem also serves as the biofilter, which eliminates the capital cost and operational expense of a separate biofilter. Granular hydroponic media such as gravel, sand, and perlite provide sufficient substrate for nitrifying bacteria and generally serve as the sole biofilter in some aquaponic systems, although they have a tendency to clog, as stated earlier. If serious clogging occurs due to organic matter overloading, gravel and sand filters can actually produce ammonia as organic matter decays, rather than remove it. If this occurs, the gravel or sand must be washed, and the system must be redesigned by installing a solids removal device prior to the granular biofilter, or the organic loading rate must be decreased by reducing fish stocking and feeding rates. McMurtry et al. (1990) relied on sand for hydroponic substrate and biofiltration while the early work at UVI (Rakocy, 1984) utilized gravel media for both functions. In some aquaponic systems, nitrification in the hydroponic component has supplemented that in the biofilter. Lewis et al. (1978) used pea gravel for hydroponic substrate in conjunction with an RBC for biofiltration. Watten and Busch (1984) used crushed-gravel hydroponic substrate and a trickling biofilter. When gravel was eliminated from UVI's experimental units, Fig. 18.2, an RBC was installed for biofiltration. When the experimental units were scaled up the first time, two RBCs were installed.

A series of three stock management experiments were conducted in the scaled-up unit, in which the surface area of each RBC was 93 m^2 and the total growing area for hydroponic lettuce was 71.4 m^2. In the first two experiments the feeding to the system ranged from 3 to 6 kg/day and good water quality was maintained. A feeding rate of 4 kg/day was nearly equivalent to the optimum design ratio (57 g of feed/day/m^2) for staggered lettuce production. In the third experiment the RBCs were eliminated to obtain a preliminary determination of the biofiltration capacity of the hydroponic component. The maximum feeding rate to the system was increased to 8 kg/day, and water quality continued to be good. Average levels of ammonia-nitrogen and nitrite-nitrogen in the rearing tank were 1.3 and 0.7 mg/L, respectively. This experiment showed that a combination of direct ammonia uptake by lettuce plants and nitrification on the hydroponic tank floor, walls, and the underside of the floating polystyrene sheets can provide sufficient nitrification to maintain good water quality at a feeding rate that is two times greater than the optimum design ratio. The total surface area in each hydroponic tank was 92.9 m^2 (not including the additional surface area provided by the plant roots), the same surface area as contained in an RBC.

An experiment was conducted in UVI's replicated systems, Fig. 18.2, to determine the waste treatment capacity of raft hydroponics (Gloger et al. 1995). Three identical systems were stocked with large *Oreohromis niloticus* (480 g/fish) so that feed could be incremented weekly until the waste treatment capacity was reached. The fish were initially underfed. Romaine lettuce (Parris Island) at a density of 29.6 plants/m^2 was produced continuously on a 4-week, staggered cycle. The results were based on the hydroponic growing area and expressed as g/m^2/day. The mean values and ranges were as follows: feeding rate, 159 (77 to 230); wet weight plant production, 338; dry weight plant production, 16.9; ammonia-nitrogen removal, 0.56 (-1.23 to 2.29); nitrite-nitrogen removal, 0.62 (-2.68 to 2.28); BOD removal, 4.78 (0.0 to 4.92); COD removal, 30.29 (-10.58 to 69.72); total nitrogen uptake by lettuce, 0.83; total phosphorus uptake by lettuce, 0.17. If the ammonia-nitrogen removal rate is divided by two to account for the surface area of the tank floor and the underside of the polystyrene sheet, the resultant value (0.28 g/m^2/day) falls within the range of removal rates reported for recirculating-system biofilters (Losordo, 1997). The maximum sustainable feeding rate was equivalent to 180 $g/m^2/day$, about three times greater than the optimum design ratio for the staggered production of leaf lettuce in aquaponic systems. Raft hydroponics not only provides adequate waste treatment for correctly-sized aquaponic systems and eliminates the need and expense of separate biofiltration units, but its excess treatment capacity ensures safe and stable water quality. After an initial acclimation period of about one month, experience with the UVI systems has shown that it is not necessary to monitor ammonia and nitrite values on a frequent basis, but weekly checks would be a prudent management practice.

Massive Root Development of Tomato Plants Supported by a Sheet of Polystyrene in UVI's Aquaponic System.

Aquaponic systems using nutrient film technique (NFT) as the hydroponic component may require a separate biofilter. NFT uses long narrow channels that are 15 to 30 cm wide. A thin layer (film) of culture water continually flows down the length of the channel, delivering water, nutrients, and oxygen to the plant roots. Compared to raft culture, the surface area is considerably smaller as there is no substantial wall area or raft underside surface area for colonization by nitrifying bacteria. The standard equations used to size biofilters do not apply to aquaponic systems because additional surface area is provided by plant roots and a considerable amount of ammonia removal is due to direct uptake by plants.

pH is an important factor for nitrification efficiency. The optimum pH range for nitrification is 6.0 to 9.0 although most studies cited by Wheaton et al. (1994) indicate that nitrification efficiency is greater at the higher end of the range. Most vegetables develop best at pH in the range of 6.0 to 7.5. pH affects nutrient solubility. Essential nutrients such as iron, manganese, copper, zinc, and boron are less available to plants at pH above 7.5 while the solubility of phosphorus, calcium, and magnesium sharply decreases at pH below 6.0. A compromise is reached by maintaining pH in the range of 7.0 to 7.5.

DO is another important factor for nitrification efficiency. Nitrification is most efficient when water is saturated with DO. The UVI commercial-scale aquaponic system maintains DO levels near 80% saturation (6 to 7 mg/L) by aerating the hydroponic tanks with numerous air stones distributed along the central axis of the tanks. Reciprocating (ebb and flow) gravel systems expose nitrifying bacteria to high atmospheric oxygen levels during the dewatering phase. The thin film of water that flows through NFT channels absorbs oxygen by diffusion, but dense plant roots and associated organic matter can block water flow and create anaerobic zones, which precludes the development of nitrifying bacteria and further requires the installation of a separate biofilter.

18.3 Hydroponic Subsystems

A wide range of hydroponic subsystems has been used in aquaponics (Rakocy and Hargreaves, 1993). Gravel hydroponic subsystems have been the most common on an experimental basis. To ensure adequate aeration of plant roots, gravel beds have been operated in a reciprocating (ebb and flow) mode, in which the beds are alternately flooded and drained (Lewis et al. 1978; Lewis et al. 1980; Sutton and Lewis, 1982; Wren, 1984; Rakocy, 1984), or in a dewatered state, in which culture water is applied continuously to the base of the individual plants through small-diameter plastic tubing (Watten and Busch, 1984). Depending on its composition, gravel can provide some nutrients for plant growth, e.g., calcium is slowly released as the gravel reacts with acid produced during nitrification (Rakocy and Nair, 1987).

Gravel beds have a number of negative aspects. The gravel is subject to clogging with suspended solids, microbial growth, and the plant roots that remain after harvest. The resulting reduction in water circulation together with decomposition of organic matter leads to the formation of anaerobic zones, which impairs or kills vegetable roots. The small plastic tubes used to irrigate gravel are also subject to clogging with biological growth. Moving and cleaning gravel substrate is difficult due to its weight. Planting in gravel is also difficult and plant stems can be damaged by abrasion in systems exposed to wind. Having high porosity, gravel retains very little water if drained. Disruption in flow will lead to the rapid onset of water stress (wilting). The sturdy infrastructure required to support the weight of the gravel and the potential of clogging impose a size limitation on gravel beds, which tend to be small.

One popular gravel-based aquaponic system uses pea gravel in 1.2 m x 2.4 m beds that are irrigated through a distribution system of PVC pipes over the gravel surface. Numerous small holes in the pipes distribute culture water on the flood cycles. The beds are allowed to drain completely between flood cycles. Solids are not removed from the culture water and organic matter does accumulate, but the beds are tilled between planting cycles, allowing some organic matter to be dislodged and discharged.

Sand has been used as hydroponic media in aquaponic systems (Ferguson, 1982; McMurtry et al. 1990). Sand is an excellent substrate for plant growth. Ferguson (1982) raised 28 types of vegetables in trays of sand that were stacked three tiers high in a commercial greenhouse. The greenhouse produced an average of 8,700 kg of vegetables during 10-week growth cycles in trays (0.6–0.9 m wide) with a total length of 1,155 m. The trays were made of parachute cloth so that water applied to the upper tray slowly percolated through it to irrigate the trays beneath it. McMurtry et al. (1990) constructed hydroponic sand beds (7.5 m x 1.5 m x 0.5 m) on sloped ground that was covered by polyethylene sheets. The beds were adjacent to in-ground rearing tanks with their floors sloping to one side. A pump in the deep end of the rearing tank was activated for 30 minutes five times daily to furrow irrigate the adjacent sand bed. The culture water percolated through the sand and returned to the rearing tank. The potential of sand substrates becoming clogged with solids can be reduced by regulating the solids loading rate. A coarse grade of sand is needed to reduce the potential for clogging over time and some solids should be removed prior to irrigation.

Perlite is another media that is starting to be used in aquaponic systems. Perlite is placed in shallow aluminum trays (8 cm deep) with a baked enamel finish. The trays vary from 20 cm to 1.2 m in width and can be fabricated to any length, but 6 m is the maximum recommended length. At intervals of 6 m, adjoining trays should be separated by 8 cm or more in elevation so that effluent drops to the lower tray and becomes reaerated. A slope 1 to 144 (0.7%) is needed for water flow. A small trickle of water enters at the top of the tray, flows trough the perlite, keeping it moist, and discharges into a trough at the lower end. Solids must be removed from the waste stream before it enters the perlite tray. Full solids loading would clog the perlite, form channels, create anaerobic zones, and lead to non-uniform plant growth. Shallow perlite trays provide minimal area for root grow and are better for smaller plants such as lettuce and herbs.

The nutrient film technique (NFT) has been successfully incorporated into aquaponic systems (Head, 1984; Burgoon and Baum, 1984). NFT consists of many narrow plastic troughs in which plant roots are exposed to a thin film of flowing nutrient solution (culture water). The troughs are lightweight, inexpensive, and versatile. Troughs can be mounted over rearing tanks to efficiently utilize vertical greenhouse space. High plant density can be maintained by adjusting the distance between troughs to provide optimum plant spacing during the growing cycle. Bioshelters, an aquaponic company in Amherst, Massachusetts, has used NFT for nearly 15 years to produce basil in conjunction with tilapia (Reid, personal communication). Inslee Fish Farm in Connerville, Oklahoma uses a modified NFT system to grow chives in an aquaponic system with tilapia (Inslee, personal communication). Aquaponic systems utilizing NFT require effective solids removal to prevent excess solids accumulation on roots, which can lead to root death and poor growth. With NFT, a disruption in the water flow can lead quickly to plant death. Small drip emitters and tubes are not recommended for water delivery to NFT troughs because they may clog. The use of a 1.3-cm PVC pipe with a plastic ball valve is recommended for water delivery because it reduces the risk of clogging and is easier and faster to clean if clogging occurs.

Raft Culture of Romaine Lettuce in an Aerated Hydroponic Tank in UVI's Aquaponic System.

A floating or raft hydroponic subsystem is ideal for the cultivation for leafy green and other types of vegetables (Zweig, 1986; Rakocy et al. 1989B). Long channels with closed-cell polystyrene sheets support vegetables at the water surface with roots suspended in the culture water (Jensen and Collins, 1985). The system provides maximum exposure of roots to the culture water and avoids clogging, although suspended solids captured by the roots can cause root death if concentrations are high (Zweig, 1986). The sheets shield the water from direct sunlight and maintain lower than ambient water temperatures. A disruption in pumping does not affect the plant's water supply as in gravel, sand, and NFT subsystems. The sheets are easily moved along the channel to a harvesting point where they can be lifted out of the water and placed on supports at an elevation that is comfortable for the workers.

A disadvantage of rafts in an aquaponic system is that roots are exposed to harmful organisms associated with aquaculture systems. For example, if tilapia fry gain access to the raft tanks, they consume plant roots and thereby severely stunt growth, although it is relatively easy to prevent the entry of tilapia by using a fine mesh screen. Similarly, blooms of zooplankton, especially ostracods, will consume root hairs and fine roots, retarding plant growth. Other pests are tadpoles and snails, which consume roots and nitrifying bacteria. These problems are surmounted by stocking some carnivorous fish that prey on pests in the hydroponic tanks. At UVI, snails are controlled with shellcracker sunfish (*Lepomis microlophus*), and zooplankton are controlled with black tetra (*Gymnocorymbus ternetzi*).

SUMP

Water flows by gravity from most hydroponic subsystems to a sump, located at the lowest point in the system. The sump contains a pump, which returns the culture water to the rearing tanks. If pea-gravel tanks, NFT troughs or perlite trays are located above the rearing tanks, the sump would be situated in front of them so that water could be pumped up to the troughs for gravity return to the rearing tanks. The sump is the only tank in the system where the water level decreases as a result of the system's overall water loss due to evaporation, transpiration, sludge removal, and splashing. An electrical or mechanic valve is used for the automatic addition of replacement water from a storage reservoir or well (municipal water should not be used unless it is dechlorinated). A water meter should be used to record additions. Unusually high readings indicate that there is a leak somewhere in the system.

The sump is the best location for the addition of base to the UVI system; UVI uses potassium hydroxide and calcium hydroxide. A well-aerated base addition tank (0.2 m^3) is located next to the sump. A small side-stream of water from the main water line continually enters the base addition tank and discharges into the sump. Base is added to this tank as needed to maintain a pH of 7.0–7.5 in the system. The base dissolves, gradually enters the sump and is pumped to the rearing tanks where it is quickly diluted in large volumes of turbulent water. Gradual addition of base avoids spikes in pH values, which are harmful to both fish and plants.

18.4 CONSTRUCTION MATERIALS

A wide range of materials has been used to construct aquaponic systems, and some have proven to be unsuitable. Budget limitations have played a major role in selecting inexpensive and often questionable materials for experimental systems. For example, vinyl-lined, steel-walled swimming pools, which were initially used at UVI, are not desirable fish rearing units. Plasticizers used in vinyl manufacture are toxic to fish. The liners must be aged in poor quality water for a few weeks before the fish can be added safely to a tank full of clean water. After a few growing periods, vinyl liners shrink upon drying, become brittle and crack, while the steel walls gradually rust. In the Virgin Islands, emerging nut sedge punctures holes in the vinyl, causing leaks. Nylon-reinforced, neoprene rubber liners proved to be equally troublesome. Holes developed at the folds as tilapia grazed on periphyton. When herbicides and soil sterilants were applied under these liners to kill nut sedge, they diffused into the culture water and killed the hydroponic vegetables.

Fiberglass is the best construction material for the rearing tanks, sump, and filter tanks. Fiberglass tanks are sturdy, durable, non-toxic, movable, and easy to plumb. An alternative to fiberglass is concrete, which is cheaper in many countries, although it lacks the flexibility of fiberglass construction. Commercially available NFT troughs, made from extruded polyethylene, are specifically designed to prevent puddling and water stagnation leading to root death and are preferable to makeshift structures (rain gutters, PVC pipes, etc.). Plastic troughs are commercially available for floating hydroponic subsystems, but they are expensive. A suitable alternative is to use of polyethylene liners and concrete-block or poured-concrete walls. Four types of liners have been tested in UVI's commercial-scale system. They are high-density polyethylene [1.5 mm (60 mil) and 0.5 mm (20 mil)], low-density polyethylene [0.5 mm (20 mil)], and a thick grade of nylon-reinforced vinyl with an under layer of high-density polyethylene. All of these liners are performing well after 5 to10 years, but it appears that 0.5-mm high-density polyethylene liners (HDPE) are best.

They are easy to install, relatively inexpensive and durable, having an expected service life of 12 to 15 years. Initially, HDPE liners were black, but recently UV-resistant white liners have been introduced. White liners are preferable in that they reflect light and do not become as hot as black liners. This is an important characteristic in the tropics and during summers in temperate climates where the goal is to avoid high water temperatures.

18.5 COMPONENT RATIOS

The design of an aquaponic system must consider two critical aspects of performance. Aquaponic systems are generally designed to meet the minimum size requirements for the suspended solids removal and biofiltration components (if a separate biofilter is used) relative to the amount of fish being raised. The other key design criterion is the ratio between the hydroponic and fish rearing components. If the ratio of fish to plants is too high, nutrients will accumulate rapidly and may reach phytotoxic levels. Higher water exchange rates will be required to prevent excessive nutrient buildup. If the ratio of fish to plants is too low, the plants will develop nutrient deficiencies and more nutrient supplementation will be required. Fortunately, hydroponic vegetables do grow well over a range of concentrations.

To date, there has been only limited research in the area of component ratios. McMurtry et al. (1993A; 1993B) studied four ratios (2.4, 3.0, 4.6 and 6.8 m^2/m^3) of hydroponic growing area to rearing tank volume, using sand hydroponics. All of the system water was essentially in the rearing tank except during irrigation periods. The sand beds (33 cm deep) were flooded and drained eight times daily during daylight hours with tilapia rearing tank water. The sand filtered out suspended solids and served as a biofilter. Tilapia stocking rate and nutrient input were the same for all ratios, but the number of plants increased in direct proportion to the component ratio in two experiments with tomatoes. As the component ratio increased, water quality improved and overall vegetable yield increased, but the yield per plant decreased, indicating declining nutritional status. The mean daily feed input per plant growing area ($g/m^2/day$) was 110, 76, 51 and 33 for the variety 'Laura' and 54, 36, 24 and 15 for the variety 'Kewalo' at ratios of 2.4, 3.0, 4.6 and 6.8 m^2/m^3, respectively.

Information on fish to plant ratios is available for gravel and raft hydroponics. A volume ratio of 1 m^3 of tilapia rearing tank to 2 m^3 of hydroponic media (3 to 6-cm pea gravel) is recommended for the reciprocating (flood and drain) aquaponic system developed by S & S Aqua Farm (P. Speraneo, personal communication). A 3.6-m^3 rearing tank with a final tilapia density of 60 kg/m^3 is supported by eight hydroponic tanks, each with a dimension of 2.4 m x 1.2 m x 30 cm. At the recommended ratio, no solids are removed from the system, and no nutrient supplementation is necessary. The hydroponic beds are stocked with red worms to help break down and assimilate the organic matter.

> **"RULE OF THUMB"**
> Pea Gravel Hydroponic Media
> 1 m^3 of fish tank volume to 2 m^3 of hydroponic media

Rakocy (1989A) evaluated six ratios between daily feed input and leaf lettuce growing area, using raft hydroponics. The variety 'Summer Bibb' was grown at a density of 23.9 plants/m^2 in 3-week, staggered production cycles so that 18 crops were harvested over a 21-week period. At ratios of 14, 29, 43, 57, 72 and 86 g/m^2/day, lettuce production was 2.3, 2.6, 2.7, 3.1, 2.8 and 2.4 kg/m^2/crop, respectively. Although the optimum ratio appeared to be 57 g/m^2/day, this result was not statistically validated because the treatments were not replicated. Recently, UVI research found with batch culture of basil that 57 g/day/m^2 was too low, and that ratios of >100 g/day/m^2 may be appropriate. Caution should be applied at this point until further research is conducted.

As a general guide for raft aquaponics, a ratio in the vicinity of 57 g of fish feed/day/m^2 of plant growing area should be used, although this ratio is subject to variation depending on many factors. This ratio is based on the staggered production of tilapia and leaf lettuce and requires that all solids are removed, with a residence time of <1 day for settleable solids (>100 microns), removed by a clarifier, and 4–7 days for non-settleable solids, removed by a filter. Potassium, calcium, and iron must be supplemented.

> **"RULE OF THUMB"**
> 57 g of fish feed per day per square meter of plant growing area for the staggered production of leaf lettuce.

The critical determinant in sizing the hydroponic component is the daily feed input to the system, which is directly proportional to the mass loading of dissolved nutrients. The total water volume is of lesser importance, but it does affect nutrient concentrations. In the UVI's commercial aquaponic system using raft hydroponics, approximately 60% of the total water volume is in the hydroponic component, whereas in aquaponic systems using gravel beds or NFT troughs, only a small portion (perhaps <10%) of the total water volume is in the hydroponic subsystem. Theoretically in systems producing the same quantity of fish and plants, a feeding rate of 57 $g/m^2/day$ would produce total nutrient concentrations more than two times higher in gravel and NFT systems, e.g., >2,000 mg/L, as compared to raft systems, e.g., 1,000 mg/L, but total nutrient mass would be equal among the systems. This example assumes that nutrient levels can be maintained at a steady state. In reality, nutrient levels fluctuate within a range of concentrations. However, if plant performance depends on the concentration of nutrients, then the optimum design ratio would vary with the type of hydroponic component. Moreover, an optimum design ratio for a particular plant in a raft system could lead to toxic concentrations in gravel or NFT systems. Other factors involved in accurately sizing the hydroponic growing area are the water exchange rate, the nutrient levels in the source water, the degree of suspended solids removal, and the type vegetable being raised. Lower rates of water exchange, higher source-water nutrient levels, incomplete solids removal, resulting in the release of more dissolved nutrients through mineralization, and slower growing plants would require larger hydroponic areas or lower feeding rates.

In properly designed aquaponic systems, the surface area of the hydroponic component is quite large compared to the surface area of the fish rearing tanks if it is stocked at commercial levels. The commercial-scale unit at UVI has a ratio of 7.3:1. The total plant growing area is 214 m^2. With diffused aeration, final tilapia densities have reached a mean of 76 kg/m^3. With pure oxygen, final densities can be increased to 120 kg/m^3 or greater. Therefore, smaller rearing tanks could be used to produce the same amount of fish and a ratio of 11.5:1 would be more appropriate.

"RULE OF THUMB"
11.5 to 1 ratio of plant beds to fish tank surface area in high density fish systems ($120\ kg/m^3$)

18.6 Plant Growth Requirements

Maximum plant growth in aquaponic systems requires 16 essential elements for proper nutrition. These nutrients are referred to below, in the order of their concentrations (ppm) in plant tissue with carbon and oxygen being the highest. The essential elements are arbitrarily divided into macronutrients, those required in relatively large quantities, and micronutrients, those required in considerably smaller amounts. Three of the macronutrients, carbon (C), oxygen (O) and hydrogen (H), are supplied by water (H_2O) and carbon dioxide gas (CO_2). All of the remaining nutrients are absorbed from the culture water. Other macronutrients include nitrogen (N), potassium (K), calcium (Ca), magnesium (Mg), phosphorus (P) and sulfur (S). The seven micronutrients include chlorine (Cl), iron (Fe), manganese (Mn), boron (B), zinc (Zn), copper (Cu), and molybdenum (Mo). All of these nutrients must be in proper balance for optimum plant growth. High levels of one nutrient can influence the bioavailability of others. For example, excessive amounts of potassium may interfere with uptake of magnesium or calcium while excessive amounts of either of the latter nutrients may interfere with the uptake of the other two nutrients (Gerber, 1985).

An elevated CO_2 level in the atmospheric environment of unventilated hydroponic structures has given dramatic increases in crop yield in northern latitudes (Jensen and Collins, 1985). Kimball (1982) summarized the data on CO_2 enrichment from more than 360 observations on 24 crops in a series of 50 reports. The study showed that doubling atmospheric CO_2 increased agricultural yields by an average of 30%. The high cost of energy to generate CO_2 has discouraged its use in conventional hydroponic systems. However, an enclosed aquaponic system is ideal for generating CO_2 due to the huge amounts that are constantly vented from the culture water.

Luther (1990, 1991A, 1991B, 1991C, 1993) cites a growing body of evidence that healthy plant development relies on a wide range of organic compounds in the root environment. These compounds, generated by complex biological processes involving microbial decomposition of organic matter, include vitamins, auxins, gibberellins, antibiotics, enzymes, coenzymes, amino acids, organic acids, hormones, and other metabolites. These compounds are directly absorbed and assimilated by the plants and stimulate growth, enhance yields, increase vitamin and mineral content, improve fruit flavor and hinder the development of pathogens. Various fractions of dissolved organic matter, e.g., humic acid, form organo-metalic complexes with Fe, Zn, and Mn, thereby increasing the availability of these micronutrients to plants (Chen and Solvitch, 1988). Luther states that although inorganic nutrients are necessary for plant survival, plants need organic metabolites from the environment to reach full hereditary potential.

Maintaining high DO levels in the culture water is extremely important for optimal plant growth, especially in aquaponic systems with their high organic loads. Hydroponic plants are subject to intense root respiration and they draw large amounts of oxygen from the surrounding water. If DO is deficient, root respiration decreases, resulting in reduced water absorption, decreased nutrient absorption, loss of cell tissue from roots and a reduction in plant growth (De Wit, 1978). Low DO levels correspond with high concentrations of carbon dioxide, conditions that promote the development of plant root pathogens. Chun and Takakura (1993) tested four nutrient-solution DO levels for hydroponic lettuce and found that root respiration, root growth, and transpiration were greatest at saturated DO levels.

Climatic factors also influence hydroponic vegetable production. Production is generally best in regions with maximum intensity and duration of light. Jensen and Collins (1985) reported that growth rates of lettuce plants in an Arizona greenhouse correlated positively with levels of available light up to the highest levels measured, although radiation levels in the Arizona desert are two to three times that of more temperate climates. When 30% shade cloth was used to cover lettuce plants in the UVI system, also in a region of intensive solar radiation, the plants elongated, the leaves twisted around the stem, and production declined. Growth slows substantially in temperate greenhouses during winter due to low solar radiation. Supplemental illumination can improve wintertime production, but it is not generally cost effective.

The optimum temperature range for most hydroponic vegetables is 20–22°C. Maintaining the optimum temperature range requires heating during the winter in temperate greenhouses and year-round cooling in tropical greenhouses. In addition to evaporative cooling of tropical greenhouses, chillers are often used to cool the nutrient solution. Summertime production declines in UVI's systems and seasonal adjustment to heat-tolerant varieties is necessary. Nevertheless, lettuce production is relatively good throughout the year because shading by the polystyrene sheets reduces water temperature, which has a greater effect on lettuce production than air temperature does.

Vegetables cultured in outdoor, aquaponic systems must be protected from strong winds, especially following transplanting when seedlings are most vulnerable to damage.

18.7 Nutrient Dynamics

The accumulation of excessive nutrients in an aquaponic system is a concern. High feeding rates, low water exchange and insufficient plant growing areas can lead to the rapid buildup of total dissolved solids to potentially phytotoxic levels (>2,000 mg/L as TDS or 3125 micromhos/cm). Since aquaponic systems have variable environmental conditions, daily feed input, solids retention, mineralization, water exchange, nutrient input from source water or supplementation, and nutrient uptake by different plant species, it is difficult to predict the exact level of total dissolved solids and how it is changing. Therefore, the culturist should purchase an inexpensive conductivity meter and periodically measure total dissolved solids using the range of 0–2000 ppm. If TDS is steadily increasing and approaching 2000 ppm, increasing the water exchange rate or lowering the fish-stocking rate and feed input will quickly and easily reduce nutrient accumulation. These quick and easy fixes are not good long-term solutions, though, because they reduce profits either by requiring more water or by decreasing fish production. Better but still costly long-term solutions involve increased solids removal, i.e., upgrade the solids removal component, or enlarged plant-growing areas.

Early work with UVI's experimental systems showed that conductivity measurements of TDS increased steadily as increasing quantities of feed were added to the system (Rakocy et al. 1993). Phytotoxic levels were reached after the addition of approximately 10 kg feed/m^3 of system volume. In an experiment to determine the optimum ratio of daily feed input to leaf lettuce growing area, conductivity as TDS increased by 147.5 g/kg of dry weight of feed at the optimum ratio of 57 g/day/m^2. However, during the first 8 months of operation of an early model of UVI's commercial-scale system, 26.9 kg of feed/m^3 of system volume was applied and the highest conductivity level was only 890 mg/L as TDS. The accumulation rate for conductivity as TDS was 26 g/kg of dry weight of feed, Table 18.2. Three factors contributed to the lower nutrient-accumulation rate. The actual mean ratio of daily feed input to plant growing area (49.5 g/day/m^2) turned out to be less than the optimum design ratio. Lettuce productivity was greater in the commercial-scale system. Romaine and leaf lettuce plants grew to a size of 250–650 g in 4 weeks (8.9–23.2 g/day) in the commercial-scale system compared to an average Bibb lettuce size of 131 g in three weeks (6.2 g/day) in the experimental system. (The average reduction of conductivity on passage through the hydroponic component during the first 8 months was 4.2 mg/L as TDS.) Substantial amounts of solids were removed by the filter tanks and consequently less mineralization may have occurred than in the experimental systems.

Table 18.2 Nutrient Accumulation (G/Kg Dry Weight Feed) of Conductivity and Major Cations and Anions from Two Experimental Aquaponic Systems and a Commercial-Scale Aquaponic System Using Raft Hydroponics for Lettuce Production

Nutrient	Exp. System 1[a]	Exp. System 2[b]	Commercial System[c]
Conductivity[d]	215.2	147.5	26.2
NO_3-N	35.6	14.9	3.7
PO_4-P	3.0	-	0.2
SO_4-S	1.9	1.8	0.6
K	66.0	-	4.2
Ca	-	7.3	2.3
Mg	1.2	1.8	0.4

[a] Supplementation with K but not Ca. Minor, one-time supplementation with P. From Rakocy et al. 1993
[b] Optimum feed to growing area ratio (57 g/day/m^2) from ratio study (Rakocy et al. 1993). K and Ca supplementation
[c] From the first 8 months of operation of an early model of the commercial-scale system. K and Ca supplementation
[d] As TDS

The major nutrients contributing to the increase in conductivity were NO_3, PO_4, SO_4, K, Ca, and Mg, Table 18.2. Among the cations, NO_3 accumulated at the highest rate. The accumulation rates of PO_4 and SO_4 were lower by an order of magnitude. Among the cations, K accumulated at the highest rate followed by Ca and then Mg. The accumulation of K and Ca was due mainly to supplementation because the levels of K and Ca generated from fish waste are insufficient for maximum plant growth (New Alchemy Institute, 1980; Head, 1984; McMurtry, 1989; Parker et al. 1990; McMurtry et al. 1993A; Rakocy et al. 1993). Potassium was added to the system in the form of potassium hydroxide (KOH) while Ca was added as calcium oxide (CaO) in experiments 1 and 2 and calcium hydroxide [$Ca(OH)_2$] in the commercial-scale system. In the commercial-scale system, KOH and $Ca(OH)_2$ were added in equal amounts. The bases were added alternately several times weekly to maintain pH at 7.0–7.5. Adding basic compounds of K and Ca serves the dual purpose of supplementing essential nutrients and neutralizing acid. All of the macronutrients with the exception of orthophosphate have a strong correlation with feed input (Rakocy et al. 1993).

The accumulation of nitrate ions is a concern with aquaponic systems. The discharge from one experimental system at UVI contained 180 mg/L as NO_3-N (Rakocy, 1994). The installation of the filter tanks in the UVI commercial-scale system provided a mechanism for controlling nitrate levels through the process of denitrification, the reduction of nitrate ions to nitrogen gas by anaerobic bacteria. Large quantities of organic matter accumulate on the orchard netting between cleanings. Denitrification occurs in anaerobic pockets that develop in the sludge. The entire water column moves through the accumulated sludge, which provides good contact between nitrate ions and denitrifying bacteria. The frequency of cleaning the netting regulates the degree of denitrification. When the netting is cleaned frequently, e.g., twice per week, sludge accumulation and denitrification are minimized, which leads to an increase in nitrate concentrations. When the netting is cleaned less frequently, e.g., once per week, sludge accumulation and denitrification are maximized, which leads to a decrease in nitrate levels. Nitrate-nitrogen levels can be regulated within a range of 1 to 100 mg/L or higher. High nitrate concentrations promote the growth of leafy green vegetables while low nitrate concentrations promote fruit development in vegetables such as tomatoes.

In a study using raft hydroponics for lettuce production, Seawright (1995) obtained similar results for macronutrient accumulation with two important exceptions: P accumulated in relation to feed input, but there was no significant relationship between feed input and N. Sodium bicarbonate ($NaHCO_3$) was added for pH control. The addition of Ca bases in the UVI system may have contributed to the precipitation of P from the culture water in the form of calcium phosphate [$Ca_3(PO_4)_2$]. Seawright removed solid waste from the system once per week and therefore denitrification may have been greater than in the UVI system where solid waste is removed at least twice per day from the clarifier and up to twice per week from the filter tanks. Although Seawright did not supplement with K, it accumulated with respect to feed input. However, plants require high levels of K and supplementation is needed in aquaponic systems. Seawright's finding that Ca is negatively correlated with feed input agrees with early work at UVI (Rakocy and Nair, 1987). Some investigators have found that Mg is limiting (Pierce, 1980; Head, 1984; Zweig, 1986). Magnesium can be supplemented by using dolomite [$CaMg(CO_3)_2$] as the base to adjust pH.

Sodium bicarbonate ($NaHCO_3$) should never be added to an aquaponic system for pH control. The accumulation of Na is a concern in aquaponic systems because high Na levels in the presence of chloride are toxic to plants (Resh, 1995). The maximum Na concentration in hydroponic nutrient solutions should not exceed 50 mg/L (Verwer and Wellman, 1980). Higher Na levels will interfere with the uptake of K and Ca (Douglas, 1985). In lettuce, reduced Ca uptake leads to tip burn, resulting in an unmarketable plant (Collier and Tibbitts, 1982). In UVI's systems, tip burn has occurred during the warmer months. Soluble salt (NaCl) levels in fish feed are relatively high. In the initial commercial-scale system, Na reached

51.0 mg/L in the 6th month and then declined to 37.8 mg/L in the 8th month, possibly due to rainfall dilution. The Na accumulation rate through the 6th month was 2.56 g/kg of dry weight feed. If Na exceeds 50 mg/L and the plants appear to be affected, a partial water exchange (dilution) may be necessary. Rainwater is used in UVI's systems because the groundwater of semiarid islands generally contains too much salt for aquaponics.

Table 18.3 Mean Concentration (Mg/L) of Micronutrients from Two Experimental Aquaponic Systems and a Commercial-Scale Aquaponic System Using Raft Hydroponics for Lettuce Production

	Micronutrient					
System	Fe	Mn	Cu	Zn	B	Mo
Exp. 1[a]	1.79	0.04	0.12	0.78	0.16	-
Exp. 2[b]	0.48	0.13	0.07	0.68	-	-
Commercial[c]	0.57	0.05	0.05	0.44	0.06	0.006
HNF[d]	5.0	0.5	0.03	0.05	0.5	0.02
HNF[e]	5.0	0.5	0.1	0.1	0.5	0.05

[a] From Rakocy et al. 1993.
[b] Optimum feed to growing area ratio (57 g/day/m^2) from ratio study (Rakocy et al. 1993).
[c] From the first 8 months of operation of a commercial-scale system.
[d] Hydroponic nutrient formulation for lettuce grown in the tropics (Resh, 1995).
[e] Hydroponic nutrient formulation for lettuce grown in Florida and California (Resh, 1995).

The micronutrients Fe, Mn, Cu, and B do not accumulate with respect to cumulative feed, although Zn does, Table 18.3, (Rakocy et al. 1993). Molybdenum does not accumulate either, based on data from the commercial-scale system, Table 18.3. The Fe derived from fish feed is insufficient for hydroponic vegetable production and must be supplemented (Lewis et al. 1980; MacKay and Van Toever, 1981; Zweig, 1986). Chelated Fe is applied to UVI's systems every 3 weeks at a rate calculated to give a concentration of 2.0 mg/L. A comparison of Mn, B, and Mo levels with standard nutrient formulations for lettuce shows that their concentrations in aquaponic systems are three to ten times lower than their initial levels in hydroponic formulations. Deficiency symptoms for Mn, B and Mo were not detected in UVI's aquaponic systems and so their concentrations appear to be adequate for normal plant growth.

Concentrations of Cu are similar in aquaponic systems and hydroponic formulations while Zn accumulates in aquaponic systems to levels that are four to sixteen times higher than initial levels in hydroponic formulations.

Nevertheless, Zn levels remained within the upper limit for safety, which is 1 mg/L in hydroponic solutions (Douglas, 1985).

Seawright (1995) worked on the development of a "designer diet" for aquaponic systems that would generate nutrients in proportion to their requirements for hydroponic plant nutrition, thereby creating stable and balanced nutrient concentrations over prolonged periods. Data on the change in nutrient concentrations in relation to dietary nutrient input was collected for the co-culture of *O. niloticus* and romaine lettuce (Jericho). This data was used to develop a mass balance model theoretically capable of predicting the nutrient inclusion rates required in fish diets to maintain stable dissolved nutrient concentrations in aquaponic systems. The model was validated by applying a specially-formulated "designer diet" to an aquaponic system and maintaining near-equilibrium concentrations of Ca, K, Mg, N and P, suitable concentrations of Mn and Cu and acceptable accumulation rates of Na and Zn. The results showed that Cu, Fe, and Mn are not good candidates for dietary manipulation because of low bioavailability. Phosphorus is a good candidate at sub-neutral pH, but it precipitates from solution at basic pH. Sulfur, B and Mo were not tested. The fish grew well on the diet, but the development of bacterial diseases indicated that elevated nutrient levels may have lowered their disease resistance.

18.8 Suitable Vegetables

Many types of vegetables have been grown in aquaponic systems. High value cash crops such as tomatoes, lettuce, and cucumbers have been cultured most frequently on an experimental scale, Tables 18.4–18.6. All of these crops proved to be suitable for aquaponic systems. Lettuce is one of the best crops for aquaponic systems because it can be produced in a short period (3 to 4 weeks in the system) and, as a consequence, pest pressure is relatively low. Unlike tomato and cucumber, a high proportion of the harvested biomass is edible. With lettuce, income per unit area per unit time is very high. Other fast-growing and high-income-generating crops are herbs such as basil and chive, which are being grown commercially in aquaponic systems (Rakocy and Hargreaves, 1993). Mint is another herb that grows luxuriantly in UVI's aquaponic system. Additional crops with potential include pak choi and Chinese cabbage, Table 18.5, bush beans (McMurtry et al. 1990) and celery (New Alchemy Institute, 1980). Recent trials at UVI have shown that okra, cantaloupe, watermelon, collard and pole bean have excellent potential. Inconclusive results have been obtained in experiments with pea (Van Toever and MacKay, 1980), green peppers

(Pierce, 1980) and squash (Lewis et al. 1980), although green peppers, hot peppers, summer squash and zucchini do moderately well in UVI's aquaponic system. Sweet potatoes have produced luxuriant growth but no tubers in aquaponic systems (Lewis et al. 1980). The cultivation of flowers has potential in aquaponic systems. Good results were obtained with marigold, zinnia and nasturtium in UVI's aquaponic system. Traditional medicinal plants and plants used for the extraction of modern pharmaceuticals have not been cultivated in aquaponic systems, but there may be potential for growing some of these plants and research is needed.

Staggered Production of Leaf Lettuce in an Early Model of UVI's Commercial-Scale Aquaponic System.

18.9 Crop Production Systems

There are three strategies for producing vegetable crops in the hydroponic component. These are staggered cropping, batch cropping, and intercropping. A staggered crop production system is one in which groups of plants in different stages of growth are cultivated simultaneously in the hydroponic subsystem. This production system allows regular harvest of produce and relatively constant uptake of nutrients from the culture water. This system is most effectively implemented where crops can be grown continuously, as in the tropics, subtropics, or temperate greenhouses with environmental control. Zweig (1986) developed a small floating system in a greenhouse in which six groups of leaf lettuce at different stages of growth were cultivated. At UVI, the production of leaf lettuce is staggered so that a crop can be harvested weekly on the same day, which facilitates marketing arrangements. Bibb lettuce reaches market size in 3 weeks from transplanting. Therefore, three growth stages of Bibb lettuce are cultivated simultaneously and one third of the crop is harvested weekly. Romaine, red leaf lettuce and green leaf lettuce require 4 weeks to reach marketable size. The cultivation of four growth stages of these lettuce varieties allows one fourth of the crop to be harvested weekly. In three years of continuous operation of UVI's commercial-scale system, 148 crops of lettuce were harvested, which demonstrated the system's sustainability. Leafy green vegetables, herbs and other crops with short production periods are well suited for continuous, staggered production systems.

Table 18.4 Varieties and Yields of Tomatoes Evaluated in Aquaponic Systems[a]. Environmental Codes: TEMP=Temperate Zone, TROP=Tropical Zone, O=Outside, GH=Greenhouse

Variety	Yield (kg/plant)	Environment	Reference
Tropic	0	TEMP/GH	Landesman, 1977
Floradel	5.4	TEMP/O	Lewis et al. 1978
Campbells 1327	4.6	TEMP/O	Lewis et al. 1978
San Marzano	4.6	TEMP/O	Lewis et al. 1978
Sweet 100	6.2	TEMP/O	Lewis et al. 1980
Better Boy	4.5	TEMP/O	Lewis et al. 1980
Rampo	3.9	TEMP/O	Lewis et al. 1980
Campbells 1327	2.3	TEMP/O	Lewis et al. 1980
Sweet 100	1.9	TEMP/GH	Baum, 1981
Spring Set	1.6	TEMP/GH	Baum, 1981
Sweet 100	0	TEMP/GH	MacKay et al. 1981
Jumbo	0	TEMP/GH	MacKay et al. 1981
Michigan Ohio Forcing	0	TEMP/GH	MacKay et al. 1981
Burpee Big Boy	0.2	TEMP/O	Markin, 1982
Floradel	8.9, 9.1	TEMP/O	Sutton and Lewis, 1982
Korala #127	6.8[b]	TEMP/GH	Burgoon and Baum, 1984
Vendor	4.1	TEMP/GH	Head, 1984
Tropic	0.5, 3.2	TROP/O	Watten and Busch, 1984
Homestead	2.7	TROP/O	Watten and Busch, 1984
Red Cherry	1.5	TROP/O	Watten and Busch, 1984
Prime Beefsteak	1.4	TROP/O	Watten and Busch, 1984
Vendor	0	TROP/O	Nair et al. 1985
Tropic	0	TROP/O	Nair et al. 1985
Jumbo	0	TROP/O	Nair et al. 1985
Perfecta	0	TROP/O	Nair et al. 1985
Laura	2.3–3.4	TEMP/GH	McMurtry, 1989
Kewalo	2.5–5.0	TEMP/GH	McMurtry, 1989
Sunny	10.1	TROP/O	Rakocy, 1989
Floradade	9.0	TROP/O	Rakocy, 1989
Vendor	3.7	TROP/O	Rakocy, 1989
Cherry Challenger	2.9	TROP/O	Rakocy, 1989
Champion	4.6[b]	TEMP/GH	McMurtry et al. 1990

[a] From Rakocy and Hargreaves, 1993
[b] kg/m^2

Table 18.5 Varieties and Yields of Leafy Greens (Lettuce, Pak Choi, Chinese Cabbage, Spinach) Evaluated in Aquaponic Systems[a]. Environment Codes. TEMP=Temperate Zone, TROP=Tropical Zone, O=Outside, GH=Greenhouse

Variety	Yield (g/plant)	Environment	Reference
Ostinata	162	TEMP/GH	Baum, 1981
Reskia	236	TEMP/GH	Burgoon and Baum, 1984
All Year Round	236	TEMP/GH	Burgoon and Baum, 1984
Karma	98	TEMP/GH	Burgoon and Baum, 1984
Ravel	45–50	TEMP/GH	Burgoon and Baum, 1984
Salina	44	TEMP/GH	Burgoon and Baum, 1984
El Captain	43	TEMP/GH	Burgoon and Baum, 1984
Bruinsma Columbus	0	TEMP/GH	Burgoon and Baum, 1984
Salad Bowl	-	TEMP/GH	Head, 1984
Pak Choi	-	TEMP/GH	Head, 1984
Winter Bloomsdale (Spinach)	-	TEMP/GH	Head, 1984
Buttercrunch	-	TEMP/GH	Zweig, 1986
Buttercrunch	193	TROP/O	Rakocy, 1989B
Summer Bibb	180	TROP/O	Rakocy, 1989B
Le Choi	508	TROP/O	Rakocy, 1989B
Pak Choi	422	TROP/O	Rakocy, 1989B
50-Day Hybrid (Chinese Cabbage)	638	TROP/O	Rakocy, 1989B
Tropical Delight (Chinese Cabbage)	589	TROP/O	Rakocy, 1989B
Summer Bibb	107–116	TEMP/GH	Parker et al. 1990
Sierra	182–340	TROP/O	Rakocy et al. 1997
Nevada	149–360	TROP/O	Rakocy et al. 1997
Jerhico	267–344	TROP/O	Rakocy et al. 1997
Parris Island	181–446	TROP/O	Rakocy et al. 1997

[a] From Rakocy and Hargreaves (1993) and some more recent data.

Table 18.6 Varieties and Yields of Cucumbers Evaluated in Aquaponic Systems[a]. Environmental Codes. TEMP=Temperate Zone, TROP=Tropical Zone, O=Outside, GH=Greenhouse

Variety	Yield (kg/plant)	Environment	Reference
Triumph	4.1	TEMP/O	Lewis et al. 1980
Patio Pik	1.6	TEMP/O	Lewis et al. 1980
Corona (Stokes)	28.6[b]	TEMP/GH	Burgoon and Baum, 1984
Bruinsma Vetomil	8.2[b]	TEMP/GH	Burgoon and Baum, 1984
Superator	4.1	TEMP/GH	Head, 1984
Sprint 4405	0.7	TEMP/GH	Wren, 1984
Burpee Hybrid II	7.3[b]	TEMP/GH	McMurtry, 1990

[a] From Rakocy and Hargreaves, 1993

[b] kg/m^2

A batch cropping system is more appropriate for crops that are grown seasonally or have long growing periods (>3 months) such as tomatoes and cucumbers. Various intercropping systems can be used in conjunction with batch cropping. For example, if lettuce is intercropped with tomatoes and cucumbers, one crop of lettuce can be harvested before tomato plant canopy development limits light penetration (Resh, 1995).

18.10 Pest and Disease Control

A number of plant pest and disease problems have been encountered in aquaponic systems. Pests observed on tomatoes include spider mite (Landesman, 1977; Nair et al. 1985), russet mite (Rakocy, unpublished data), hornworm (Lewis et al. 1978; Sutton and Lewis, 1982; Nair et al. 1985), western locust (Sutton and Lewis, 1982), fall armyworm, pinworm, aphid, and leaf minor (Nair et al. 1985). Diseases observed on tomatoes include blight (Lewis et al. 1980) and bacteria wilt (McMurtry et al. 1990). In UVI's systems, lettuce is affected by fall armyworm, corn earworm, and two species of pathogenic root fungus (*Pythium dissotocum* and *P. myriotylum*). The root diseases that plague conventional hydroponics may be a threat to aquaponics. Four viral, two bacterial and 20 fungal pathogens have been associated with root diseases in hydroponically grown vegetables (Stanghellini and Rasmussen, 1994). Most of the destructive root diseases in hydroponics have been attributed to the fungal genera *Pythium, Phytophthora, Plasmopara, Olpidium* and *Fusarium*.

Pesticides should not be used to control insects on aquaponic plant crops. Even pesticides that are registered would pose a threat to the fish and would not be permitted in a fish culture system. Similarly, most therapeutants for treating fish parasites and diseases should not be used either. Vegetables may absorb and concentrate them. Even the common practice of adding salt to treat fish diseases or reduce nitrite toxicity would be deadly to vegetables. Non-chemical methods of plant pest and disease control are required such as biological control (resistant cultivars, predators, antagonistic organisms, pathogens), physical barriers, traps, treatment of the nutrient solution (filtration, UV sterilization), manipulation of the physical environment and other specialized cultural practices. Opportunities for biological control methods are greater in enclosed greenhouse environments than exterior installations. McMurtry (1989) used *Encarsia formosa* and *Chrysopa carnea* to control greenhouse white fly (*Trialeurodes vaporariorum*) and *Hippodamia convergens* to control potato aphid (*Macrosiphum euphorbiac*). In UVI's systems, caterpillars are effectively controlled by twice-a-week spraying with *Bacillus thuringiensis*, a bacterial pathogen that is specific to caterpillars. The fungal root pathogens that are encountered in summer dissipate in winter in response to lower water temperature and manipulation of suspended solids levels. An outbreak of *Pythium* coincided with a period during which the efficiency of suspended solids removal was dramatically increased.

Prohibition on the use of pesticides makes crop production in aquaponic systems more difficult. However, this restriction assures that crops from aquaponic systems will be raised in an environmentally sound manner, free of pesticide residues. A major advantage of aquaponic systems is that crops are less susceptible to attack from soil-borne diseases. It also appears that aquaponic systems may be more resistant to diseases that affect standard hydroponics. This resistance may be due to the presence of some organic matter in the culture water which creates a stable, ecologically-balanced, growing environment with a wide diversity of microorganisms, some of which may be antagonistic to plant root pathogens.

18.11 Vegetable Yields

Vegetable yields from aquaponic systems have been quite variable, ranging from 0 to 10.1 kg/plant for tomatoes, Table 18.4, 0 to 786 g/plant for leaf lettuce (Table18.5 and more recent data) and 0.7 to 4.1 kg/plant for cucumbers, Table 18.6. This variability reflects vegetable production under a wide range of environments in many experimental systems. As in all pioneering efforts, some of these initial system designs produced unfavorable environments and disappointing results. Nevertheless, several investigators determined that the yields from aquaponic systems were greater than local field crops or yields from plants grown in soil under the same environmental conditions (Lewis et al. 1978; Lewis et al. 1980; Watten and Busch, 1984; McMurtry et al. 1990). In an experiment at UVI using raft hydroponics, aquaponic production of leaf lettuce was compared to hydroponic production using a commercial nutrient formulation (Rakocy, 1989A). There was no significant difference in total production, average plant size at harvest, or survival. Under ideal conditions, UVI's aquaponic systems have produced exceptional lettuce growth. Romaine (Parris Island) and red leaf (Sierra) lettuce reached mean harvest weights of 660 g and 532 g, respectively, in 4 weeks while crisphead lettuce (Montello) grew to an average size of 786 g in 5 weeks. The planting density (16 plants/m^2) was four times greater than the density (4 plants/m^2) typically used in field production. An aquaponic system can produce 13 lettuce crops per year as opposed to four or five field crops in areas that allow year-round growth. In UVI's commercial-scale system, marketable lettuce production has averaged 26 cases per week and ranged from 13 to 38 cases (24–30 heads/cs.). Annual production has averaged 1,350 cases (Rakocy et al. 1997). Equivalent or better tomato yields (Sutton and Lewis, 1982; Rakocy, 1989B) have been obtained from aquaponic systems when compared to the average yield (8.2–9.1 kg/plant) in soil less culture (Resh, 1995), Table 18.4.

18.12 Prospects For The Future

Aquaponics is still in its infancy. There are only a few small commercial operations in the U.S. Hydroponic growers generally do not consider aquaculture as a nutrient source for their operations. Aquaculturists, on the other hand, frequently mention the possibility of incorporating hydroponics into their closed recirculating systems to mitigate waste discharge and earn extra income. Data from successful, large-scale trials is needed to attract investor capital and spur commercial development.

The immediate potential for aquaponic systems appears to be niche markets where consumers are willing to pay a higher price for consistently high-quality fish and specialty crops that are available on a regular basis throughout the year. Environmental control, e.g., heated greenhouses, allows aquaponic systems to be located adjacent to urban markets, substantially reducing transportation costs.

Aquaponic systems also have potential in arid and semiarid areas due to their water reuse efficiency and conservation characteristics. If the economics are favorable, aquaponic systems can be operated with desalinated water. In desert areas, the effluent from aquaponic systems is a valuable resource for field crop production.

As production and marketing studies on St. Croix show, there is considerable potential for aquaponic systems in tropical and subtropical island communities. Over fishing and the constraint of unreliable water supplies and rugged, non-arable terrain for crop production make many islands dependent on fish and produce imports, but the long transportation links and time in transit substantially increase the price of imported food and contribute to poor quality. Aquaponic systems are ideal for resource-limited areas such as islands and would afford a degree of food self-sufficiency.

The intensive and sustainable production of aquaponic systems and their lower requirements for resources such as nutrients, land, and water will undoubtedly provide the impetus for continued aquaponic research, development, and commercialization.

Leafy Green Vegetables Such as Collard Thrive on the Rich Nutrient Concentrations in Aquaculture Effluent.

REFERENCES

Adams, C.R., 1982. Folic acid, thiamin and pyridoxine. Vitamins – The Life Essentials. National Feed Ingredient Institute. NFIA. Ames, Iowa.

Ahmad, T., Boyd, C., 1988. Design and performance of paddlewheel aerators. Aquacultural Engineering 7:39–62.

Alabaster, J.S., Herbert, D.W.M., Memens, J., 1957. The survival of rainbow trout (*Salmo gairdneri Richardson*) and perch (*Perca fluviatilis L.*) at various concentrations of dissolved oxygen and carbon dioxide. Ann. Appl. Biol. 45:177–188.

Alabaster, J.S., Lloyd, R., 1984. Water Quality Criteria For Freshwater Fish (2nd Ed.). London, Butterworths.

Al-Layla, M.A., Ahmad, S., Middlebrooks, E.J., 1980. Handbook of wastewater collection and treatment. Principles and Practice, Chapter 13, Sedimentation, Garland STPM Press, NY, 177–209.

Amend, D.F., Conte, F.S., 1982. Disinfection: Necessary preventive maintenance for healthy fish. Aquaculture Magazine, 25–29.

American Public Health Association, 1989. Standard Methods for the Examination of Water and Wastewater, 17th Ed. American Public Health Association, Washington, DC.

American Public Health Association, 1995. Standard Methods for the Examination of Water and Wastewater, 19th Ed. American Public Health Association, Washington, DC, 1108 p.

American Society of Civil Engineers (ASCE), 1959. Sewage treatment plant design. Manual of Engineering Practice, No. 36, 375 p.

Anthonisen, A.C., Lehr, R.C., Prakasam, T.B.S., Srinath, E.G., 1976. Inhibition of nitrification of ammonia and nitrous acid. J. Water Pollut. Control Fed. 48(5), 835–852.

Antoniou, P., Hamilton, J., Koopman, B., Jain, R., Holloway, B., Lyberatos, G., Svoronos, S.A., 1990. Effect of temperature and pH on the effective maximum specific growth rate of nitrifying bacteria. Wat. Res. 24(1):97–101.

Arceivala, S.J., 1983. Hydraulic modeling for waste stabilization ponds. Journal of Environmental Engineering 109(5):265–268.

Arimoto M., Sato J., Maruyama K., Mimura G. and Furusawa I., 1996. Effect of chemical and physical treatment of striped jack nervous necrosis virus (SJNNV). Aquaculture 143:15–22.

Asato Y., 1976. Ultraviolet light inactivation and photoreactivation of AS-1 cyanophage in *Anacystis nidulans*. J. Bacteriol. 126:550–552.

Ashley, L.M., 1970. A symposium on diseases of fishes and shellfishes. Am. Fish. Soc. Spec. Pub. 5:366–379.

Austic, R.E., Jiao, R.R., Rumsey, G.L., Ketola, H.G., 1989. Monovalent minerals in the nutrition of salmonid fishes. Proceedings of the 1989 Cornell Nutrition Conference for Feed Manufacturers, October 24–26, Syracuse Marriott, East Syracuse, NY, 25–30.

Austreng, E. 1978. Digestibility determination in fish using chromic oxide marking and analysis of contents from different segments of the gastrointestinal tract. Aquaculture 13:265–272.

Bablon, G., Bellamy, M.-M. Bourbigot, F. B. Daniel, M. Doré, F. Erb, G. Gordon, B. Langlais, A. Laplanche, B. Legube, G. Martin, W. J. Masschelein, G. Pacey, D. A. Reckhow, and C. Ventresque, 1991. Fundamental aspects. In B. Langlais, D. A. Reckhow and D. R. Brink, editors. Ozone in water treatment: application and engineering. American Water Works Association Research Foundation. Denver, Colorado, pp. 11–132.

Balarin, J.D., Haller, R.D., 1982. The intensive culture of tilapia in tanks, raceways, and cages. In: J.F. Muir, R.J. Roberts (Eds.), Recent Advances in Aquaculture. Westview Press, Boulder, CO, pp. 265–356.

Bangs, D.H., 1998. The Business Planning Guide: Creating a Plan for Success in Your Own Business. Upstart Publishing Company, Chicago, IL.

BASF Corporation, 1986. Effect of pelleting on crystal and ethyl cellulose-coated ascorbic acid assay levels of poultry feed. BASF Animal Nutrition Research. RA873. BASF Corporation: Animal Nutrition Research, 100 Cherry Hill Road, Parsippany, NY.

BASF Corporation, 1991. Vitamins—One of the most important discoveries of the century. Documentation Animal Nutrition 4th Ed. 36 p. BASF Corporation: Animal Nutrition Research. RA873, 100 Cherry Hill Road, Parsippany, NY.

BASF Corporation, 1992. Vitamin stability in premixes and feeds: A practical approach. KC 9138, 2nd Revised Ed. BASF Corporation: Animal Nutrition Research, 100 Cherry Hill Road, Parsippany, NY.

Baum, C.M., 1981. Gardening in fertile waters. New Alchemy Quarterly 5:3–8.

Bell, G.R., 1962. Design criteria for diatomite filters. Journal of AWWA Vol. 10:1241–1256.

Bellamy, W. D., F. Damez, B. Langlais, A. Montiel, K. L. Rakness, D. A. Reckhow, C. M. Robson, 1991. Engineering aspects. In: B. Langlais, D. A. Reckhow and D. R. Brink, editors. Ozone in water treatment: application and engineering. American Water Works Association Research Foundation. Denver, Colorado, pp. 317–468.

Bergheim A., Åsgård T., 1996. Waste production from aquaculture. In: Baird D., Beveridge, M.C.M., Kelly L.A., Muir, J.F. (Eds.). Aquaculture and Water Resource Management), Blackwell Science Ltd, Oxford, UK, 50–80.

Bessems, E., 1998. The effect of practical conditions on the efficacy of disinfectants. Internat. Biodeter. Biodegr. 41:177–183.

Bishop, T.M., Smalls, A., Wooster, G.A., Bowser, P.R., 2002. Aerobiological (airborne) dissemination of the fish pathogen *Ichthyopthirius multifiliis* and the implications in fish health management. In: Lee, C.S., and O'Bryen, P., Editors. Biosecurity in Aquaculture Production Systems: Exclusion of Pathogens and Other Undesirables. The World Aquaculture Society, Baton Rouge, Louisiana, United States (in press).

Bisogni, J.J., Timmons, M.B., 1994. Control of pH in closed cycle aquaculture systems. In: M.B. Timmons and T.S. Losordo (Eds.), Aquaculture Water Reuse Systems: Engineering Design and Management, Elsevier Science, Amsterdam, NY, pp. 235–245.

Bondi, Aron A. 1987. Animal Nutrition. John Wiley and Sons, New York, pp. 298–299.

Boroughs, H., Townsley, S.J., Hiatt, R.W., 1957. The metabolism of radionuclides by marine organisms. III. Uptake of Ca in solution by marine fish. Limnol. Oceanogr. 2:28–32.

Boyd, C.E., Watten, B.J., 1989. Aeration systems in aquaculture. CRC Crit. Rev. Aquat. Sci., 1:425–472.

Braaten, B., 1991. Impact of pollution from aquaculture in six Nordic countries. Release of nutrients, effects, and wastewater treatment. In: De Pauw, N., Joyce, J.N. (Eds.), Aquaculture and the Environment. European Aquaculture Society Special Publication 16, Gent, Belgium, pp. 79–101.

Brandi, G., Sisti, M., Amagliani, G., 2000. Evaluation of the environmental impact of microbial aerosols generated by wastewater treatment plants utilizing different aeration systems. J. Appl. Microbiol. 88(5):845–852.

Brazil, B. L., 1996. Impact of ozonation on system performance and growth characteristics of hybrid striped bass (Morone chrysops X M. saxatilis) and tilapia hybrids (*Sarotherodon sp.*) reared in recirculating aquaculture systems. Ph.D. Dissertation, Virginia Polytechnic Institute and State University, Blacksburg, VA.

Brill Corporation, 1995. Agricultural Software.

Brune, D.E., Tomasso, J.R., Editors, 1991. Aquaculture and water quality. The World Aquaculture Society, Louisiana State University, Baton Rouge, LA 70803.

Bullock, G.L., Stuckey, H.M., 1977. Ultraviolet treatment of water for destruction of five gram-negative bacteria pathogenic to fishes. J. Fish. Res. Board Can. 34:1244–1249.

Bullock, G.L., Summerfelt, S.T., Noble, A.C., Weber, A.L., Durant, M.D., Hankins, J.A. 1997. Ozonation of a recirculating rainbow trout culture system. I. Effects on bacterial gill disease and heterotrophic bacteria. Aquaculture. 158:43–55.

Burgoon, P.S., Baum, C., 1984. Year round fish and vegetable production in a passive solar greenhouse. Proc. Intern. Soc. Soilless Cult. 6:151–172.

Burnstock, G. 1959. The morphology of the gut of the brown trout. Quart. J. Microscop. Sd. 100:183–198.

Burrows, R., Chenoweth, H., 1955. Evaluation of three types of fish rearing ponds. Research Report 39, U.S. Department of the Interior, Fish and Wildlife Service, Washington, D.C., 29 p.

Burrows, R., Chenoweth, H., 1970. The rectangular circulating rearing pond. Prog. Fish-Cult. 32:67–80.

Buterbaugh, G.L. and H. Willoughby. 1967. A feeding guide for brook, brown, and rainbow trout. Progr. Fish-Cult. 29:210–215.

Carlins, J. J, and R. G. Clark, 1982. Ozone generation by corona discharge. In: R. G. Rice and A. Netzer, editors. Handbook of ozone technology and applications, Volume I. Ann Arbor Science Publishers, Ann Arbor, Michigan, pp. 41–76.

Chang P.S., Chen L.J., Wang Y.C., 1998. The effect of ultraviolet irradiation, heat, pH, ozone, salinity and chemical disinfectants on the infectivity of white spot syndrome baculovirus. Aquaculture 166:1–17.

Chapman, P.E., Popham, J.D., Griffin, J., Michaelson, J., 1987. Differentiation of physical from chemical toxicity in solid waste fish bioassay. Water, Air, and Soil Pollution 33:295–308.

ChemGen Corporation, 1998. Exogenous Digestive Enzyme Blends and Custom Formulations.

Chen, J., 1990. Technical Service Internal Reports. BASF Corporation, Wyandotte, Michigan.

Chen, S., 1991. Theoretical and experimental investigation of foam separation applied to aquaculture, Ph.D. dissertation, Cornell University, Ithaca, New York.

Chen, S., Timmons, M.B., Aneshansley, D.J., Bisogni, J.J., 1992. Protein in foam fractionation applied to recirculating systems. Prog. Fish-Cult. 55(2):76–82.

Chen, S., Coffin, D.E., Malone, R.F., 1993A. Production, characteristics, and modeling of aquaculture sludge from a recirculating aquaculture system using a granular media biofilter. In: J.K. Wang (Ed.), Techniques for modern aquaculture. St. Joseph, MI, American Society of Agricultural Engineers, pp. 16–25.

Chen, S., Timmons, M.B., Aneshansley, D.J. Bisogni, J.J., 1993B. Suspended solids characteristics from recirculating aquacultural systems and design implications. Aquaculture 112:143–155.

Chen, S., Timmons, M.B., Aneshansley, D.J., Bisogni, J.J. Jr., 1994C. Suspended solids characteristics from recirculating aquacultural systems and design implications. Aquaculture 112:143–155.

Chen, S., Timmons, M.B., Bisogni, J.J., Aneshansley, D.J., 1994A. Modeling surfactant removal in foam fractionation I: Theoretical development. Aquacultural Engineering 13:101–120.

Chen, S., Timmons, M.B., Bisogni, J.J., Aneshansley, D.J., 1994B. Modeling surfactant removal in foam fractionation II: Experimental investigations. Aquacultural Engineering 13:121–138.

Chen, Y., Solovitch, T., 1988. Effects of humic substances on plant growth. Acta. Hort. 221:412.

Chenoweth, H.H., Larmoyeux, J.D., Piper, R.G., 1973. Evaluation of circular tanks for salmonid production. Prog. Fish-Cult. 35:122–131.

Chesness, J.L., Poole, W.H., Hill, T.K., 1975. Settling basin design for raceway effluent. Trans. Am. Soc. Agr. Engr.19:192–194.

Chiang, H.C., Lee, J.C., 1986. Study of treatment and reuse of aquacultural wastewater in Taiwan. Aquacultural Engineering 5:301–312.

Cho, C.Y., Slinger, S.J., 1979. Apparent digestibility measurement in feedstuffs for rainbow trout. In: Halver, J.E., Tiews, K., (Eds.). Finfish Nutrition and Fish Food Technology, Vol. II, Heenemann Verlagsgsellschaft, Berlin., 239–247.

Cho, C.Y., J.D. Hynes, K.R. Wood, Yoshida, H.K., 1994. Development of high-nutrient-dense, low pollution diets and prediction of aquaculture wastes using biological approaches. Aquaculture 124:293–305.

Christian, L.D., 1983. Vitamin stability in mineral mixes formulated with different calcium phosphates stored at two temperatures and relative humidities. Proceedings AFIA Nutrition Council, 22 p.

Chun, C., Takakura, T., 1993. Control of root environment for hydroponic lettuce production: rate of root respiration under various dissolved oxygen concentrations. Paper presented at the 1993 International Summer Meeting of the American Society of Agricultural Engineers, Spokane, WA.

Cobb, W.W., Titcomb, J.W., 1930. A circular pond with central outlet for rearing fry and fingerlings of the salmonidae. Trans. Am. Fish. Soc. 60:121–123.

Coelho, M.B., 1991. Fate of vitamins in premixes and feeds: vitamin stability. Feed Management (Oct.) 42(10):24–35.

Colberg, P.J., Lingg, A.J., 1978. Effect of ozonation on microbial fish pathogens, ammonia, nitrate, nitrite, and BOD in simulated reuse hatchery water. J. Fish. Res. Board Can. 35:1290–1296.

Collier G.F., Tibbitts, T.W., 1982. Tipburn of lettuce. Hort. Rev. 4:49–65.

Colt, J., 1984. Computation of dissolved gas concentrations in water as functions of temperature, salinity, and pressure. American Fisheries Society Special Publication, 14, Bethesda, Maryland.

Colt, J.E., Tchobanoglous, G., 1976. Evaluation of the short-term toxicity of nitrogenous compounds to channel catfish, *Ictalurus punctatus*. Aquaculture 8(3):209–224.

Colt, J., Bouck, G., 1984. Design of packed columns for degassing. Aquacultural Engineering 3:251–273.

Colt, J. and B. Watten, 1988. Application of pure oxygen in fish culture. Aquacultural Engineering 7:397–441.

Colt, J.E., Orwicz, K., Bouck, G., 1991. Water quality considerations and criteria for high-density fish culture with supplemental oxygen. In: Colt, J., White, R.J. (Eds.), Fisheries BioEngineering Symposium 10, American Fisheries Society, Bethesda, MD, pp. 372–385.

Commercial 3-242-R, Penn. Fish Commission, 345 p.

Cowey, C.B., Sargent, J.R., 1972. Fish nutrition. Adv. Mar. Biol. 10:383–492.

Crampton, V., Kreiberg, H., Powell, J., 1990. Feed control in salmonids. Aquaculture International Congress Proceedings, Vancouver, B.C., September, pp. 99–108.

Cripps, S.J., 1992. Characterization of an aquaculture effluent based on water quality and particle size distribution data. Aquaculture '92 Conference, May 21–25, 1992, Orlando, FL, 72 p.

Cripps, S.J., 1993. The application of suspended particle characterization techniques to aquaculture systems. In: J.K. Wang (Ed.), Techniques for modern aquaculture. St. Joseph, MI, American Society of Agricultural Engineers, pp. 26–34.

Cripps, S.J., 1995. Serial particle size fractionation and characterization of an aquacultural effluent. Aquaculture 133:323–339.

Cripps, S.J., Poxton, M.G., 1992. A review of the design and performance of tanks relevant to flatfish culture. Aquacultural Engineering 11:71–91.

Crites, R., Tchobanoglous, G., 1998. Small and Decentralized Wastewater Management Systems. New York, McGraw-Hill.

Cryer, E., 1992. Recent applications of ozone in freshwater fish hatchery systems. In: Blogoslawski, W.J (Ed.). Proceedings of the 3rd International Symposium on the Use of Ozone in Aquatic Systems. International Ozone Association, Pan American Committee, Stamford, Connecticut, 134–154.

Davenport, M.T., Timmons, M.B., Vinci, B.J., Crum, M.K., 2001. Experimental evaluation of low head oxygenators. Aquacultural Engineering 24:245–256.

Davies D.A., Arnold C.R., 1997. Tolerance of the rotifer *Brachionus plicatilis* to ozone and total oxidative residuals. Ozone Sci. Eng. 19:457–469.

De Antonis, K.M., P.R. Brown, Z. Yi, Maugle, P.D., 1993. High-Performance liquid chromatography with ion paring and electrochemical detection for the determination of the stability of two forms of vitamin C. J. of Chromatography 632: 91–96.

de Veer I, Moriske H.J., Ruden, H., 1994. Photochemical decomposition of organic compounds in water after U.V.-irradiation: investigation of positive mutagenic effects. Toxicol. Lett. 72:113–119.

De Wit, M.C.J., 1978. Morphology and function of roots and shoot growth of crop plants under oxygen deficiency. In: D.D. Hook and R.M.M., Eds. Plant Life and Anaerobic Environments. Ann Arbor Science. Ann Arbor, MI, pp. 330–350.

Dimitriou, M. A., 1990. Design guidance manual for ozone systems. International Ozone Association, Pan American Committee. Norwalk, Connecticut.

Douglas, J.S., 1985. Advanced Guide to Hydroponics. Pelham Books, London.

Dove, C.R. and R.C. Ewan, 1986. The effect of diet composition on the stability of natural and supplemental vitamin E. Swine Research Report AS-580-J. Iowa State University, Ames, Iowa.

Downey, P.C., Klontz, G.W., 1981. Aquaculture techniques: Oxygen (pO_2) requirements for trout quality. Publication status unknown.

Downing, A.L., Truesdale, G.A., 1955. Some factors affecting rate of solution of oxygen in water. J. Applied Chemistry 5:570–581.

Dwyer, W.P., Peterson, J.E., 1993. Evaluation of a low head oxygenator at Giant Springs State Fish Hatchery, Great Falls, Montana. The Progressive Fish-Culturist 55:121–124.

Easter, C., 1992. System component performance and water quality in a recirculating system produce hybrid striped bass. Paper presented at Aquacultural Expo V. New Orleans, January 11–16, 1992.

Ebeling, J.M., 1993. Engineering experiences during the design and construction of an aquaculture research and demonstration center at The Ohio State University. In: Wang, J.K. (Ed.). Techniques for Modern Aquaculture, ASAE, St. Joseph, MI, pp. 204–210.

Ebeling, J.M., 1994. Monitoring and Control. In: Timmons, M.B., Losordo, T.M. (Eds.). Aquaculture Water Reuse Systems: Engineering Design and Management, Developments in Aquaculture and Fisheries Science, Vol. 27. Elsevier, NY, pp. 307–324.

Ebeling, J.M., 1995. Engineering design and construction details of distributed monitoring and control systems for aquaculture. In: Aquacultural Engineering and Waste Management, NRAE-90, Northeast Regional Aquaculture Engineering Service, Ithaca, NY, pp. 1–22.

Ebeling, J.M., 1997. Monitor/Control for recirculating aquaculture systems based on a centralized computer using LabVIEW and a stand-alone module. In: Advances in Aquaculture Engineering, NRAE-105, Northeast Regional Aquaculture Engineering Service, Ithaca, NY, pp. 87–95.

Ebeling, J.M., 1998. Development and evaluation of a feedback control system for dynamic control of dissolved oxygen in intensive recirculating aquaculture systems. In: G. Libey, (Ed.). Proceedings of the 2nd International Conference on Recirculating Aquaculture, Northeast Regional Agriculture Engineering Service, Ithaca, NY, pp. 366–367.

Ebeling, J.M., 2000. Kinetic Reaction Rate Analysis of Nitrifying Bead Filters in Aquaculture. PhD Thesis, University of Maryland.

Eikebrokk B., Ulgenes Y., 1993. Characterization of treated and untreated effluents from land based fish farms. In: Reinertsen H., Dahle L.A., Jørgensen L., Tvinnereim K. (Eds.). Fish Farming Technology. A.A. Balkema, Rotterdam, The Netherlands, pp. 361–374.

Eker, A.P.M., Formenoy L., Wit, L.E.A., 1991. Photoreactivation in the extreme halophilic archaebacterium *Halobacterium cutirubrum*. Photochem. Photobiol. 53:643–651.

Environmental Protection Agency (EPA), 1975. Process design manual for suspended solids removal. U.S. EPA Technology Transfer Publication.

Farooq S., Engelbrecht R.S., Chian E.S.K., 1977. Influence of temperature and U.V. light on disinfection with ozone. Wat. Res. 11:737–741.

Federal Joint Subcommittee on Aquaculture. 1994. Guide to Drug, Vaccine and Pesticide Use in Aquaculture. Texas Agricultural Extension Service, Texas A&M University System, College Station, Texas.

Ferguson, O., 1982. Aqua-ecology: the relationship between water, animals, plants, people and their environment. Rodale's NETWORK, Summer, Rodale Press, Emmaus, PA.

FIFAC, 1980. Symposium on new developments in the utilization of heated effluent and recirculation systems for intensive aquaculture. EIFAC, 11th Session, Stavanger, Norway, May 28–30th.

Fitt, P.S., Sharma N., Castellanos, G., 1983. A comparison of liquid holding recovery and photoreactivation in halophilic and non-halophilic bacteria. Biochim. Biophys. Acta 739:73–78.

Fløgstad, H., Torgersen, Y., Schei, I., Røttereng, P.J. 1991. Disinfection of effluent from slaughterhouses. SINTEF report STF60 A91096 (In Norwegian), Trondheim, Norway.

Friedberg, E.C., Walker G.C., Siede, W., 1995. DNA Repair and Mutagenesis. ASM Press, Washington D.C., VA.

Fry, F.E.J., Hart, J.S., 1948. The relation of temperature to oxygen consumption in the goldfish. Biol. Bull. 94:66–77.

Gerber, N.N., 1979. Volatile substances from actinomycetes: their role in the odor pollution of water. Critical Reviews in Microbiology 7:191–194.

Gerber, N.N., 1985. Plant growth and nutrient formulas. In: A.J. Savage, Ed. Hydroponics Worldwide: State of the Art in Soilless Crop Production. International Center for Special Studies, Honolulu, HI, pp. 58–69.

Gjessing, E., Kallqvist, T., 1991. Algicidal and chemical effect of u.v.-radiation of water containing humic substances. Wat. Res. 25:491–494.

Gloger, K.G., Rakocy, J.E., Cotner, J.B., Bailey, D.S., Cole W.M., Shultz, K.A., 1995. Waste treatment capacity of raft hydroponics in a closed recirculating fish culture system. Book of Abstracts, World Aquaculture Society, Baton Rouge, LA.

Goldsmith, D.L., Wang, J.-K., 1993. Hydromechanics of settled solids collection using rotating flow. In: J.-K. Wang (Ed.), Techniques for Modern Aquaculture (Proceedings). June 21–23, 1993, Spokane, Washington. American Society of Agricultural Engineers, Saint Joseph, MI, pp. 467–480.

Golz, W.J., Rusch, K.A., Malone, R.F., 1999. Modeling the major limitations on nitrification in floating-bead filters. Aquacultural Engineering 20:43–61.

Grace, G.R., Piedrahita, R.H., 1994. Carbon dioxide control. In: M.B. Timmons and T.S. Losordo (Eds.), Aquaculture Water Reuse Systems: Engineering Design and Management, Elsevier Science, Amsterdam, NY, pp. 209–234.

Graham N.J.D., 1999. Removal of humic substances by oxidation/ biofiltration processes – a review. In: Ødegaard H. (Ed.). Removal of humic substances from water. Tapir, Trondheim, Norway, pp. 151–158.

Gregory, R., Zabel, T.F., 1990. Sedimentation and Flotation. In Pontius, F.W. (Ed.), Water quality and treatment (4th Ed. New York, McGraw-Hill, Inc., pp. 367–453.

Grasso, D. and Weber, J. W. Jr., 1988. Ozone-induced particle destabilization. J. Am. Water Works Assn. 80(8):73–81.

Hackney, G.E., Colt, J., 1982. The performance and design of packed column aeration systems for aquaculture. Aquacultural Engineering 1:275–295.

Hagopian, D.S, Riley, J.G., 1998. A closer look at the bacteriology of nitrification. Aquacultural Engineering 18:223–244.

Halas, E.S., M.J. Halon, Sandstead, H.H., 1975. Intrauterine nutrition and aggression. Nature 257:221–222.

Halver, J.E. (Ed.). 1989. Fish Nutrition (2nd Ed. Won). New York, NY, Academic Press, 798 p.

Halver, J.E. ed., 1989. Fish Nutrition, 2nd edition. Academic Press, New York, pp. 405–408, 540–541.

Hardy, R.W., 1998. Phytate. Aquaculture Magazine. Vol. 24(6):77–80.

Harm W., 1968. Dark repair of photorepairable u.v. lesions in *Escherichia coli*. Mutat. Res. 6:25–35.

Harman, J.P., 1978. Characterization, treatment and utilization of the effluent from an intensive fish farm. Ph.D. dissertation, University of Aston in Birmingham.

Harms, C.A., 1993. Quarantine procedures for fish aquaria. Ann. Proc. Amer. Assoc. Zoo Vet., pp. 70–74.

Harris G.D., Adams V.D., Sorensen D.L., Curtis M.S., 1987. Ultraviolet inactivation of selected bacteria and viruses with photoreactivation of the bacteria. Wat. Res. 21:687-692.

Haug, R.T., McCarty, P.L., 1972. Nitrification with submerged filters. J. Water Pollut. Control Fed. 44:2086.

Hazen, A., 1904. On sedimentation. Transactions ASCE 53:45–88.

Head, W., 1984. An assessment of a closed greenhouse aquaculture and hydroponic system. Ph. D. Dissertation. Oregon State University, Corvallis, OR.

Heinen, J.M., Hankins, J.A., Adler, P.R., 1996. Water quality and waste production in a recirculating trout culture system with feeding of a higher-energy or a lower-energy diet. Aquaculture Research 27:699–710.

Heinzel, M., 1998. Phenomena of biocide resistance in microorganisms. Internat. Biodeter. Biodegr. 41:225–234.

Henderson, J.P., Bromage, N.R., 1988. Optimizing the removal of suspended solids from aquaculture effluents in settlement lakes. Aquacultural Engineering 7:167–188.

Hilton, J.W., Slinger, S.J., 1981. Nutrition and Feeding of Rainbow Trout. Canadian Special Publication of Fisheries and Aquatic Sciences 55, 15 p.

Hoar, W.S., 1966. General and Comparative Physiology. Englewood Cliffs, NJ, Prentice-Hall, Inc., 243 p.

Hochheimer, J.N. 1990. Trickling filter model for closed system aquaculture. Ph.D. Dissertation, University of Maryland, College Park, Maryland.

Hoigne, J., 1988. The chemistry of ozone in water. In: Stucki, S. (Ed.). Process Technologies for Water Treatment. Plenum Press, New York, NY, pp. 121–141.

Hopkins, T.A., Manci, W.E., 1989. Feed conversion, waste and sustainable aquaculture, the fate of the feed. Aquaculture Magazine 15(2):32–36.

Hughes, S.G., Rumsey, G.L., Nesheim, M.C., 1983. Dietary requirements for essential branched-chain amino acids by lake trout. Trans. Ant. Fish. Soc. 112:812–817.

Huguenin, J.E., Colt, J., 1989. Design and Operating Guide for Aquaculture Seawater Systems. Elsevier Interscience, Amsterdam, The Netherlands, 264 p.

Hunter, G. L., W. J. O'Brien, R. A. Hulsey, K. E. Carns, and R. Ehrhard, 1998. Emerging disinfection technologies: medium-pressure ultraviolet lamps and other systems are considered for wastewater applications. Water Environment and Technology 10(6):40–44.

IDEQ (Idaho Division of Environmental Quality), 1988. Idaho Waste Management Guidelines for Aquaculture Operations. Idaho Department of Health and Welfare, Division of Environmental Quality, Twin Falls, ID.

Iwana, G.K., 1991. Interaction between aquaculture and the environment. Critical Reviews in Environmental Control, CRC Press 21(2):177–216.

Jagger, J. 1967. Introduction To Research in Ultraviolet Photobiology. Prentice-Hall Inc., Engelwood Cliffs, NJ.

Jensen, M.H., Collins, W.L., 1985. Hydroponic vegetable production. Hort. Rev. 7:483–558.

Jones, R.D., Morita, R.Y., 1985. Low temperature growth and whole cell kinetics of marine ammonium oxidizer. Marine Eco. Prog. Series 21:239–243.

Josse, M., Remacle, C., Dupont, E., 1989. Trout "body building" by ichthyodrome. In: DePauw, N., Jaspers, E., Ackefors, H., Wilkins, N. (Eds.), Aquaculture--A Biotechnology in Progress. European Aquaculture Society, Bredena, Belgium, pp. 885–984.

Juell, J.E., 1991. Hydroacoustic detection of food waste--a method to estimate maximum food intake of fish populations in sea cages. Aquacultural Engineering 10:207–217.

Juttner, F., 1983. Volatile odorous excretion products of algae and their occurrence in the natural aquatic environment. Water Science and Technology 15:247–257.

Kaiser, G.E., Wheaton, F.W., 1983. Nitrification filters for aquatic culture systems: state of the art. J. World Maricult. Soc. 14:302–324.

Kanazawa, A., Teshizna, S-T., Sakamoto, M., Awal, MA., 1980. Requirement of *Tilapia zili* for essential fatty acids. Bull. Japan Soc. Sd. Fish. 46:1353–1356.

Katzenelson, E., Kletter, B., Shuval, H.I., 1974. Inactivation kinetics of viruses and bacteria in water by ozone. J. Am. Water Works Assoc. 66:725–729.

Kaushik, S.J. Medale, F., 1994. Energy requirements, utilization and dietary supply to Salmonids. Aquaculture. 124:81–97.

Kelly, L.A., Bergheim, A., Stellwagen, J., 1997. Particle size distribution of wastes from freshwater fish farms. Aquaculture International 5:65–87.

Kelner, A., 1961. Historical background of the study of photoreactivation. In: Christensen B.C., Buchmann, B. (Eds.). Progress in Photobiology. Elsevier Publishing Company, Amsterdam, The Netherlands, pp. 276–278.

Ketola, H.G., 1976. Quantitative nutritional requirements of fishes for vitamins and minerals. Feedstuffs 48(7):42–44.

Ketola, H.G., 1979. Influence of dietary zinc on cataracts in rainbow trout *(Salmo gairdneri*). J. Nuir. 109(6):965–969.

Ketola, H.G., 1983. Requirement for dietary lysine and arginine by fry of rainbow trout. An. Sci. 56:101–107.

Ketola, H.G., Smith, C.E., Kindschi, G.A., 1989. Influence of diet and oxidative rancidity on fry of Atlantic and coho salmon. Aquaculture 79:417–423.

Ketola, H.G., Richmond, M.E., 1994. Requirement of rainbow trout for dietary phosphorus and its relationship to the amount discharged in hatchery effluents. Trans. Am. Fish. Soc. 123(4):587–594.

Kimball, B.A., 1982. Carbon dioxide and agricultural yield: an analysis of 360 prior observations. Agr. Res. Service, USDA.

Kimura T., Yoshimizu M., Tajima K., Ezura Y., Sakai M., 1976. Disinfection of hatchery water supply by ultraviolet irradiation – Susceptibility of some fish pathogenic bacteria and microorganisms inhabiting pond waters (In Japanese with English summary). Bull. Jap. Soc. Sci. Fish. 42:207–211.

Klapsis, A., Burley, R., 1984. Flow distribution studies in fish rearing tanks. Part 1--Design constraints. Aquacultural Engineering 3:103–118.

Klapsis, A., Burley, R., 1985. Flow distribution studies in fish rearing tanks. Part 1--Analysis of hydraulic performance of 1 m square tanks. Aquacultural Engineering 4:118–134.

Klontz, G.W., P.C. Downey, Foclit, R.L., 1985. A Manual for Trout and Salmon Production. Sterling H. Nelson and Sons, Inc., Murray Elevators Division, 118 West 480 South, Murray, UT 84107, 23 p.

Komanapalli, I.R., Lau B.H.S. 1996. Ozone-induced damage of *Esherichia coli* K-12. Appl. Microbiol. Biotechnol. 46:610–614.

Kostenbauder, H.B., 1991. Physical factors influencing the activity of antimicrobial agents. In: Block S.S. (Ed.). Disinfection, Sterilization, and Preservation. Lea & Febiger, Philadelphia, PA., pp. 59–71.

Kumar, S., 1984. Closed recirculating nitrification systems for soft-shelled crabs. Louisiana State University, Baton Rouge. Masters Thesis, 99 p.

Landesman, L., 1977. Overwintering tilapia in a recirculating system. In: Essays on Food and Energy, Howard Community College and Foundation for Self-Sufficiency, Inc., pp. 121–127.

Larmoyeux, J.D., Piper, R.G., Chenoweth, H.H., 1973. Evaluation of circular tanks for salmonid production. Prog. Fish-Cult. 35:122–131.

Lawson, T.B., 1995. Fundamentals of Aquacultural Engineering. Chapman & Hall, N.Y., 355 p.

Lee, P.G. 1998. The use of process control software for the monitoring and control of aquaculture systems. In: G. Libey, (Ed.). Proceedings of the 2nd International Conference on Recirculating Aquaculture, Virginia Tech., VA, pp. 48–58.

Lee, P.G., 1995. A review of automated control systems for aquaculture and design criteria for their implementation. Aquacultural Engineering 14:205–227.

Lemlich, R., 1972. Principles of foam fractionation and drainage. In: Lemlich, R. (Ed.). Absorptive Bubble Separation Techniques. Academic Press, NY.

Lewis, W.K., Whitman, W.C., 1924. Principles of gas adsorption. Industrial and Chemical Engineering Chemistry 16:1215–1220.

Lewis, W.M., Yopp, J.H., Brandenburg, A.M., Schnoor, K.D., 1980. On the maintenance of water quality for closed fish production systems by means of hydroponically grown vegetable crops. In: K. Tiews, Ed. Symposium on New Developments in the Utilization of Heated Effluents and Recirculation Systems for Intensive Aquaculture, EIFAC, Stravanger, Norway, pp. 121–129.

Lewis, W.M., Yopp, J.H., Schram, Jr., H.L., Brandenburg, A.M., 1978. Use of hydroponics to maintain quality of recirculated water in a fish culture system. Trans. Am. Fish. Soc. 197:92–99.

Li, J.C. 1964. Statistical Inference (Volume I). Edwards Brothers, Inc., Ann Arbor, Michigan. pp. 146–147 (t-test).

Liao, P.B., Mayo, R.D., 1974. Intensified fish culture combining water reconditioning with pollution abatement. Aquaculture 3:61–85.

Libey, G.S. and M.B. Timmons, Eds., 1996. Successes and Failures in Commercial Recirculating Aquaculture. Northeast Regional Agricultural Engineering Service - 98, Ithaca, NY.

Libey, G.S., 1993. Evaluation of a drum filter for removal of solids from a recirculating aquaculture system. In: J.K. Wang (Ed.), Techniques for modern aquaculture. St. Joseph, MI, American Society of Agricultural Engineers, pp. 519–532.

Liltved H., 1997. Characterisation of wastewater from fillets and shrimp processing industry. Norwegian Institute for Water Research, Report LNR 3631–97, Oslo, Norway.

Liltved H., Hektoen, H., Efraimsen, H., 1995. Inactivation of bacterial and viral fish pathogens by ozonation or UV irradiation in water of different salinity. Aquacultural Engineering 14:107–122.

Liltved H., Landfald, B., 1995. Use of alternative disinfectants, individually and in combination, in aquacultural wastewater treatment. Aquaculture Res. 26:567–576.

Liltved, H., Landfald, B., 1996. Influence of liquid holding recovery and photoreactivation on survival of ultraviolet-irradiated fish pathogenic bacteria. Wat. Res. 30:1109–1114.

Liltved, H., Cripps, S., 1999. Removal of particle-associated bacteria by prefiltration and ultraviolet irradiation. Aquaculture Res. 30:445–450.

Liltved, H., Landfald, B., 2000. Effects of high intensity light on ultraviolet-irradiated and non-irradiated fish pathogenic bacteria. Wat. Res. 34:481–486.

Losordo, T.M., Westers, H., 1994. System carrying capacity and flow estimation. In: Timmons, M.B., Losordo, T.M. (Eds.), Aquaculture Water Reuse Systems: Engineering Design and Management. Elsevier Science LTD, New York, pp. 9–60.
Losordo, T.M. 1997. Tilapia culture in intensive recirculating systems. In: B.A. Costa-Pierce and J.E. Rakocy, (Eds.). Tilapia Aquaculture in the Americas. World Aquaculture Society, Baton Rouge, LA.
Lovelace, F.E Podoliak, H.A., 1952. Absorption of radioactive calcium by brook trout. Progr. Fish-Cult. 14:154–158.
Loveless, J.E., Painter, H.A., 1968. The influence of metal ion concentration and pH value on the growth of a *Nitrosomonas* strain isolated from activated sludge. J. Gen. Micro. 52: 1–14.
Lovell, T. 1989. Diet and fish husbandry. Chapter 10, In: Halver, J.E. (Ed.), 1989. Fish Nutrition, Academic Press, New York, NY, 798 p.
Lovell, T. 1989. Nutrition and Feeding of Fish. (An AVI Book). New York, NY, Van Nostrand Reinhold, 260 p.
Loyless, C.L., Malone, R.F., 1997. A sodium bicarbonate dosing methodology for pH management in freshwater-recirculating aquaculture systems. Prog. Fish Cult. 59:198–205.
Lund V., Hongve, D., 1994. Ultraviolet irradiated water containing humic substances inhibits bacterial metabolism. Wat. Res. 28:1111–1116.
Lunde, T., Skybakmoen, S., Schei, I., 1997. Particle trap. US Patent 5, 636, 595.
Luther, T., 1990. Bioponics - organic hydroponic gardening. The Growing EDGE 1(3):40–43.
Luther, T., 1991A. Bioponics part 2. The Growing EDGE 2(2):37–43, 65.
Luther, T., 1991B. Bioponics part 3 - temperature and pH as limiting factors. The Growing EDGE 2(3):40–43, 61.
Luther, T., 1991C. Bioponics part 4. The Growing EDGE 2 (4):35–40, 59.
Luther, T., 1993. Bioponics part 5 - enzymes for hereditary potential. The Growing EDGE 4(2):36–41.
MacKay, K.T., Van Toever, W., 1981. An ecological approach to a water recirculating system for salmonids: preliminary experience. In: L.J. Allen and E.C. Kinney, Eds. Proceedings of the Bio-Engineering Symposium for Fish Culture, Fish Culture Section of the American Fisheries Society, Bethesda, MD, pp. 249–258.
MacMillan, R., 1992. Economic implications of water quality management for a commercial trout farm. In: National Livestock, Poultry, and Aquaculture Waste Management St. Joseph, MI, American Society of Agricultural Engineers, pp. 185–190.
MAFES, 1982. Protein and Amino Acid Nutrition for Channel Catfish. Information Bulletin 25. Mississippi Agricultural and Forestry Experiment Station, Mississippi State, MS. 18 p.
Magor, B.G., 1988. Gill histopathology of juvenile *Oncorhynchus kisutch* exposed to suspended wood debris. Can. J. Zool 66:2164–2169.
Mäkinen, T., Lindgren, S., Eskelinen, P., 1988. Sieving as an effluent treatment method for aquaculture. Aquacultural Engineering 7:367–377.

Malone, R.F., Rusch, K.A., Burden, D.G., 1990. Kemp's ridley sea turtle waste characterization study: precursor to a recirculating holding system design. World Aqua. Soc. 21(2):137–144.

Malone, R.F., Chitta, B.S., Drennan, D.G., 1993. Optimizing nitrification in bead filters for warmwater recirculating aquaculture systems. In: J. Wang (Ed.), Techniques for modern aquaculture. St. Joseph, MI, American Society of Agricultural Engineers, pp. 315–325.

Malone, R.F., Beecher, L.E., DeLosReyes, A.A. Jr., 1998. Sizing and management of floating bead bioclarifiers. In: Libey, G.S. (Ed.), Second International Conference on Recirculating Aquaculture (Proc.), 16–19 July 1998, Roanoke, VA. Northeast Regional Agricultural Engineering Service, Ithaca, NY, pp. 319–341.

Malone, R.F., Beecher, L.E., 2000. Use of floating bead filters to recondition recirculating waters in warmwater aquaculture production systems. Aquacultural Engineering 22:57–73.

Manthe, D.P., Malone, R.F., Kumar, S., 1988. Submerged rock filter evaluation using oxygen consumption criterion for closed recirculating systems. Aquacultural Engineering 7:97–111.

Marine, D., Lenhart, C.H., 1911. Further observations and experiments on the so-called thyroid carcinoma of brook trout and its relation to endemic goiter. J. Exp. Med. 13:45–460.

Markin, J., 1982. Tomato production in gravel beds as a means of maintaining water quality and recovering waste nutrients in recirculating catfish production systems. M.S. Thesis. Auburn University, Auburn, AL.

Masschelein, W. J., 1998. Ozone generation: Use of air, oxygen, or air simpsonized with oxygen. Ozone Science & Engineering 20:191–203.

Maugle, P.D., 1982. Digestive Enzymes in Shrimp. Ph.D. Dissertation. University of Rhode Island, Kingston, Rhode Island.

Maugle, P.D., 1993. Simultaneous determination of three vitamin C forms. Bulletin Aqua. Asc. of Canada. No. 93-4, pp. 44–46, Pirquet, K.T. (Ed.)) ISSN: 0840-5417.

Mayo, R.D., 1976. A technical and economic review of the use of reconditioned water in aquaculture. FAO Tech. Conference on Aquaculture, Kyoto, Japan, R. 30.

McCay, C.M., Tunison, A.V., Crowell, M., Paul, H., 1936. The calcium and phosphorus content of the body of the brook trout in relation to age, growth, and food. J. Biol. Chem. 114:259–263.

McLaughlin, T.W., 1981. Hatchery effluent treatment. In: L.J. Allen, Kinney, E.C. (Eds.), Bio-Engineering symposium for fish culture (FCS Publ. 1) Bethesda, MD, American Fisheries Society, pp. 167–173.

McMurtry, M.R., 1989. Performance of an integrated aqua-olericulture system as influenced by component ratio. Ph. D. Dissertation. North Carolina State University, Raleigh, North Carolina.

McMurtry, M.R., Nelson, P.V., Sanders, D.C. Hughes, L., 1990. Sand culture of vegetables using recirculating aquaculture effluents. J. Appl. Agric. Res. 5(4):280–284.

McMurtry, M.R., Sanders, D.C. Nelson, P.V., Nash, A., 1993A. Mineral nutrient concentration and uptake by tomato irrigated with recirculating aquaculture water as influenced by quantity of fish waste products supplied. J. Plant Nutrition 16(3):407–419.

McMurtry, M.R., Sanders, D.C., Patterson, R.P., Nash, A., 1993B. Yield of tomato irrigated with recirculating aquacultural water. J. Prod. Agric. 6(3):428–432.

Meade, J.W., 1985. Allowable ammonia for fish culture. Prog. Fish.Cult. 47(3):135–145.

Metcalf and Eddy Inc. 1991. Wastewater Engineering: Treatment, Disposal, Reuse. 3rd Ed., New York, McGraw Hill, 929 p.

Mia, R.I., 1996. In situ nitrification in high density recirculating aquaculture production systems. Master's Thesis, Louisiana State University, Baton Rouge, LA, 90 p.

Millamena, O.M., 1992. Ozone treatment of slaughterhouse and laboratory wastewaters. Aquacultural Engineering 11:23–31.

Miller, R.V., Jeffrey, W., Mitchell, D., Elasri, M., 1999. Bacterial responses to ultraviolet light. ASM News, 65:535–541.

Momayama, K., 1989. Inactivation of baculoviral mid-gut gland necrosis (BMN) virus by ultraviolet irradiation, sunlight exposure, heating and drying. Fish Pathol. 24:115–118.

Monaghan, T.J., DelosReyes, A.A., Jeansonne, T.M., Malone, R.F., 1996. Effects of media size on nitrification in fluidized sand filters. Aquaculture America '96 Book of Abstracts. World Aquaculture Society, Baton Rouge, Louisiana.

Montgomery, J.M. (Consulting Engineers, Inc.), 1985. Water Treatment Principles and Design, John Wiley and Sons, 696 p.

Moss, S.H, Davies, J.G., 1974. Interrelationship of repair mechanisms in ultraviolet-irradiated *Escherichia coli*. J. Bacteriol. 120:15–23.

Mudrak, V.A., 1981. Guidelines for economical commercial fish hatchery wastewater treatment systems. In: L.J. Allen, Kinney, E.C. (Eds.). Proceedings of the Bio-engineering Symposium for Fish Culture. American Fisheries Society, Fish Culture Section, Bethesda, MD, pp. 174–182

Mudrak, V.A., Stark, K.R., 1981. Guidelines for economical commercial hatchery wastewater treatment systems. U.S. Dept. of Commerce, NOAA, NMFS Project No.

Muir, J.F., 1978. Aspects of water treatment and reuse in intensive fish culture. PhD. Thesis. University of Stratchclyde, Glasgow, 451 p.

Muir, J.F., 1982. Recirculated system in aquaculture, In: Muir, J.F., Roberts, R.J. (Eds.). Recent Advances in Aquaculture, Vol. 1, Croom Helm and Westview Press, London, 453 p.

Naegel, L.C.A., 1977. Combined production of fish and plants in recirculating water. Aquaculture 10:17–24.

Nair, A., Rakocy, J.E., Hargreaves, J.A., 1985. Water quality characteristics of a closed recirculating system for tilapia culture and tomato hydroponics. In: R. Day and T.L. Richards, Eds. Proceedings of the Second International Conference on Warm Water Aquaculture - Finfish, Laie, HI, pp. 223–254.

Needham, J., 1988. Salmon smolt production. In: Laird and Needham, (Eds.). Salmon and Trout Farming. Halsted Press, Div. of John Wiley and Sons.

New Alchemy Institute, 1980. Assessment of a semi-closed, renewable resource-based aquaculture system. Progress Report No. 5 (OPA77-16790 AO3), Appropriate Technology Program, National Science Foundation.

Noble, A.C., Summerfelt, S.T., 1996. Diseases encountered in rainbow trout cultured in recirculating systems. Ann. Rev. Fish Dis. 6:65–92.

NRC (National Research Council, National Academy of Science), 1983. Nutrient requirements of domestic animals. Nutrient requirements of warmwater fishes and shellfishes. Washington, D.C., National Academy Press.

NRC. (National Research Council, Committee on Animal Nutrition, Board of Agriculture), 1993. Nutrient requirements of fish. Washington, D.C., National Academy Press.

Ogino, C., 1980. Requirements of carp and rainbow trout for essential amino acids. Bull. Japan. Soc. Sci. Fish. 46:171–174.

Oliver, B.G., Carey, J.H., 1976. Ultraviolet disinfection: an alternative to chlorination. J. Wat. Pollut. Contr. Fed. 48:2619–2624.

Ozawa, T., Yotsumoto H., Sasaki, T., Nakayama, S., 1991. Ozonation of seawater – applicability of ozone for recycled hatchery cultivation. Ozone Sci. Eng. 13:697–710.

Page, J.W., Andrews, J.W., 1974. Chemical composition of effluents from high density culture of channel catfish. Water, Air and Soil Pollution 3, 365.

Paller, M.H., Lewis, W.M., 1988. Use of ozone and fluidized-bed biofilters for increased ammonia removal and fish loading rates. Progressive Fish-Culturist 50:141–147.

Pankratz, T.M., 1995. Screening Equipment Handbook For Industrial and Municipal Water and Wastewater Treatment, 2nd Ed. Technomic Publishing Co., Lancaster, Pennsylvania, 282 p.

Parker, D., Anouti A., Dickerson, G., 1990. Experimental results: integrated fish/plant production systems. Report N. ERL 90-34. University of Arizona, Environmental Research Laboratory, Tucson, AZ.

Parker, N.C., 1981. An air-operated fish culture system with water-reuse and subsurface silos. In: L.J. Allen and E.C. Kinney (Eds.), Bio-Engineering symposium for fish culture (FCS Publ. 1) Bethesda, MD, American Fisheries Society, pp. 131–137.

Paterson, W.D., Horne, M., Zinn, J., 1991. Prevention and control of fish diseases: Some proven strategies. Bull. Aqua. Assoc. Can. 91:8–15.

Paul, T.C., Sayal, S.K., Sakhuja, V.S., Dhillon, G.S., 1991. Vortex-settling basin design considerations. J. Hydraul. Eng., 117:172–189.

Pecor, C.H., 1978. Intensive culture of tiger muskellunge in Michigan during 1976 and 1977. In: R.L. Kendall (Ed.). Selected coolwater fishes of Northern America. Spec. Publ. No. 11, Am. Fish Soc., pp. 246–253.

Perry, R.H., C.H. Chilton. 1973. Chemical Engineers Handbook, Fifth Ed., McGraw-Hill Book Company, New York, New York.

Person, P.E., 1979. Notes on muddy flavor. IV. Variability of sensory response to 2-methylisoborneol. Aqua Fennica 9:48–52.

Phillips, M.J., Beveridge, M.C.M., Clarke, R.M., 1991. Impact of aquaculture on water resources. In: Brune, D.E., Tomasso, J.R. (Eds.). Aquaculture and water quality. World Aquaculture Society, Advances in Aquaculture Vol. 3, Baton Rouge, LA, USA, pp. 568–591.

Piedrahita, R.H., Seland, A. 1995. Calculation of pH in fresh and seawater aquaculture systems. Aquacultural Engineering 14:331–346.

Pierce, B.A., 1980. Water reuse aquaculture systems in two greenhouses in northern Vermont. Proc. World Maric. Soc. 11:118–127.

Piper, R.E., McElwain, I.B., Orme, L.E., McCraren, J.P., Fowler, L.G Leonard, J.R., 1982. Fish Hatchery Management. U.S. Fish and Wildlife Service, Washington, DC.

Popma, T.J., 1982. Digestibility of selected feedstuffs and naturally occurring algae by tilapia. Ph.D. dissertation. Auburn University. Auburn, Alabama.

Poston, HA., R.C. Riis, G.L. Rumsey, H.G. Ketola., 1977. The effect of supplemental dietary amino acids, minerals and vitamins on salmonids fed cataractogenic diets. Cornell Veterinarian 67(4):472–509.

Poston, H.A., Rumsey, G.L., 1983. Factors affecting dietary requirements and deficiency signs of L-tryptophan in rainbow trout. J. Nutr. 113(2):2568–2577.

Potter, A., 1997. Alteration of the mechanical properties of fish waste via dietary components. Master of Science Thesis, Cornell University, Ithaca, NY.

Qualls, R.G., Flynn M.P., Johnson J.D., 1983. The role of suspended particles in ultraviolet disinfection. J. Wat. Poll. Contr. Fed. 55:1280–1285.

Quillere, I., Marie, D., Roux, L. Gosse, F., Morot-Gaudry, J.F., 1993. An artificial productive ecosystem based on fish/bacteria/plant association. I. Design and management. Agric., Ecosystems and Environment 7:13–30.

Rakocy, J.E., 1984. A recirculating system for tilapia culture and vegetable hydroponics. In: R.C. Smitherman and D. Tave (Eds.), Proceedings of the Auburn Symposium on Fisheries and Aquaculture, Auburn University, Auburn, AL., pp. 103-114.

Rakocy, J.E., 1989A. Hydroponic lettuce production in a recirculating fish culture system. Virgin Islands Agricultural Experiment Station, Island Perspectives 3:4–10.

Rakocy, J.E., 1989B. Vegetable hydroponics and fish culture: a productive interphase. World Aqua. 20(3):42–47.

Rakocy, J.E., 1995. Aquaponics: the integration of fish and vegetable culture in recirculating systems. In: M.C. Palada and C.C. Clarke, Eds. Proceedings of the 30th Annual Meeting of the Caribbean Food Crops Society, St. Thomas, USVI, pp. 101–108.

Rakocy, J.E., 1994. Waste management in integrated recirculating systems. Proceedings of the 21st United States-Japan Meeting on Aquaculture. Bulletin of National Research Institute of Aquaculture, Supplement 1:75–80.

Rakocy J.E., Nair, A., 1987. Integrating fish culture and vegetable hydroponics: problems and prospects. Virgin Islands Perspective - Agriculture Research Notes 2:16–18.

Rakocy, J.E., Hargreaves, J.A., 1993. Integration of vegetable hydroponics with fish culture: a review. In: J.-K. Wang, Ed. Techniques for Modern Aquaculture. American Society of Agricultural Engineers, St. Joseph, MI, pp. 112–136.

Rakocy, J.E., Hargreaves, J.A., Bailey, D.S., 1991. Comparative water quality dynamics in a recirculating system with solids removal and fixed-film or algal biofiltration. J. World Aqua. Soc. 22(3):49A.

Rakocy, J.E., Hargreaves, J.A., Bailey, D.S. 1993. Nutrient accumulation in a recirculating aquaculture system integrated with vegetable hydroponic production. In: J.-K. Wang, Ed. Techniques for Modern Aquaculture. American Society of Agricultural Engineers, St. Joseph, MI, pp. 148–158.

Rakocy, J.E., D.S., Bailey, K.A., Shultz and W.M., Cole, 1997. Evaluation of a commercial-scale aquaponic unit for the production of tilapia and lettuce. In K. Fitzsimmons, ed. Tilapia Aquaculture: Proceedings of the Fourth International Symposium on Tilapia in Aquaculture, Orlando, Florida, pp. 357–372.

Reckhow, D. A., J. K. Edzwald, and J. E. Tobiason, 1993. Ozone as an aid to coagulation and filtration. American Water Works Association, Denver, Colorado.

Rice, R. G., C. M. Robson, G. W. Miller, and A. G. Hill, 1981. Uses of ozone in drinking water treatment. Journal American Water Works Association 73:1–44.

Reid, D.F., Townsley, J.J., Ego, W.T., 1959. Uptake of Sr and Ca through epithelia of freshwater and seawater adapted *Tilapia mossambica.* Proc. Hawaii Acad. Sd. 34, 32 (abstr.).

Resh, H.M., 1995. Hydroponic food production: a definitive guidebook of soilless food-growing methods. Woodbridge Press Publishing Company, Santa Barbara, CA.

Robertson, W.D., 1992. Assessment of filtration and sedimentation systems for total phosphorus removal at the Lake Utopia Fish Culture Station. Canadian Fish Culture Operations, Aquaculture Division, Connors Bros., Limited, Blacks Harbour, New Brunswick, Canada.

Rodriguez, J., and S. Gagnon, 1991. Disinfection: liquid purification by UV radiation, and its many applications. Ultrapure Water 8(6): 26–31.

Roselund, B. D., 1975. Disinfection of hatchery influent by ozonation and the effects of ozonated water on rainbow trout. In: W. J. Blogoslawski and R. G. Rice, editors. Aquatic applications of ozone. International Ozone Institute, Stamford, Connecticut, pp. 59–69.

Rosenthal, H., 1981. Ozonation and sterilization. In: K. Tiens (Ed.), Proceedings from the World Symposium on Aquaculture in Heated Effluents and Recirculation System Vol. I, Heenemann, Berlin, Germany, pp. 219–274.

Roustan, M., Debellefontaine H., Do-Quang Z., Duguet J., 1998. Development of a method for the determination of ozone demand of a water. Ozone Sci. Eng. 20:513–520.

Ruane, R.J., Chu, T.Y.J., Vandergriff, V.E., 1977. Characterization and treatment of waste discharged from high-density catfish cultures. Water Research. 11:789–800.

Rumsey, G.L, Page, J.W., Scott, M.L., 1983. Methionine and cystine requirements of rainbow trout. Prog. Fish-Cult. 45:139- 143.

Rusten, G., McCoy, M., Proctor, R., Siljudalen, J., 1998. The innovative moving bed biofilm reactor/solids contact reaeration process for secondary treatment of municipal wastewater. Water Environment Research 70:1083–1089.

Sako, H., Sorimachi, M., 1985. Susceptibility of fish pathogenic viruses, bacteria and fungus to ultraviolet irradiation and the disinfectant effect of U.V.-ozone water sterilizer on the pathogens in water (In Japanese with English summary). Bull. Nat. Res. Inst. Aquaculture 8:51-58.

Satoh, S., Takeuchi, T., Watanabe, T., 1987. Requirement of tilapia for alpha-tocopherol. Nippon Suisan Gakkaishi 53:119.

Schaaf, R.L., 1990. Quality Assurance Internal Reports. BASF Corporation, Wyandotte, Michigan.

Schneider, J., 1988. Technical Service Internal Reports, BASF AG Animal Nutrition, Ludwigshafen, Germany.

Scott, K.R., Allard, L., 1984. A four-tank water recirculation system with a hydrocyclone prefilter and a single water reconditioning unit. Progressive Fish-Culturist 46:254–261.

Seawright, D.E., 1995. Integrated aquaculture-hydroponic systems: nutrient dynamics and designer diet development. Ph.D. Dissertation, University of Washington, Seattle, WA.

Severin, B.F., Suidan M.T., Engelbrecht S., 1983. Effects of temperature on ultraviolet light disinfection. Environ. Sci. Tecnol. 17:717–721.

SFT, 1997. Classification of environmental quality in freshwater (In Norwegian). Veiledning 97, 04, Statens forurensningstilsyn, Oslo, Norway.

Shehadeh, Z.H., Gordon, M.S., 1969. The role of the intestine in salinity adaptation of the rainbow trout, *Salmo gairdneri*. Comp. Biochem. Physiol. 30:397–418.

Shelton, G., 1970. The regulation of breathing. In: Hoar and Randall, (Eds.) Fish Physiology Vol. IV. The nervous system, circulation and breathing. Academic Press, New York, pp. 293–352.

Singh. R.P., Nose, T., 1967. Digestibility of carbohydrates in young rainbow trout. Bull. Freshwater Fish. Res. Lab. (Tokyo) 17(1):21–25.

Skybakmoen, S., 1989. Impact of water hydraulics on water quality in fish rearing units. In: Conference 3--Water Treatment and Quality. Proceedings of AquaNor 89, August 11–16, 1989. AquaNor, Trondheim, Norway, pp. 17–21.

Slater, G.P., Blok, V.C., 1983. Volatile compounds of the *cyanophyceae*-a review. Water Science and Technology 165:181–190.

Smith, H.W. 1930. The absorption and excretion of water and salts by marine teleosts. Am. J. Physiol. 93:480–505.

Smith, L.S., 1980. Digestive functions in teleost fishes (Chapter 11). In: Fish Nutrition, Second Ed., J.E. Halver, Ed., New York, NY, Academic Press.

Smith, R.R., 1971. A method for measuring digestibility and metabolizable energy of fish feeds. Progr. Fish-Cult. 33(3):132–134.

Soliman, A.K., Jauncey, K., Roberts, R.J., 1986. The effect of varying forms of dietary ascorbic acid on the nutrition of juvenile Tilapia (*Oreochromis niloticus*), Aquaculture, 52.

Solomon D.J., Brafield, A.G., 1972. The energetics of feeding, metabolism and growth of perch. J. Anim. Ecol. 41:699–718.

Spotte, S., 1979. Fish and Invertebrate Culture. Wiley-Interscience Publication, John Wiley and Sons, NY, NY.

Stanghellini, M.E., Rasmussen, S.L., 1994. Hydroponics: a solution for zoosporic pathogens. Plant Disease 78(12):1129–1138.

Stechey, D.M., Trudell, Y., 1990. Aquaculture wastewater treatment: Wastewater characterization and development of appropriate treatment technologies for the New Brunswick smolt production industry. Prepared for New Brunswick Department of the Environment and Local Government, PO Box 6000, Fredericton, New Brunswick, Canada, E3B 5H1. 101 p.

Stenstrom, M.K., Gilbert, R.G., 1981. Effect of alpha, beta, and theta factors upon the design, specification, and operation of aeration systems. Water Res.15:43–54.

Stickney, R.R. and R.T. Lovell (Eds.), 1977. Nutrition and Feeding of Channel Catfish. Southern Cooperative Bulletin 218. Alabama Agricultural Experiment Station, Auburn University, Auburn, AL (October), 67 p.

Stickney, R.R., 1979. Principles of Warmwater Aquaculture. New York, NY, J. Wiley and Sons, 375 p.

Stickney, R.R., McGeachin, R.B., Lewis, D.H., Marks, J., Sis, R.F., Robinson, E.H., W. Warts, 1984. Response of *Tilapia aurea* to dietary vitamin C. J. World Mariculture Society 15:179.

Stover, E.L., Haas C.N., Rakness K.L., Scheible O.K., 1986. Design Manual. Municipal Wastewater Disinfection. EPA/625/1-86/021. U.S. Environmental Protection Agency, Center for Environmental Research Information, Cincinnati, OH.

Streeter, V.L., 1966. Fluid Mechanics. McGraw-Hill Inc., New York.

Stumm, W., Morgan, J.J., 1981. Aquatic Chemistry. 2nd Edition. Wiley and Sons, New York, NY, 780 p.

Sugita, H., Asai, T., Hayashi, K., Mitsuya, T., Amanuma, K., Maruyama, C., Deguchi, Y., 1992. Application of ozone disinfection to remove *Enterococcus seriolicida*, *Pasteurella piscicida*, and *Vibrio anguillarum* from seawater. Appl. Environ. Microbiol. 58:4072-4075.

Summerfelt, S.T., 1996. Engineering design of a water reuse system. In: R.C. Summerfelt (Ed.), Walleye culture manual. Ames, IA, North Central Regional Aquaculture Center Publication Office, Iowa State University, pp. 277–309.

Summerfelt, S.T., 1999. Waste-handling systems. In: CIGR (Series Ed.) and F. Wheaton (Volume Ed.), CIGR handbook of agricultural engineering: Volume II, Aquacultural Engineering St. Joseph, MI, American Society of Agricultural Engineers, pp. 309–350.

Summerfelt, S.T., Cleasby, J.L., 1996. A review of hydraulics in fluidized-bed biological filters. ASAE Transactions 39:1161–1173.

Summerfelt, S.T., Hankins J.A., Durant, M.D., Goldman, J.N., 1996. Removing obstructions: Modified pipe-lateral flow distribution mechanism reduces backflow in fluidized-sand biofilters. Water Environment & Technology 8(11):39–43.

Summerfelt, S.T., Hankins J.A., Weber A.L., Durant M.D., 1997. Ozonation of a recirculating rainbow trout culture system. II. Effects on microscreen filtration and water quality. Aquaculture 158:57–67.

Summerfelt, S.T., Hochheimer J.N., 1997. Review of ozone processes and application as an oxidizing agent in aquaculture. Prog. Fish-Cult. 59:94–105.

Summerfelt, S.T., Wade, E.M., 1997. Recent advances in water treatment processes to intensify fish production in large recirculating systems. In: Advances in Aquacultural Engineering, Proceedings from Aquacultural Engineering Society Technical Sessions at the 4th International Symposium on Tilapia in Aquaculture, November 9–12, Orlando, FL, NRAES-105, pp. 350–367.

Summerfelt, S.T., Davidson, J., Waldrop, T., Tsukuda, S., 2000A. A partial-reuse system for coldwater aquaculture. In: G.S. Libey and M.B. Timmons (Eds.), Proceedings of the third international conference on recirculating aquaculture. Blacksburg, VA, Virginia Polytechnic Institute and State University. 167–175.

Summerfelt, S.T., Davidson, J., Timmons, M.B., 2000B. Hydrodynamics in the 'Cornell-type' dual-drain tank. In: G.S. Libey and M.B. Timmons (Eds.), Proceedings of the third international conference on recirculating aquaculture Blacksburg, VA, Virginia Polytechnic Institute and State University, pp. 160–166.

Summerfelt, S.T., Vinci, B.J. and R.H. Piedrahita, 2000. Oxygenation and carbon dioxide control in water reuse systems. Aquacultural Engineering 22 (1-2):87–108.

Summerfelt, S.T., Bebak-Williams, J., Tsukuda, S., 2001. Controlled Systems: Water Reuse and Recirculation. In: G. Wedemeyer (Ed.), Fish Hatchery Management, Second Ed. Bethesda, MD, American Fisheries Society, pp. 285–395.

Surber, E.W., 1933. Observations on circular pool management. Trans. Am. Fish. Soc. 63:139–143.

Surber, E.W., 1936. Circular rearing pools for trout and bass. Prog. Fish-Cult. 21:1–14.

Sutton, R.J., Lewis, W.M., 1982. Further observations on a fish production system that incorporates hydroponically grown plants. Prog. Fish Cult. 44:55–59.

Swenson P.A., 1976. Physiological responses of *Escherichia coli* to far-ultraviolet light. In: K.C. Smith (Ed.), Photochemical and Photobiological Reviews, Vol. 1, Plenum Press, New York. NY, pp. 269–387.

Takeuchi, T., Watanabe, T., 1982. Effects of various polyunsaturated fatty acids on growth and fatty acid composition of rainbow trout *Salmo gairdneri,* coho salmon *Onchorhynchus kisutch,* and chum salmon *Onchorhynchiss keta.* Bull. Japan Soc. Sci. Fish. 48:1745–1752.

Takeuchi, T., Satoh, S., Watanabe, T., 1983. Requirement of *Tilapia nilotica* for essential fatty acids. Bull. Japan Soc. Sci. Fish. 49:1127–1134.

Tchobanoglous, G., Burton, F.L., 1991. Wastewater Engineering: Treatment, Disposal, and Reuse, Third Ed. New York, McGraw-Hill, Inc.

The Task Committee on Design of Wastewater Filtration Facilities, 1986. Tertiary filtration of wastewaters, Journal of Environmental Engineering 112(6):1008–1025.

Theisen, D.D., Stansell, D.D., Woods, L.C., 1998. Disinfection of nauplii of *Artemia franciscana* by ozonation. Prog. Fish-Cult. 60:149–151.

Thoesen, J.C., 1994. Suggested procedures for the detection and identification of certain finfish and shellfish pathogens. Version 1. Fish Health Section, American Fisheries Society, Bethesda, MD.

Thomasson, M.P., 1991. Nitrification in fluidized-bed sand filters for use in recirculating aquaculture systems. Master's thesis, Louisiana State University, Baton Rouge, Louisiana.

Timmons, M.B., 2000. Aquacultural engineering–experiences with recirculating aquaculture system technology, Part 1. Aquaculture Magazine 26(1):52–56.

Timmons, M.B., Youngs, W.D., 1991. Considerations on the design of raceways. In: Aquaculture Systems Engineering, Proceedings of World Aquaculture Society and American Society of Agricultural Engineers, June 16–20, 1991, San Juan, Puerto Rico (ASAE Publication 02-91), American Society of Agricultural Engineers, St. Joseph, Michigan.

Timmons, M.B., Youngs, W.D., Ley, D., 1993. The Northern Fresh Fish Cooperative: Formation and initial results and technology description with plans. Cornell Agricultural Engineering Extension Bulletin Number 465.

Timmons, M.B., Losordo, T.S., 1994. Aquaculture water reuse systems: engineering design and management. Elsevier Science, The Netherlands.

Timmons, M.B., Chen, S. Weeks, N.C., 1995. Mathematical model of foam fractionators used in aquaculture. Journal of World Aquaculture 26(3):225–233.

Timmons, M.B., Summerfelt, S.T., 1998. Application of fluidized-sand biofilters to aquaculture. In: Proceedings Second International Conference on Recirculating Aquaculture, Available from NRAES as Publication VA-1, pp. 342–355.

Timmons, M.B., Summerfelt, S.T., Vinci, B.J., 1998. Review of circular tank technology and management. Aquacultural Engineering 18:51–69.

Timmons, M.B., Aho, P.W., 1998. Comparison of aquaculture and broiler production systems. In: Proceedings Second International Conference on Recirculating Aquaculture, Available from NRAES as Publication VA-1, pp. 190–200.

Timmons, M.B., Vinci, B.J., Davenport, M.T., Crum, M.K., 2001. A mathematical model of low-head oxygenators. Aquacultural Engineering 24:257–277.

Tomasso, J.R., Simco, B.A., Davis, K., 1979. Chloride inhibition of nitrite induced methemoglobinemia in channel catfish (*Ictalurus punctatus*). J. Fish. Res. Board of Canada 36:1141–1144.

Torgersen, Y., 1998. Physical and chemical inactivation of the infectious salmon anaemia (ISA) virus. In: Proceedings of the New England Farmed Fish Health Workshop, Washington County Technical College, Eastport, ME, 2–10.

Tsivoglou, J.R., O'Connel, R.L., Walter, C.M., Godsil, P.J., Logsdon, G.S., 1965. Tracer measurements of atmospheric respiration. 1. Laboratory studies. Journal Water Pollution Control Federation 37:1343–1362.

Tsukuda, S.M., Hankins, J.A., Marshall, C.P., Summerfelt, S.T., Bullock G.L., Sawyer, T.K., 1997. Effects of sand size on fluidized-bed biofilter performance in cold water systems. In: M.B. Timmons and T.M. Losordo (Eds.) Recent Advances in Aquacultural Engineering (Proceedings). Northeast Regional Agricultural Engineering Service, Ithaca, NY, pp. 368–380.

Tsukuda, S.M., J.M. Ebeling, Watten, B.J., 1998. Application of industrial monitoring and control for an experimental carbon dioxide stripper in a recirculating aquaculture system. In: G. Libey, (Ed.). Proceedings of the 2nd International Conference on Recirculating Aquaculture, Virginia Tech., VA, pp. 371–372.

Tucker, C.S., Robinson, E.H., 1990. Channel Catfish Farming Handbook. New York, Van Nostrand Reinhold.

Tvinnereim, K., Skybakmoen, S., 1989. Water exchange and self-cleaning in fish rearing tanks. In: De Pauw, N., Jaspers, E., Ackefors, H., Wilkens, N. (Eds.), Aquaculture: A Biotechnology in Progress. European Aquaculture Society, Bredena, Belgium, pp. 1041–1047.

Twarowska, J.G., Westerman, P.W., Losordo, T.M., 1997. Water treatment and waste characterization evaluation of an intensive recirculating fish production system. Aquacultural Engineering 16:133–147.

U.S. EPA, 1975. Process design manual for nitrogen control. A design manual prepared for the Office of Technology Transfer of the U.S. EPA.

Van Gorder, S., 1991. Optimizing production by continuous loading of recirculating systems. In: Workshop on Design of High Density Recirculating Systems, Baton Rouge, LA, pp. 17–26.

Van Toever, W., MacKay, K.T., 1980. A modular recirculating hatchery and rearing system for salmonids utilizing ecological design principles. In: K. Tiews, Ed. Symposium on New Developments in the Utilization of Heated Effluents and Recirculating Systems for Intensive Aquaculture. EIFAC, Stravanger, Norway, pp. 403–414.

Vaughn, J.M., Chen Y.S., Lindburg K., Morales D., 1987. Inactivation of human and simian rotaviruses by ozone. Appl. Environ. Microbiol. 53:2218–2221.

Venosa, A.D., Petrasek, A.C., Brown, D., Spark, H.L. and Allen, D.M., 1984. Disinfection of secondary effluent with ozone/UV. J. Wat. Poll. Contr. Fed. 56:137–142.

Verwer, F.L., Wellman, J.J.C., 1980. The possibilities of Grodan rockwool in horticulture. Int. Congress on Soilless Culture 5:263–278.

Vinci, B.J., Watten, B.J., Timmons, M.B., 1997. Modeling gas transfer in a spray tower oxygen absorber. Aquacultural Engineering 16:91–105.

Vinci, B., Summerfelt, S., Bergheim, A., 2001. Solids control. In: Summerfelt, S.T. (Ed.), Presentation Notebook: Aquacultural Engineering Society Workshop: Intensive Fin-fish Systems and Technologies. Aquaculture 2001, International Triannual Conference and Exposition of the World Aquaculture Society, January 23, Disney's Coranado Springs Resort, Orlando, FL.

Wagner, E.J., Bosakowski, T., Miller, S.A., 1995. Evaluation of the absorption efficiency of the low head oxygenation system. Aquacultural Engineering 14:49–57.

Walker G.C., 1984. Mutagenesis and inducible responses to deoxyribonucleic acid damage. *Escherichia coli*. Microbiol. Rev. 48:60–93.

Warrer-Hansen, I., 1982. Evaluation of matter discharged from trout farming in Denmark. In: Alabaster, J.E., (Ed.), Report of the E1FAC Workshops on Fish Farm Effluents, Silkegorg, Denmark, Technical Paper No.4.

Watanabe, T., Takeuchi, T., Ogino, C., 1975. Effect of dietary methyl linoleate and linolenate on growth of carp II. Bull. Japan Soc. Sci. Fish. 41:263–269.

Watten, B.J., 1989. Multiple Stage Gas Absorber. United States Patent Number 4, 880, 445.

Watten, B.J., 1990. Design of packed columns for commercial oxygen addition and dissolved nitrogen removal based on effluent criteria. Aquacultural Engineering 9:305–328.

Watten, B.J., Busch, R.L., 1984. Tropical production of tilapia (*Sarotherodon aurea*) and tomatoes (*Lycopersicon esculentum)* in a small-scale recirculating water system. Aquaculture 41:271–283.

Watten, B.J., Boyd, C.E., 1990. Gas transfer within a multi-stage packed column oxygen absorber: Model development and application. Aquacultural Engineering 9:33–59.
Watten, B.J., Johnson, R.P., 1990. Comparative hydraulics in rearing trial performance of a production scale cross-flow rearing unit. Aquacultural Engineering 9:245–266.
Watten, B.J., Honeyfield, D.C., 2000. Hydraulic characteristics of a rectangular mixed-cell rearing unit. Aquacultural Engineering 24:59–73.
Wedekind, H., Knösche, R., Göthling, U., 1995. Treatment of trout farm effluents by drum filtration with different mesh sizes. In: C.S. Lee (Ed.), Aquaculture in Eastern European Countries. Stara Zagora, Bulgaria.
Wedemeyer, G.A., 1996. Physiology of fish in intensive culture. International Thompson Publishing, New York.
Wedemeyer G.A., Nelson N.C., 1977. Survival of two bacterial fish pathogens (*Aeromonas salmonicida* and the enteric redmouth bacterium) in ozonated, chlorinated, and untreated waters. J. Fish. Res. Board Can. 34:429–432.
Wedemeyer G.A., Nelson N.C., Smith C.A., 1978. Survival of the salmonid viruses infectious hematopoietic necrosis (IHNV) and infectious pancreatic necrosis (IPNV) in ozonated, chlorinated, and untreated waters. J. Fish. Res. Board Can. 35:875–879.
Wedemeyer, G.A., Nelson, N.C., Yasutake, W.T., 1979. Physiological and biochemical aspects of ozone toxicity to rainbow trout (*Salmo gairdneri*). J. Fish. Res. Board Can. 36:605–614.
Weeks, N.C., Timmons, M.B. and S. Chen. 1992. Feasibility of using foam fractionation for the removal of dissolved and suspended solids from fish culture water. Aquacultural Engineering 11:251–265.
Weiss, R.F., 1970. The solubility of nitrogen, oxygen, and argon in water and seawater. Deep-sea Res. 17:721–735.
Weiss, R.F., 1974. Carbon dioxide in water and seawater: the solubility of a non-ideal gas. Mar. Chem. 2:203–215.
Westers, H. 1979. Principles of intensive fish culture. Mich. Dept. of Nat. Res. 108 p.
Westers, H. 1987. Feeding levels for fish fed formulated diets. Progr. Fish-Cult. 49:87–92.
Westers, H. 1989. Water quality impacts of aquaculture. A review, (manuscript), 34 p. Michigan Department of Natural Resources.
Wheaton, F.W., 1977. Aquacultural Engineering. Robert E. Krieger Publishing Co., Malabar, FL, 708 p.
Wheaton, F.W., 1985. Aquacultural Engineering, 2nd Ed., Wiley-Interscience, New York. 708 p.
Wheaton, F.W., Hochheimer, J.N., Kaiser, G.E., 1991. Principles of biological filtration. In: M.B. Timmons (Ed.) Engineering Aspects of Intensive Aquaculture. Northeast Regional Agricultural Engineering Service, Cooperative Extension, Ithaca, NY, pp. 1–31.

Wheaton, F.W., Hochheimer, J.N., Kaiser, G.E., Krones, M.J., Libey G.S., Easter, C.C., 1994. Nitrification Principles. In: M.B. Timmons and T.M. Losordo, Eds. Aquaculture Reuse Systems: Engineering Design and Management. Development in Aquaculture and Fisheries Sciences, Vol. 27, Elsevier Science B.V., Amsterdam, The Netherlands, pp.101–126.

Whitby, G.E. and Palmateer, G., 1993. The effect of UV transmission, suspended solids and photoreactivation on micro-organisms in wastewater treated with UV. Wat. Sci. Tech. 27:379-386.

Wickramanayake, G.B., 1991. Disinfection and sterilization by ozone. In: Disinfection, Sterilization, and Preservation 4th Ed. (Edited by Block S.S.), Lea & Febiger, Philadelphia, PA, pp. 182–190.

Williams, R.C., S.G., Hughes, and G.L. Rumsey. 1982. Use of ozone in a water reuse system for salmonids. The Progressive Fish-culturist 44:102–105.

Wilson, R.P. and W. E. Poe. 1985. Apparent digestibility of protein and energy in feed ingredients for channel catfish. Prog. Fish-Cult. 47:154–158.

Wimberly, Douglas M., 1990. Development and evaluation of a low-density media biofiltration unit for use in recirculating finfish culture systems, Master's Thesis, Louisiana State University, Baton Rouge, Louisiana.

Windell, J.T., J. W. Foltz, and J.A. Sarokon, 1978. Effect of body size, temperature and ration size on the digestibility of a dry pelleted diet by rainbow trout. Trans. Am. Fish. Soc. 107:613–616.

Wong, K.B., and Piedrahita, R.H., 2000. Settling velocity characterization of aquacultural solids. Aquacultural Engineering 21:233–246.

Wooster, G.A., Bowser, P.R., 1996. The aerobiological pathway of a fish pathogen: Survival and dissemination of *Aeromonas salmonicida* in aerosols and its implications in fish health management. Journal of the World Aquaculture Society 27(1):7–14.

Wortman, B., Wheaton, F.W., 1991. Temperature effects on biodrum nitrification. Aquacultural Engineering 10:183–205.

Wren, S.W., 1984. Comparison of hydroponic crop production techniques in a recirculating fish culture system. M.S. Thesis, Texas A&M University, College Station, TX.

Yoshimizu M., Takizawa H. and Kimura T., 1986. UV-susceptibility of some fish pathogenic viruses (in Japanese with English summary) Fish Pathol. 21:47–52.

Zhuge, Q. and C.F. Klopfenstein, 1986. Factors affecting storage stability of vitamin A, riboflavin and niacin in a broiler diet premix. Poult. Sci. 65, 987.

Zweig, R.D., 1986. An integrated fish culture hydroponic vegetable production system. Aqua. Mag. 12(3):34–40.

APPENDIX

Table A-17 Size and Properties of PVC Pipe Schedule 40 and 80

Table A-18 PVC Pipe Friction Chart for Pipe and Length Loss for Schedule 40

Table A-19 PVC Pipe Friction Chart for Pipe and Length Loss for Schedule 80

Table A-20 Tank Volumes for Various Depths and Diameters

Table A-21 Temperature Equivalents Between Celsius and Fahrenheit

Table A-22 Area Under the Normal Distribution Curve

Table A-23 Percentage Points of the Student's t-Distribution

Table A-1 Conversion Factors for Commonly Used Terms in Aquaculture[a]

Multiply Number of	By	To Obtain
microns	$3.93 \cdot 10^{-5}$	inches (in)
microns	$1.0 \cdot 10^{-6}$	meters (m)
feet (ft)	0.3048	m
ft^2	0.0929	m^2
gallons (US)	0.1337	ft^3
gallons (US)	3.7854	liters (L)
gallons (US)	0.0037854	m^3
gallons (US)	8.34	lbs water at 60°F
264 gallons	1	m^3
gallons/ft^2	40.7	L/min
ft^3	0.02832	m^3
ft^3/s	28.317	L/s (liters per second)
ft^3/s	0.02832	m^3/s
gallons/minute (gpm)	0.06309	L/s
gpm	3.78	L/minute (Lpm)
gpm	0.00006309	m^3/s
gpm/ft^2	40.7	Lpm/m^2
gpm/ft^2	58.646	m^3/day•m^2
ft^3/s	448.83	gpm
pounds force (lb)	4.448	newtons (N)
lb/in^2 (psi)	6895	N/m^2 (Pascals)
lb/ft^2	47.88	N/m^2
lb/in^2 (psi)	2.246	ft seawater head (50°F, 35 ppt)
	2.307	ft freshwater head (60°F, 0 ppt)
lb/gpm	0.1198	kg/Lpm
lb/gpm	0.001997	kg/Lps
lb/ft^2	4.8824	kg/m^2
lb/ft^3	16.018	kg/m^3
lb/gal	120.1	kg/m^3
British Thermal Unit (BTU)	1055.1	joule (J)
BTU	252.0	gram calories
Horsepower (hp)	0.707	BTU/s
hp	745.7	watts (W)
W	0.239	gram-calories/s
To Obtain	Divide By	Starting With

[a]Units on the right column may be obtained by starting with units on the left column, multiply by the given number. To go the other way, from right to left, divide by the given value. Example: watts = 745.7·HP and HP = watts / 745.7

Table A-2 Fish Health Conversion Factors

1 ppm (mg/L)	=	0.38 grams per 100 gallons of water
		3.8 milligrams per gallon of water
		0.0283 grams per cubic foot of water
		0.38 milliliters per 100 gallons of water
		2.72 pounds per acre-foot of water
		1 milligrams per liter of water (mg/L)
		1 grams per cubic meter of water (g/m^3)
		0.001 milliliters per liter of water (mL/L)
1 acre-foot	=	43,560 cubic feet
1 acre-foot	=	325,850 gallons
1 acre foot of water	=	2,718,144 pounds
1 cubic foot of water	=	7.48 gallons
1 cubic foot of water	=	62.4 pounds
1 cubic foot of water	=	28.3 liters
1 cubic foot of water	=	28.3 kilograms
1 cubic meter of water	=	1,000 liters
1 cubic meter of water	=	35.2 cubic feet of water
1 cubic meter of water	=	2.205 pounds of water
1 gallon of water	=	8.34 pounds
1 gram	=	0.0353 ounces
1 kilogram	=	2.2 pounds
1 pound	=	454 grams
1 gallon	=	3.785 liters
1 gallon of water	=	3,785 grams
1 liter	=	0.26 gallons
1 liter	=	1,000 cubic centimeters
1 liter	=	1,000 milliliters
1 liter of water	=	1,000 grams
1 ounce	=	28.35 grams
1 gallon	=	128 fluid ounces
1 fluid ounce	=	28.4 grams
1 inch	=	2.54 centimeters
1 foot	=	30.48 centimeters
1 cubic centimeter of water	=	1 gram
1 cubic centimeter of water	=	1 milliliter
1 hectare	=	10,000 square meters
1 hectare	=	2.47 acre
1 acre	=	0.405 hectare
1 acre	=	43,560 square feet
Percent Solution		
For 1 percent solution add:		
38 grams per gallon		1.3 ounces per gallon
10 grams per liter		38 cc per gallon
10 cc per liter		
Temperature Conversion		
Centigrade to Fahrenheit	=	(°C X 9/5) + 32
Fahrenheit to Centigrade	=	(°F - 32) X 5/9

Table A-3 Broad List of Conversion Factors: Inch:Pound (IP) to System International (SI)

To Convert IP to SI......

Value	in IP Units of	= Value	in SI Units of
Length, Thickness, Width, Depth, Height			
1	chain (= 66 feet)	2.012E+01	m
1	chain (= 66 feet)	2.012E-02	km
1	foot	3.048E+02	mm
1	foot	3.048E+01	cm
1	foot	3.048E-01	m
1	furlong (= 1/8 mile)	2.012E+02	m
1	furlong (= 1/8 mile)	2.012E-01	km
1	hand (= 4 inches)	1.016E+02	mm
1	hand (= 4 inches)	1.016E+01	cm
1	hand (= 4 inches)	1.016E-02	m
1	inch	2.540E+01	mm
1	inch	2.540E+00	cm
1	inch	2.540E-02	m
1	mil (= 0.001 inch)	2.540E-02	mm
1	mile	1.609E+03	m
1	mile	1.609E+00	km
1	nautical mile (= 6076.115 feet)	1.852E+03	m
1	nautical mile (= 6076.115 feet)	1.852E+00	km
1	rod	5.029E+00	m
1	rod	5.029E-03	km
1	yard	9.144E+02	mm
1	yard	9.144E+01	cm
1	yard	9.144E-01	m
Area			
1	acre	4.047E+03	m^2
1	acre	4.047E-01	ha
1	$foot^2$	9.290E+04	mm^2
1	$foot^2$	9.290E-02	m^2
1	$inch^2$	6.542E+02	mm^2
1	$inch^2$	6.452E+00	cm^2
1	$inch^2$	6.452E-04	m^2
1	$mile^2$	2.590E+06	m^2

1 mile^2	2.590E+02 ha
1 mile^2	2.590E+00 km^2
1 square (= 100 ft^2)	9.290E+00 m^2
1 yard^2	8.361E+03 cm^2
1 yard^2	8.361E-01 m^2

Volume, Capacity, Section Modulus

1 acre-ft	1.233E+03 m^3
1 acre-ft	1.028E+02 m^3
1 barrel (42 US gallons, petroleum)	1.590E+02 L
1 barrel (42 US gallons, petroleum)	1.590E-01 m^3
1 board ft (actual)	2.360E-03 m^3
1 board ft (actual)	2.360E+03 cm^3
1 bushel (= 2150.42 in3; 32 US dry quarts)	3.524E+01 L
1 bushel (= 2150.42 in3; 32 US dry quarts)	3.524E-02 m^3
1 cord (of wood)	3.625E+00 m^3
1 cup	2.366E+02 mL
1 cup	2.366E+02 cm^3
1 cup	2.366E-01 L
1 fluid oz	2.957E+01 cm^3
1 fluid oz	2.957E+01 mL
1 fluid oz	2.957E-02 L
1 ft^3	2.832E+04 cm^3
1 ft^3	2.832E+01 L
1 ft^3	2.832E-02 m^3
1 gallon, Imperial	4.546E+00 L
1 gallon, Imperial	4.546E-03 m^3
1 gallon, US (= 231 in^3)	3.785E+00 L
1 gallon, US (= 231 in^3)	3.785E-03 m^3
1 hogshead	2.385E+02 L
1 inch^3	1.639E+04 mm^3
1 inch^3	1.639E+01 cm^3
1 inch^3	1.639E+01 mL
1 pint	4.732E+02 mL
1 pint	4.732E+02 cm^3
1 pint	4.732E-01 L
1 quart, Imperial (= 1.2 US liquid quarts)	1.136E+03 mL
1 quart, Imperial (= 1.2 US liquid quarts)	1.136E+03 cm^3
1 quart, Imperial (= 1.2 US liquid quarts)	1.136E+00 L
1 quart, Imperial (= 1.2 US liquid quarts)	1.136E-03 m^3
1 quart, US (dry, = 67.204 in^3)	1.101E+03 mL

1 quart, US (dry, = 67.204 in^3)	1.101E+03 cm^3
1 quart, US (dry, = 67.204 in^3)	1.101E+00 L
1 quart, US (dry, = 67.204 in^3)	1.101E-03 m^3
1 quart, US (liquid, = 57.749 in^3)	9.464E+02 mL
1 quart, US (liquid, = 57.749 in^3)	9.464E+02 cm^3
1 quart, US (liquid, = 57.749 in^3)	9.464E-01 L
1 quart, US (liquid, = 57.749 in^3)	9.464E-04 m^3
1 tablespoon (approximate)	1.479E+01 mL
1 tablespoon (approximate)	1.479E+01 cm^3
1 tablespoon (approximate)	1.479E-02 L
1 teaspoon (approximate)	4.929E+00 mL
1 teaspoon (approximate)	4.929E+00 cm^3
1 teaspoon (approximate)	4.929E-03 L
1 $yard^3$	7.646E-01 m^3

Mass (and weight in IP system)

1 cwt	4.536E+01 kg
1 cwt	4.536E-02 tonne
1 grain (= 1/7000pound)	6.480E+01 mg
1 grain (= 1/7000pound)	6.480E-02 g
1 ounce (liquid, US)	2.960E+01 mL
1 ounce (liquid, US)	2.960E-02 L
1 ounce (mass, avoirdupois)	2.835E+01 g
1 ounce (mass, avoirdupois)	2.835E-02 kg
1 penneyweight (= 1/20 ounce)	1.555E+00 g
1 pound (avoirdupois)	4.536E+02 g
1 pound (avoirdupois)	4.536E-01 kg
1 stone (= 14 pounds)	6.350E+00 kg
1 ton	9.072E+02 kg
1 ton	9.072E-01 tonne
1 ton, long (= 2240 pounds)	1.016E+03 kg

Mass per unit area, application rate, yield, seeding rate

1 cwt/acre	1.121E-02 kg/m^2
1 cwt/acre	1.121E+02 kg/ha
1 cwt/acre	1.121E-01 tonne/ha
1 lb/acre	1.121E-04 kg/m^2
1 lb/acre	1.121E+00 kg/ha
1 lb/acre	1.121E-03 tonne/ha
1 lb/ft^2	4.882E-01 g/cm^2
1 lb/ft^2	4.882E+03 g/m^2

1 lb/ft^2	4.882E+00 kg/m^2
1 lb/yd^2	5.425E-02 g/cm^2
1 lb/yd^2	5.425E-01 kg/m^2
1 oz/ft^2	3.052E-02 g/cm^2
1 oz/ft^2	3.052E+02 g/m^2
1 oz/ft^2	3.052E-01 kg/m^2
1 oz/yd^2	3.391E-03 g/cm^2
1 oz/yd^2	3.391E+01 g/m^2
1 oz/yd^2	3.391E-02 kg/m^2
1 ton/acre	2.242E-01 kg/m^2
1 ton/acre	2.242E+03 kg/ha
1 ton/acre	2.242E+00 tonne/ha
Soybeans, alfalfa, clover, wheat (60#/bu)	
1 bu/acre	6.726E-03 kg/m^2
1 bu/acre	6.726E+01 kg/ha
1 bu/acre	6.726E-02 tonne/ha
Shelled corn	
1 bu/acre	6.277E-03 kg/m^2
1 bu/acre	6.278E+01 kg/ha
1 bu/acre	6.278E-02 tonne/ha
Ear corn, husked	
1 bu/acre	3.924E-03 kg/m^2
1 bu/acre	3.924E+01 kg/ha
1 bu/acre	3.924E-02 tonne/ha
Oats (32#/bu)	
1 bu/acre	3.587E-03 kg/m^2
1 bu/acre	3.587E+01 kg/ha
1 bu/acre	3.587E-02 tonne/ha
Sudan grass (40#/bu)	
1 bu/acre	4.484E-03 kg/m^2
1 bu/acre	4.484E+01 kg/ha
1 bu/acre	4.484E-02 tonne/ha

Volume per unit area, application rate, yield

1 bu/acre	7.484E-02 m^3/ha
1 bu/acre	7.484E-06 m^3/m^2
1 gal/acre	9.355E+03 cm^3/ha
1 gal/acre	9.355E-01 cm^3/m^2
1 gal/acre	9.355E+00 L/ha
1 gal/acre	9.355E-04 L/m^2
1 gal/acre	9.355E-03 m^3/ha

1 gal/acre	9.355E-07 m^3/m^2
1 pt/acre	1.169E+03 cm^3/ha
1 pt/acre	1.169E-01 cm^3/m^2
1 pt/acre	1.169E+00 L/ha
1 pt/acre	1.169E-04 L/m^2
1 pt/acre	1.169E-03 m^3/ha
1 pt/acre	1.169E-07 m^3/m^2
1 qt/acre	2.339E+03 cm^3/ha
1 qt/acre	2.339E-01 cm^3/m^2
1 qt/acre	2.339E+00 L/ha
1 qt/acre	2.339E-04 L/m^2
1 qt/acre	2.339E-03 m^3/ha
1 qt/acre	2.339E-07 m^3/m^2

Mass per unit volume, density, dilution

1 grain/gallon	1.710E+01 mg/L
1 lb/ft^3	1.600E+04 kg/m^3
1 lb/ft^3	1.602E+01 g/L
1 lb/gal	1.199E+02 kg/m^3
1 lb/gal	1.199E+02 g/L
1 oz/gal	7.491E+00 kg/m^3
1 oz/gal	7.491E+00 g/L
1 ton/ft^3	3.203E+04 kg/m^3
1 ton/yd^3	1.186E+03 kg/m^3

Mass per unit mass, concentration

1 grain/lb	1.430E-01 g/kg
1 grain/lb	1.430E-04 kg/kg

Velocity, speed

1 ft/s	3.048E-01 m/s
1 ft/s	1.097E+00 km/hr
1 ft/min	5.080E-03 m/s
1 ft/min	1.829E-02 km/hr
1 mile/hr	4.470E-01 m/s
1 mile/hr	1.609E+00 km/hr

Power

1 Btu/hr	2.931E-01 W
1 Btu/hr	2.931E-04 kW
1 Btu/min	1.759E+01 W

1 Btu/min	1.759E-02 kW
1 ft-lb/hr	3.767E-04 W
1 ft-lb/hr	3.767E-07 kW
1 ft-lb/min	2.260E-02 W
1 ft-lb/min	2.260E-05 kW
1 ft-lb/s	1.356E+00 W
1 ft-lb/s	1.356E-03 kW
1 horsepower (boiler)	9.810E+03 W
1 horsepower (boiler)	9.810E+00 kW
1 horsepower (mechanical)	7.457E+02 W
1 horsepower (mechanical)	7.457E-01 kW
1 ton (refrigeration, 12,000 Btu/hr)	3.517E+03 W
1 ton (refrigeration, 12,000 Btu/hr)	3.517E+00 kW
1 J/s	1.00E+00 W

Energy

1 Btu	1.055E+00 kJ
1 ft-lb	1.356E+00 J
1 horsepower-hr	2.685E+00 MJ
1 kWh	3.600E+00 MJ
1 calorie (gram)	4.187E+00 J
1 therm	1.055E+02 MJ
1 langley	4.186E+00 J/cm^2
1 langley	4.186E+01 kJ/m^2
1 quad (US)	1.055E+12 MJ

Volumetric flow rate, ventilation rate

1 ft^3/s	2.832E-02 m^3/s
1 ft^3/s	2.832E+01 L/s
1 ft^3/min (cfm)	4.719E-04 m^3/s
1 ft^3/min (cfm)	4.719E-01 L/s
1 gal/min	6.309E-05 m^3/s
1 gal/min	6.309E-02 L/s
1 gal/hr	1.052E-06 m^3/s
1 gal/hr	1.052E-03 L/s
1 acre-ft/s	1.233E+03 m^3/s
1 million gallon/day	4.383E-02 m^3/s
1 million gallon/day	4.383E+01 L/s

Mass per unit time, flow rate

1 lb/s	4.536E-01 kg/s
1 lb/min	7.560E-03 kg/s

1 lb/hr	1.260E-04 kg/s
1 lb/hr	5.515E-04 tonne/hr
1 ton/hr	2.520E-01 kg/s
1 ton/hr	1.103E+00 tonne/hr

Pressure, stress

1 atmosphere	1.013E+05 Pa
1 atmosphere	1.013E+00 bar
1 inches mercury	3.377E+03 Pa
1 inches mercury	3.377E-02 bar
1 inches water	2.488E+02 Pa
1 kip/in^2 (ksi)	6.895E+06 Pa
1 lb/ft^2 (psf)	4.788E+01 Pa
1 lb/in^2 (psi)	6.895E+03 Pa

Force, weight

1 cwt	4.448E+02 N
1 kip (= 1000 lb)	4.448E+03 N
1 lb	4.448E+00 N
1 ounce	2.780E-01 N
1 stone	6.227E+01 N
1 ton	9.964E+03 N

Area per unit volume, coverage, specific area

1 acre/pint	8.553E+06 m^2/m^3
1 acre/pint	8.553E+02 ha/m^3
1 acre/pint	8.553E-01 ha/L
1 acre/pint	8.553E+03 m^2/L
1 acre/quart	4.276E+06 m^2/m^3
1 acre/quart	4.276E+02 ha/m^3
1 acre/quart	4.276E-01 ha/L
1 acre/quart	4.276E+03 m^2/L
1 acre/gallon	1.069E+06 m^2/m^3
1 acre/gallon	1.069E+02 ha/m^3
1 acre/gallon	1.069E-01 ha/L
1 acre/gallon	1.069E+03 m^2/L
1 ft^2/pint	1.963E+02 m^2/m^3
1 ft^2/pint	1.963E-02 ha/m^3
1 ft^2/pint	1.963E-05 ha/L
1 ft^2/pint	1.963E-01 m^2/L
1 ft^2/quart	9.816E+01 m^2/m^3

1 ft^2/quart	9.816E-03 ha/m^3
1 ft^2/quart	9.816E-06 ha/L
1 ft^2/quart	9.816E-02 m^2/L
1 ft^2/gallon	2.454E+01 m^2/m^3
1 ft^2/gallon	2.454E-03 ha/m^3
1 ft^2/gallon	2.454E-06 ha/L
1 ft^2/gallon	2.454E-02 m^2/L
1 acre/bu	1.336E+05 m^2/m^3
1 acre/bu	1.336E+01 ha/m^3
1 acre/bu	1.336E-02 ha/L
1 acre/bu	1.336E+02 m^2/L

Area per unit mass, coverage

1 acre/cwt	8.922E+00 ha/tonne
1 acre/cwt	8.922E+01 m^2/kg
1 acre/cwt	8.922E-03 ha/kg
1 acre/lb	8.922E+02 ha/tonne
1 acre/lb	8.922E+03 m^2/kg
1 acre/lb	8.922E-01 ha/kg
1 acre/ton	4.461E-01 ha/tonne
1 acre/ton	4.461E+00 m^2/kg
1 acre/ton	4.461E-04 ha/kg
1 ft^2/lb	2.048E-02 ha/tonne
1 ft^2/lb	2.048E-01 m^2/kg
1 ft^2/lb	2.048E-05 ha/kg
1 ft^2/oz	3.277E-01 ha/tonne
1 ft^2/oz	3.277E+00 m^2/kg
1 ft^2/oz	3.277E-04 ha/kg

Area per unit time

1 acre/hr	4.047E-01 ha/hr
1 acre/hr	1.124E+00 m^2/s

Moment of inertia (second moment of area)

1 in^4	4.162E+05 mm^4
1 in^4	4.162E+01 cm^4

Egg weights

1 oz/dozen	2.363E+00 g/egg

Fan efficiency

1 cfm/watt	4.719E-01 m^3-s/kW
1 cfm/watt	4.719E-04 m^3-s/W

Thermal conductivity

1 Btu/hr-ft-F	1.731E+00 W/m-K
1 Btu-in/hr-ft^2-F	1.442E-01 W/m-K

Thermal conductance

1 Btu/hr-ft^2-F	5.678E+00 W/m^2-K

Thermal resistance

1 hr-ft^2-F/Btu	1.761E-01 m^2-K/W

Thermal flux, solar radiation

1 Btu/hr-ft^2	3.169E+00 W/m^2
1 langley.min	6.980E+02 W/m^2
1 langley	4.186E+04 J/m^2
1 langley	4.186E+01 kJ/m^2
1 Btu/ft^2	1.140E+04 J/m^2
1 Btu/ft^2	1.140E+01 kJ/m^2
1 Btu/ft^2	1.140E-02 MJ/m^2
1 MBtu/ft^2	1.140E+01 GJ/m^2

Light level, illumination

1 foot-candle	1.076E+01 Lux

Photosynthetically Active Radiation (PAR)

1 Btu/hr-ft^2 (solar)	6.565E+00 mmol/m^2-s (solar)
1 Btu/ft^2 (solar)	2.363E-02 mol/m^2 (solar)

Acceleration

1 ft/s^2	3.048E-01 m/s^2

Bending moment, torque

1 foot-lb	1.355E+00 N-m
1 inch-lb	1.130E-01 N-m
1 ounce-inch	7.060E-03 N-m

Water hardness

1 grain	1.710E+01 ppm

Fuel consumption

1 mile/gallon	4.251E-01 km/L

Heat value (as of fuels)

1 Btu/ft^3	3.726E+01 kJ/m^3
1 Btu/gal	2.785E-01 kJ/L
1 Btu/lb	2.326E+00 kJ/kg

Heat capacity

1 Btu/lb-F	4.187E+03 J/kg-K
1 Btu/lb-F	4.187E+00 kJ/kg-K
Btu/lb-F	0.000E+00 MJ/kg-K

Ventilation on a live-weight basis

1 cfm/lb	1.040E-03 m^3/s-kg
1 cfm/1000 lb	1.040E-06 m^3/s-kg
1 cfm/1000 lb	5.203E-04 m^3/s-500 kg

Unit cost

1 $/lb	2.205E+00 $/kg
1 $/cwt	2.205E-02 $/kg
1 $/ton	9.070E-01 $/tonne
1 $/ft	3.281E+00 $/m
1 $/yard	1.094E+00 $/m
1 $/ft^2	1.076E+01 $/m^2
1 $/yard^2	1.196E+00 $/m^2
1 $/acre	2.471E+00 $/ha
1 $/quart	1.056E+00 $/L
1 $/gallon	2.640E-01 $/L
1 $/ft^3	3.531E+01 $/m^3
1 $/yard^3	1.308E+00 $/m^3

Concentration, molar

	mg/kg

Other

centipoise (dynamic viscosity, m)	0.000E+00 mPa-s
centistokes (kinematic viscosity, n)	0.000E+00 mm^2/s

	centistokes (kinematic viscosity, n)	0.000E+00	m^2/s
1	clo	1.550E-01	m^2-K/W
1	decacards	1.000E+01	cards
1	$dyne/cm^2$	1.000E-01	Pa
1	EER	2.930E-01	COP
1	ft water/100 ft of pipe (pressure loss)	9.810E+01	Pa/m
1	ft water/100 ft of pipe (pressure loss)	9.810E-02	kPa/m
1	in. water/100 ft of duct (pressure loss)	8.175E+00	Pa/m
1	in. water/100 ft of duct (pressure loss)	8.175E-03	kPa/m

Table A-4 (Table 2.1) Physical Properties of Water

Temp. C	Density kg/m^3	Kinematic Viscosity (m^2/s)·E-06	Vapor Pressure mm Hg
0	999.84	1.79	4.8
1	999.90	1.73	5.1
2	999.94	1.68	5.4
3	999.97	1.62	5.8
4	1000.00	1.57	6.2
5	999.97	1.52	6.6
6	999.94	1.48	7.0
7	999.90	1.43	7.5
8	999.85	1.39	8.0
9	999.78	1.35	8.5
10	999.70	1.31	9.1
11	999.61	1.27	9.7
12	999.50	1.23	10.3
13	999.38	1.20	11.0
14	999.25	1.17	11.7
15	999.10	1.13	12.5
16	998.94	1.10	13.4
17	998.78	1.07	14.3
18	998.60	1.05	15.2
19	998.41	1.02	16.2
20	998.21	0.99	17.3
21	997.99	0.97	18.4
22	997.77	0.95	19.7
23	997.54	0.93	21.0
24	997.30	0.90	22.4
25	997.05	0.88	23.9
26	996.79	0.87	25.5
27	996.52	0.85	27.2
28	996.24	0.83	29.0
29	995.95	0.81	30.9
30	995.65	0.80	33.0
31	995.34	0.78	35.2
32	995.03	0.77	37.5
33	994.71	0.75	40.0
34	994.38	0.74	42.7
35	994.04	0.72	45.5
36	993.69	0.71	48.5
37	993.33	0.70	51.8
38	992.97	0.68	55.2
39	992.60	0.67	58.9
40	992.22	0.66	62.8

Table A-5 (Table 2.5) Percentage of Free Ammonia (as NH_3) in Freshwater at Varying pH and Water Temperature, (Spotte, 1979)

pH	10°C (50°F)	15°C (59°F)	20°C (68°F)	25°C (77°F)
7.0	0.19	0.27	0.40	0.55
7.1	0.23	0.34	0.50	0.70
7.2	0.29	0.43	0.63	0.88
7.3	0.37	0.54	0.79	1.10
7.4	0.47	0.68	0.99	1.38
7.5	0.59	0.85	1.24	1.73
7.6	0.74	1.07	1.56	2.17
7.7	0.92	1.35	1.96	2.72
7.8	1.16	1.69	2.45	3.39
7.9	1.46	2.12	3.06	4.24
8.0	1.83	2.65	3.83	5.28
8.1	2.29	3.32	4.77	6.55
8.2	2.86	4.14	5.94	8.11
8.3	3.58	5.16	7.36	10.00
8.4	4.46	6.41	9.09	12.27
8.5	5.55	7.98	11.18	14.97

Table A-6 (Table 2.3) Dissolved Oxygen (mg O_2 per Liter, ppm) at Saturation in Freshwater, Brackish Water, and Seawater at Different Temperatures

Temperature	Chlorinity (0/00)[1]					
(C)	0	4	8	12	16	20
1	14.24	13.54	12.91	12.29	11.70	11.15
2	13.84	13.18	12.56	11.98	11.40	10.86
3	13.45	12.84	12.25	11.68	11.12	10.59
4	13.09	12.51	11.93	11.38	10.83	10.34
5	12.75	12.18	11.63	11.09	10.57	10.10
6	12.44	11.86	11.33	10.82	10.32	9.86
7	12.13	11.58	11.06	10.56	10.07	9.63
8	11.85	11.29	10.80	10.32	9.84	9.40
9	11.56	11.02	10.54	10.07	9.61	9.20
10	11.29	10.77	10.30	9.84	9.40	9.00
11	11.05	10.53	10.07	9.63	9.20	8.80
12	10.80	10.29	9.84	9.41	9.00	8.61
13	10.56	10.07	9.63	9.21	8.81	8.42
14	10.33	9.86	9.41	9.01	8.62	8.25
15	10.10	9.64	9.23	8.83	8.44	8.09
16	9.89	9.44	9.03	8.64	8.28	7.94
17	9.67	9.26	8.85	8.47	8.11	7.78
18	9.47	9.07	8.67	8.31	7.97	7.64
19	9.28	8.88	8.50	8.15	7.08	7.49
20	9.11	8.70	8.32	7.99	7.66	7.36
21	8.93	8.54	8.17	7.84	7.52	7.23
22	8.75	8.38	8.02	7.69	7.39	7.11
23	8.60	8.22	7.87	7.55	7.26	6.99
24	8.44	8.07	7.72	7.42	7.13	6.86
25	8.27	7.92	7.58	7.29	7.01	6.75
26	8.12	7.78	7.45	7.16	6.89	6.63
27	7.98	7.64	7.32	7.03	6.78	6.52
28	7.84	7.51	7.19	6.92	6.66	6.40
29	7.69	7.38	7.08	6.82	6.55	6.29
30	7.56	7.25	6.96	6.70	6.45	6.19

[1]Conversion Chlorinity (CI) to Salinity (S) is: S = 1.80655 x CI
Normal seawater is 35 0/00 or 35 ppt S (19.37 ppt CI)

Table A-7 Hardness Conversion to Other Units of Measure

Units of Measure	mg/L $CaCO_3$	English gr/gal (Imperial) $CaCO_3$	American gr/gal (US) $CaCO_3$	French parts/ 100,000 $CaCO_3$	German parts/ 100,000 CaO	meq/L	g/L CaO	lb/cu ft $CaCO_3$
mg/L $CaCO_3$	1.0	0.07	0.058	0.1	0.056	0.02	$5.64x10^{-4}$	$6.23x10^{-5}$
English gr/gal (Imperial) $CaCO_3$	14.3	1.0	0.83	1.43	0.83	0.286	$8.0x10^{-3}$	$8.91x10^{-4}$
American gr/gal (US) $CaCO_3$	17.1	1.2	1.0	1.72	0.96	0.343	$9.66x10^{-3}$	$1.07x10^{-3}$
French parts/100,000 $CaCO_3$	10.0	0.7	0.58	1.0	0.56	0.2	$5.6x10^{-3}$	$6.23x10^{-4}$
German parts/100,000 CaO	17.9	1.25	1.04	1.79	1.0	0.358	$1.0x10^{-2}$	$1.12x10^{-3}$
meq/L	50	3.5	2.9	5.0	2.8	1.0	$2.8x10^{-2}$	$3.11x10^{-3}$
g/L CaO	1,790	125	104.2	179	100	35.8	1.0	0.112
lb/cu ft $CaCO_3$	16,100	1,123	935	1,610	900	321	9.0	1.0

Table A-8 (Table 3.7) Standard U.S. Atmospheric Pressure at Different Altitudes

Altitude		Atmospheric Pressure				
M	ft	mm Hg	kPa abs	psia	m H_2O	ft H_2O
-200	-656	778	103.7	15.0	10.6	34.8
0	0	760	101.3	14.7	10.3	33.8
200	656	742	98.9	14.3	10.1	33.1
400	1,312	725	96.6	14.0	9.9	32.5
600	1,968	707	94.3	13.7	9.6	31.5
800	2,625	690	92.0	13.3	9.4	30.8
1,000	3,281	674	89.8	13.0	9.2	30.2
4,000	13,123	462	61.6	8.9	6.3	20.7

Table A-9 (Table 2.6) Alkalinity Supplement Properties (Bisogni and Timmons, 1991)

Chemical Formula	Common Name(s)	Equivalent Wt. (gm/eq.)	Solubility	Rate of solubilization
NaOH	sodium hydroxide	40	high	rapid
Na_2CO_3	sodium carbonate soda ash	53	high	rapid
$NaHCO_3$	sodium bicarbonate baking soda	83	high	rapid
$CaCO_3$	calcium carbonate calcite	50	moderate	moderate
CaO	slaked lime	28	high	moderate
$Ca(OH)_2$	calcium hydroxide hydrated lime	37	high	moderate
$CaMg(CO_3)_2$	dolomite	46	moderate	slow
$MgCO_3$	magnesium carbonate magnesite	42	moderate	slow
$Mg(OH)_2$	magnesium hydroxide brucite	29	moderate	slow

Note: Na compounds are highly soluble in water while Mg compounds have poor solubility. Ca compounds are intermediate. Mg compounds tend to dissolve very slowly, so may have application in situations requiring a long-term application. Na compounds may prove to be the most expensive.

Based on 100% pure compound. To calculate for impurities, divide the tabulated value by the pure fraction ({100%-impurities %}/100 gives pure fraction) to get the true value.

Table A-10 (Table 7.6) Opening Sizes of U.S. Sieve Series Designation Number (Perry and Chilton, 1973)

Sieve Designation Number†	Size of Opening, (mm)	Sieve Designation Number†	Size of Opening (mm)
4	4.76	35	0.500
5	4.00	40	0.420
6	3.36	45	0.354
7	2.83	50	0.297
8	2.38	60	0.250
10	2.00	70	0.210
12	1.68	80	0.177
14	1.41	100	0.149
16	1.19	120	0.125
18	1.00	140	0.105
20	0.841	170	0.088
25	0.707	200	0.074
30	0.595	230	0.063

† Number of meshes per inch.

Table A-11 (Table 8.1) Dry Air Components

Species	% volume	% mass	Molecular Wt.
Nitrogen	78.084	75.600	28.0
Oxygen	20.946	23.200	32.0
CO_2	0.032	0.048	44.0
Argon	0.934	1.300	39.9
Air	100.000	100.000	29.0

Table A 12 (Table 8.2) Solubility of Four Major Gases in Water

Gas Species	Solubility in Air* (mg/L)	Solubility of pure gas* (mg/L)
Oxygen	10.08	48.14
Nitrogen	16.36	20.95
Argon	0.62	65.94
Carbon dioxide	0.69	1992.00

*At 15 Deg. C.

Table A-13 (Table 17.1) Maximum and Minimum Monthly Average Outside Temperatures for Selected USA Locations in Degrees °F (Kreider, J. F., & Kreith, F. ,1975, Solar Heating And Cooling: Engineering, Practical Design, And Economics, Washington, DC: Hemisphere Publishing)

Location	Max	Min		Location	Max	Min
Mobile, AL	82.6	53.0		Great Falls, MT	69.4	22.1
Phoenix, AZ	89.8	49.7		Omaha, NE	78.5	22.3
Little Rock, AR	81.9	40.6		Reno, NV	67.7	30.4
Los Angeles, CA	69.1	54.4		Concord, NH	69.6	21.2
Denver, CO	72.9	28.5		Atlantic City, NJ	75.1	34.7
Hartford, CT	73.4	26.0		Albuquerque, NM	78.5	35.0
Wilmington, DE	76.0	33.4		Buffalo, NY	69.8	24.1
Washington, DC	78.2	36.9		Raleigh, NC	77.9	41.6
Miami, FL	82.3	66.9		Bismark, ND	71.7	9.9
Atlanta, GA	78.9	44.7		Columbus, OH	74.8	29.9
Honolulu, HI	79.4	72.4		Oklahoma City, OK	82.8	37.0
Boise, ID	75.2	29.1		Portland, OR	67.2	38.4
Chicago, IL	75.6	26.0		Pittsburgh, PA	72.1	28.9
Indianapolis, IN	75.2	29.1		Providence, RI	72.1	29.2
Des Moines, IA	76.3	19.9		Columbia, SC	81.6	46.4
Wichita, KS	80.9	32.0		Sioux Falls, SD	74.3	15.2
Louisville, KY	77.6	35.0		Memphis, TN	81.3	41.5
New Orleans, LA	81.9	54.6		Dallas, TX	85.0	45.9
Portland, ME	68.1	21.8		Salt Lake City, UT	76.9	27.2
Baltimore, MD	76.8	34.8		Burlington, VT	69.0	16.2
Boston, MA	73.7	29.9		Richmond, VA	78.1	38.7
Detroit, MI	74.4	26.9		Seattle, WA	64.9	38.3
Minneapolis, MN	72.3	12.4		Charleston, WV	74.9	36.6
Jackson, MS	82.3	47.9		Milwaukee, WI	68.7	20.6
St. Louis, MO	78.1	31.9		Cheyenne, WY	70.0	25.4

Table A-14 (Table 17.2) Unit Area Thermal Resistance's (R-Values) of Typical Building Construction and Insulation Materials (ASHRAE, 1985, Handbook of Fundamentals. Atlanta, GA: American Society of Heating, Refrigerating, and Air-Conditioning Engineers)

Building Material	R-value °F-ft^2·hr/BTU (W/m^{2}K)
Plywood (Douglas Fir)	
0.25 inch thick	0.3 (0.05)
0.50 inch thick	0.6 (0.11)
Particleboard (Medium Density)	
0.25 inch thick	0.3 (0.05)
0.50 inch thick	0.5 (0.09)
Spruce-Pine-Fir – 1.75 inch thick	2.4 (0.42)
Cinder Block	
4 inch thick	1.1(0.19)
8 inch thick	1.7 (0.30)
12 inch thick	1.9 (0.33)
Asphalt shingles	0.4 (0.07)
Blanket and Batt Mineral Fiber Insulation	
4 inch thick	11.0 (1.94)
6.5 inch thick	19.0 (3.35)
7 inch thick	22.0 (3.87)
9 inch thick	30.0 (5.28)
12 inch thick	38.0 (6.69)
Expanded Polystyrene Insulation Board	
1 inch thick	3.4 (0.60)
2 inch thick	6.7 (1.18)
Convection Coefficients for Wall Surfaces[1]	
Winter time 15 mph (24 kph) wind	0.17 (0.029)
Summer time 7.5 mph (12 kph) wind	0.25 (0.044)
Still air (less than 100 ft/min or 0.5 m/s)	1.00 (0.176)

[1]Convection coefficients are the inverse of these R values;
to convert (°F-ft^2-hr/BTU) to W/m^{2}K divide by 5.678

Table A-15 (Table 17.4) Allowable Concentrations of Carbon Dioxide at Standard Temperature and Pressure to the Nearest 25 ppm (ASHRAE, 1991, Handbook - Applications. Atlanta, GA: American Society of Heating, Refrigerating, and Air-Conditioning Engineers)

Toxicity Standard	Allowable Concentration (ppm)
Immediately dangerous to life and health[a]	50,000
Acceptable ceiling concentration, not to be exceeded during an 8 hour shift[b]	30,000
Time-weighted average, not to be exceeded in any 8 hour shift of a 40 hour week[b]	5,000
Atmospheric concentration	350

[a]National Institute for Occupational Safety and Health standard
[b]Occupational Safety and Health Administration standard

TABLE A-16 PLASTIC PROPERTIES FOR VARIOUS TYPES OF PIPES

PVC

Polyvinyl Chloride, Type 1, Grade 1. This pipe is strong, rigid and resistant to a variety of acids and bases. Some solvents and chlorinated hydrocarbons may damage the pipe. PVC is very common, easy to work with and readily available at commercial vendors. Maximum useable temperature is 140°F (60°C) and pressure ratings start at a minimum of 125 to 200 psi (8 ATM to 13 ATM) (check for specific ratings stamped on the pipe). PVC can be used with water, gas, and drainage systems, but **NOT** for hot water or pressurized gases.

ABS

Acrylonitrite Butadiene Styrene, Type 1. This pipe is strong and rigid and resistant to a variety of acids and bases. Some solvents and chlorinated hydrocarbons may damage the pipe. ABS is very common, easy to work with and readily available from commercial vendors. Maximum useable temperature is 160°F (71°C) at low pressure. It is most commonly used as a DWV pipe.

CPVC

Chlorinated polyvinyl chloride. Similar to PVC, but designed specifically for piping water at up to 180°F (82°C), although it can withstand 200°F for a limited time. Pressure rating is 100 psi (7 ATM). Note diameter and fittings are not interchangeable with PVC pipe and fitting.

PE

Polyethylene. A flexible pipe for pressurized water systems, not for hot water.

PB

Polybutylene. A flexible pipe for pressurized water systems, both hot and cold. ONLY compression and banded type joints can be used.

Table A-17 Size and Properties of PVC Pipe Schedule 40 and 80

		PVC Schedule 40			PVC Schedule 80		
Nominal Size (in)	Actual OD (in)	Inside Diameter (in)	Wall Thickness (in)	Weight (lbs/ft)	Inside Diameter (in)	Wall Thickness (in)	Weight (lbs/ft)
¼	0.540				0.302	0.119	0.10
½	0.840	0.622	0.109	0.16	0.546	0.147	0.21
¾	1.050	0.824	0.113	0.22	0.742	0.154	0.28
1	1.315	1.049	0.133	0.32	0.957	0.179	0.40
1–1/4	1.660	1.380	0.140	0.43	1.278	0.191	0.57
1–1/2	1.900	1.610	0.145	0.52	1.500	0.200	0.69
2	2.375	2.067	0.154	0.70	1.939	0.218	0.95
2–1/2	2.875	2.469	0.203	1.10	2.323	0.276	1.45
3	3.500	3.068	0.216	1.44	2.900	0.300	1.94
4	4.500	4.026	0.237	2.05	3.826	0.337	2.83
6	6.625	6.065	0.280	3.61	5.761	0.432	5.41
8	8.625	7.981	0.322	5.45	7.625	0.500	8.22
10	10.750	10.020	0.365	7.91	9.564	0.593	12.28
12	12.750	11.938	0.406	10.35	11.373	0.687	17.10

Table A-18 (Table 3.3) PVC Pipe Friction Chart for Pipe and Length Loss for Schedule 40

GPM	Velocity (fps)	Head Loss (ft)	Head Loss (psi)	Velocity (fps)	Head Loss (ft)	Head Loss (psi)	Velocity (fps)	Head Loss (ft)	Head Loss (psi)	Velocity (fps)	Head Loss (ft)	Head Loss (psi)	Velocity (fps)	Head Loss (ft)	Head Loss (psi)	Velocity (fps)	Head Loss (ft)	Head Loss (psi)	Velocity (fps)	Head Loss (ft)	Head Loss (psi)
	1/2 inch			**3/4 inch**			**1 inch**			**1-1/4 inch**			**1-1/2 inch**			**2 inch**			**3 inch**		
1	1.13	2.08	0.90	0.63	0.51	0.22															
2	2.26	4.16	1.80	1.26	1.02	0.44	0.77	0.55	0.24	0.44	0.14	0.06									
5	5.64	23.44	10.15	3.16	5.73	2.48	1.93	1.72	0.75	1.11	0.44	0.19	0.81	0.22	0.09						
7	7.90	43.06	13.64	4.43	10.52	4.56	2.72	3.17	1.37	1.55	0.81	0.35	1.13	0.38	0.17						
10				6.32	20.04	8.68	3.86	6.02	2.61	2.21	1.55	0.67	1.62	0.72	0.31	0.98	0.21	0.09			
15				9.48	42.46	18.39	5.79	12.77	5.53	3.31	3.28	1.42	2.42	1.53	0.66	1.46	0.45	0.19			
20							7.72	21.75	9.42	4.42	5.59	2.42	3.23	2.61	1.13	1.95	0.76	0.33	0.88	0.11	0.05
25										5.52	8.45	3.66	4.04	3.95	1.71	2.44	1.15	0.50	1.10	0.17	0.07
30	**4 inch**									6.63	11.85	5.13	4.85	5.53	2.39	2.93	1.62	0.70	1.33	0.23	0.10
35										7.73	15.76	6.82	5.66	7.36	3.19	3.41	2.15	0.93	1.55	0.31	0.13
40	1.02	0.11	0.05							8.84	20.18	8.74	6.47	9.43	4.08	3.90	2.75	1.19	1.77	0.40	0.17
45	1.15	0.13	0.06	**5 inch**									7.27	11.73	5.08	4.39	3.43	1.49	1.99	0.50	0.22
50	1.28	0.16	0.07										8.08	14.25	6.17	4.88	4.16	1.80	2.21	0.60	0.26
60	1.53	0.22	0.10	0.97	0.07	0.03										5.85	5.84	2.53	2.65	0.85	0.37
70	1.79	0.30	0.13	1.14	0.10	0.04	**6 inch**									6.83	7.76	3.36	3.09	1.13	0.49
75	1.92	0.34	0.15	1.22	0.11	0.05										7.32	8.82	3.82	3.31	1.28	0.55
80	2.05	0.38	0.16	1.30	0.13	0.06	0.90	0.05	0.02							7.80	9.94	4.30	3.53	1.44	0.62
90	2.30	0.47	0.20	1.46	0.16	0.07	1.01	0.06	0.03							8.78	12.37	5.36	3.98	1.80	1.78
100	2.56	0.58	0.25	1.62	0.19	0.08	1.12	0.08	0.04	**8 inch**						9.75	15.03	6.51	4.42	2.18	1.94
125	3.20	0.88	0.38	2.03	0.29	0.13	1.41	0.12	0.05										5.52	3.31	1.43
150	3.84	1.22	0.53	2.44	0.40	0.17	1.69	0.16	0.07	0.97	0.04	0.02							6.63	4.63	2.00
175	4.48	1.63	0.71	2.84	0.54	0.24	1.97	0.22	0.10	1.14	0.06	0.02	**10 inch**						7.73	6.16	2.67
200	5.11	2.08	0.90	3.25	0.69	0.30	2.25	0.28	0.12	1.30	0.07	0.03							8.83	7.88	3.41
250	6.40	3.15	1.36	4.06	1.05	0.45	2.81	0.43	0.19	1.63	0.11	0.05	1.03	0.04	0.02	**12 inch**					
300	7.67	4.41	1.91	4.87	1.46	0.63	3.37	0.60	0.26	1.94	0.16	0.07	1.23	0.05	0.02						
350	8.95	5.87	2.55	5.69	1.95	0.85	3.94	0.79	0.34	2.27	0.21	0.09	1.44	0.07	0.03	1.01	0.03	0.01			
400	10.23	7.52	3.26	6.50	2.49	1.08	4.49	1.01	0.44	2.59	0.27	0.12	1.64	0.09	0.04	1.16	0.04	0.02			
450				7.31	3.09	1.34	5.06	1.26	0.55	2.92	0.33	0.14	1.85	0.11	0.05	1.30	0.05	0.02			
500				8.12	3.76	1.63	5.62	1.53	0.66	3.24	0.40	0.17	2.05	0.13	0.06	1.45	0.06	0.03			
750							8.43	3.25	1.41	4.86	0.85	0.37	3.08	0.28	0.12	2.17	0.12	0.05			
1000							11.24	5.54	2.40	6.48	1.45	0.63	4.11	0.48	0.21	2.89	0.20	0.09			
1250										8.11	2.20	0.95	5.14	0.73	0.32	3.62	0.31	0.13			
1500										9.72	3.07	1.33	6.16	1.01	0.44	4.34	0.43	0.19			
2000													8.21	1.72	0.74	5.78	0.73	0.32			
2500													10.27	2.61	1.13	7.23	1.11	0.49			
3000																8.68	1.55	0.67			

(Shaded area recommended flow rates to minimize settlement of solids and avoid scouring of walls and junctions)

Table A-19 PVC Pipe Friction Chart for Pipe and Length Loss for Schedule 80

GPM	Velocity (fps)	Head Loss (ft)	Head Loss (psi)	Velocity (fps)	Head Loss (ft)	Head Loss (psi)	Velocity (fps)	Head Loss (ft)	Head Loss (psi)	Velocity (fps)	Head Loss (ft)	Head Loss (psi)	Velocity (fps)	Head Loss (ft)	Head Loss (psi)	Velocity (fps)	Head Loss (ft)	Head Loss (psi)	Velocity (fps)	Head Loss (ft)	Head Loss (psi)
	1/2 inch			**3/4 inch**			**1 inch**			**1-1/4 inch**			**1-1/2 inch**			**2 inch**			**3 inch**		
1	1.48	4.02	1.74	0.74	0.86	0.37															
2	2.95	8.03	3.48	1.57	1.72	0.74	0.94	0.88	0.38	0.52	0.21	0.09	0.38	0.10	0.04						
5	7.39	45.23	19.59	3.92	9.67	4.19	2.34	2.75	1.19	1.30	0.66	0.29	0.94	0.30	0.13	0.56	0.10	0.04			
7	10.34	83.07	35.97	5.49	17.76	7.69	3.28	5.04	2.19	1.82	1.21	0.53	1.32	0.55	0.24	0.78	0.15	0.07			
10				7.84	33.84	14.65	4.68	9.61	4.16	2.60	2.30	1.00	1.88	1.04	1.45	1.12	0.29	0.13			
15				11.76	71.70	31.05	7.01	20.36	8.82	3.90	4.87	2.11	2.81	2.20	0.95	1.68	0.62	0.27			
20							9.35	34.68	15.02	5.20	8.30	3.59	3.75	3.75	1.62	2.23	1.06	0.46	1.00	0.15	0.07
25							11.69	25.43	22.70	6.50	12.55	5.43	4.69	5.67	2.46	2.79	1.60	0.69	1.25	0.22	0.10
30	**4 inch**									7.80	17.59	7.62	5.63	7.95	3.44	3.35	2.25	0.97	1.49	0.31	0.13
35										9.10	23.40	10.13	6.57	10.58	4.58	3.91	2.99	1.29	1.74	0.42	0.18
40	1.15	0.14	0.06							10.40	29.97	12.98	7.50	13.55	5.87	4.47	3.83	1.66	1.99	0.54	0.23
45	1.29	0.17	0.07	**5 inch**						11.70	37.27	16.14	8.44	16.85	7.30	5.03	4.76	2.07	2.24	0.67	0.29
50	1.43	0.21	0.09							13.00	45.30	19.61	9.38	20.48	8.87	5.58	5.79	2.51	2.49	0.81	0.35
60	1.72	0.30	0.13	1.08	0.10	0.04							11.26	28.70	12.43	6.70	8.12	3.52	2.99	1.14	0.49
70	2.01	0.39	0.17	1.26	0.13	0.06	**6 inch**									7.82	10.80	4.68	3.49	1.51	0.65
75	2.15	0.45	0.19	1.35	0.14	0.06										8.38	12.27	5.31	3.74	1.72	0.74
80	2.29	0.50	0.22	1.44	0.16	0.07	1.00	0.07	0.03							8.93	13.83	5.99	3.99	1.94	0.84
90	2.58	0.63	0.27	1.62	0.20	0.09	1.13	0.08	0.04							10.05	17.20	7.45	4.48	2.41	1.04
100	2.87	0.76	0.33	1.80	0.24	0.10	1.25	0.10	0.05	**8 inch**						11.17	20.90	9.05	4.98	2.93	1.27
125	3.59	1.16	0.50	2.25	0.37	0.16	1.57	0.16	0.07										6.23	4.43	1.92
150	4.30	1.61	0.70	2.70	0.52	0.23	1.88	0.22	0.10	1.07	0.05	0.02							7.47	6.20	2.68
175	5.02	2.15	0.93	3.15	0.69	0.30	2.20	0.29	0.12	1.25	0.08	0.03	**10 inch**						8.72	8.26	3.58
200	5.73	2.75	1.19	3.60	0.88	0.38	2.51	0.37	0.16	1.43	0.09	0.04							9.97	10.57	4.58
250	7.16	4.16	1.81	4.50	1.34	0.58	3.14	0.56	0.24	1.79	0.14	0.06	1.14	0.05	0.02	**12 inch**					
300	8.60	5.83	2.52	5.40	1.87	0.81	3.76	0.78	0.34	2.14	0.20	0.09	1.36	0.07	0.03						
350	10.03	7.76	3.36	6.30	2.49	1.08	4.39	1.04	0.45	2.50	0.27	0.12	1.59	0.09	0.04	1.12	0.04	0.02			
400	11.47	9.93	4.30	7.19	3.19	1.38	5.02	1.33	0.58	2.86	0.34	0.15	1.81	0.11	0.05	1.28	0.05	0.02			
450				8.09	3.97	1.72	5.64	1.65	0.71	3.21	0.42	0.18	2.04	0.14	0.06	1.44	0.06	0.03			
500				8.99	4.82	2.09	6.27	2.00	0.87	3.57	0.51	0.22	2.27	0.17	0.07	1.60	0.07	0.03			
750							9.40	4.25	1.84	5.36	1.08	0.47	3.40	0.36	0.16	2.40	0.15	0.07			
1000							12.54	7.23	3.13	7.14	1.84	0.80	4.54	0.61	0.26	3.20	0.26	0.11			
1250										8.93	2.78	1.20	5.67	0.92	0.40	4.01	0.40	0.17			
1500										10.71	3.89	1.68	6.8	1.29	0.56	4.81	0.55	0.24			
2000													9.07	2.19	0.95	6.41	0.94	0.41			
2500													11.34	3.33	1.44	8.01	1.42	0.62			
3000																9.61	1.99	0.86			

(Shaded area recommended flow rates to minimize settlement of solids and avoid scouring of walls and junctions)

Table A-20 Tank Volumes for Various Depths and Diameters

Diameter	Depth	Volume		Diameter	Depth	Volume	
ft	ft	gallons	m^3	m	m	gallons	m^3
3	1	53	0.2	1	0.25	52	0.2
	2	106	0.4		0.5	104	0.4
	3	159	0.6		0.75	156	0.6
4	1	94	0.4	1.5	0.25	52	0.2
	2	188	0.7		0.5	104	0.4
	3	283	1.1		0.75	156	0.6
6	2	424	1.6	2	0.5	104	0.4
	3	636	2.4		0.75	156	0.6
	4	848	3.2		1	208	0.8
8	2	753	2.9	2.5	0.5	104	0.4
	3	1130	4.3		0.75	156	0.6
	4	1507	5.7		1	208	0.8
10	3	1766	6.7	3	0.75	156	0.6
	4	2354	8.9		1	208	0.8
	5	2943	11.1		1.25	260	1.0
12	3	2543	9.6	4	0.75	156	0.6
	4	3390	12.8		1	208	0.8
	5	4238	16.0		1.25	260	1.0
15	3	3973	15.0	5	0.75	156	0.6
	4	5297	20.1		1	208	0.8
	5	6622	25.1		1.25	260	1.0
20	3	7063	26.7	6	0.75	156	0.6
	4	9418	35.6		1	208	0.8
	5	11772	44.6		1.25	260	1.0
30	4	21190	80	10	4	832	3.1
	6	31784	120		6	1248	4.7
	8	42379	160		8	1663	6.3
36	4	30513	115	12	4	832	3.1
	6	45769	173		6	1248	4.7
	8	61026	231		8	1663	6.3
42	4	41531	157	13.5	4	832	3.1
	6	62297	236		6	1248	4.7
	8	83063	314		8	1663	6.3
48	4	54245	205	15	4	832	3.1
	6	81368	308		6	1248	4.7
	8	108490	411		8	1663	6.3

To calculate the Volume of a Tank:

$$Volume(\,ft^3\,) = \frac{\pi D^2}{4} \cdot Depth$$

where Depth — Tank depth, ft
D — Tank diameter, ft

Conversions: 1 gal=7.481 ft^3; 1m^3=264.2 gal

Table A-21 Temperature Equivalents Between Celsius and Fahrenheit

C	F	C	F	C	F	C	F
0.0	32.0	**10.0**	50.0	**20.0**	68.0	**30.0**	86.0
0.6	33.0	10.6	51.0	20.6	69.0	30.6	87.0
1.0	33.8	**11.0**	51.8	**21.0**	69.8	**31.0**	87.8
1.1	34.0	11.1	52.0	21.1	70.0	31.1	88.0
1.7	35.0	11.7	53.0	21.7	71.0	31.7	89.0
2.0	35.6	**12.0**	53.6	**22.0**	71.6	**32.0**	89.6
2.2	36.0	12.2	54.0	22.2	72.0	32.2	90.0
2.8	37.0	12.8	55.0	22.8	73.0	32.8	91.0
3.0	37.4	**13.0**	55.4	**23.0**	73.4	**33.0**	91.4
3.3	38.0	13.3	56.0	23.3	74.0	33.3	92.0
3.9	39.0	13.9	57.0	23.9	75.0	33.9	93.0
4.0	39.2	**14.0**	57.2	**24.0**	75.2	**34.0**	93.2
4.4	40.0	14.4	58.0	24.4	76.0	34.4	94.0
5.0	41.0	**15.0**	59.0	**25.0**	77.0	**35.0**	95.0
5.6	42.0	15.6	60.0	25.6	78.0	35.6	96.0
6.0	42.8	**16.0**	60.8	**26.0**	78.8	**36.0**	96.8
6.1	43.0	16.1	61.0	26.1	79.0	36.1	97.0
6.7	44.0	16.7	62.0	26.7	80.0	36.7	98.0
7.0	44.6	**17.0**	62.6	**27.0**	80.6	**37.0**	98.6
7.2	45.0	17.2	63.0	27.2	81.0	37.2	99.0
7.8	46.0	17.8	64.0	27.8	82.0	37.8	100.0
8.0	46.4	**18.0**	64.4	**28.0**	82.4		
8.3	47.0	18.3	65.0	28.3	83.0		
8.9	48.0	18.9	66.0	28.9	84.0		
9.0	48.2	**19.0**	66.2	**29.0**	84.2		
9.4	49.0	19.4	67.0	29.4	85.0		

Table A-22 Area Under the Normal Distribution Curve

Z	0.00	0.01	0.02	0.03	0.04	0.05	0.06	0.07	0.08	0.09
0	.5000	.5040	.5080	.5120	.5160	.5199	.5239	.5279	.5319	.5359
0.1	.5398	.5438	.5437	.5517	.5557	.5596	.5636	.5675	.5714	.5753
0.2	.5793	.5832	.5871	.5910	.5948	.5987	.6026	.6064	.6103	.6141
0.3	.6179	.6217	.6255	.6293	.6331	.6368	.6406	.6443	.6480	.6517
0.4	.6554	.6591	.6628	.6664	.6700	.6736	.6772	.6808	.6844	.6879
0.5	.6915	.6950	.6985	.7019	.7054	.7088	.7123	.7157	.7190	.7224
0.6	.7257	.7291	.7324	.7357	.7389	.7422	.7454	.7486	.7517	.7549
0.7	.7580	.7611	.7642	.7673	.7704	.7734	.7764	.7794	.7823	.7852
0.8	.7881	.7910	.7939	.7967	.7995	.8023	.8051	.8079	.8106	.8133
0.9	.8159	.8186	.8212	.8238	.8264	.8289	.8315	.8340	.8365	.8389
1.0	.8413	.8438	.8461	.8485	.8508	.8531	.8554	.8577	.8599	.8621
1.1	.8643	.8665	.8686	.8708	.8729	.8749	.8770	.8790	.8810	.8830
1.2	.8849	.8869	.8888	.8907	.8925	.8944	.8962	.8980	.8997	.9015
1.3	.9032	.9049	.9066	.9082	.9099	.9115	.9131	.9147	.9162	.9177
1.4	.9192	.9207	.9222	.9236	.9251	.9265	.9279	.9292	.9306	.9319
1.5	.9332	.9345	.9357	.9370	.9382	.9394	.9406	.9418	.9429	.9441
1.6	.9452	.9463	.9474	.9484	.9495	.9505	.9515	.9525	.9535	.9545
1.7	.9554	.9564	.9573	.9582	.9591	.9599	.9608	.9616	.9625	.9633
1.8	.9641	.9649	.9658	.9664	.9671	.9678	.9686	.9693	.9699	.9706
1.9	.9713	.9719	.9726	.9732	.9738	.9744	.9750	.9756	.9761	.9767
2.0	.9773	.9778	.9783	.9788	.9793	.9798	.9803	.9808	.9812	.9817
2.1	.9821	.9826	.9830	.9834	.9838	.9842	.9846	.9850	.9854	.9857
2.2	.9861	.9864	.9868	.9871	.9875	.9878	.9881	.9884	.9887	.9890
2.3	.9893	.9896	.9898	.9901	.9904	.9906	.9909	.9911	.9913	.9916
2.4	.9918	.9920	.9922	.9925	.9927	.9929	.9931	.9932	.9934	.9936
2.5	.9938	.9940	.9941	.9943	.9945	.9946	.9948	.9949	.9951	.9952
2.6	.9953	.9955	.9956	.9957	.9959	.9960	.9961	.9962	.9963	.9964
2.7	.9965	.9966	.9967	.9968	.9969	.9970	.0971	.9972	.9973	.9974
2.8	.9974	.9975	.9976	.9977	.9977	.9978	.9979	.9979	.9980	.9981
2.9	.9981	.9982	.9983	.9983	.9983	.9984	.9985	.9985	.9986	.9986
3.0	.99865	.99869	.99874	.99878	.99882	.99886	.99889	.99893	.99896	.99900
3.1	.99903	.99906	.99910	.99913	.99915	.99918	.99921	.99924	.99926	.99929
3.2	.99931	.99934	.99936	.99938	.99940	.99942	.99944	.99946	.99948	.99950
3.3	.99952	.99953	.99955	.99957	.99958	.99960	.99961	.99962	.99964	.99965
3.4	.99966	.99967	.99969	.99970	.99971	.99972	.99973	.99974	.99975	.99976
3.5	.99977	.99978	.99978	.99979	.99980	.99981	.99981	.99982	.99983	.99983

Z	0.00	0.01	0.02	0.03	0.04	0.05	0.06	0.07	0.08	0.09
-3.5	.00023	.00022	.00022	.00021	.00020	.00019	.00019	.00018	.00017	.00017
-3.4	.00034	.00033	.00031	.00030	.00029	.00028	.00027	.00026	.00025	.00024
-3.3	.00048	.00047	.00045	.00043	.00042	.00040	.00039	.00038	.00036	.00035
-3.2	.00069	.00066	.00064	.00062	.00060	.00058	.00056	.00054	.00052	.00050
-3.1	.00097	.00094	.00090	.00087	.00085	.00082	.00079	.00076	.00074	.00071
-3.0	.00135	.00131	.00126	.00122	.00118	.00114	.00111	.00107	.00104	.00010
-2.9	.0019	.0018	.0017	.0017	.0016	.0016	.0015	.0015	.0014	.0014
-2.8	.0026	.0025	.0024	.0023	.0023	.0022	.0021	.0021	.0020	.0019
-2.7	.0035	.0034	.0033	.0032	.0031	.0030	.0029	.0028	.0027	.0026
-2.6	.0047	.0045	.0044	.0043	.0041	.0040	.0039	.0038	.0037	.0036
-2.5	.0062	.0060	.0059	.0057	.0055	.0054	.0052	.0051	.0049	.0048
-2.4	.0082	.0080	.0078	.0075	.0073	.0071	.0069	.0068	.0066	.0064
-2.3	.0107	.0104	.0102	.0099	.0096	.0094	.0091	.0089	.0087	.0084
-2.2	.0139	.0136	.0132	.0129	.0125	.0122	.0119	.0116	.0113	.0110
-2.1	.0179	.0174	.0170	.0166	.0162	.0158	.0154	.0150	.0146	.0143
-2.0	.0228	.0222	.0217	.0212	.0207	.0202	.0197	.0192	.0188	.0183
-1.9	.0287	.0281	.0274	.0268	.0262	.0256	.0250	.0244	0.239	.0233
-1.8	.0359	.0351	.0344	.0336	.0329	.0322	.0314	.0307	.0301	.0294
-1.7	.0446	.0436	.0427	.0418	.0409	.0401	.0392	.0384	.0375	.0367
-1.6	.0548	.0537	.0526	.0516	.0505	.0495	.0485	.0475	.0465	.0455
-1.5	.0668	.0655	.0643	.0630	.0618	.0606	.0594	.0582	.0571	.0559
-1.4	.0808	.0793	.0778	.0764	.0749	.0735	.0721	.0708	.0694	.0681
-1.3	.0968	.0951	.0934	.0918	.0901	.0885	.0869	.0853	.0838	.0823
-1.2	.1151	.1131	.1112	.1093	.1075	.1057	.1038	.1020	.1003	.0985
-1.1	.1357	.1335	.1314	.1292	.1271	.1251	.1230	.1210	.1190	.1170
-1.0	.1587	.1562	.1539	.1515	.1492	.1469	.1446	.1423	.1401	.1379
-0.9	.1841	.1814	.1788	.1762	.1736	.1711	.1685	.1660	.1635	.1611
-0.8	.2119	.2090	.2061	.2033	.2005	.1977	.1949	.1922	.1894	.1867
-0.7	.2420	.2389	.2358	.2327	.2297	.2266	.2236	.2207	.2177	.2148
-0.6	.2743	.2709	.2676	.2643	.2611	.2578	.2546	.2514	.2483	.2451
-0.5	.3085	.3050	.3015	.2981	.2946	.2912	.2877	.2843	.2810	.2776
-0.4	.3446	.3409	.3372	.3336	.3300	.3264	.3228	.3192	.3156	.3121
-0.3	.3821	.3783	.3745	.3707	.3669	.3632	.3594	.3557	.3520	.3483
-0.2	.4207	.4168	.4129	.4090	.4052	.4013	.3974	.3936	.3897	.3859
-0.1	.4602	.4562	.4522	.4483	.4443	.4404	.4364	.4325	.4286	.4247
-0	.5000	.4960	.4920	.4880	.4840	.4801	.4761	.4721	.4681	.4641

Table A-23 Percentage Points of the Students t-Distribution

Degrees of Freedom	Percentage Points of Student's t-Distribution (value is area in one-tail of distribution curve)						
	0.250	0.200	0.100	0.050	0.025	0.010	0.005
1	2.414	3.078	6.314	12.706	25.452	63.656	127.321
2	1.604	1.886	2.920	4.303	6.205	9.925	14.089
5	1.301	1.476	2.015	2.571	3.163	4.032	4.773
10	1.221	1.372	1.812	2.228	2.634	3.169	3.581
20	1.185	1.325	1.725	2.086	2.423	2.845	3.153
30	1.173	1.310	1.697	2.042	2.360	2.750	3.030
40	1.167	1.303	1.684	2.021	2.329	2.704	2.971
50	1.164	1.299	1.676	2.009	2.311	2.678	2.937
60	1.162	1.296	1.671	2.000	2.299	2.660	2.915
70	1.160	1.294	1.667	1.994	2.291	2.648	2.899
80	1.159	1.292	1.664	1.990	2.284	2.639	2.887
90	1.158	1.291	1.662	1.987	2.280	2.632	2.878
100	1.157	1.290	1.660	1.984	2.276	2.626	2.871
200	1.154	1.286	1.653	1.972	2.258	2.601	2.838
500	1.152	1.283	1.648	1.965	2.248	2.586	2.820
1,000	1.151	1.282	1.646	1.962	2.245	2.581	2.813
100,000	1.150	1.282	1.645	1.960	2.241	2.576	2.807
1.00E+09	1.150	1.282	1.645	1.960	2.241	2.576	2.807

GLOSSARY

Calculating Volume of Tanks

Volume measurements are required to determine the proper concentration of chemicals and to calculate holding and transport densities.

Rectangular Tank Volume:

Volume = length · width · depth or L · W · H

When measuring a tank, take inside measurements of length and with and the depth at the appropriate water level. If a standpipe or other overflow drain is present, then the height of the standpipe is the correct depth measurement. If the bottom of the tank is sloped toward the drain, an average depth measurement should be determined. To find the average depth, take measurements at the two ends and in the middle, add together and divide by 3.

To convert from cubic feet or cubic inches, the conversion table included in the appendix can be used.

Circular Tank Volume:

Circular tank volume is determined by the formula:

Volume = 3.14 · $(\text{radius})^2$ · depth or $\pi\, r^2 \cdot H$

The radius is measured as ½ of the inside diameter of the tank, at its base. The radius is squared or multiplied by itself.

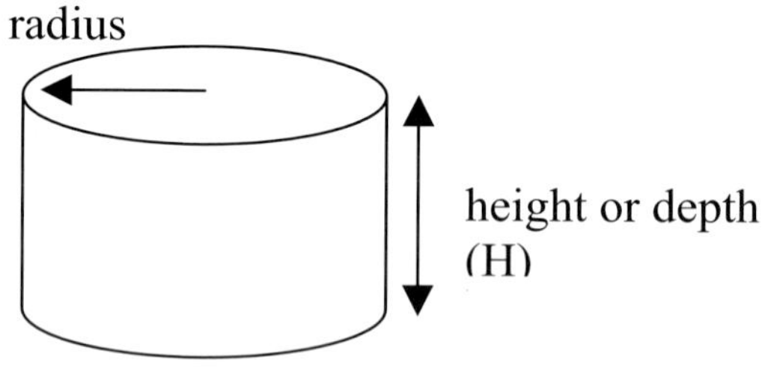

Useful conversions:

1 gal = 7.481 ft^3	1 ft^3 = 0.1337 gal
1 m^3 = 264.2 gal	1 gal = 0.0038 m^3
1 ft^3 = 1,728 in^3	

DEFINITION OF TERMS USED IN AQUACULTURE

Absolute growth - This is growth measured in grams/day.

Acclimation - The process of metabolic compensation in response to change in an environmental factor.

Aerobic - This refers to chemical or biological processes which take place in the presence of oxygen.

Allele - One of two or more forms of a given gene.

Ammonia - NH_3. Sometimes this term is loosely used to refer to both forms of total ammonia present in the water (NH_3 and NH_4^+).

Ammonia volatization - This is the loss of ammonia (NH_3) directly into the atmosphere resulting from aeration (called Air stripping). Ammonia is volatilized more readily at high pH values.

Ammonia-nitrogen - This is the amount of ammonia expressed in terms of nitrogen concentration. The molecular weight of ammonia (NH_3) is 17 and the molecular weight of nitrogen is 14. Therefore, 1 mg/L of ammonia-nitrogen = 1 x 17/14 or 1.2 mg/L of ammonia.

Ammonification - This is an aerobic or anaerobic heterotrophic process that deaminates proteins and produces ammonia-nitrogen.

Ammonium - NH_4^+. This is the predominant form of the ammonia equilibrium.

Anaerobic - A chemical or biological process that takes place in the absence of oxygen.

Aquaculture - The controlled production of a crop in water.

Autosome - A chromosome other than the sex chromosome.

Backcross - Crossing an F1 hybrid heterozygote back to one of the parental types.

Button-up - Fingerling closes yolk sac.

Broodfish - Fish used to produce fish that will be cultured.

Carrying capacity - Biomass in a given area where growth stops (lbs/ft^3, kg/m^3).

Clarification - The process of removing suspended solids from water.

Cohort - A group of fish of the same age or size stocked in the system at the same time.

Critical standing crop - The inflection point on a growth curve indicating the presence of a growth limiting factor.

Denitrification - This is the anaerobic process by which chemotropic bacteria convert nitrates (NO_3^-) to nitrogen gas (N_2), N_2O or ammonia (NH_3).

Digestible Carbohydrate- A value that indicates the utilizable amount of carbohydrates. For salmonids the following are digestibility values for different types of carbohydrates: simple sugars 100%; complex sugars 90%, cooked starch 60%; raw starch 30%; fiber 0%. (Piper et al. 1989, 219)

FDA - United States Food and Drug Administration

Feed conversion ratio - This is the ratio of the amount of feed fed to the amount of weight gained fro a group of fish.

Flashing - Fish rolling over upside-down and exposing their ventral side. Fish swimming at the surface appearing to gasp for air.

Genotype - Genetic determination of sex.

Grading - The process of sorting fish by size.
Gross yield - Total weight of fish harvested for a volume of water.
Gynogenesis - Producing all female fish.
Hapa - A net enclosure used to hold fish that allows water circulation. It is usually suspended in a tank or pond.
Harvest - The removal of harvestable sized fish or the removal of fish from any stage of the production cycle.
Haploid - Half the normal number of chromosomes.
Heterogametic - Fish whose sex determining alleles are different, i.e., XY or WY.
Homogametic - Fish whose sex determining alleles are the same, i.e., XX or WW.
Hybrid vigor - Exceptional growth ability of the first generation (F1) resulting from a cross of two different species.
Low Regulatory Priority (LRP) - Drugs with this status are not officially approved by the FDA. The FDA is "unlikely to object" to these compounds if used under the exact conditions specified. The FDA has not made any statements regarding the safety or efficacy of these products for the purpose under which they are Low Regulatory Priority.
Meiosis - The process by which chromosomes replicate, form homologous pairs, and then segregate into different nuclei to produce haploid cells.
Mortality rate - This is the calculation of the number of fish that die versus the number of fish stocked for any part (or all) of the production cycle.
Mitosis - A process of cell division in which the chromosomes replicate and divide equally so that identical daughter cells are produced.
Net yield - Total weight harvested minus total weight stocked for a volume of water (lbs/ft^3, kg/m^3).
Necropsy - A postmortem internal and external examination of fish.
Nitrate-nitrogen - This is the nitrite concentration expressed in terms of its equivalent nitrogen concentration. The molecular weight of nitrite is 62 and the molecular weight of nitrogen is 14. Therefore, 1 mg/L of nitrite-nitrogen = 1 x (62/14) or 4.4 mg/L of nitrate.
Nitrification - The aerobic process by which bacteria convert ammonium (NH_4^+) to nitrates ($NO_3^=$). Nitrification is most rapid at pH of 7–8 and at temperatures of 25–30°C. Nitrification causes waters to decrease in pH.
Nitrobacter - The nitrifying bacteria responsible for converting nitrites (NO_2^-) to nitrates ($NO_3^=$).
Nitrosomonas - The nitrifying bacteria responsible for converting ammonium (NH_4^+) to nitrites (NO_2^-).

Osmotic regulation - This is a process by which freshwater fish are able to excrete excess water yet retain ions and substances vital to normal function. There is a constant flow of water into their bodies caused by the difference in ion concentrations between the inside and outside of their body. Saltwater fish have the opposite problem. They are continually losing water from their bodies and must work to retain water needed for normal function.

Phenotype - Secondary characteristics (not genetic makeup) that indicate sex, i.e., genitals, size, color.

Polishing filter - These are secondary filters that remove fine solid materials from water that has already passed through a primary filter (ozone, ultraviolet filters, cartridge filter, drum filter, foam fractionator).

Polyploidy - Other than the normal number of chromosomes, i.e., triploids, tetraploids.

Purging - A process that removes any "off flavor" in harvestable sized fish by holding fish unfed in clean water for a few days.

Relative fecundity - This is a measure of egg production from broodfish. It is the number of eggs received per spawn for each kilogram of broodfish weight.

Relative growth - Growth measured as a percent of the fishs body weight.

Sac Fry - Early stage fingerlings, still on egg sac.

Settleable solids - This is the volume or weight of solid material that will settle in one hour from one liter of water confined in a cone shaped cylinder called an Imhoff cone.

Sex chromosome - A chromosome, usually X or Y that plays a major role in the determination of sex.

Species - A population or populations of individuals that share a common gene pool, are phenotypically similar, and are reproductively isolated from other species.

Standing crop - Biomass in a given volume of water at a particular point in time (lbs/ft^3, kg/m^3).

Stress - A physiological response caused by an external stimulus (stressor) that results in an energetic cost to the organism.

Supermale - A male carrying only male determining genetic information (YY).

Survival rate - The calculation of the number of fish that remains the number of fish stocked for any part or all of the production cycle.

System - A system refers to a complete aquaculture production unit that includes: production tanks, filtration and circulation.

Tetraploids - Double the normal chromosome number.

Total dissolved solids - This is the total residue (mg/L) left after evaporation of a filtered water sample.

Toxic ammonia - This refers to ammonia (NH_3), which is lethal to fish at fairly low concentrations.

Triploid - 1.5 the normal number of chromosomes.

Laboratory Safety Procedures

(Adapted from MCRA Operators Manual, Freshwater Institute, 2000)

Knowing and utilizing safe research procedures is your responsibility!

This section provides an overview of safety practices that one is expected to follow when conducting laboratory activities. Providing specific information on every potential safety hazard associated is impossible; however, you should customize this document to reflect precautions necessary for your specific facility.

Emergency Evacuation Procedure

The emergency evacuation procedure outlined in the following paragraphs is applicable to any life-threatening situation you may face at your workplace. For example, fire, extremely hazardous laboratory chemical spills, major pesticide spills, major fuel spills, or propane gas leaks.

It is critical to know the location of **ALL** fire alarms and fire extinguishers throughout your building and work area. When a fire is discovered, be sure to alert all coworkers. If an extinguisher is readily available and the fire can be easily controlled in this manner, do so! If not, pull the nearest fire alarm and evacuate the building through the nearest exit. All employees should exit, meet at a predetermined location, and report immediately to his/her supervisor. **DO NOT** reenter the building until directed to do so by the Station Safety Officer, Director, or maintenance staff.

If time permits after the alarm is sounded, secure any hazardous materials or research procedures you may be involved with, turn off all heat or open flames, and close all doors. If you are disabled, be sure to inform your supervisor so that assistance can be provided upon time of evacuation.

Emergency Telephone Numbers

This is an abbreviated list. You should also have numbers for the facility manager, electric company, gas supplier, and other important contacts posted in central locations.

Local Fire, Ambulance, Police	**9-1-1**
National Poison Control Center	**1-800-222-1222**
Chemtrec (chemical spills)	**1-800-424-9300**

ELECTRICAL POWER FAILURE (BLACK OUT)

Fume hoods are an important part of the fresh air circulation systems in laboratory buildings. During a black out, fresh air supply is interrupted, foul air remains in the building, and health hazards may persist. If a black out occurs while you are working with a toxic chemical under the fume hood, cap the container, close the hood sash, and contact your supervisor or director of the laboratory.

EMERGENCY RESPONSE EQUIPMENT

- Know fire alarm and fire extinguisher locations
- Know how to use an extinguisher
- Know which extinguishers are made for which types of fires
- Know where emergency showers and eye wash stations are available—any water source may be used to flush the eyes, i.e., drinking fountains, water taps, etc.
- Know where First-Aid kits are located and make sure items are kept up-to-date
- Never be embarrassed to use emergency equipment; they are provided to keep each employee safe and are intended to keep serious problems to a minimum

BASIC RULES FOR LABORATORY SAFETY

- Keep laboratory aisles and doorways free of boxes, carts, equipment, etc.
- Keep cabinet doors closed and equipment away from edge of lab benches
- Do not store glassware, metal instruments, or heavy items on overhead shelves or on top of cabinets
- **NEVER** place broken glass, sharps, or syringe needles in the household laboratory trash receptacles. Make sure these items are placed in appropriately labeled receptacles

- **NEVER** mouth-pipette any chemical or any substance, including water; ALWAYS use mechanical suction devices
- **ALWAYS** wear proper protective clothing when conducting research (closed shoes, gloves, lab coats, safety goggles, face, shields, respirators, etc.)
- Label all prepared solution containers with contents, date of preparation, and any hazard warnings
- **STUDENTS** are to be under direct supervision when working after normal working hours and on weekends
- Smoking is not permitted in any research facility; smoke only in designated areas outside of the building
- **NEVER** store, prepare, or consume any food or drink in any laboratory where hazardous materials are being utilized

LABORATORY CHEMICAL SAFETY

USE AND HANDLING

It is the responsibility of employees and cooperators to know hazards, properties, and precautions for safe handling of chemicals they are using. ALWAYS READ THE LABELS before using any type of chemical. MATERIAL SAFETY DATA SHEETS (MSDS) should be consulted to find complete information on a chemical or pesticide's properties, hazards, and safe handling methods. Hardcopy MSDS information should be available at the entrance of any laboratory.

Storage of Chemicals

All containers holding prepared solutions or materials should be labeled with a hazard warning if applicable, contents, and preparation date. Large quantities (>1gal) of flammable solvents or concentrated acids/bases should not be stored in your lab; keep only enough for your daily use requirements. Any flammable liquids requiring refrigeration should be done so only in specifically approved refrigerators/freezers. Storage cabinets are provided for the compatible storage of all bulk chemicals and are labeled either by contents (acids, bases) or by scientist's name (solvents). Glass bottles of solvents should never be stored near paper, books or boxes, or above head level or on the floor where they are most susceptible to breakage. Store compatible chemicals (solvents, acids, and bases) together using secondary chemical-resistant spill containment, but separate them into individual spill containment areas.

NEVER STORE GLACIAL AND NITRIC ACIDS TOGETHER!

Disposal of Waste Chemicals - Liquid Wastes

Never dispose of any chemical solution; containing even traces of a toxic chemical, by flushing down the sink drain.

ALL HAZARDOUS LIQUID CHEMICAL WASTE MUST BE CAPTURED AND PROPERLY STORED IN LABELED CONTAINERS.

- Plastic-coated, shatterproof glass bottles or DOT-approved polyethylene drums can be obtained from the safety officer to serve as collection containers for hazardous liquid waste.

- All hazardous waste containers are to be stored in an identified area of your laboratory marked "satellite accumulation area."

Disposal of Waste Chemicals - Solid Wastes

HAZARDOUS MATERIALS SHOULD NEVER BE PLACED IN THE NORMAL LABORATORY TRASH!

- Empty chemical bottles and some pesticide containers must be washed clean before disposal.
- All contaminated gloves, disposable clothing, syringes, pipettes, paper goods, gels, etc. must be discarded as a hazardous waste.
- Package all contaminated materials in labeled containers and dispose of by appropriate means.

LABORATORY CHEMICAL SPILLS

All laboratories should have spill neutralizing and spill control materials readily available to respond to "incidental spills"—releases of hazardous substances where the substance can be absorbed, neutralized, or otherwise controlled at the time of release by employees in the immediate release area, or by maintenance personnel. These materials should be disposed of properly through a waste disposal program.

PERSONAL PROTECTIVE EQUIPMENT

Many operations conducted in the laboratory are dangerous. The hazard potential from explosions, implosions, ruptures; liquid splash or generation of toxic vapors should always be given special consideration. Carefully select appropriate eyewear, gloves, respirators, clothing, and safety shield. Always dress sensibly—avoid open-toed shoes, long, loose hair, and loose-fitting clothes.

Eye And Face Protection

- Wear protective equipment to prevent the hazard of flying objects, liquid splash, dust generation, injurious radiation, etc.
- Select equipment that has been designed and tested in accordance with the American National Standard for Occupational and Educational Eye and Face Protection.
- Wear flexible-fitting goggles with hooded ventilation or safety glasses with side shields and a full-face shield to protect against chemical splash and/or flying glass.
- Never wear contact lenses while handling chemicals. This increases the potential for eye injuries by increasing the concentration of chemicals in contact with the cornea in the case of a chemical splash, laminar flow air can cause dehydration of the eye and cause corneal abrasions, and water-soluble chemical vapors can be absorbed by soft lenses creating exposure to the eye.

Emergency Eye Treatment Program

- Know where all eye wash stations are located; treat an eye injury within 15 seconds of occurrence.
- When washing eyes, get under eyelid and roll eyeballs around in order to wash the entire surface.
- **NEVER** use neutralizing solutions including boric acid and out-of-date eye wash solutions for a chemical splash treatment.
- Apply clean, cool, wet, soft pads to eyes after washing until follow-up help is obtained.
- Do not attempt to remove embedded objects; seek assistance from trained medical personnel.
- See Administrative Officer to fill out an Accident Report.

- Flush eye lavages weekly to assure the water is clear, and then report any problems to maintenance staff.

Hand Protection

- Wear gloves to protect against chemical absorption into skin, highly toxic chemicals, and physical hazards such as heat or sharp objects.
- Use gloves designed for chemical protection made from various materials such as rubber, neoprene, nitrile, and PVC.
- Do not use asbestos gloves; select heat-resistant gloves made from ceramic (alumina silica) fiber.
- Use disposable gloves for handling toxic materials, especially carcinogens and tetragens.
- Removal of gloves should be done so as to eliminate any chance of contamination—using fingers of one gloved hand, pinch the material of the other glove at the base of the palm and peel off glove; continue to hold glove; with the ungloved hand reach under the other glove on the palm side of the wrist, pinch and peel off the other glove.
- Thoroughly rinse reusable gloves after use.
- Discard torn or punctured gloves.

Respiratory Protection

- Wear respiratory protection whenever toxic chemicals are handled in areas without adequate ventilation.
- Consider the toxicity and concentration of airborne contaminants, the contaminant's warning properties (odor, taste, and irritation), and proper face piece fit when selecting respiratory equipment.
- DO NOT use an air-purifying respirator if airborne contaminants do not have good warning properties.
- Wear full-face piece respirators when airborne contaminants cause eye irritation.
- Select appropriate cartridge/filter.
- NEVER leave respirators uncovered in work area; cartridges/filters and inner surface of face piece may become saturated with contaminants.
- Inspect the exhalation valve system and headbands of respirators and replace any damaged parts.

- Remove cartridges/filters before cleaning.
- Wash face pieces with a warm aqueous solution of a germicidal detergent; rinse in clean, warm water and air dry.
- Store respirators in plastic bags and keep in station respirator storage cabinet or in a clean, dry, cool area.

USE OF t-TEST TO DETERMINE STATISTICAL DIFFERENCES

Determination of the statistical significance of a difference between two sets of data, for example the digestibilities of phosphorus for two diets, may be done using the t-test (as discussed by Li, 1964).

A commonly tested hypothesis is that the difference between two means (μ_1 - μ_2) is equal to zero. In testing this hypothesis, the statistic t is calculated from some number of sample observations (n, with y being used to designate the value associated with a particular sample observation), ($n_1 + n_2$ -2) degrees of freedom, and the observed sample means (Y_1 and Y_2) from each population

$$t = \frac{(Y_1 \ and \ Y_2)}{\sqrt{S_P^2\left(\frac{1}{n_1}+\frac{1}{n_2}\right)}}$$

If t is approximately equal to zero, the indication is that $\mu_1 = \mu_2$ (as indicated by our estimates of these true means using sample means, Y_1=Y_2). If t is too much larger than zero, the indication is that μ_1 is greater than μ_2. If t is too much smaller than zero, the indication is that μ_1 is less than μ_2. How large is too large or how small is too small is determined by the significance level obtained from a t distribution table, Table A-23.

The procedure of testing the hypothesis that the digestibilities of phosphorus in two diets (#1 and #2) are equal ($\mu_1 = \mu_2$) is illustrated by the following example: Three measurements (n=3) of percentage digestibility (y) for diet #1 are 72.6, 71.9 and 73.3. For diet #2, the digestibilities were determined to 76.2, 75.2, and 77.4. The statistical question is to determine whether the mean digestibilities for the two diets are equal.

The procedure to answer this question is as follows:

1. Null Hypothesis: The two diet means are equal, that is, $\mu_1 = \mu_2$ (a 2-tailed t-test).
2. Alternative hypothesis: The alternative hypothesis is that (a) $\mu_1 < \mu_2$ or (b) $\mu_1 > \mu_2$.

3. Assumptions: The samples are drawn at random from normal populations with the same variance (σ^2). The assumption of randomness is by far the most important assumption.
4. Level of significance: The 95% confidence level is commonly chosen to determine significance, which means a difference of this magnitude occurs only 5% of the time by chance alone when there is no significant or real difference.
5. Critical *t* value: The critical *t* value for a 2-tailed *t* test with (n_1+n_2 -2) or 4 degrees of freedom (in this example since we have 3 samples for each group) is the table value (3.182) found in the *t*-Distribution table (A-22). For a 2-tailed *t*-test with these data, one uses a table value found in the column with 2.5%/tail.
6. Computation of *t:* The details of the computation of *t* are given in table of calculations (see table below). The computed observed *t*-value is -4.9514 with 4 degrees of freedom.

Definitions of terms for the calculation of *t*.

Observation data values: y

Observed difference between sample means: (Y_1 and Y_2)

Uncorrected sums of squares: $\sum_{i=1}^{n} y_i^2$

Correction term: $\dfrac{\left(\sum_{n=1}^{n} y_i\right)^2}{n}$

Corrected sums of squares: $SS = \sum_{n=1}^{n} y^2 - \dfrac{\left(\sum_{n=1}^{n} y\right)^2}{n}$

Pooled corrected sums of squares: Pooled SS = $SS_1 + SS_2$

Pooled degrees of freedom: Pooled d.f. = $df_1 + df_2$

Pooled variance: $S_P^2 = \dfrac{Pooled\ SS}{Pooled\ d.f.}$

Observed *t* value: $t = \dfrac{(Y_1\ and\ Y_2)}{\sqrt{S_P^2\left(\dfrac{1}{n_1} + \dfrac{1}{n_2}\right)}}$

7. Conclusion: Because the absolute value of the observed *t* (4.78) is equal to or greater than the table value (3.182), this means we can conclude that the means are significantly different ($\mu_1 \neq \mu_2$) at the 95% level of confidence ($P < 0.05$). (Note that *t* is a pure number and not influenced by a change of the unit of measurement.)

Calculation of t-value

Item	Diet 1	Diet 2	Combination	Explanation
Observations	$y_1 = 72.9$	$y_4 = 76.4$		
	$y_2 = 71.9$	$y_5 = 75.2$		
	$y_3 = 73.3$	$y_6 = 77.4$		
Calculations				
Σy	218.18	229.0		
n	3	3		
Means (Y)	72.700	76.333	-3.633	$(Y_1 - Y_2)$
$(\Sigma y)^2$	47,567.61	52,441.		
$(\Sigma y)^2/n$	15,855.87	17,480.33		Correction terms
Σy^2	15,856.91	17,482.76		Uncor. SS
SS	1.0400	2.4267	3.4067	Pooled SS
d.f.	2	2	4	Pooled d.f.
s^2			0.8667	s_p^2 =3.4067/4
1/n	0.3333	0.3333	0.6667	$1/n_1 + 1/n_2$
			0.5778	$[s_p^2 (1/n_1 + 1/n_2)]$
			0.7602	$[s_p^2 (1/n_1 + 1/n_2)]^{0.5}$
			-4.779	t (observed)

(Table adapted from page 147 in Li, 1964).

Conclusion: Because the absolute values of the observed *t* is greater than the table *t* (2.776), we conclude that μ_1 and μ_2 are significantly different at the 95% confidence level ($P < 0.05$).

If a 1-tailed *t*-test were made, the null hypothesis could be $\mu_1 = \mu_2$ or $\mu_1 > \mu_2$ verses the alternative hypothesis that $\mu_2 > \mu_1$ (meaning digestibility of phosphorus in diet 2 is not greater than that for diet 1 *vs* the digestibility of phosphorus in diet 2 is greater than that for diet 1). In that case the critical table *t* value (2.132) would be found in the column with 5%/tail with 4 degrees of freedom. The conclusion would be that we reject the null hypothesis and accept the alternative hypothesis ($\mu_2 > \mu_1$) or that the digestibility of phosphorus in diet 2 is significantly greater than that in diet 1.

INFORMATION RESOURCES FOR AQUACULTURE

REGIONAL AQUACULTURE CENTERS

Northeastern Regional Aqua Center Univ. of Mass. Dartmouth, Research 201, 285 Old Westport Rd., College of Arts and Sciences, Dept. of Biology, Southeastern Mass. Univ., North Dartmouth, MA 02747-2300
TEL: 508/999-8157 FAX: 508/999-8590

North Central Reg. Aqua Center College of Agr. & Natural Resources, Dept. of Fisheries & Wildlife, Michigan State Univ., Rm. 13, Natural Resources Building, East Lansing, MI 48824
TEL: 517/353-1962, FAX: 517/353-7181

Southern Regional Aqua Center Delta Branch Experiment Station, PO Box 197, Stoneville, MS 38776
TEL: 601/686-9311, FAX: 601/686-3569

The Tropical & Subtropical, Regional Aquaculture Center, Univ. of Hawaii & Oceanic Inst., Makapuu Point, 41-202 Kahenianeole, Waimanalo, HI 96795
TEL: 808/259-7951, FAX: 808/259-8395

The Western Regional Aqua Center School of Fisheries, Box 357980, Univ. of Washington, Seattle, WA 98195-7980
TEL: 206/543-4290, FAX: 206/685-4674

Other Resource Centers

National Sea Grant Depository
Pell Library Bldg., Univ. of Rhode Island, Narragansett Bay Campus, Narragansett, RI 02882-1197
TEL: 401/874-6114, FAX: 401/874-6160
Lending library of all pub. funded by Sea Grant.

USDA/CSREES/PAS
Dr. Merle Broussard, Principal Aqua Scientist, Dr. Gary Jensen, Aqua Program Leader, Stop 2220, Washington, DC 20250-2220
TEL: 202/401-5787, FAX: 202/401-6869
Internet E-mail Addresses:
Dr. Broussard - mbroussard@reeusda.gov
Dr. Jensen - giensen@reeusda.gov

U.S. Fish & Wildlife Service
1 Arlington Square, Rm. 820, Arlington, Virginia 22203
TEL: 703/358-1715, FAX: 703/358-2210

ORGANIZATIONS

Advanced Aquaculture
PO Box 429, Syracuse, IN 46567
TEL: 219/457-5802, FAX: 219/457-5887

American Fisheries Society
5410 Grosvenor Ln., Suite 110, Bethesda, MD 20814 -2199
TEL: 301/897-8616, FAX: 301/897-8096
Student $37, Individual $74

American Tilapia Association
111 West Washington St., Charles Town, WV 25414-1529
TEL: 304/728-2175
www.tilapia.org

Aquacultural Engineering Society
Brian J. Vinci, Freshwater Institute, PO Box 1746, Shepherdstown, WV 25443
TEL: 304/876-2815, FAX: 304/870-2208
Student $20, Individual $75
www.aesweb.org

Aquaculture Assoc. of Canada
Box 1987, St. Andrews, New Brunswick, EOG 2X0 Canada
TEL: 506/529-4766, FAX: 506/529-4609
Student $40, Individual $50

Catfish Farmers of America
1100 Highway 82 East, Suite 202, Indianola, MS 38751
TEL: 601/887-2699, FAX: 601/887-6857
Individual $40

Division of Fisheries, US Fish & Wildlife
1 Federal Dr., Fort Snelling, MN 55337
TEL: 612/725-3447, FAX: 612/725-3026

National Aquaculture Association
111 West Washington St., Suite 1, Charles Town, WV 25414-1529
TEL: 800/626-3301, FAX: 304/728-2196

Pond Dynamics/Aquaculture CRSP
Office of International Research & Development, Oregon State University, Snell Hall 400, Corvallis, OR 97331-1641
TEL: 541/737-6416, FAX: 541/737-3447

US Trout Farmers Assoc.
111 West Washington St., Charles Town, WV 25414-1529
TEL: 304/728-2189
www.ustfa.org

US Aquaculture Suppliers Assoc.
PO Box 24866, Little Rock, AR 72221
TEL: 909/652-2612, FAX: 909/652-2492

World Aquaculture Society
143 JM Parker Coliseum, LSU, Baton Rouge, LA 70803
TEL: 504/388-3137, FAX: 504/388-3493
Student $40, Individual $60
www.was.org

MAGAZINES/NEWSLETTERS

Aquaculture Magazine
PO Box 2329, Asheville, NC 28802
TEL: 704/254-7334
Monthly trade magazine, $19/year

Aquaculture
American Elsevier Scientific Pub. Co., 52 Vanderbilt Ave., New York, NY 10017. 32 times/year. Technical journal covering current research, $640.

Aquacultural Engineering
Elsevier Science Publishing Co., Inc., 655 Avenue of the Americas, New York, NY 10010
TEL: 212/989-5800, FAX: 212/633-3990
6 issues/year. Technical journal covering design and development of effective. Aquaculture systems, $230.00/year or join Aquaculture Engineering Society (receive journal as part of membership privileges).

The Aquaculture News
PO Box 416, Jonesville, LA 71343
TEL: 318/339-4660, FAX: 318/339-4664
Monthly trade newspaper, issues that affect catfish industry, $30/year.

Aquaculture: Situation and Outlook Report
ERS-NASS, PO Box 1608, Rockville, MD 20849-1608
Quarterly Report, $12/year

The Catfish Journal
PO Box 55648, Jackson, MS 39296
TEL: 601/981-0807, FAX: 601/981-0506
Monthly trade newspaper

Freshwater and Marine Aquarium
PO Box 487, Sierra Madre, CA 91025
TEL: 626/355-1476, FAX: 626/355-6415
Monthly Magazine, $23/year

Fish Farming International
Livingston Road, Fessle, HU13 0EE, North Humberside, England
TEL: 011-44-1482-642302
Monthly newspaper covering Aquaculture around the world.

Infofish International
1st Floor, Wisma PKNS, Janal Raja Laut, 50350 Kuala Lumpar, Malaysia
TEL: 011-603291-4466, FAX: 011-603291-6804
Bi-monthly magazine FAO/WHO covering fisheries, mostly in Asia.

Northern Aquaculture
4623 William Head Rd., Victoria, British Columbia, V9C 3Y7 Canada
TEL: 250/478-9209, FAX: 250/478-1184
Trade magazine covering trout and salmon in North America, $25/year.

Progressive Fish-Culturist
American Fisheries Society, 5410 Grosvenor Ln., Suite 110,
Bethesda, MD 20814-2199
TEL: 301/897-8616, FAX: 301/897-8096
Quarterly technical journal of applied research, $25/year.

Salmonid Magazine
U.S. Trout Farmers Assoc., 111 West Washington St.,
Charles Town, WV 25414-1529
TEL: 304/728-2189
Quarterly trade magazine covering trout and salmon industry, free to members.

Seafood International
19 Scarbrook Rd., Croydon, CR9 1QH, Surrey, England
TEL: (0044) 0181 277-5527, FAX: (0044) 0181-2775-440
Monthly magazine covering marketing of wild and aquacultured fish around the world.

Tropical Fish Hobbyist
One TFH Plaza, 3rd & Union Ave., Neptune City, NJ 07753
TEL: 732/988-8400, FAX: 732/988-5466
Monthly Magazine, $40/year

World Aquaculture Magazine
World Aquaculture Society, 143 JM Parker Coliseum, LSU Baton Rouge, LA 70803
TEL: 504/388-3137, FAX: 504/388-3493
Quarterly magazine for WAS members, free to members.
$40/year

Books

Recirculating Aquaculture Systems, 2nd Edition. M.B. Timmons, J.E. Ebeling, F.W. Wheaton, S.T. Summerfelt, and B.J. Vinci, 2002. Cayuga Aqua Ventures, Ithaca, NY 14850. ISBN 0-9712646-1-9. 800 p.

Aquaculture Water Reuse Systems: Engineering Design and Management. M.B. Timmons and T.M. Losordo. Elsevier Science. ISBN 0-444-89585-X. 346 p. $210.00.

Fundamentals of Aquacultural Engineering. Lawson, T.B., 1995. Chapman & Hall, N.Y., 355 p. ISBN 0-412-06511-8.

Third Report to the Fish Farmers - Status of Warm Water Fish Farming & Progress in Fish Farming Research. Dupree, H.K. and J.V. Hunter, Eds. 1984. U.S. Government Printing Office; Superintendent of Documents; Washington, DC 20402. U.S. Fish And Wildlife Service. Stock # 024-010-00654-4. $8.00.

Aquacultural Engineering. Wheaton, F. 1977. Robert E. Krieger Publishing Co., Malabar, FL, 708 p. $72.00.

Aquaculture: Principals and Practices. Pillay, T.V.R. 1990. Fishing News Books Ltd./Div. of Blackwell Scientific, 591 p. $100.00.

Aquaculture: The Farming and Husbandry of Freshwater & Marine Organisms. Bardach, J., J. Ryther, and W. McLamey, 1972. Wiley-Interscience, Box 092, Somerset, NJ 08873, 868 p. $49.00.

Channel Catfish Farming Handbook. Tucker, C. and E. Robinson. 1990. Van Nostrand, Reinhold, NY, 464 p. $59.00.

Culture and Propagation of Striped Bass and Its Hybrids. Harrell, R., J. Kerby, and R. Minton, Eds. 1990. American Fisheries Society. 79 p. $50.00.

Fish Hatchery Management. Piper, et al. 1983. American Fisheries Society and the U.S. Fish and Wildlife Service, Washington, DC. 517 p. $54.00; and 2nd Edition by G.A. Wedemeyer (Editor) 2001.

The Freshwater Aquaculture Book: A Handbook for Small Scale Fish Culture. McLamey, W., 1984. Harley and Marks, Inc. 583 p. $37.00.

Introduction to Fish Technology. Regenstein, J.M. and C.E. Regenstein. 1991. Van Nostrand, Reinhold, NY. 270 p. $54.00.

Principals and Practices of Pond Aquaculture. Lannan. J., R. Smitherman, and G. Tchobangous. Eds. 1986. Oregon State Univ. Press, Corvallis, OR 252 p. $45.00.

Principals of Warmwater Aquaculture. Stickney R. 1979. John Wiley and Sons, Inc. 375 p. $59.00.

Water Quality in Ponds for Aquaculture. Boyd, C.E. 1990. Alabama Agricultural Experiment Station, Auburn Univ., AL 482 p. $9.60.

***Trout and Salmon Culture* (Hatchery Methods).** 1980. 197 p. Publications, Div. of Agriculture and Natural Resources, Univ. of California, 6701 San Pablo Ave., Oakland, CA 94608-1239. $6.50
TEL: 510/642-2431, FAX: 510/643-5470

Water Quality & Pond Soil Analyses for Aquaculture. Tucker, C.S. and C. Boyd. Alabama Agriculture Experiment Station, Auburn Univ., AL. $8.00.

DIRECTORIES

Aquaculture Magazine Buyer's Guide, PO Box 2329, Asheville, NC 28802
TEL: 704/254-7334, FAX: 704/253-0677, $16.00.

Canadian Aquaculture Buyers Guide
Harrison House Publishers, 4623 William Head Rd., Victoria, British Columbia, V9C 3Y7, CANADA
TEL: 250/478-9209, FAX: 250/478-1184, $7.00.

WORLD WIDE WEB

Alternative Farming Systems Information Center (AFSIC)
http://www.nal.usda.gov/afsic/
The AFSIC is one of 10 information centers at the National Agricultural Library. AFSIC serves as a national clearinghouse for aquaculture information and provides materials for aquafamers, consumers, et al.

American Tilapia Association
http://www.tilapia.org

Aquacultural Engineering Society
http://www.aesweb.org/

AquaNIC - Aquaculture Network Information Center
http://aquanic.org/
An excellent clearinghouse of aquaculture information available on the internet. It also provides information on how to obtain aquaculture information that is not on the internet.

Cornell Aquaculture Resources and Short Course
http://www.bee.cornell.edu/extension/aquaculture

FINS
http://www.actwin.com/fish/
An archive of information about aquariums. It covers both freshwater and marine, tropical and temperate.

Fish-Vet
http://users.jaguNET.com/~fishvet/
Information on Fish-Vet (tm), a multimedia program for fast and accurate determination of abnormal conditions in fish keeping, including disease, environmental, and nutritional problems, and the veterinary consultancy service of Fish-Vet, Inc.

National Fisheries Institute
http://www.nfi.org
National Fisheries Institute is the largest trade association serving the US seafood industry. NFI has always been a clearinghouse of business-related information. This website offers dozens of links to government statistics, seafood companies and other web resources. A members-only section adds promotional materials, HACCP information and the latest news on import alerts, legislative updates and more.

Natural Resource, Agricultural, and Engineering Service (NRAES)
http://www.nraes.org

Seafood NIC
http://www-seafood.ucdavis.edu/home.htm
Hosted by the Sea Grant Extension Program at UC Davis, the Seafood Network Information Center or Seafood NIC, is an on-line home to the HACCP Alliance. It offers page after page of seafood-safety information, as well as training materials, seminar schedules and the FDA's official fishery hazard and control guides. The site provides generic HACCP plans for scores of seafood products and processes.

U.S. Trout Farmers Association
http://www.ustfa.org/
Trout information for both the consumer and industry. Information for the consumer includes: Tips and handling, farm-raised trout, trout tips, handling how-to's, prep pointers, and nutritional values. Industry information includes: Salmonid Magazine, Quality Assurance, membership info, and other internet resources, as well as a trout recipe book.

World Aquaculture Society (WAS)
http://www.was.org/
An international nonprofit society with over 4,000 members in 94 countries. Founded in 1970, its primary focus is to improve communication and information exchange within the diverse global aquaculture community.

INDEX